Biotechnology: Innovative Technologies

Biotechnology: Innovative Technologies

Edited by Sophia Martin

MURPHY & MOORE
www.murphy-moorepublishing.com

Published by Murphy & Moore Publishing,
1 Rockefeller Plaza,
New York City, NY 10020, USA
www.murphy-moorepublishing.com

Biotechnology: Innovative Technologies
Edited by Sophia Martin

International Standard Book Number: 978-1-63987-079-0 (Hardback)

Cataloging-in-Publication Data

Biotechnology : innovative technologies / edited by Sophia Martin.
p. cm.
Includes bibliographical references and index.
ISBN 978-1-63987-079-0
1. Biotechnology. 2. Genetic engineering. I. Martin, Sophia.
TP248.2 .B563 2022
660.6--dc23

TABLE OF CONTENTS

PREFACE

The world is advancing at a fast pace like never before. Therefore, the need is to keep up with the latest developments. This book was an idea that came to fruition when the specialists in the area realized the need to coordinate together and document essential themes in the subject. That's when I was requested to be the editor. Editing this book has been an honour as it brings together diverse authors researching on different streams of the field. The book collates essential materials contributed by veterans in the area which can be utilized by students and researchers alike.

Biotechnology is referred to as the utilization of biological processes for industrial and other purposes. It involves the genetic manipulation of microorganisms for the production of antibiotics and hormones. Innovative biotechnology provides advanced products and techniques to combat devitalizing and rare diseases and develop greener, and more efficient industrial manufacturing processes. Biotechnology has applications across multiple industries such as food, pharmaceutical, chemical, bio-products, textiles, medicine, nutrition, and animal sciences. A few of its major branches are animal, medical, industrial, environmental, and plant biotechnology. This book attempts to understand the multiple branches that fall under the discipline of biotechnology and how such concepts have practical applications. The various studies that are constantly contributing towards innovative technologies and evolution of this field are examined in detail. As this field is emerging at a fast pace, this book will help the readers to better understand the concepts of biotechnology.

Each chapter is a sole-standing publication that reflects each author´s interpretation. Thus, the book displays a multi-facetted picture of our current understanding of application, resources and aspects of the field. I would like to thank the contributors of this book and my family for their endless support.

Editor

1

New α- and SIN γ-retrovectors for safe transduction and specific transgene expression in pancreatic β cell lines

Olivier Albagli[1*], Alicia Maugein[1], Lukas Huijbregts[1], Delphine Bredel[1,3], Géraldine Carlier[1], Patrick Martin[2] and Raphaël Scharfmann[1]

Abstract

Background: Viral vectors are invaluable tools to transfer genes and/or regulatory sequences into differentiated cells such as pancreatic cells. To date, several kinds of viral vectors have been used to transduce different pancreatic cell types, including insulin-producing β cells. However, few studies have used vectors derived from « simple » retroviruses, such as avian α- or mouse γ-retroviruses, despite their high experimental convenience. Moreover, such vectors were never designed to specifically target transgene expression into β cells.

Results: We here describe two novel α- or SIN (Self-Inactivating) γ-retrovectors containing the RIP (Rat Insulin Promoter) as internal promoter. These two retrovectors are easily produced in standard BSL2 conditions, rapidly concentrated if needed, and harbor a large multiple cloning site. For the SIN γ-retrovector, either the VSV-G (pantropic) or the retroviral ecotropic (rodent specific) envelope was used. For the α-retrovector, we used the A type envelope, as its receptor, termed TVA, is only naturally present in avian cells and can efficiently be provided to mammalian β cells through either exogenous expression upon cDNA transfer or gesicle-mediated delivery of the protein. As expected, the transgenes cloned into the two RIP-containing retrovectors displayed a strong preferential expression in β over non-β cells compared to transgenes cloned in their non-RIP (CMV- or LTR-) regulated counterparts. We further show that RIP activity of both retrovectors mirrored fluctuations affecting endogenous *INSULIN* gene expression in human β cells. Finally, both α- and SIN γ-retrovectors were extremely poorly mobilized by the BXV1 xenotropic retrovirus, a common invader of human cells grown in immunodeficient mice, and, most notably, of human β cell lines.

Conclusion: Our novel α- and SIN γ-retrovectors are safe and convenient tools to stably and specifically express transgene(s) in mammalian β cells. Moreover, they both reproduce some regulatory patterns affecting *INSULIN* gene expression. Thus, they provide a helpful tool to both study the genetic control of β cell function and monitor changes in their differentiation status.

Keywords: α-Retrovector, SIN γ-retrovector, Pancreatic β cell lines, EndoC-β H2 cells, Rat insulin promoter, BXV1 xenotropic retrovirus, TVA gesicles

* Correspondence: olivier.albagli-curiel@inserm.fr
[1]INSERM U1016, CNRS UMR8104, Institut Cochin, Université Paris Descartes, 123 Boulevard de Port-Royal, 75014 Paris, France
Full list of author information is available at the end of the article

Background

Transferring DNA sequences through viral vectors represents an invaluable tool to study the development and physiology of the pancreas, and explore new therapeutic strategies [1–5], and references therein]. Huge efforts have been put into targeting specific pancreatic cell types, especially among the endocrine populations clustered in islets of Langerhans [6–8]. Hence, numerous viral vectors were designed with specific, and often short, cis-regulatory DNA sequences. The goal usually is to achieve tissue-specific expression or, less frequently, to introduce a read-out for a given signaling pathway [9]. In each case, the viral vector must both achieve efficient transduction and keep the proper specificity/responsiveness of the host-derived DNA sequence.

To date, the most widely used viral vectors to transduce pancreatic cells are based on either adenoviruses, adeno-associated viruses or lentiviruses (generally HIV-1, Human Immunodeficiency Virus-1). These viral vectors provide efficient tools that allow successful and even specific gene expression in different pancreatic exocrine or endocrine sub-populations, including insulin-producing β cells [6–8, 10–13]. However, the use of vectors based on « simple » retroviruses is much rarer, and, to the best of our knowledge, was never described for specific expression in β cells. The aim of this study is thus to provide and validate new convenient and harmless viral vectors for stable transduction and specific transgene expression in β cells. We used the well known α – and γ- « simple » retroviruses as backbones and describe here two new α – and SIN (Self-Inactivating) γ-retrovectors respectively based on ASLV(A) (Avian Sarcoma and Leukosis Virus subgroup A) and MLV (Murine Leukemia Virus) retroviruses. Both harbor the RIP (Rat Insulin Promoter) as internal (transgene regulating) promoter. These retrovectors are safe, easily produced in standard BSL2 conditions and can rapidly be concentrated if needed. Importantly, in both retroviral contexts, the RIP activity is much stronger in β cells than in non-β cells and displays fluctuations that parallel those affecting the endogenous *INSULIN* gene expression.

Results

A SIN γ-retrovector for specific transgene expression in pancreatic β cells

pPRIZ, pSERS RIP MCS and pSERS SF MCS retrovectors

The pPRIZ vector has been previously described [14]. It is a non-SIN γ-retrovector harboring the Sh ble gene (thereafter referred to as ZeoR) conferring resistance to zeocin downstream of an IRES element (Fig. 1). Once retro-transcribed and integrated, the vector transcription is driven by its intact 5′ LTR, which displays a fairly ubiquitous expression. In order to design a pancreatic β-cell specific γ-retrovector, we started from the MLV-based retrovector pSERS11 SF GFP pre, a gift from Axel Schambach and Christopher Baum (Hannover Medical School, Germany) [15]. pSERS11 SF GFP pre is a SIN γ-retrovector, as its 3′ LTR (becoming 5′ LTR after retro-transcription and integration) lacks the U3 region [15]. The SF internal promoter (U3 region of SFFV LTR, Spleen Focus Forming Virus LTR), which is active in most cells, was replaced by the 405 bp RIP promoter [6, 10]. We further added several cloning sites to create the pSERS RIP MCS vector (Fig. 1). These cloning sites were also inserted into the pSERS11 SF pre GFP vector to obtain the pSERS SF MCS derivative (Fig. 1).

Transduction of β and non-β cell lines with the pSERS RIP SIN γ-retrovector

Human non-β (293 T) and β cells (EndoC-βH2 [6]) were transduced with retroviral supernatants containing pSERS RIP encoding either a nuclearized and HA-tagged YFP (3HAH2BYFP) or GFP with a co-expressed selectable marker (GFP-IRES-ZeoR). pPRIZ GFP served as a control, as it also encodes a GFP-IRES-ZeoR transcript driven by the viral LTR (Fig. 1). All transductions were carried out in parallel under the same conditions, using VSV-G as envelope. After 4 days (without selection), FACS analysis showed a much stronger expression of the transgenes encoded by the RIP-containing SIN γ-retrovector in EndoC-βH2 compared to 293 T cells (Fig. 2). In contrast, transduction with pPRIZ GFP (without selection) led to a higher level of fluorescence in 293 T compared to EndoC-βH2 cells (Fig. 2 and Additional file 1). Stable populations of highly fluorescent β cells transduced with either pSERS RIP GFP-IRES-ZeoR or pPRIZ GFP were readily obtained upon zeocin selection (Additional file 2 and Fig. 8b). Transduction with pSERS SF GFP-IRES-ZeoR (without selection) led to a strong fluorescence level in both 293 T and EndoC-βH2 cells (Additional file 1). As pSERS RIP GFP-IRES-ZeoR and pSERS SF GFP-IRES-ZeoR only differ by their internal promoter, this result indicates that β cell specific expression of pSERS RIP transgene solely depends on the RIP sequence, rather than on any other feature of the construct. Finally, pSERS RIP (as well as pPRIZ and pSERS SF) also led to efficient transgene expression in transduced mouse MIN6 β insulinoma cells (Additional file 3). Thus, a SIN γ-retrovector with an internal RIP allows a stronger and preferential transgene expression in β cells compared to non-β cells (see also Additional file 10).

pSERS retrovectors are not mobilized by the BXV1 xenotropic retrovirus

We recently reported that EndoC-βH2 cells are productively infected by BXV1 (XMV-43), a mouse xenotropic

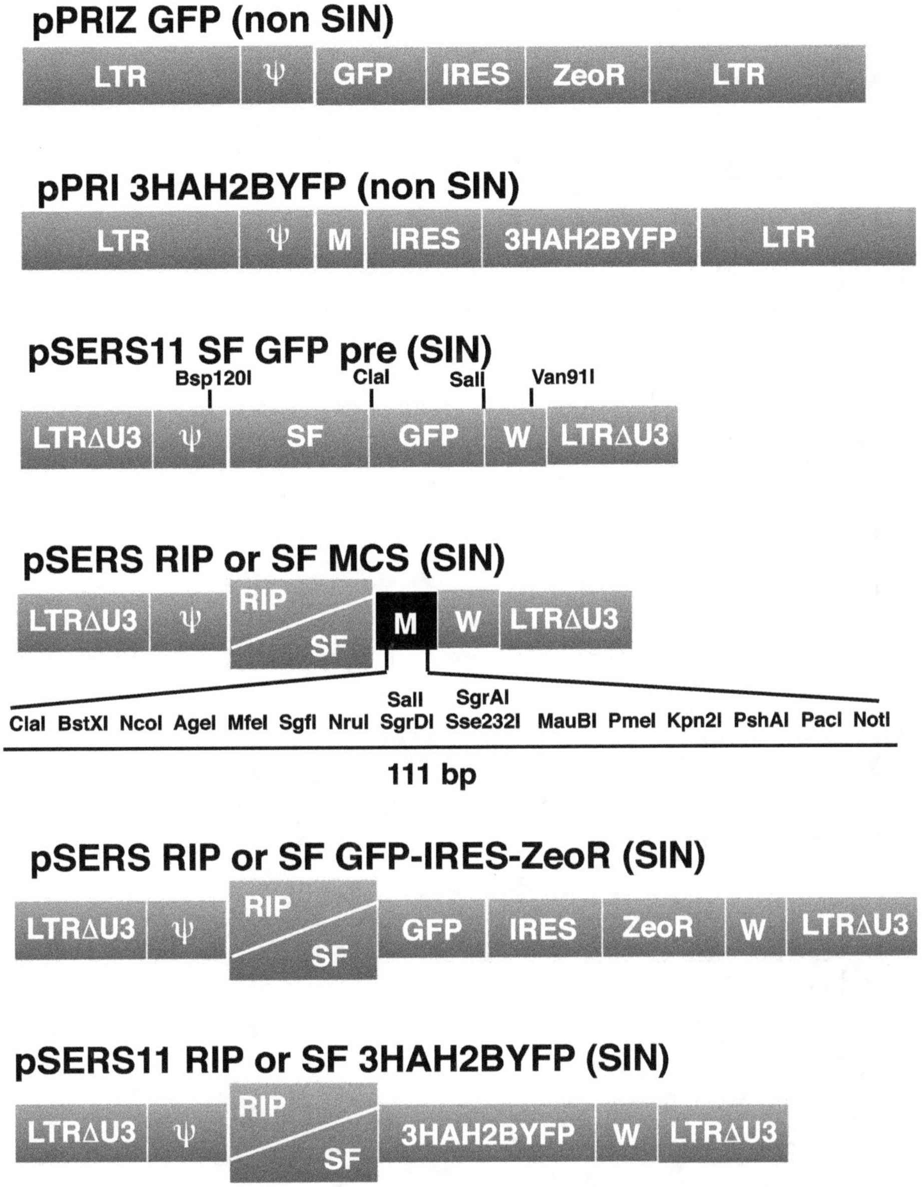

Fig. 1 Schematic diagram of the γ-retrovectors. The integrated « proviral » form is depicted. The pSERS11 GFP pre [15] was used as a starting vector to generate a β-specific SIN γ-retrovector. LTR**Δ**U3: Long Terminal Repeat without the U3 region (SIN deletion) Ψ: packaging sequence. RIP: 405 bp promoter of rat insulin II gene. SF: U3 enhancer/promoter region from Spleen Focus Forming Virus. M: Multi Cloning Sites are depicted in two colors (grey or black) because they are different in pPRI 3HAH2BYFP and in pSERS RIP/SF MCS (only the latter is detailed). The cDNAs subsequently inserted in these derivatives are indicated. W: Woodchuck hepatitis virus post-transcriptional regulatory element. 3HAH2BYFP: YFP fluorescent protein fused in 5′ with 3 copies of the HA epitope followed by the H2B histone. The position of the restriction sites relevant for subsequent constructs are indicated (see Methods). The different parts of the diagrams are not at scale

endogenous γ-retrovirus [16]. As a consequence, defective γ-retrovectors are « mobilized », i.e. trans-complemented and released as xenotropic particles, raising concerns about their use in EndoC-βH2 cells [16]. In principle, such trans-complementation depends on the presence of the Ψ packaging signal within the γ-retrovector-encoded transcripts. Upon transient transfection, we previously verified that Ψ-deleted constructs are indeed no more trans-complemented by BXV1 [16]. However, by definition, such deleted constructs can not be used to produce viral particles.

The SIN (U3) deletion is aimed to dampen the transcriptional activity of the LTR [15] and hence the synthesis of Ψ-containing transcripts (Fig. 1). Therefore, most pSERS encoded mRNA should be « ignored » by the BXV1 proteins, thereby preventing trans-complementation in pSERS-

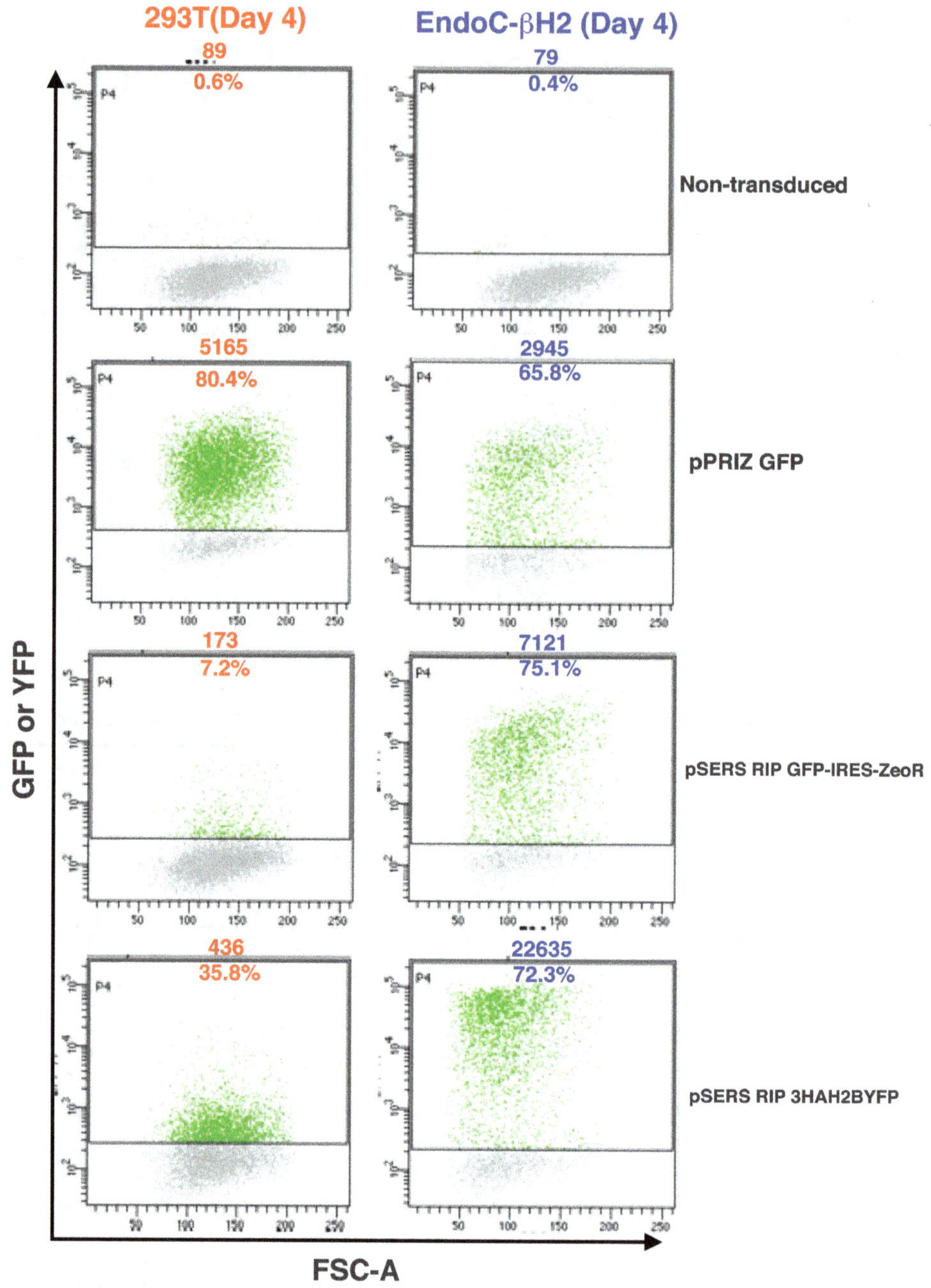

Fig. 2 A SIN γ-retrovector mediating preferential transgene expression in β cells. Human β (EndoC-βH2) and non-β (293 T) cells were transduced in parallel with the indicated VSV-G pseudotyped γ-retrovectors. Cells were harvested 4 days later (without selection) and the level of GFP or YFP fluorescence was analyzed by flow cytometry. The number above each diagram indicates the mean of fluorescence of the whole population of analyzed cells. The number in the diagram is the percentage of fluorescent cells. This experiment was performed once, the two different pSERS RIP constructs giving similar results. Note that the positive threshold level is slightly higher for 293 T cells transduced with pPRIZ GFP to take into account an increase in the autofluorescence level of negative cells (lower cloud). FSC-A: Forward-Scatter-Area

transduced EndoC-βH2. To test this prediction, EndoC-βH2 cells were stably transduced with either pSERS SF GFP-IRES-ZeoR or with the non-SIN γ-retrovector, pPRIZ GFP (Fig. 1). After zeocin selection, the two EndoC-βH2 transduced populations showed similar GFP fluorescence levels (Fig. 3a). « Naive » 293 T cells exposed to the conditioned medium of EndoC-βH2 cells transduced with pPRIZ-GFP generated many zeocin-resistant (and GFP fluorescent) foci (more than 100 in two experiments) while cells exposed to the conditioned medium of EndoC-βH2 cells transduced with pSERS SF GFP-IRES-ZeoR gave much fewer zeocin-resistant foci (0 or 2 in two

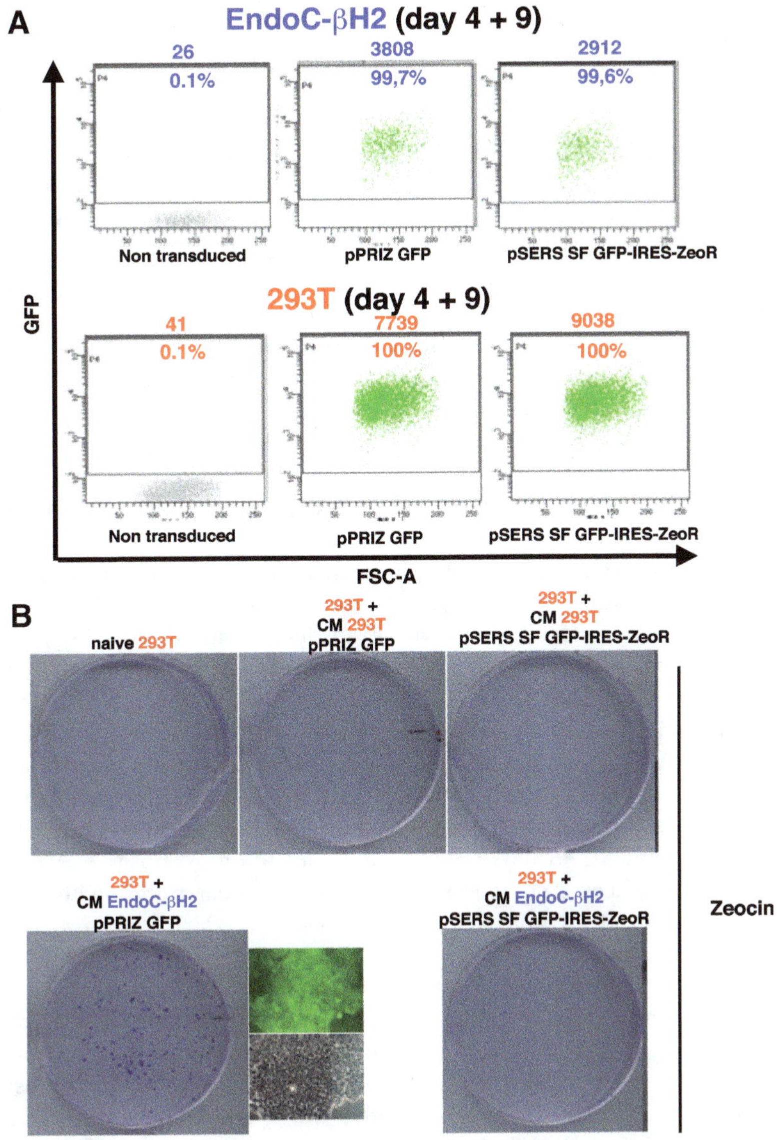

Fig. 3 pSERS SF SIN γ-retrovector is not mobilized by the BXV1 xenotropic γ – retrovirus in EndoC-βH2 cells. **a**. GFP fluorescence level in transduced 293 T and EndoC-βH2 cells. 4 days after transduction with the indicated non-SIN (pPRIZ GFP) or SIN (pSERS SF GFP-IRES-ZeoR) γ-retrovector, 293 T and EndoC-βH2 cells were selected in zeocin for 9 days (day 4 + 9), then harvested and analyzed for GFP fluorescence by flow cytometry. The number above each diagram indicates the mean of fluorescence of the whole population of analyzed cells. The number in the diagram is the percentage of fluorescent cells. **b**. Conditioned medium (CM) of the transduced 293 T or EndoC-βH2 cells shown in A was added to « naive » 293 T cells. Transmission of zeocin resistance as a measure of γ-retrovector mobilization/trans-complementation was assessed by selecting the exposed 293 T to zeocin. GFP fluorescence was detected at various levels in most of the 293 T foci observed after their exposure to the CM of EndoC-βH2-pPRIZ GFP, one of them is shown. This experiment was done in duplicates, which gave similar results

experiments) (Fig. 3b). As expected, both the supernatants of pPRIZ GFP- and pSERS SF GFP-IRES-ZeoR-transduced 293 T failed to transfer the zeocin-resistance to naive 293 T cells [16] (Fig. 3b). We also transduced EndoC-βH2 with both pSERS SF GFP-IRES-ZeoR and another non-SIN γ-retrovector, pPRIHy TVA, encoding hygromycin-resistance [16] and then exposed naive 293 T cells to the conditioned medium of this doubly transduced EndoC-βH2 population. Under these conditions, the number of hygromycin-resistant foci was well higher than that of zeocin-resistant foci (3 and 12 versus more than 500, in two experiments, respectively), confirming that non SIN γ-retrovectors are much more readily trans-complemented by BXV1 than SIN γ-retrovectors in EndoC-β H2 cells (Additional file 4). These results indicate that the SIN γ-retrovector pSERS SF MCS, and by extension, pSERS RIP MCS, can be used in EndoC-βH2 cells, or in any cell contaminated with a xenotropic γ-retrovirus, without the safety concerns raised by usual, non-SIN, γ-retrovectors.

An α-retrovector for specific transgene expression in pancreatic β cells

RCASBP(a) MCS, RCANBP(a) CMV MCS and RCANBP(a) RIP MCS retrovectors

RCASBP(A) and RCANBP(A) retrovectors are derived from the Rous Sarcoma Virus-A, an avian α-retrovirus belonging to ASLV(A) α-retroviridae [17]. Different pancreatic subpopulations, including β cells, have been selectively targeted with RCASBP(A) retrovectors. This was achieved through tissue-specific expression of their avian (generally quail) receptor, termed TVA, in transgenic mice [18, 19]. However, no ASLV(A)-based retrovector endowed with an intrinsic specificity for any pancreatic cell has been described to date.

To design an α-retrovector for specific transgene expression in β cells, we started from RCANBP(A) CMV GFP (a gift from Stephen Hughes and Andrea Ferris, NIH/NCI, Fredericks, USA). We first removed GFP and added a new multiple cloning sites to generate RCANBP(A) CMV MCS (Fig. 4). The CMV enhancer/promoter of RCANBP(A) CMV MCS was then replaced by RIP, leading to the RCANBP(A) RIP MCS derivative (Fig. 4). Finally, the same MCS was added to the original RCASBP(A) retrovector (another gift from Stephen Hughes and Andrea Ferris) to design RCASBP(A) MCS (Fig. 4). Of note, all these constructs keep a complete 5′ LTR and should thus be all regarded as non-SIN α-retrovectors.

Transduction of β and non-β cell lines with RCANBP(a) RIP retrovector

ASLV(A)-based retrovectors are highly efficient on chicken or quail cells. In contrast, mammalian cells are resistant because they are devoid of the ASLV(A) receptor, termed TVA. Accordingly, only mammalian cells engineered to express quail or chicken TVA can be transduced by ASLV(A)-based retrovectors [17–19]. We used two methods to express the short isoform of quail TVA in mammalian cells: 1) stable expression of the tva encoding sequence through transduction with a γ-retrovector, pPRIHy TVA, followed by hygromycin selection; 2) transient transfer of TVA through TVA-gesicles, a method adapted from the previously described transfer of the ecotropic receptor (mCAT-1) into several human cells [14, 20].

EndoC-βH2 and 293 T cells were transduced with RCANBP(A) RIP containing either 3HAH2BYFP or GFP-IRES-ZeoR using TVA-gesicles. As controls, both cell lines were also transduced with RCANBP(A) CMV MCS and RCASBP(A) MCS containing the same transgenes (Fig. 4). Both fluorescent reporters showed a much stronger expression in EndoC-βH2 cells transduced with RCANBP(A) RIP retrovector compared to 293 T cells (Fig. 5). As another comparison, while both RCANBP(A) RIP constructs were more expressed than RCASBP(A) constructs in EndoC-βH2 cells, it was the other way around in 293 T cells (Fig. 5). Moreover, while the fluorescence emitted by RCANBP(A) RIP-transduced EndoC-βH2 is in the same range to that of RCANBP(A) CMV-transduced EndoC-βH2 cells (about 1/2 to 1/3), the difference is dramatically greater in 293 T cells (about 1/50 to 1/100 fold) (Fig. 5, see also Additional files 5 and 6). Similar results were obtained in MIN6-TVA cells (Additional file 3 and Additional file 10). Thus, when delivered by an α-retrovector, RIP is much more active in β cell lines than in a non-β cell line. Of note, RIP is also barely active in chicken DF1 fibroblasts (Additional files 6 and 9), a useful feature to overcome the problem raised by the production of replicative retrovectors harboring a cytostatic or cytotoxic transgene.

Interestingly, the few fluorescent 293 T cells that were transduced with RCANBP(A) RIP 3HAH2BYFP mainly showed an evenly distributed intracellular signal, while the same construct gave the expected strictly nuclear signal in mouse or human β cells (Additional file 5). This suggests that the vector has somehow undergone « aberrant » events upon or after integration (e. g. rearrangement and/or abnormal splicing / transcriptional initiation) in most positive 293 T cells.

It may appear puzzling that despite being very dimly fluorescent, 293 T cells transduced with RCANBP(A) RIP GFP-IRES-ZeoR were resistant to zeocin (Fig. 5 and Additional file 6). Indeed, a RIP-driven bicistronic GFP-IRES-ZeoR transcript should preferentially be translated into GFP given the 5′ (Cap) position of its coding sequence [21]. This uncoupling between the expression of the two markers suggests that the resistance to zeocin does not arise from RIP-driven transcripts, but rather from the LTR-driven « genomic » transcript (5'R-U5-Gag-Pol-Env-RIP-GFP-IRES-ZeoR-U3-R3') translated into ZeoR thanks to the IRES. This hypothesis is further supported by the fact that 293 T-TVA cells transduced with RCANBP(A) RIP ZeoR are sensitive to zeocin while cells transduced with RCASBP(A) ZeoR or RCANBP(A) CMV ZeoR are resistant (Additional file 6). Thus, RIP-driven transcription is not sufficient to confer resistance to zeocin in 293 T cells. These data confirmed that RIP brought by an α-retrovector is barely active in non-β cells.

α-Retrovectors are not trans-complemented by BXV1

ASLV(A) and MLV, although sometimes both referred to as « onco-retroviruses », are actually prototypes of two phylogenetically distant genus of the retroviridae family [22]. We thus anticipated that BXV1, a γ-retrovirus,

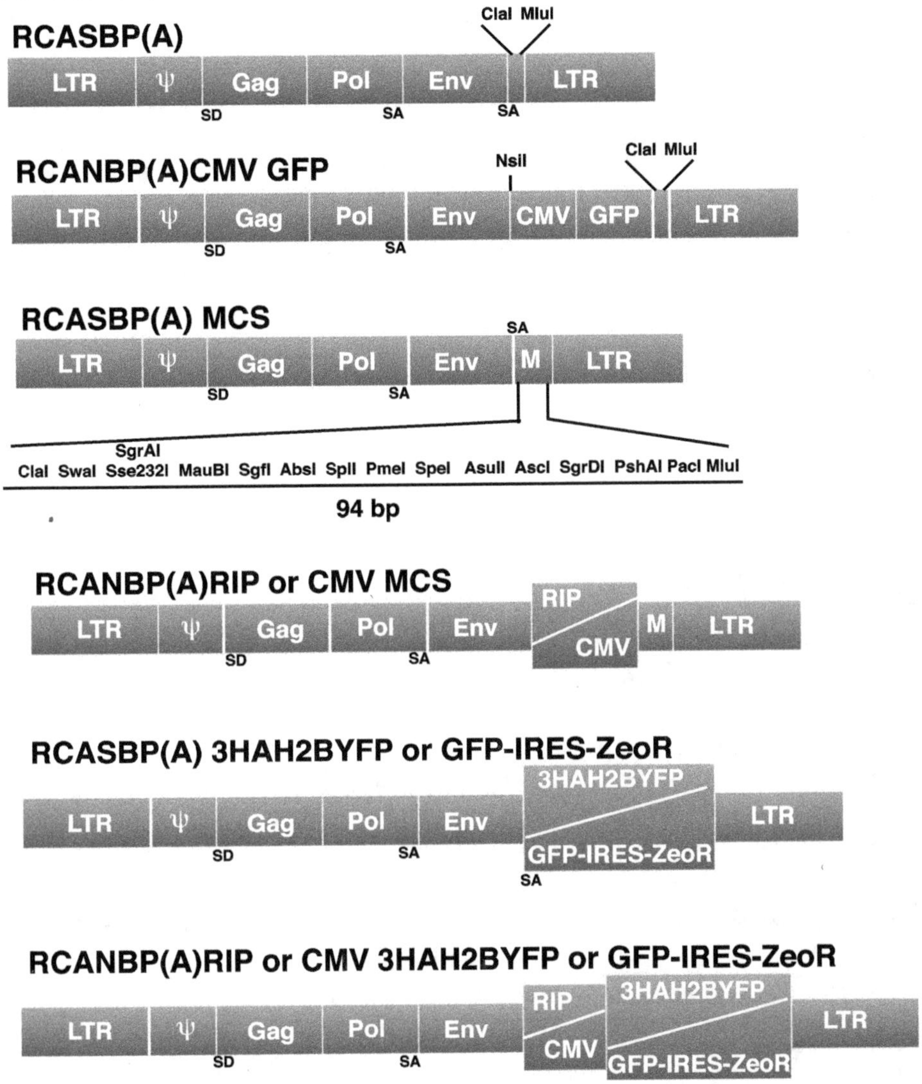

Fig. 4 Schematic diagram of the α-retrovectors. The RCASBP(A) and RCANBP(A) CMV GFP [17–19] were used as starting vectors to generate three derivatives. All these derivatives were equipped with a large MCS (M, detailed in RCASBP(A) MCS) and one of them, RCANBP(A) RIP MCS, contains the RIP sequence as internal promoter. The cDNAs subsequently inserted in these derivatives are indicated. SD: donor splice site. SA: acceptor splice site. The different parts of the diagrams are not at scale

would, at most, barely trans-complemente α-retrovectors. To test this hypothesis, we generated EndoC-βH2 cells sequentially transduced, first with the non-SIN γ-retrovector pPRIHy-TVA, and then with RCANBP(A) CMV GFP-IRES-ZeoR. While this derivative is doubly resistant and GFP positive (see Fig. 7b), its conditioned medium endowed naive 293 T cells with resistance to hygromycin but not to zeocin (62 and 77 versus 0 focus, respectively, in two experiments) (Fig. 6a). Concordant results were obtained when we compared the conditioned media of EndoC-βH2 cells transduced with either RCANBP(A) CMV GFP-IRES-ZeoR or, as a positive control, the non-SIN γ-retrovectors pPRIZ GFP (0 versus more than 100 in two experiments, respectively) (Fig. 6b). Thus, non-SIN α-retrovectors can be used in EndoC-βH2 cells without the safety concerns raised by non-SIN γ-retrovectors.

Transduction at defined retroviral copy number

We next performed transductions at defined retroviral copy number of the RIP retrovectors to verify that their preferential expression in β cells is not due to higher viral titer and/or integration efficiency. pSERS RIP transgene was much more expressed in EndoC-βH2-TVA than in 293 T-TVA cells at a low and similar mean retroviral copy number per cell (1 vs. 1.5 respectively, Fig. 7a). Importantly, this specificity was observed according to two criterias: i) the percentage of fluorescent cells (Fig. 7a) and ii) the mean fluorescence intensity (MFI) fold increase,

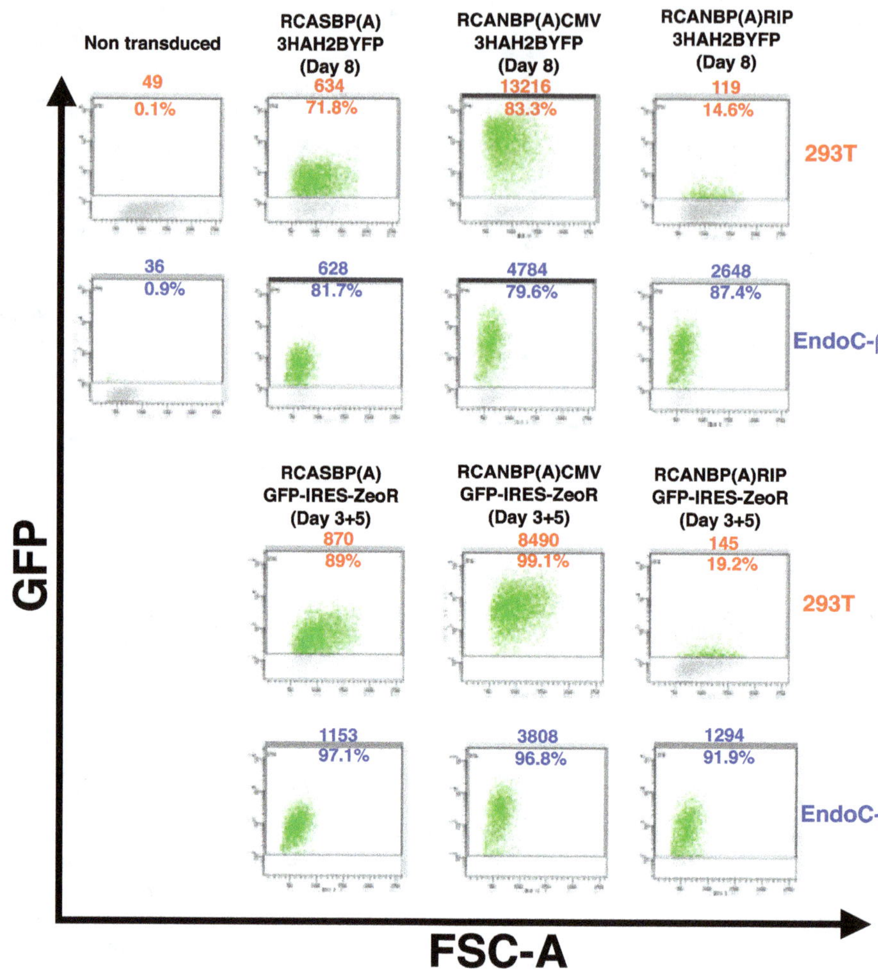

Fig. 5 An α-retrovector mediating preferential transgene expression in β cells. EndoC-βH2 and 293 T cells were transduced with the indicated α-retrovectors using TVA-gesicles. Cells were harvested 8 days later and GFP fluorescence was analyzed by flow cytometry. For cells transduced with a selectable construct (GFP-IRES-ZeoR), 8 days corresponds to 3 days without zeocin and 5 days with zeocin (3 + 5). The number above each diagram indicates the mean of fluorescence of the whole population of analyzed cells. The number in the diagram is the percentage of fluorescent cells. This experiment was performed once, with two distinct reporters in the three α-retrovectors giving similar results

defined as the fluorescence intensity measured in all cells exposed to the retroviral supernatant divided by that measured in control cells that were not exposed to the retroviral supernatant (Fig. 7b). A preferential expression, albeit less pronounced, was also observed with RCANBP(A) RIP at a higher retroviral mean copy number per cell (10,7 in EndoC-βH2-TVA vs 8,8 in 293 T-TVA, Fig. 7a and b). Conversely, the same transgenes carried by the corresponding non-RIP retrovectors (pSERS SF and RCANBP(A) CMV) were more strongly expressed in 293 T-TVA cells than in EndoC-βH2-TVA cells (Fig. 7a and b). Finally, transgene expression is remarkably stable over time (38 days for pSERS RIP and 48 days for RCAN RIP) in human β cells without any selection, as demonstrated by the percentage of fluorescent cells or fluorescence intensity (Fig. 7a and b).

RIP activity in α- and SIN γ-retroviral contexts faithfully reflects regulations affecting endogenous INSULIN gene expression

EndoC-βH2 cells have been generated through lentivector-mediated transfer of two immortalizing transgenes, hTERT and SV40 T, which can be eliminated upon sustained expression of the CRE recombinase [6]. Under these conditions, EndoC-βH2 cells undergo several changes akin to terminal differentiation: i) cessation of proliferation; ii) withdrawal from the cell cycle; and iii) increased expression of some key β cell genes, including IAPP (AMYLIN) and *INSULIN*, most likely at the transcriptional level [6, 14]. Most of these changes can be recapitulated upon transient knock-down of SV40 T (Fig. 8a and our unpublished data, [23]). Interestingly, knock-down of SV40 T also led to increased GFP fluorescence intensity in both pSERS RIP GFP-

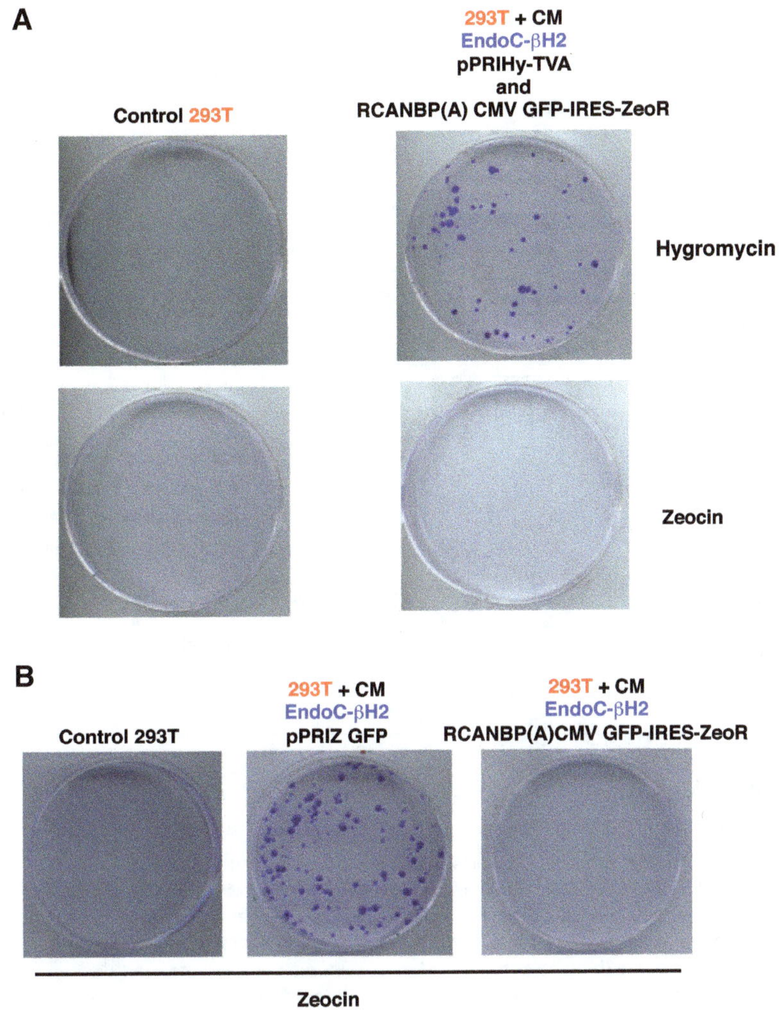

Fig. 6 α-retrovectors are not mobilized by the BXV1 xenotropic γ – retrovirus in EndoC-βH2 cells. **a** EndoC-βH2 cells transduced with the non SIN γ-retrovector pPRIHy TVA encoding a selectable marker (HygroR) and the TVA receptor (EndoC-βH2-TVA) were further transduced with an α-retrovector, (RCANBP(A) CMV GFP-IRES-ZeoR) encoding a distinct selectable marker (ZeoR). Naive 293 T cells were exposed to the conditioned medium (CM) of the doubly transduced EndoC-βH2 cells and next divided and cultured into two petri dishes in presence of either hygromycin or zeocin. After 10 days, exposed and selected 293 T cells were fixed and colored with crystal violet. The same doubly transduced EndoC-βH2 derivative (EndoC-βH2-TVA RCANBP(A) CMV GFP-IRES-ZeoR) is used in Fig. 7**b**, showing that it is heavily positive for GFP expression. This experiment was done in duplicate, with similar results, and further confirmed using the CM of EndoC-βH2- transduced with both pPRIHy TVA and RCASBP(A) GFP-IRES-ZeoR) (not shown). **b** Same experiments as in (a), but naive 293 T cells were exposed to the conditioned medium (CM) of EndoC-βH2 harboring only one retrovector, as indicated, each encoding ZeoR. This experiment was done in duplicate, which gave similar results

IRES-ZeoR and RCANBP(A) RIP GFP-IRES-ZeoR-transduced EndoC-βH2 cells (Fig. 7b).

The neutralization of the RB family of « pocket proteins » (RB, P107(RBL1) and P130(RBL2)) through direct physical interaction is one of the most important mechanisms underlying SV40 T induced-immortalization [24]. Quite surprisingly, the pocket proteins activity was not completely abrogated by a RIP-driven SV40 T transgene in a mouse model of insulinoma. Accordingly, Rb knock-out promotes proliferation and accelerates tumorigenesis in this experimental setting [25]. As EndoC-βH2 cells also harbor a RIP-driven SV40 T transgene [6], we examined whether pocket proteins depletion in these cells might cause changes reciprocal to those observed upon SV40 T depletion. Indeed, the transient knock-down of both RB and P130 mRNA (the two most highly expressed pocket proteins encoding transcripts in EndoCβH2 cells) reduced INS and IAPP expression (Fig. 8a). Similarly, RIP activity in the α- and SIN γ-retrovectors is reduced upon RB or RB plus P130 knock-down, respectively (Fig. 8b). Thus RIP activity again mirrored the fluctuations affecting the endogenous INS gene expression (Fig. 8b).

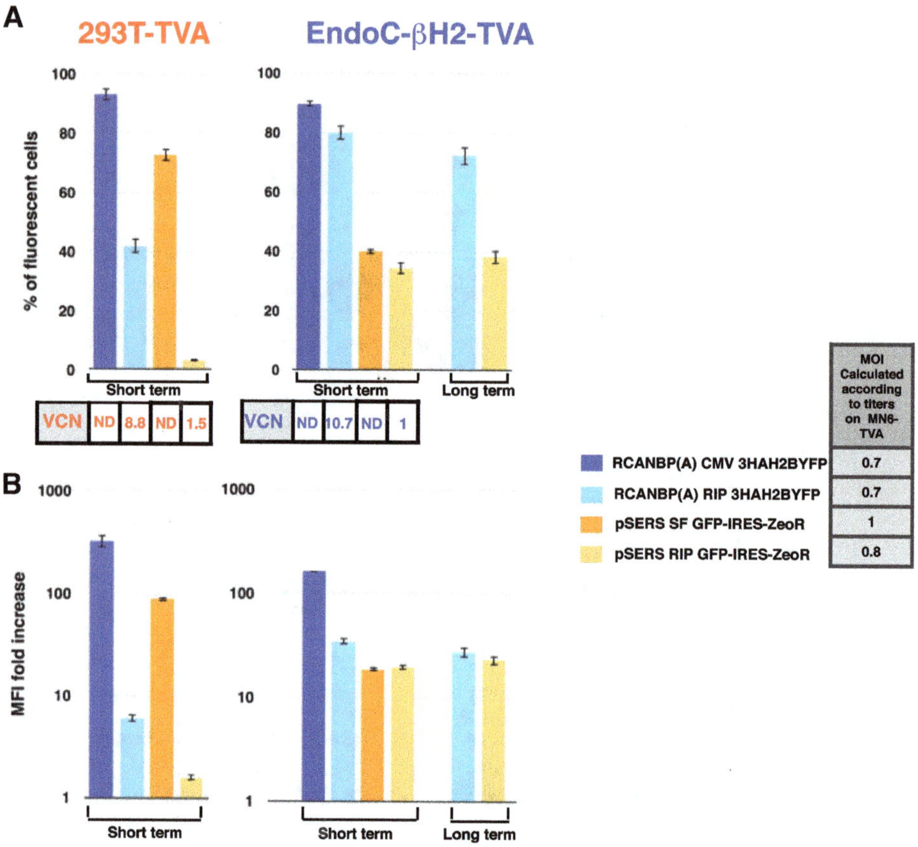

Fig. 7 Transgene expression level and stability upon transduction with titrated RIP and non-RIP retrovectors. Two α- (RCANBP(A) CMV or RIP 3HAH2BYFP) and two γ- (pSERS SF or RIP GFP-IRES-ZeoR) retroviral supernatants, harboring either an ubiquitous (CMV or SFFV U3) or a β-specific (RIP) promoter were titrated in mouse MIN6-TVA insulinoma cells. Two human cells lines, either 293 T-TVA (left) or EndoC-βH2-TVA (right) cells were next transduced with the indicated retroviral supernatant, at a similar MOI (see internal legend) according to this titration. The percentage of fluorescent cells (**a**) and the MFI fold increase (**b**) were determined by flow cytometry 6 days after transduction (short term). The histogram shows the means and SD of the measures in three independent populations. The estimated mean viral copy number per cell (VCN) is indicated below each corresponding column. For clarity, VCN values have been mentioned only below the upper histogram. The same three populations of EndoC-βH2-TVA cells transduced with RCANBP(A) RIP 3HAH2BYFP or pSERS RIP GFP-IRES-ZeoR were cultured without selection and re-analyzed by flow cytometry for MFI fold increase and percentage of transduced cells 48 or 38 days after transduction, respectively, (long term). The MFI fold increase and percentage of fluorescent cells correspond to the mean and SD of the values measured in the three independent populations. ND: not done

In summary, the variations of RIP activity in both α- and SIN γ-retroviral contexts faithfully reflected both an up-regulation and a down-regulation of INS gene expression. This tight correlation is specific to the RIP since the transfection of the same siRNA did not alter the activity of other promoters tested in similar retroviral contexts, with the exception of CMV in α-retrovector upon SV40 T depletion (Fig. 8b).

Discussion

Our present work describes new retrovectors, termed pSERS RIP MCS and RCANBP(A) RIP MCS, designed to selectively express transgenes in pancreatic β cells. They are easy to produce and to concentrate, and harbor a large Multi Cloning Sites to facilitate DNA cloning. Both allow a strong preferential transgene expression in mammalian (mouse or human) β cells lines over non-β cell lines. Importantly, reaching a transgene expression level within the range of abundantly expressed endogenous mRNA usually is a desirable setting for practical applications. Satisfactorily, after more than 50 days without selection, transgene mRNA expression level in human β-cells transduced with either pSERS RIP or RCANBP(A) RIP was in the range of CYCLOPHILIN A (CYPA/PPIA) and INSULIN expression. Moreover, short selection of pSERS RIP GFP-IRES-ZeoR transduced cells in zeocin further increased transgene expression (Additional file 8). Our new retrovectors thus appear as convenient and promising tools to study the genetic control of β cell development or physiology. Accordingly, transgene in pSERS RIP and, to an even greater extent, in

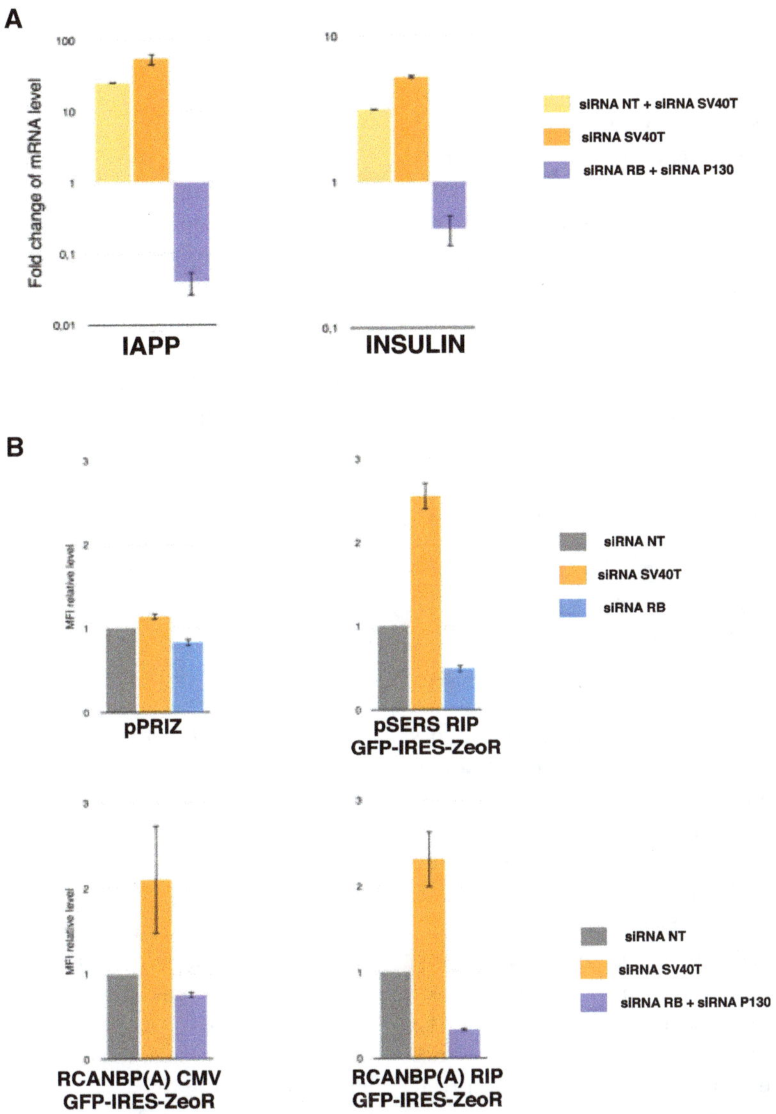

Fig. 8 Modulations of the RIP activity in both SIN γ- and α-retrovectors tightly correlate with those affecting endogenous INSULINgene expression. a EndoC-βH2 cells or their derivatives stably expressing fluorescent reporters were transfected with the indicated siRNA. 6 days later, cells were harvested and changes in INSULIN and IAPP mRNA levels were determined through qRT-PCR analyses, and represented as fold change over control siRNA only (siRNA NT, Non Targeting) transfected cells. Each co-transfection (either siRNA NT + siRNA SV40T or siRNA RB + siRNA P130) was done with an equimolar mix (final concentration of 40 nM for each siRNA), while single transfection (either siRNA NT, siRNA SV40T only) was done at a final concentration of 80 nM. The results represent the mean and standard deviation (SD) of at least 3 experiments for each mRNA level analyzed, each point made in duplicate **b** EndoC-βH2 cells or their derivatives expressing the TVA receptor were stably transduced with the indicated γ – or α – retrovector and then transiently transfected with the indicated siRNA at a final concentration of 80 nM (80 nM for the sIRNA in single transfections, 40 nM for each siRNA in co-transfections). GFP fluorescence level was analyzed by flow cytometry 6 days after transfection. Results were expressed as fold variation in MFI (mean of fluorescence intensity) with the value in cells transfected with the siRNA arbitrarily taken as 1. The histograms show the means and SD of three (upper panels) of four (bottom panels) experiments

RCANBP(A) RIP, also display a strong selectivity for β cell expression upon transduction of dissociated pancreatic buds from E11-E12 mouse embryo (our unpublished data, manuscript in preparation). Although we have transferred DNA sequences encoding protein(s) in the present study, both MLV- and ASLV(A)-based retrovectors are also convenient for loss-of-function experiments [18, 19, 26–28].

Importantly, these new retrovectors are not, or extremely faintly, mobilized by the BXV1 xenotropic γ-retrovirus. Presumably, SIN γ-retrovectors are not trans-complemented because they mostly encode transcripts devoid of the Ψ packaging sequence, while α-retrovectors are too loosely related to BXV1 to use its gag- and pol-encoded proteins. Unlike non-SIN γ-retrovectors, those newly constructed retrovectors can therefore be serenely used to transduce EndoC-βH2 cells or any cells potentially or certainly contaminated with a xenotropic γ-retrovirus. Arguably, ASLV(A)-based retrovectors are even safer than SIN γ-retrovectors under this criteria since the formers are also very unlikely to recombine with any mammalian endogenous retroviral elements [29]. The safety of RCASBP(A)/RCANBP(A) vectors stems also from the observation that they appear defective in mammalian cells, even though they harbor the complete set of viral genes that support their efficient replication in avian cells [18, 19]. Yet, presumably because our experiments favored heavy transduction, a faint production of RCANBP(A) CMV particles was detected in transduced mammalian β or non-β cells, including cell lines, such as 293 T (Additional file 7) or rat INS1 (data not shown), not infected with xenotropic γ-retrovirus (Fig. 3b and [16]). In EndoC-βH2 cells, this faint production may cause the release of a few « xenotropic » RCANBP(A) CMV particles, as it bypasses the need for a true trans-complementation (i.e sharing of gag- and pol-encoded proteins) and only requires the pseudotyping of budding particles by the xenotropic Env protein. Accordingly, we sometimes observed a very limited number of zeocin-resistant foci in the experiments described in Fig. 6 (at most two) when the conditioned medium was harvested on very dense cell cultures (data not shown).

We also report here two methods to make mammalian β cells permissive to ASLV(A)-based vectors. The first method is based on transient gesicles-mediated delivery of the quail TVA receptor. We previously reported the efficiency of this method to transfer the ecotropic receptor (mCAT-1) to human β cells [14]. The second method is the stable expression of the quail tva cDNA, as previously described in transgenic mice or in other cell lines [17–19], It works also very well in human, mouse or rat β cells (Figs. 6a, 7b, Additional files 3, 5, 8, 9 and data not shown). Although more time consuming, this method allows TVA-expressing cells to stably co-express a selectable resistance gene (HygroR in our experiments). This resistance can be exploited in co-culture experiments to transduce mammalian β cells followed by a selective elimination of RCASBP(A)/RCANBP(A)-producing chicken DF1 fibroblasts (Additional file 9).

Transduction by α- and γ-retrovectors were consistently reported to be dependent on the proliferation of the target cells [18, 19]. This raises an obvious limitation for their use on mammalian β cells, which mainly are quiescent, and even probably post-mitotic in adult [30]. However, several situations throughout mammalian lifespan are associated with a significant proliferation of β cells, such as during gestation, in response to obesity (at least in rodents) and during the perinatal period (in rodents and humans) [30, 31]. β cells under such metabolic challenges should therefore be permissive to transduction with α- as well as γ-retrovectors. Experimental manipulations, such as treatment with growth hormone, could also mimic these situations and stimulate β cell proliferation [32]. Moreover, it actually is not so clear that quiescent cells are fully refractory to α-retrovirus transduction. Indeed, this assumption partly stems from observations primarily made on MLV, which requires mitosis for nuclear entry and integration of viral DNA, and was extrapolated to all « onco-retroviruses » [33]. Interestingly, early works reported that α-retrovirus (RSV) integration occurs in the absence of mitosis. In fact, either artificially arrested cell lines as well as, most importantly, naturally arrested in vitro differentiated mouse embryonic neurons, remain remarkably permissive for ASLV transduction [34]. Overall, the efficiency of α-retrovectors on non-dividing cells appears higher to that of γ-retrovectors, even though lower to that of lentivectors [33, 34]. Thus, α-retrovectors may work to some extent on differentiated (quiescent) rodent and human β cells, especially on those that have « recently » exited the cell cycle [34].

Finally, we show here that genetic manipulations that either enhance (SV40 T depletion) or decrease (RB and p130 depletion) endogenous human *INSULIN* expression, elicit similar modulations of the RIP activity in both α- and SIN γ-retroviral contexts. Transducing EndoC-βH2 cells with either pSERS RIP GFP-IRES-ZeoR or RCAN RIP GFP-IRES-ZeoR thus provide a convenient tool to monitor *INSULIN* gene regulation. Consequently, these derivatives can be used in high throughput screens for molecules modulating human *INSULIN* expression. A similar reporter cell line was recently designed in mouse β cells (MIN6) through the lentiviral transfer of a destabilized GFP under the control of the proximal human *INSULIN* promoter [35], indicating that its regulation can also be studied using lentivectors. However, it is worth noting that α-retroviruses integrate quite randomly throughout the genome, while lentiviruses tend to target active transcription units [36], which might not only distort normal host cell functions, but also transgene regulation. This feature of α-retroviruses is of particular interest to transfer and study short regulatory DNA regions. We thus guess that the regulation of other specific promoters of chosen pancreatic subpopulations will also be preserved upon transfer with α-retrovectors, although this remains to be tested.

Interestingly, the two means we have used to modulate *INSULIN* gene expression are unlikely to only act (directly or indirectly) on insulin (endogenous and exogenous) promoters. Indeed, the observed modulations are accompanied by variations in the expression of other β markers or in the proliferative activity of the cells. Therefore, the depletion of SV40 T or RB/P130 triggers a more global (and seemingly reciprocal) change in EndoC-βH2 cell phenotype of which *INSULIN* mRNA expression (or RIP activity) is only one aspect. Hence, our two RIP-GFP-transduced EndoC-βH2 derivatives could provide convenient read-out to follow up changes affecting not only *INSULIN* expression but also human β cell differentiation status.

Conclusions

In the present study, we designed and validated one SIN (Self-Inactivating) γ-retrovector (MLV-based) and one α-retrovector (ASLV(A)-based), each harboring the Rat Insulin Promoter for specific transgene expression in pancreatic insulin-producing β cells. We show that: 1) these two retrovectors efficiently transduce mammalian (mouse or human) β cell lines, and we describe here a rapid method to transiently transfer the (avian) ASLV(A) receptor TVA into mammalian cells; 2) the transgenes they harbor display a strong preferential expression in β cells compared to non-β cells; 4) these two retrovectors contain a large Multi Cloning Sites, and retroviral supernatants can easily be prepared and concentrated if necessary; 5) in contrast to non-SIN γ-retrovectors, they are extremely faintly mobilized by BXV1, and can thus safely be used in human EndoC-βH β cell lines or in any cells contaminated with a mouse xenotropic γ-retrovirus; 5) the activity of the Rat Insulin Promoter in the context of both retrovectors is co-regulated with that of endogenous human *INSULIN* gene in various experimental conditions. We conclude that these new retrovectors are convenient and promising tools to study the genetic control of β cell development or physiology.

Methods

Constructs

γ-Retrovectors

We started from the SIN γ-retrovector pSERS11 SF GFP pre (a gift from Axel Schambach and Christopher Baum). Its composite 5′ promoter (RSV, Rous Sarcoma Virus U3 region fused to the SV40 virus enhancer) increases rate and processivity of transcription. Accordingly, transiently transfected 293 T cells with pSERS11 SF GFP pre produce high levels of full-length (genomic) retroviral RNA resulting in high titer supernatants [15]. We replaced part of the Ψ region, the internal SF promoter (U3 region from the SFFV retrovirus) and the GFP sequence by a synthetically synthetized DNA sequence (Eurogentec) comprising the same part of the Ψ region, the RIP 405 promoter [6, 10], and a new MCS. This linker was digested by Bsp120I and XhoI (all the restrictions enzymes were purchased from ThermoFischer or New England Biolabs) and introduced into pSERS11 SF GFP pre-digested by Bsp120I and SalI (see Fig. 1 for the location of the relevant restriction sites). pSERS SF MCS was next constructed by taking a fragment containing the MCS from pSERS RIP MCS through a ClaI and Van91I digestion (Fig. 1) which was then cloned by the same enzymes into pSERS11 SF GFP pre (which removed GFP). The GFP-IRES-ZeoR cassette was taken from RCASBP(A) GFP-IRES-ZeoR (see below) by a MreI and PacI digestion and cloned into the pSERS RIP or SF MCS digested by AgeI and PacI. pSERS RIP and SF 3HAH2BYFP were constructed by cloning a BstXI PacI fragment from pPRI 3HAH2BYFP (see below) into the vectors digested by the same enzymes. peGFP-N1 (Clontech) was digested by BamHI and NotI to harvest the GFP sequence which was inserted by the same enzymes into pPRIZ 204, a derivative of pPRIZ [14] containing a single PacI site 3′ to ZeoR. pSERS RIP and SF GFP-IRES-ZeoR were constructed by a MreI and PacI digestion of RCAN GFP-IRES-ZeoR (see below) cloned into AgeI and PacI digested vectors. pPRI 3HAH2BYFP was constructed by cloning a synthetically synthetized new 5′ part of 3HAH2BYFP containing 5′ ClaI and BstXI sites and a 3′ AgeI site (Eurogentec) into a ClaI and AgeI digested pPRIPu 3HAH2BYFP. The resulting 3HAH2-BYFP sequence was digested by ClaI and DraI and inserted into RCASBP(A) MCS digested by ClaI and PshAI. Finally, 3HAH2BYFP was digested by BstXI and PacI to be cloned into pPRIG 204 (a derivative of pPRIG containing a single PacI site 3′ to GFP) which exactly replaced GFP by 3HAH2BYFP to give pPRI 3HAH2BYFP 204 (termed pPRI 3HAH2BYFP in text for simplicity). All the sequences of the final γ-retrovectors are available upon request.

α-Retrovectors

RCANBP(A) CMV GFP and RCASBP(A) (two gifts from Stephen Hughes and Andrea Ferris) were digested by ClaI and MluI and filled with a ClaI and MluI compatible double strand primer (Eurogentec) containing several new uniques cloning sites (Fig. 4). This step provided the RCASBP(A) MCS and RCASBP(A) CMV MCS derivatives. Next, RCANBP(A) CMV MCS was digested with NsiI and ClaI (Fig. 4) and filled with a synthetically synthetized RIP405 promoter (Eurogentec) giving rise to RCANBP(A) RIP MCS. 3HAH2BYFP was constructed by cloning a double strand DNA encoding 3HA linker (Eurogentec) compatible with ClaI in 5′ and BglII in 3′ and H2BYFP from pCMV H2BYFP (a gift from Dr. Chantal Desdouets, Institut Cochin, Paris, France) digested by BglII and MunI into ClaI and MunI digested pPRIZ 204. The resulting vector (pPRIZ 204 3HAH2BYFP) was next digested by ClaI and DraI to harvest 3HAH2BYFP, which was cloned into ClaI and

PshAI digested RCASBP(A) MCS, RCANBP(A) RIP MCS and RCANBP(A) CMV MCS. RCASBP(A) GFP-IRES-ZeoR was constructed by a NarI and PacI digestion of pPRIZ 204 GFP to harvest GFP-IRES-ZeoR cloned into ClaI and PacI digested retrovector. The same cassette was inserted through a SgsI and PacI digestion into MauBI and PacI digested RCANBP(A) CMV MCS and RCANBP(A) RIP MCS. All the sequences of the final α-retrovectors are available upon request.

Expression vector

pCMV-TVA was constructed by taking the tva cDNA through a SmaI (a site present just 5′ to the tva cDNA in the starting vector, pENTER-TVA, a gift from Philippe Ravassard, IMC, Paris) and BglII digestion from pPRIHy-TVA [16] cloned into Eco47III and BamHI-digested peGFP-C1 (Clontech) to replace GFP by TVA.

Production of retroviral supernatants and TVA gesicles

Ecotropic and VSV-G (Vesicular Stomatis Virus Glycoprotein)-pseudotyped γ-retrovectors were prepared as previously described [14, 16].

α-retrovectors supernatants were prepared as follows. First, 1 µg of retrovector plasmid was transfected into 30–40% confluent DF1 chicken fibroblasts (ATCC #CRL-12203) plated on six-well plates using 3 µl of X-tremeGENE 9 transfection reagent (Roche). DF1 cells were next amplified for about 8 days (following the extension of the fluorescent and/or selection marker). After this period, DF1 cells were grown to high confluency. α-retrovector supernatants were then harvested, filtered on 0.45 µm (Millipore) and concentrated (if needed) exactly like 293 T-produced γ-retrovector supernatants. Several other collects were made, if needed, during the ten following days, with an intervening 1/3 or 1/4 passage when required.

TVA gesicles were produced exactly like mCAT-1-gesicles [14] except that the pCMV-mCAT-1 expression vector was replaced by a pCMV-TVA expression vector.

All the supernatants, containing either retroviral particles or TVA-gesicles were concentrated, if necessary, on Amicon Ultra-15 (Merck) according to the recommandations of the supplier.

Cell culture

Cell lines

293 T and derivatives were cultured as previously described [14]. EndoC-βH2 cells (and derivatives) were cultured in a Advanced-DME/F12 (Thermo Fischer) -based medium supplemented with BSA fraction V (Roche) 9,8 g/500 ml; 2 β-mercapto-ethanol (Sigma) 50 µM; sodium selenite (Sigma) 6,7 ng/ml; transferrin (Sigma) 1,2 µg/ml; nicotinamide 10 mM and antibiotics (Thermo Fischer).

The EndoC-βH2-TVA, 293 T-TVA, MIN6-TVA and INS1-TVA derivatives were previously described as the non-SIN γ-retrovector pPRIHy TVA encoding TVA and a selectable marker (HygroR) was used to assess trans-complementation by BXV1 in all these cell lines [16]. All of them were continuously grown in presence of hygromycin. MIN6-TVA cells were cultured in a RPMI 1640-based medium (Thermo Fischer) supplemented with glutamax (Thermo Fischer); Fetal Calf Serum (Eurobio, 10%), 2-β-mercapto-ethanol (50 microM), sodium pyruvate (Thermo Fischer, 10 mM) and antibiotics (Thermo Fischer). Chicken DF1 fibroblasts were cultured in a F12/DMEM + Glutamine and Hepes based medium (Thermo Fischer) supplemented with fetal calf serum (Eurobio, 10%) chicken serum (Biowest 2%), sodium pyruvate (Thermo Fischer, 10 mM) and antibiotics. For selection and maintenance of selected cells, zeocin (Thermo Fischer) and hygromycin (Calbiochem) were used at 50–100 microg/ml for human β cells and at 100–200 microg/ml for rodent β cells, DF1 and 293 T cells.

Transduction

To compare the efficiency of γ-retrovectors on β and non-β cells, EndoC-βH2 or 293 T cells were plated at 10^6 cells in 25 cm^2 flasks. The day after, cells were exposed to 6 ml of γ-retroviral supernatants (pseudotyped by the VSV-G enveloppe) in presence of polybrene (5 microg/ml) for 4–6 h, and then grown with or without zeocin for the time lengths indicated in the main text and figures.

To compare the efficiency of the different γ-retrovectors on MIN6-TVA cells, 10^5 cells were plated on six well plates and exposed the day after to 10 µl of concentrated retroviral ecotropic supernatant in presence of polybrene (5 µg/ml) and then grown with or without zeocin for the time lengths indicated in the main text and figures.

To compare the efficiency of α-retrovectors on β and non-β cells, EndoC-βH2, 293 T or MIN6-TVA cells were plated at 3×10^5 cells in six-well plates. The day after, cells were exposed to 50 µl of concentrated retroviral supernatants, along with 20 µl of concentrated TVA-gesicles and polybrene (5 µg/ml) for TVA-negative cells. Transduced cells were then grown with or without zeocin for the time length indicated in the main text and figures.

For all transductions of mammalian cells with α-retrovectors, the TVA receptor was either transiently provided through gesicles (in « parental » EndoC-βH2 or 293 T cells), or stably expressed through prior transduction of the cells with the γ-retrovector pPRIHyTVA. In the latter cases, the cells are always denoted X-TVA (e.g. EndoC-βH2-TVA, MIN6-TVA, 293 T-TVA).

Trans-complementation/mobilization assays

Trans-complementation, or retrovector mobilization, by BXV1 was assessed by exposing naive 293 T cells (10^5 in six-well plates) to 1–2 ml of the tested conditioned medium (filtered on 0.45 μm, Millipore) for 4–6 h in presence of polybrene (5 μg/ml). The day after, cells were transferred in duplicate (and divided in two for cells transduced with two retrovectors) into the selective medium (containing zeocin or hygromycin) in 6 or 10 cm dishes to be and fixed and colored with crystal violet after 10–12 days.

Knock-down experiments

EndoC-βH2 cells were plated at 40 × 10^3 cells/cm^2 in either six well-plates or 25cm^2 flasks. The day after, they were transfected with the siRNA at the final concentration of 80 nM using Lipofectamine RNAiMax (Thermo-Fischer) according to the manufacturer's instructions. siRNA RB, siRNA P130 (RBL2) or siRNA control (NT: Non targeting) were genome SMARTpool from Dharmacon(M-003296-03-0005, M-003299-03-0005, D-001206-13-20). siRNA SV40T (referred to as siRNA 2047 in [37]) was synthetized by Thermo-Fischer. Transfected cells were harvested for RNA and/or GFP fluorescence analysis 6 days after transfection.

Cytometry

Cytometry analyzes to measure the intensity of GFP/YFP fluorescence (expressed as mean of fluorescence) were carried out as previously described [14].

qRT-PCR analyses

Quantitative RT-PCR experiments were performed, normalized and plotted as previously described [14]. All the primers used were previously detailed [6].

Titration of retroviral supernatants

Titration of four retroviral supernatants (pSERS SF or RIP GFP-IRES-ZeoR pseudotyed by VSV-G and RCANBP(A) CMV or RIP 3HAH2BYFP), was carried out on a reference cell line, mouse MIN6-TVA insulinoma, which efficiently expresses transgenes under the control of both ubiquitous (SF, CMV) and β-specific (RIP) promoters. Either fresh and non-concentrated (pSERS SF or RIP GFP-IRES-ZeoR) or concentrated and frozen (RCANBP(A) CMV or RIP 3HAH2BYFP) retroviral supernatant was titrated. Three doses of supernatants (25, 100 or 500 μl plus polybrene for pSERS SF or RIP retrovectors; 1, 3 or 10 μl for RCANBP(A) CMV or RIP retrovectors) were added to 2 × 10^5 MIN6-TVA cells in six-well plates. The percentage of transduced (fluorescent) cells in each plate was measured by flow cytometry 3 days later, giving in each case two points of near-linearity. The titer was thus deduced from the value observed at the intermediate dose and used to transduce 2 × 10^5 EndoC-βH2-TVA or 293 T-TVA cells in six-well plates at a defined MOI. To reach the indicated MOI (see Additional file 8), 400 μl of (fresh) pSERS SF or RIP GFP-IRES-ZeoR plus polybrene or 5 μl of (concentrated/frozen) RCANBP(A) CMV or RIP 3HAH2BYFP supernatant was added to the cells. The retroviral copy number was consistent with that expected from the MOI for the transduction of 293 T-TVA and EndoC-βH2-TVA cells with pSERS RIP but higher than expected for RCANBP(A) RIP 3HAH2BYFP (see below and Fig. 7). This indicates that the titration on MIN6-TVA cells is accurate for pSERS RIP retrovectors but underestimates the efficiency of the RCANBP(A) RIP retrovector on 293 T-TVA and EndoC-βH2-TVA cells. Yet, this discrepancy does not preclude the comparison of transgene expression level per retroviral copy in 293 T-TVA and EndoC-βH2-TVA transduced by RCANB(A) RIP 3HAH2BYFP, as this number of copy is similar in both cell lines (see below and Fig. 7).

Determination of mean retroviral copy number per cell

Two EndoC-βH2-TVA or 293 T-TVA populations transduced with either pSERS RIP GFP-IRES-ZeoR or RCANBP(A) RIP 3HAH2BYFP were randomly chosen out of three populations generated for each vectors in each cell line. The genomic DNA of these four populations was extracted from using QIAamp DNA Micro (Qiagen). Determination of retroviral copy number was then carried out at Texcell (Evry, France). Genomic PCR was performed in duplicates for each population using the following GFP/YFP primers firstly validated on plasmids containing either GFP or YFP: CTTC AAGATCCGCCACAACATC (Fw) and GGGTCTTT GCTCAGGGCGGAC (Rev). For each PCR, the Ct (cycle threshold) value was converted to an absolute number of transgene copies using a standard curve drawn from serial dilutions of the control vectors. To evaluate the mean copy number per cell, this value was divided by the number of genomes estimated in two manners: i) the quantity of genomic DNA in the sample (50 ng) (TG/DNA); ii) the Ct obtained for the amplification of an endogenous gene, β2-microglobulin (Fw: AATT TCCTGAATTGCTATGT; Rev.: ACGGCAGGCATAYT-CATC; TaqMan probe TTCAGCAARGACTGGT CTTTCTATC) in the same genome using prior calibration on defined quantities of a control plasmid (TG/β2-m). These two methods gave quite concordant results, assuming that the genome of 293 T and EndoC-βH2 as triploid and diploid, respectively [6, 38] and two copies of β2-microglobulin per cell in both cell lines: 0.7 (TG/DNA) and 1.5 (TG/β2-m) retroviral copy per cell for 293 T-TVA pSERS RIP GFP-IRES-ZeoR; 4,6 (TG/DNA) and 8.8 (TG/β2-m) for 293 T-TVA RCANBP(A)

3HAH2BYFP; 1 and 1 for EndoC-βH2-TVA pSERS RIP GFP-IRES-ZeoR; 8.3 and 10.7 for EndoC-βH2-TVA RCANBP(A) 3HAH2BYFP. Given this concordance and the partial tetra- or penta-ploidisation of 293 T genome [38], which may lead to somewhat underestimate the retroviral copy number calculated through TG/DNA in these cells, only the TG/β2-m results (as VCN: mean viral copy number per cell) were mentioned in the figure (Fig. 7) and text.

Statistical analyses

Means and standard deviations were calculated using the NUMBERS software (Apple) and Student's t tests were performed using XLSTAT v2014 (Addinsoft, Brooklyn, NY) in Additional file 10.

Additional files

Additional file 1: Transgenes harbored by the pSERS SF γ-retrovector are efficiently expressed in both β and non-β cells. 293 T and EndoC-βH2 cells were transduced in parallel with either pSERS SF GFP-IRES-ZeoR or, as a control, pPRIZ GFP and GFP fluorescence was analyzed 4 days later by flow cytometry. The number above each diagram indicates the mean of fluorescence of the whole population of analyzed cells. The number in the diagram is the percentage of fluorescent cells. Note that the positive threshold level is slightly higher for 293 T cells transduced with pPRIZ GFP and pSERS SF GFP-IRES-ZeoR to take into account an increase in the autofluorescence level of negative cells (lower cloud). This experiment was done once. (DOCX 79 kb)

Additional file 2: Stable EndoC-βH2 populations transduced with either pPRIZ or pSERS RIP γ-retrovector. The two stable populations described in Fig. 1 (EndoC-βH2 transduced with either pPRIZ GFP or pSERS RIP GFP) were continuously cultured in presence of zeocin for 23 days (i. e.27 days after transduction: day 4 + 23) and GFP fluorescence was analyzed by flow cytometry. The two populations display a similar level of fluorescence intensity and are entirely constituted of fluorescent cells. Non-transduced cells are below the horizontal bar (see Fig. 3). The number above each diagram indicates the mean of fluorescence of the whole population of analyzed cells. The number in the diagram is the percentage of fluorescent cells. This experiment was done once. (DOCX 48 kb)

Additional file 3 Transduction of mouse MIN6 β cells with α- and γ-retrovectors. A. MIN6 -cell line derivative stably expressing the TVA receptor (MIN6-TVA) was transduced with the indicated γ – or α-retrovector and harvested at the indicated time for GFP fluorescence analysis by flow cytometry. When the retrovector contains a selectable transgene (ZeoR), transduced cells were selected for 3 days, starting 2 days after their exposure to the retroviral supernatant (day 2 + 3). The number above each diagram indicates the mean of fluorescence of the whole population of analyzed cells. The number in the diagram is the percentage of fluorescent cells. Note that the positive threshold level is slightly higher for cells transduced with pPRI 3HAH2BYFP to take into account an increase in the autofluorescence level of negative cells (lower cloud). These experiments were done once. (DOCX 79 kb)

Additional file 4: BXV1 efficiently mobilizes non-SIN, but not SIN, γ-retrovector in doubly transduced EndoC-βH2 cells. A. EndoC-βH2 cells transduced with the non-SIN γ-retrovector pPRIHy-TVA (EndoC-βH2-TVA) were further transduced with a SIN γ-retrovector (pSERS SF GFP-IRES-ZeoR). Each γ-retrovector harbors a distinct selectable marker (HygroR and ZeoR, respectively). Naive 293 T cells were next exposed to the conditioned medium (CM) of the doubly transduced EndoC-βH2 cells, then cultured in presence of either hygromycin or zeocin for 10 days, and finally fixed and colored with crystal violet. This experiment was done in duplicate, which gave similar results. B. The doubly transduced EndoC-βH2 cells display high GFP expression. This, together with their resistance to zeocin, confirms that they harbor pSERS SF GFP-IRES-ZeoR and express the corresponding transgenes. (DOCX 162 kb)

Additional file 5: Abnormal ubiquitous localization of 3HAH2BYFP in 293 T cells transduced with RCANBP(A) RIP 3HAH2BYFP. Some foci of fluorescent cells can be found in 293 T cells transduced with RCANBP(A) RIP 3HAH2BYFP. Cells within these foci often display an evenly distributed fluorescence, i.e. detectable in both the cytosol and the nucleus of the cells (left panel, upper image). In contrast, β cells (EndoC-βH2 or MIN6-TVA) transduced with the same retrovector display the expected nuclear fluorescence (left panel, middle and bottom images) while the same transgene in RCANBP(A) CMV-transduced 293 T also gives a strictly nuclear signal (right image). (DOCX 229 kb)

Additional file 6: RCANBP(A) RIP GFP-IRES-ZeoR transduced 293 T cells are resistant to zeocin thanks to non-RIP driven transcripts. A 293 T-TVA cells were transduced with the indicated α-retrovectors and cultured in presence of zeocin. The three retrovectors harbor ZeoR 3' to an IRES sequence (GFP-IRES-ZeoR). After selection, cells were analyzed for GFP fluorescence intensity by flow cytometry. 293 T-TVA cells transduced with RCANBP(A) RIP GFP-IRES-ZeoR are extremely dimly fluorescent compared to the two other populations. MFI: Mean of Fluorescence Intensity. All the analyzed cells were taken into account to determine the MFI for each population, including the negative cells in RCANBP(A) RIP GFP-IRES-ZeoR-transduced 293 T-TVA cells. This experiment was done twice, which gave concordant results. B. The same three transduced 293 T-TVA cells populations (left image, right column) as well as three other populations transduced in parallel with the corresponding non-IRES containing retrovectors (left image, left column) were cultured in presence of zeocin for 7 days, then fixed and colored with crystal violet. While RCANBP(A) RIP GFP-IRES-ZeoR confers zeocin-resistance, RCAN RIP ZeoR does not. Similar observations were made in DF1 chicken fibroblasts producing the tested supernatants (middle image). In contrast, the same supernatants are equally efficient to confer zeocin resistance to MIN6-TVA β (right image). The repartition of the six populations for each six-well plate is the same, and indicated below the images. This experiment was done twice, which gave concordant results. (DOCX 172 kb)

Additional file 7: Non-SIN α-retrovectors are not totally defective in mammalian cells. 293 T-TVA cells were transduced with the non-SIN α-retrovector RCAN CMV GFP-IRES-ZeoR and cultured 12 days in presence of zeocin. Then (that is 14 days and three passages after transduction), the conditioned medium (CM) of the 293 T-TVA RCANBP(A) CMV GFP-IRES-ZeoR cells were added to naive 293 T-TVA cells, which were next exposed to a zeocin-containing medium. 2 weeks later, the receiving cells were fixed and colored with crystal violet. Some zeocin-resistant foci can be observed (middle image). As a control, naive 293 T cells (not TVA) were subjected in parallel to the same CM exposure and zeocin selection. With these cells, no zeocin resistant foci were observed (bottom image) indicating that the transmission of zeocin resistance by the conditioned medium is dependent on TVA expression in the receiving cells. As a negative control, the conditioned medium of 293 T-TVA transduced with RCASBP(A) 3HAH2BYFP (devoid of ZeoR) was also tested under the same conditions on 293 T-TVA cells (upper image). This experiment was done once and its results were confirmed using the conditioned medium of RCASBP(A) GFP-IRES-ZeoR-transduced 293 T cells (not shown). (DOCX 119 kb)

Additional file 8: Transgene expression level compared to that of two endogenous mRNA. Transgene expression level in populations of EndoC-βH2-TVA cells transduced with either RCANBP(A) RIP 3HAH2BYFP or pSERS RIP GFP-IRES-ZeoR (the same three populations as in Fig. 7) was analyzed through qRT-PCR. Transgene mRNA level is expressed as the cycle threshold (Ct). Cells transduced with pSERS RIP GFP-IRES-ZeoR were not exposed to zeocin (middle panel), except for one aliquot during 12 days before harvesting (right panel). The transgene expression level was compared to that of two endogenous mRNA, *CYCLOPHILIN A* and *INSULIN*, in the same transduced cells, also shown as Ct. These analyses were performed on long term transduced cells (> 50 days after transduction) and the data are the means and SD of results obtained on the three independent population for each retrovector, with each point of the qRT-PCR done in duplicate. (DOCX 37 kb)

Additional file 9: Co-culture experiments to transduce TVA-expressing mammalian β cells with RCANBPB(A) RIP retrovector. DF1 chicken fibroblasts infected with RCAN RIP GFP-IRES-ZeoR were co-cultured with either EndoC-β H2-TVA or MIN6-TVA cells (day 0). These two derivatives express TVA together with HygroR. As the RIP is very barely active in DF1 cells, almost no fluorescent cells can be found at this time point. 2 days later, the co-culture were exposed to both zeocin, to remove untransduced β cells, and to hygromycin, to remove DF1 cells. The cells were photographed at the indicated times. After 8–10 days in hygromycin and with an intervening passage, almost all DF1 cells were eliminated. This experiment was done once and EndoC-βH2-TVA were transduced through co-culture with other RCASBP(A) or RCANBP(A) derivatives (not shown). (DOCX 91 kb)

Additional file 10: Comparison of RIP and non-RIP retrovector expression in β and non-β cells. Non-β 293 T-TVA and MIN6-TVA insulinoma cells were transduced with the indicated retrovectors, in each case a pair of corresponding α- (RCANBP(A)) or SIN γ – (pSERS) retrovectors harboring either a strong ubiquitous (either CMV or SF) or RIP as internal promoter. Cells were then cultured for 4 days without selection after transduction and fixed and analyzed for eGFP fluorescence by flow cytometry. Each transduction was done in triplicates. Upper panels: the percentages of fluorescent cells (means and SD) are shown. Bottom panel: for each pair of retrovectors, the relative efficiency corresponds to the percentage of fluorescent cells upon transduction with the RIP retrovector divided by the percentage observed in the same cell line transduced with the corresponding non-RIP retrovectors. Means, SD are given. For each pair of retrovectors, the ratio is significantly higher in β than in non-β cells according to a two-tailed Student's t test. (DOCX 51 kb)

Abbreviations

3HAH2BYFP: Fusion protein with three HA epitopes (3HA), Histone H2B (H2B) and Yellow Fluorescent Protein (YFP); AAV: Adeno Associated Virus; ASLV: Avian Sarcoma and Leukosis Virus; BXV1: B10 xenotropic virus 1 (also known as XMV-43, endogenous Xenotropic Mouse Virus 43); CMV: Cyto Megalo Virus; hTERT: human Telomerase Reverse Transcriptase; IAPP: Islet Amyloid Poly Peptide; IRES: Internal Ribosome Entry Site; LTR: Long Terminal Repeat; mCAT-1: mouse Cationic Amino acid Transporter 1 (also known as Slc7a1, SoLute Carrier family 7, member 1); MCS: Multi Cloning Site; MLV: Murine Leukemia Virus; MOI: multiplicity of infection; RCANBP(A): Replication-Competent ASLV long terminal repeat (LTR) No splice acceptor Bryan Polymerase subgroup A; RCASBP(A): Replication-Competent ASLV long terminal repeat (LTR) with a Spliceacceptor Bryan Polymerase subgroup A; RIP: Rat Insulin Promoter; RSV: Rous Sarcoma Virus; SFFV: Spleen Focus Forming Virus; SIN: Self-Inactivating; SV40: Simian Virus 40; TVA: Tumor Virus A; VSV-G: Vesicular Stomatitis Virus G Glycoprotein

Acknowledgements

We thank Axel Schambach and Chistopher Baum for the gift of the pSERS11 SF GFP pre plasmid, Stephen Hughes and Andrea Ferris for the gift of RCASBP(A) and RCANBP(A) CMV GFP plasmids, and to Philippe Ravassard for sharing the pENTER-TVA vector. We are also indebted to Karine Bailly (Cochin's CYBIO facility), and Chantal Bécourt for their help in flow cytometry, Romain Fragnoud (Texcell, Evry, France) for assistance in retroviral copy number determination, and Marc Diedisheim for help in statistical analyses.

Fundings

This work was supported by ANR-16-CE18–0005-01 (BromoB). The RS laboratory is supported by The Foundation Bettencourt Schueller, belongs to the Laboratoire d'Excellence consortium Revive and to the Departement Hospitalo-Universitaire (DHU) Autoimmune and Hormonal disease. The funders had no role in study design, data collection and analysis, decision to publish, or preparation of the manuscript.

Authors' contribution

OA wrote the paper with the help of RS and LH for the text and AM for the figures; OA conceived the constructs with the help of PM; OA conceived the experiments; OA constructed the vectors and performed the experiments with the help of AM, LH, DB and GC; OA analyzed the results with the help of RS. All authors read and approved the final manuscript.

Ethics approval and consent to participate

Not applicable.

Consent for publication

Not applicable.

Competing interests

The authors declare that they have no competing interests.

Author details

[1]INSERM U1016, CNRS UMR8104, Institut Cochin, Université Paris Descartes, 123 Boulevard de Port-Royal, 75014 Paris, France. [2]Université Côte d'Azur, CNRS UMR7277 INSERM U1099, iBV (Institut de Biologie Valrose), Université Nice Sophia Antipolis, Bâtiment Sciences Naturelles; UFR Sciences, Parc Valrose, 28, avenue Valrose, 06108 Nice Cedex 2, France. [3]Present Address: Laboratoire de Recherche Translationnelle en Immunothérapie, Institut Gustave Roussy, 114 Rue Edouard Vaillant, 94800 Villejuif, France.

References

1. Arda HE, Li L, Tsai J, Torre EA, Rosli Y, Peiris H, Spitale RC, Dai C, Gu X, Qu K, Wang P, Wang J, Grompe M, Scharfmann R, Snyder MS, Bottino R, Powers AC, Chang HY, Kim SK. Age-dependent pancreatic gene regulation reveals mechanisms governing human β cell function. Cell Metab. 2016;23:1–12. 27133132. https://doi.org/10.1016/j.cmet.2016.04.002.
2. George NM, Boerner BP, Mir SU, Guinn Z, Sarvetnick NE. Exploiting expression of hippo effector, yap, for expansion of functional islet mass. Mol Endocrinol. 2015;29:1594–607. 26378466. https://doi.org/10.1210/me.2014-1375.
3. Mallol C, Casana E, Jimenez V, Casellas A, Haurigot V, Jambrina C, Sacristan V, Morró M, Agudo J, Vilà L, Bosch F. AAV-mediated pancreatic overexpression of Igf1 counteracts progression to autoimmune diabetes in mice. Mol Metab. 2017;6:664–80. 28702323. https://doi.org/10.1016/j.molmet.2017.05.007.
4. Xu X, D'Hoker J, Stangé G, Bonné S, De Leu N, Xiao X, Van de Casteele M, Mellitzer G, Ling Z, Pipeleers D, Bouwens L, Scharfmann R, Gradwohl G, Heimberg H. B cells can be generated from endogenous progenitors in injured adult mouse pancreas. Cell. 2008;132:197–207. 18243096. https://doi.org/10.1016/j.cell.2007.12.015.
5. Wang AY, Ehrhardt A, Xu H, Kay MA. Adenovirus transduction is required for the correction of diabetes using Pdx-1 or Neurogenin-3 in the liver. Mol Ther. 2007;15:255–63. 17235302. https://doi.org/10.1038/sj.mt.6300032.
6. Scharfmann R, Pechberty S, Hazhouz Y, von Bülow M, Bricout-Neveu E, Grenier-Godard M, Guez F, Rachdi L, Lohmann M, Czernichow P, Ravassard P. Development of a conditionally immortalized human pancreatic β cell line. J Clin Invest. 2014;124:2087–98. 24667639. https://doi.org/10.1172/JCI72674.
7. Pauerstein PT, Park KM, Peiris HS, Wang J, Kim SK. Research resource: genetic labeling of human islet A cells. Mol Endocrinol. 2016;30:248–53. 26745668. https://doi.org/10.1210/me.2015-1220.
8. Shuai H, Xu Y, Yu Q, Gylfe E, Tengholm A. Fluorescent protein vectors for pancreatic islet cell identification in live-cell imaging. Pflugers Arch. 2016; 468:1765–77. https://doi.org/10.1007/s00424-016-1864-z PMID: 27539300.
9. Orabi AI, Sah S, Javed TA, Lemon KL, Good ML, Guo P, Xiao X, Prasadan K, Gittes GK, Jin S, Husain SZ. Dynamic imaging of pancreatic nuclear factor κB (NF-κB) activation in live mice using adeno-associated virus (AAV) infusion and bioluminescence. J Biol Chem. 2015;290:11309–20. https://doi.org/10.1074/jbc.M115.647933 PMID: 25802340.
10. Castaing M, Guerci A, Mallet J, Czernichow P, Ravassard P, Scharfmann R. Efficient restricted gene expression in β cells by lentivirus-mediated gene transfer into pancreatic stem/progenitor cells. Diabetologia. 2005;48:709–19. https://doi.org/10.1007/s00125-005-1694-6 PMID: 15759109.
11. Jimenez V, Ayuso E, Mallol C, Agudo J, Casellas A, Obach M, Muñoz S, Salavert A, Bosch F. In vivo genetic engineering of murine pancreatic β cells mediated by single-stranded adeno-associated viral vectors of serotypes 6,

8 and 9. Diabetologia. 2011;54:1075–86. https://doi.org/10.1007/s00125-011-2070-3 PMID: 21311856.

12. Guo P, Xiao X, El-Gohary Y, Criscimanna A, Prasadan K, Rymer C, Shiota C, Wiersch J, Gaffar I, Esni F, Gittes GK. Specific transduction and labeling of pancreatic ducts by targeted recombinant viral infusion into mouse pancreatic ducts. Lab Investig. 2013;93:1241–53. https://doi.org/10.1038/labinvest.2013.113 PMID: 24100509.
13. Morró M, Teichenne J, Jimenez V, Kratzer R, Marletta S, Maggioni L, Mallol C, Ruberte J, Kochanek S, Bosch F, Ayuso E. Pancreatic transduction by helper-dependent adenoviral vectors via intraductal delivery. Hum Gene Ther. 2014;25:824–36. https://doi.org/10.1089/hum.2013.182 PMID: 25046147.
14. Carlier G, Maugein A, Cordier C, Pechberty S, Garfa-Traoré M, Martin P, Scharfmann R, Albagli O. Human Fucci pancreatic B cell lines: new tools to study B cell cycle and terminal differentiation. PLoS One. 2014;9:e108202. https://doi.org/10.1371/journal.pone.0108202 PMID: 25259951 PMCID: PMC4178124.
15. Schambach A, Mueller D, Galla M, Verstegen MM, Wagemaker G, Loew R, Baum C, Bohne J. Overcoming promoter competition in packaging cells improves production of self-inactivating. Gene Ther. 2006;13(21):1524–33. https://doi.org/10.1038/sj.gt.3302807 PMID: 16763662.
16. Kirkegaard JS, Ravassard P, Ingvarsen S, Diedisheim M, Bricout-Neveu E, Grønborg M, Frogne T, Scharfmann R, Madsen OD, Rescan C, Albagli O. Xenotropic retrovirus Bxv1 in human pancreatic β cell lines. J Clin Invest. 2016;126:1109–13. https://doi.org/10.1172/JCI83573 PMID: 26901817.
17. Hughes SH, Greenhouse JJ, Petropoulos CJ, Sutrave P. Adaptor plasmids simplify the insertion of foreign DNA into helper-independent retroviral vectors. J Virol. 1987;61:3004–12 PMCID: PMC255873. PMID: 3041020.
18. von Werder A, Seidler B, Schmid RM, Schneider G, Saur D. Production of avian retroviruses and tissue-specific somatic retroviral gene transfer in vivo using the RCAS/TVA system. Nat Protoc. 2012;7:1167–83. https://doi.org/10.1038/nprot.2012.060 PMID: 22635109.
19. Ahronian LG, Lewis BC. Using the RCAS-TVA system to model human cancer in mice. Cold Spring Harb Protoc. 2014;2014:1128–35. https://doi.org/10.1101/pdb.top069831 PMID: 25368315.
20. Mangeot P-E, Dollet S, Girard M, Ciancia C, Joly S, Peschanski M, Lotteau V. Protein transfer into human cells by VSV-G-induced nanovesicles. Mol Ther. 2011;19:1656–66. https://doi.org/10.1038/mt.2011.138 PMID: 25259951 PMCID: PMC4178124.
21. Mizuguchi H, Xu Z, Ishii-Watabe A, Uchida E, Hayakawa T. IRES-dependent second gene expression is significantly lower than cap-dependent first gene expression in a bicistronic vector. Mol Ther. 2000;1:376–82. https://doi.org/10.1006/mthe.2000.0050 PMID: 10933956.
22. Weiss RA. The discovery of endogenous retroviruses. Retrovirology. 2006;3:67. https://doi.org/10.1186/1742-4690-3-67 PMID: 17018135 PMCID: PMC1617120.
23. Arda HE, Li L, Tsai J, Torre EA, Rosli Y, Peiris H, Spitale RC, Dai C, Gu X, Qu K, Wang P, Wang J, Grompe M, Scharfmann R, Snyder MS, Bottino R, Powers AC, Chang HY, Kim SK. Age-dependent pancreatic gene regulation reveals mechanisms governing human β cell function. Cell Metab. 2016;23(5):909–20. https://doi.org/10.1016/j.cmet.2016.04.002 PMID: 27133132 PMCID: PMC4864151.
24. Ahuja D, Sáenz-Robles MT, Pipas JM. SV40 large T antigen targets multiple cellular pathways to elicit cellular transformation. Oncogene. 2005;24:7729–45. https://doi.org/10.1038/sj.onc.1209046 PMID: 16299533.
25. Casanova O, Hager JH, Hong Chun MG, Hanahan D. Incomplete inhibition of the Rb tumor suppressor pathway in the context of inactivated p53 is sufficient for pancreatic islet tumorigenesis. Oncogene. 2005;24:6597–604.
26. Hemann MT, Fridman JS, Zilfou JT, Hernando E, Paddison PJ, Cordon-Cardo C, Hannon GJ, Lowe SW. An epi-allelic series of p53 hypomorphs created by stable RNAi produces distinct tumor phenotypes in vivo. Nat Genet. 2003;33:396–400. https://doi.org/10.1038/ng1091 PMID: 12567186.
27. Bromberg-White JL, Webb CP, Patacsil VS, Miranti CK, Williams BO, Holmen SL. Delivery of short hairpin RNA sequences by using a replication-competent avian retroviral vector. J Virol. 2004;78:4914–6 PMID: 15078973 PMCID: PMC387706.
28. Seidler B, Schmidt A, Mayr U, Nakhai H, Schmid RM, Schneider G, Saur D. A Cre-loxP-based mouse model for conditional somatic gene expression and knockdown in vivo by using avian retroviral vectors. Proc Natl Acad Sci U S A. 2008;105:10137–42. https://doi.org/10.1073/pnas.0800487105 PMID: 18621715 PMCID: PMC2481330.
29. Zheng XH, Hughes SH. An avian sarcoma/leukosis virus-based gene trap vector for mammalian cells. J Virol. 1999;73(8):6946–52 PMID: 10400793 PMCID: PMC112780.
30. Kushner JA. The role of aging upon β cell turnover. J Clin Invest. 2013;123:990–5. https://doi.org/10.1172/JCI64095 PMID: 23454762 PMCID: PMC3582123.
31. Meier JJ, Butler AE, Saisho Y, Monchamp T, Galasso R, Bhushan A, Rizza RA, Butler PC. B-cell replication is the primary mechanism subserving the postnatal expansion of β-cell mass in humans. Diabetes. 2008;57:1584–94. https://doi.org/10.2337/db07-1369 PMID: 18334605 PMCID: PMC3697779.
32. Nielsen JH, Linde S, Welinder BS, Billestrup N, Madsen OD. Growth hormone is a growth factor for the differentiated pancreatic β-cell. Mol Endocrinol. 1989;3(1):165–73. https://doi.org/10.1210/mend-3-1-165 PMID: 2644530.
33. Katz RA, Greger JG, Skalka AM. Effects of cell cycle status on early events in retroviral replication. J Cell Biochem. 2005;94:880–9. https://doi.org/10.1002/jcb.20358 PMID: 15669021.
34. Greger JG, Katz RA, Tagano K, Rall GF, Skalka AM. Transduction of terminally differentiated neurons by avian sarcoma virus. J Virol. 2004;78:4902–6. https://doi.org/10.1128/JVI.78.9.4902–4906.2004 PMID: 15078971 PMCID: PMC387698.
35. Ku GM, Pappalardo Z, Luo CC, German MS, McManus MT. An siRNA screen in pancreatic β cells reveals a role for Gpr27 in insulin production. PLoS Genet. 2012;8(1):e1002449. https://doi.org/10.1371/journal.pgen.1002449 PMID: 22253604 PMCID: PMC3257298.
36. Mitchell RS, Beitzel BF, Schroder AR, Shinn P, Chen H, Berry CC, Ecker JR, Bushman FD. Retroviral DNA integration: ASLV, HIV, and MLV show distinct target site preferences. PLoS Biol. 2004;2(8):E234. https://doi.org/10.1371/journal.pbio.0020234 PMID: 15314653 PMCID: PMC509299.
37. Ihler F, Vetter EV, Pan J, Kammerer R, Debey-Pascher S, Schultze JL, Zimmermann W, Enders G. Expression of a neuroendocrine gene signature in gastric tumor Cells from CEA 424-SV40 large T antigen- transgenic mice depends on SV40 large T antigen. PLoS One. 2012;7(1):e29846. https://doi.org/10.1371/journal.pone.0029846 PMID: 22253802 PMCID: PMC3258231.
38. Lin YC, Boone M, Meuris L, Lemmens I, Van Roy N, Soete A, Reumers J, Moisse M, Plaisance S, Drmanac R, Chen J, Speleman F, Lambrechts D, Van de Peer Y, Tavernier J, Callewaert N. Genome dynamics of the human embryonic kidney 293 (HEK293) lineage in response to cell biology manipulations. Nat Commun. 2014;5:4767. https://doi.org/10.1038/ncomms5767 PMID: 25182477 PMCID: PMC4166678.

2

Development of a novel human phage display-derived anti-LAG3 scFv antibody targeting $CD8^+$ T lymphocyte exhaustion

Alessandro Ascione[1*], Claudia Arenaccio[1], Alessandra Mallano[1], Michela Flego[1], Mara Gellini[1], Mauro Andreotti[1], Craig Fenwick[2], Giuseppe Pantaleo[2], Stefano Vella[1] and Maurizio Federico[1]

Abstract

Background: Lymphocyte-activation gene (LAG)3 is a 498 aa transmembrane type I protein acting as an immune inhibitory receptor. It is expressed on activated lymphocytes, natural killer cells and plasmacytoid dendritic cells. In activated lymphocytes, LAG3 expression is involved in negative control of cell activation/proliferation to ensure modulation and control of immune responses. In view of its deregulated expression in tumor-infiltrating lymphocytes, LAG3, together with the additional immune checkpoint inhibitors CTLA4 and PD1, is considered a major target in order to reverse the immunosuppression typically mounting in oncologic diseases. Since many patients still fail to respond to current immune checkpoints-based therapies, the identification of new effective immune inhibitors is a priority in the ongoing fight against cancer.

Results: We identified a novel human single-chain variable fragment (scFv) Ab against a conformational epitope of LAG3 by in vitro phage display technology using the recombinant antigen as a bait. This scFv (referred to as F7) was characterized in terms of binding specificity to both recombinant antigen and human LAG3-expressing cells. It was then rebuilt into an IgG format pre-optimized for clinical usage, and the resulting bivalent construct was shown to preserve its ability to bind LAG3 on human cells.

Next, we analyzed the activity of the anti-LAG3 scFvF7 using two different antigen-specific $CD8^+$ T lymphocyte clones as target cells. We proved that the reconstituted anti-LAG3 F7 Ab efficiently binds the cell membrane of both cell clones after peptide-activation. Still more significantly, we observed a striking increase in the peptide-dependent cell activation upon Ab treatment as measured in terms of IFN-γ release by both ELISA and ELISPOT assays.

Conclusions: Overall, the biotechnological strategy described herein represents a guiding development model for the search of novel useful immune checkpoint inhibitors. In addition, our functional data propose a novel candidate reagent for consideration as a cancer treatment.

Keywords: LAG3, Single-chain variable fragment, Phage display, Reconstituted scFv, $CD8^+$ T lymphocytes, Lymphocyte exhaustion

* Correspondence: alessandro.ascione@iss.it
[1]National Center for Global Health, Istituto Superiore di Sanità (ISS), Viale Regina Elena 299, 00161 Rome, Italy
Full list of author information is available at the end of the article

Background

Lymphocyte activate gene (LAG)3 (also known as CD223) is a transmembrane protein showing about 20% identity with CD4 (for a recent review, see [1]). Like CD4, LAG3 binds MHC Class II molecules, albeit at a distinct site and with a greater affinity [2, 3]. LAG3 is not expressed by resting T cells, although it is upregulated on activated T lymphocytes (both $CD4^+$ and $CD8^+$ subpopulations), B lymphocytes, natural killer cells, and plasmacytoid dendritic cells (DCs). Besides MHC Class II molecules, LAG3 has been found to bind Galectin-3 [4, 5], LSECtin [6] and, in central nervous system, α-synuclein [7]. LAG3 engagement leads to a negative regulation of T-cell activation, despite the LAG3 signal transduction mechanism being still largely unknown. A couple of unique, yet largely conserved domains, i.e., the EP [8] and KIEELE [9] motifs, seem important for the transmission of intracellular inhibitory signals. In addition, LAG3 shows a bidirectional signaling activity, as inferred by the modulation of DC activation/maturation observed upon interaction with LAG3 expressing lymphocytes [10, 11].

LAG3 has been identified as one of the major inhibitory receptors (IR) involved in the T lymphocyte exhaustion typically occurring in oncologic diseases [12]. For this reason, specific anti-LAG3 mAbs are currently under investigation in the attempt to reverse cancer-associated immunosuppression. Together with other IRs (i.e. CTLA4 and PD1) LAG3 is now considered a major target of the so called immune checkpoint blockade (ICB) antitumor strategy based on delivery of mAbs against IRs [13]. This therapeutic design has been positively adopted in the field of oncology, providing persistent clinical benefits for a significant number of patients with advanced cancer. Unfortunately, however, many patients fail to respond to ICB-based therapies, and the hyper-immune activation can associate with immune-related adverse events affecting several organs, including skin, gut, heart, lungs, and bone [14]. Therefore, nowadays the emphasis is on the evaluation of new immune checkpoint inhibitors and a combination thereof possibly increasing efficacy while also reducing toxicity [15], all resulting in a considerable interest in LAG3.

IRs maintain self-tolerance and modulate the duration and amplitude of physiological immune responses. It is now clear that tumors could co-opt certain regulatory pathways as a major mechanism of immune resistance, particularly in $CD8^+$ T lymphocytes, and that this dysfunction could be reversed by anti-IRs antagonist mAbs [16]. Meanwhile, awareness is emerging that multiple IRs can be involved in lymphocyte exhaustion [17–19], so paving the way for more effective combinatorial approaches [20, 21]. A first confirmation of this hypothesis came from the fact that the combination of two IR blockers, namely anti-PD1 and anti-CTLA4 mAbs, resulted in a higher rate of response than that observed with single mAb treatments. In solid tumors, LAG3 is upregulated on tumor-infiltrating lymphocytes (TILs), while blockade of LAG3 can enhance antitumor T cell responses. On this basis, several reagents targeting LAG3 are currently being tested in various stages of preclinical and clinical development, thus attesting to the broad interest of this IR in the field.

Monoclonal Abs (mAbs) are molecules largely used in therapy, diagnostic and biotechnology. Among the various recombinant mAb formats, the single chain variable fragment (scFv) is one of the simplest and most versatile constructs that retains the antigen-binding activity [22]. This molecule is formed by the variable regions of both heavy and light chains joined by a peptide linker. While scFvs have the significant advantage that they can be expressed by a single open reading frame, their low stability and short pharmacokinetic may limit their applications. The scFv fusion with CH2 and CH3 domains of the Ig Fc region can overcome such drawbacks [23].

To avoid undesirable immune responses, human Abs represent the most suitable reagents for Ab-based therapeutic purposes. Human scFvs can be readily isolated through the well consolidated phage display technique. In this paper we describe the identification of a novel human anti-LAG3 scFv Ab recovered from a fully human naive phage library by means of an in vitro bio-panning strategy. Upon Fc-rebuilding, this Ab was shown to maintain its binding activity, while strongly increasing the activation of antigen-specific $CD8^+$ T lymphocytes. We propose our anti-LAG3 Ab, in both scFv and reconstituted forms, as a candidate for novel biotechnology and clinic applications.

Results

Isolation of a novel human anti-LAG3 scFv ab by phage display technology

Human recombinant scFvs against LAG3 were obtained by in vitro affinity enrichment process (bio-panning) as previously described [24], and depicted in Fig. 1a. In essence, an aliquot of phage-displayed scFvs representing the complexity of the naive IORISS1 library [25] was subjected to cycles of binding to recombinant LAG3 protein used as a bait and immobilized on immunotubes, followed by elution. Outputs after each round of selection were characterized by a progressive increase in phage titers (Table 1). Five rounds of selection were necessary to isolate a phage population detectable in anti-LAG3 ELISA. Soluble scFvs derived from single colonies were obtained after plating bacteria TG1 infected with the selected phages. Several of these scFvs were demonstrated to bind the recombinant LAG3 protein (Fig. 1b), whereas they were negative against an irrelevant antigen (GO, glucose oxidase) used as a negative control (data not shown). The scFv coding DNA was isolated from the positive clones, and the sequencing revealed a common genetic identity (Fig. 2).

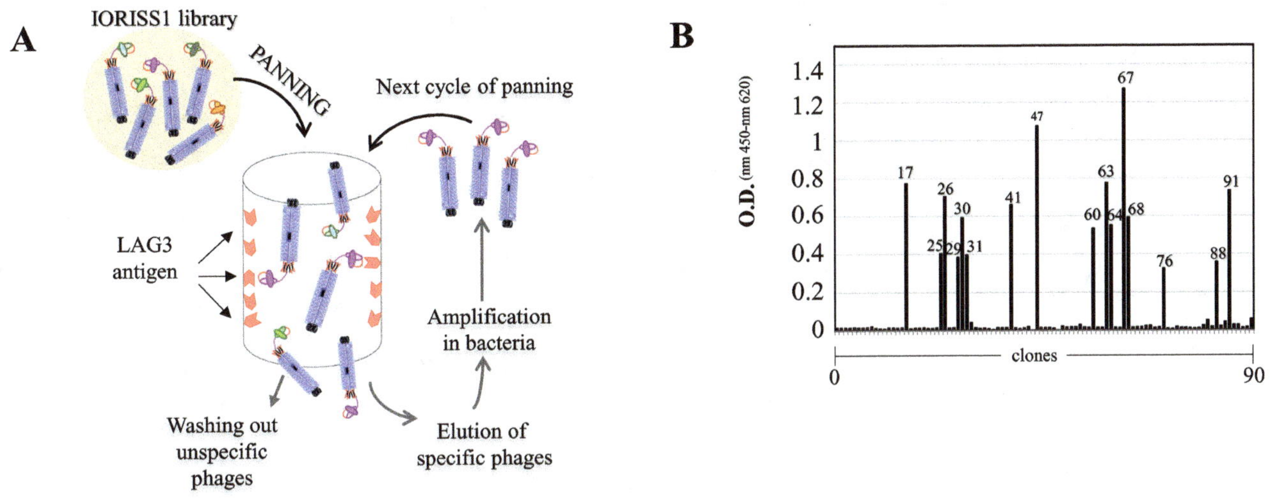

Fig. 1 Isolation and screening of anti-LAG3 scFvs by phage display technology. **a** Scheme of the LAG3 affinity-based phage selection starting from the IORISS1 library. **b** Anti-LAG3 ELISA carried out on supernatants of 90 bacterial colonies isolated after five rounds of panning cycles. The plate was coated with 0.5 μg of antigen per microwell. The numbers identify each positive clone. O.D. nm 450 ref. 620: optical density

A representative clone, referred to as scFvF7, was produced in bacteria and purified by immobilized metal affinity *chromatography* using 6 × histidine-tag located at its C-terminus. SDS-PAGE analysis showed a single 31 kilodalton (kDa) band, at the expected size of the Ab fragment, with a negligible presence of both contaminate and degraded products (Fig. 3a).

The purified scFv was then tested for binding with recombinant LAG3 antigen in ELISA. The binding assay generated a specific dose-dependent association with LAG3 and a lack of binding against GO protein used as an irrelevant antigen (Fig. 3b).

Concerning the binding activity of scFvF7 on LAG3-expressing eukaryotic cells, flow cytometric analysis on $CD4^+$ T lymphocytes carried out following a 3-day stimulation with phytohaemagglutinin (PHA) showed that scFvF7 binds native LAG3 with a comparable profile to that of a commercial murine anti-LAG3 Ab used as a control (Fig. 3c). Therefore, scFvF7 is capable of recognizing LAG3 antigen when it retains its native conformation, namely in ELISA on recombinant protein and in cytometry on live LAG3-expressing cells. Conversely, scFvF7 failed to react against its target in SDS-PAGE western blot assay (Fig. 3d), whose reducing/denaturing conditions stretch the amino acid structure of protein; this suggests a conformation-dependent epitope. An ELISA conducted on the heat-denatured LAG3 protein (by boiling) and a western blotting assay performed in non-reducing conditions strengthen this hypothesis (see Additional file 1: Figure S3D[II]).

Engineering of divalent Fc-scFvF7 Ab

The use of single-domain Abs overcomes some limitations typical of tetrameric Abs, e.g., unspecific uptake, restricted access to tissues of interest. On the other hand, scFvs have been shown to have low in vivo stability as a consequence of their efficient blood clearance. Hence, to increase the expected in vivo stability of scFvF7 in view of possible clinical

Table 1 Selection of native IORISSI Ab phage library againts LAG3 recombinant protein[a]

Panning cycle	Ag μg/mL	Input	Output	Ratio (O/I)	Enrichment	ELISA (O.D.)
1	20	5×10^{12}	2×10^{4}	0.4×10^{-8}	1	-
2	10	1×10^{12}	6×10^{5}	1.6×10^{-7}	3×10	0.018
3	5	1×10^{12}	1.4×10^{7}	1.4×10^{-5}	7×10^{2}	0.0275
4	5	1×10^{12}	4×10^{9}	4×10^{-3}	2×10^{5}	0.175
5	2.5	1×10^{12}	7.5×10^{10}	7.5×10^{-2}	3.7×10^{6}	0.536

[a]both Phage input (I) and output (O) were determined by titration in terms of colony forming units in infected *E. coli* TGI cells before and after each round of selection. Enrichment was calculated as ratio between outputs from each cycle and the output from the first one

V_H	1	ATG	GCC	CAG	GTG	CAG	CTG	GTG	GAG	TCG	GGG	GGA	GGC	TTG	GTA	CAG	45
	1	M	A	Q	V	Q	L	V	E	S	G	G	G	L	V	Q	15
	46	CCT	GGG	GGG	TCC	CTG	AGA	CTC	TCC	TGT	GCA	GCC	TCT	**GGA**	**TTC**	**ACC**	90
	16	P	G	G	S	L	R	L	S	C	A	A	S	**G**	**F**	**T**	30
	91	**TTT**	**AGC**	**AGC**	**TAT**	**GCC**	ATG	AAC	TGG	GTC	CGC	CAG	GCT	CCA	GGG	AAG	135
	31	**F**	**S**	**S**	**Y**	**A**	M	N	W	V	R	Q	A	P	G	K	45
	136	GGG	CTG	GAG	TGG	GTC	TCA	GGT	**ATT**	**AGT**	**GCT**	**GGT**	**GGT**	**ACT**	**GGC**	**ACA**	180
	46	G	L	E	W	V	S	G	**I**	**S**	**A**	**G**	**G**	**T**	**G**	**T**	60
	181	TAC	TAC	GCA	GAC	TCC	GTG	AAG	GGC	CGG	TTC	ACC	ATC	TCC	AGA	GAC	225
	61	Y	Y	A	D	S	V	K	G	R	F	T	I	S	R	D	75
	226	AAT	TCC	AAG	AAC	ACG	CTG	TAT	CTA	CAA	ATG	AAC	AGC	CTG	AGA	GCC	270
	76	N	S	K	N	T	L	Y	L	Q	M	N	S	L	R	A	90
	271	GAG	GAC	ACG	GCC	GTA	TAT	TAC	TGT	**GCG**	**AAA**	**GAT**	**CAG**	**GAC**	**GGT**	**GGT**	315
	91	E	D	T	A	V	Y	Y	C	**A**	**K**	**D**	**Q**	**D**	**G**	**G**	105
	316	**AAG**	**AAG**	**ACT**	**GAT**	**GCT**	**TTT**	**GAT**	**ATC**	TGG	GGC	CAA	GGG	ACA	ATG	GTC	360
	106	**K**	**K**	**T**	**D**	**A**	**F**	**D**	**I**	W	G	Q	G	T	M	V	120
	361	ACC	GTC	TCT	TCA	GGT	GGA	GGT	GGA	TCT	GGC	GGC	GGC	GGC	TCA	GGC	405
	121	T	V	S	S	G	G	G	G	S	G	G	G	G	S	G	135
	406	GGA	GGA	GGT	TCC	GAT	ATT	GTG	ATG	ACG	CAG	TCT	CCA	TCC	CTC	CTG	450
	136	G	G	G	S	D	I	V	M	T	Q	S	P	S	L	L	150
	451	TCT	GCA	TCT	GTA	GGA	GAC	AGA	GTC	ACC	ATC	ACT	TGC	CGG	GCC	AGT	495
	151	S	A	S	V	G	D	R	V	T	I	T	C	R	A	S	165
	496	**CAG**	**GGC**	**ATT**	**AAC**	**TAT**	**TAT**	TTA	GCC	TGG	TAT	CAG	CAA	AAA	CCA	GGG	540
	166	**Q**	**G**	**I**	**N**	**Y**	**Y**	L	A	W	Y	Q	Q	K	P	G	180
	541	AAA	GCC	CCT	AAG	CTC	CTG	ATC	TAT	**GCT**	**GCA**	**TCC**	ACT	TTG	CAA	AGT	585
	181	K	A	P	K	L	L	I	Y	**A**	**A**	**S**	T	L	Q	S	195
V_L	586	GGG	GTC	CCA	TCA	AGG	TTC	AGC	GGC	AGT	GGA	TCT	GGG	ACA	GAA	TTC	630
	196	G	V	P	S	R	F	S	G	S	G	S	G	T	E	F	210
	631	ACT	CTC	ACA	ATC	AGC	AGC	CTG	CAG	CCT	GAA	GAT	TTT	GCA	ACT	TAT	675
	211	T	L	T	I	S	S	L	Q	P	E	D	F	A	T	Y	225
	676	TAC	TGT	**CAG**	**CAG**	**ACT**	**GGT**	**TAT**	**TAC**	**CCT**	**CTC**	**ACT**	TTC	GGC	CCT	GGG	720
	226	Y	C	**Q**	**Q**	**T**	**G**	**Y**	**Y**	**P**	**L**	**T**	F	G	P	G	240
	721	ACC	AAA	GTG	GAT	ATC	AAA	GCG	GCC	GCA	GAT	GAC	GAT	TCT	GAC	GAT	765
	241	T	K	V	D	I	K	A	A	A	D	D	D	S	D	D	255
	766	GAC	TAC	AAG	GAC	GAC	GAC	GAC	CAG	CAC	CAT	CAC	CAT	CAC	CAT	TAG	810
	256	D	Y	K	D	D	D	D	Q	H	H	H	H	H	H	*	270

Fig. 2 Nucleotide and amino acid sequences of scFvF7. The whole scFvF7 sequence, including tags, is reported. CDR1, CDR2 and CDR3 regions of both VH and VL chains are indicated in bold. The linker region is reported in red. The flag tag is indicated in violet, whereas the 6 × histidine region is in blue

use, its fusion with CH2 and CH3 Fc domains was carried out. We generated a divalent scFvF7-Fc construct resulting in a recombinant Ab incorporating two scFvF7 molecules, each of them joined with CH2 and CH3 constant immunoglobulin domains. Similar to classic Abs, their spontaneous dimerization results in two anti-LAG3 binding sites on the same molecule, with the important advantage that the divalent molecule can be expressed by a single open reading frame (Fig. 4a). The presence of a hinge region between the scFv and the constant domains guarantees a flexibility similar to that of natural immunoglobulins. In addition, LALA mutations within the CH2 constant regions [26] would be particularly useful for a possible use in vivo since they are expected to lower the risk of unwanted Fc-dependent effects, e.g., Ab-dependent cell cytotoxicity. The scFvF7 open reading frame was cloned in frame with CH2 and CH3 Fc sequences in the context of the pFUSEss-CHIghGI vector which also expresses the human IgG1 secretory leader peptide sequence (Fig. 4a). The resulting molecular construct was transfected in CHO cells in order to verify the proper synthesis and secretion of the construct in an eukaryotic system. The western blot assay performed under non-reducing conditions on the supernatants of the transfected cells using anti-human Fc as detector showed a dominant signal compatible with the expected molecular weight (~ 140 kDa) of the reconstituted anti-LAG3 Ab (Fig. 4b). These supernatants efficiently and specifically reacted in LAG3 ELISA (Fig. 4c).

Next, the Fc-reconstituted F7 Ab was produced on a large scale by transient transfection in CHO cells, purified, and formulated in sterile PBS. The purified divalent Ab construct preserved its specificity as evidenced by both its reactivity in ELISA against recombinant LAG-3 (Fig. 5a), and its binding on human HEK 293 cells stably expressing LAG3 (Fig. 5b). Finally, FACS assays proved that the reconstituted F7 Ab binds human PHA-activated $CD8^+$ T cells specifically, and in a concentration-dependent manner (Fig. 5c).

Fc-reconstituted F7 Ab binds peptide-activated, antigen-specific $CD8^+$ T lymphocytes

Besides CTLA4 and PD1, LAG3 is now considered a major target for the mAb-based cancer immunotherapy.

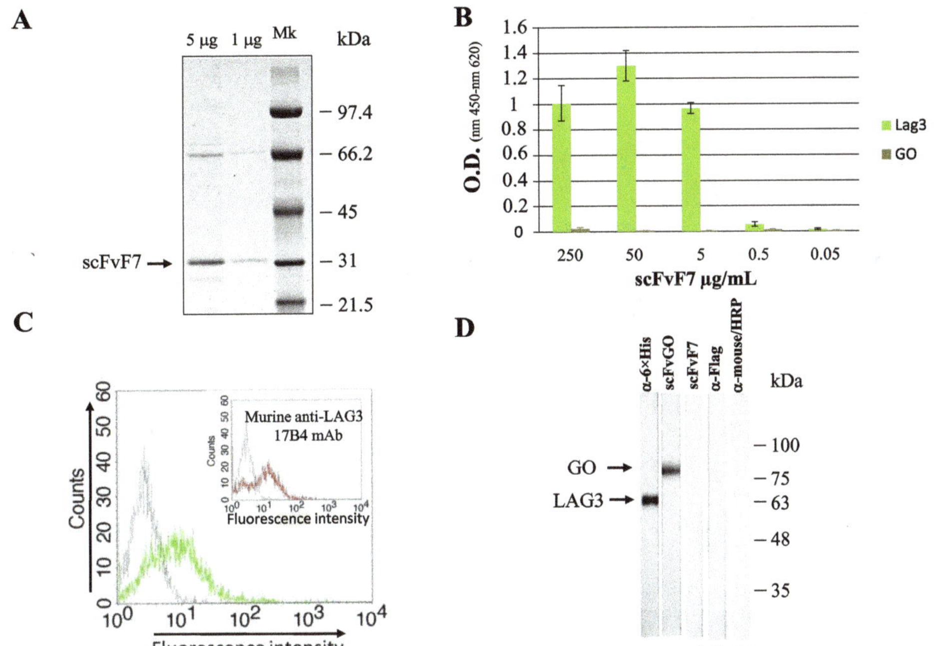

Antigen (0,5 µg/well)		Antibodies: α-6×His	scFvGO	scFvF7	α-Flag	α-mouse/HRP
ELISA O.D. (nm 450- nm 620)	GO	0.009	1.3	0.007	0.012	0.007
	LAG3	1.8	0.052	0,93	0.005	0.007

Fig. 3 Characterization of the anti-LAG3 scFvF7. **a** SDS-PAGE analysis on 1 and 5 µg of scFvF7. Molecular markers in kilodaltons (kDa) are reported on the right. **b** ELISA for the detection of LAG3 antigen carried out with decreasing doses of purified scFvF7. GO: glucose oxidase . O.D.: optical density. The mean values ±SEM are shown as calculated from results of three independent assays. **c** FACS analysis carried out with scFvF7 on either unstimulated (grey line) or PHA-stimulated (green line) human CD4^{+} T lymphocytes. Inset: results from the same assay carried out with the murine anti-LAG3 17B4 mAb are reported. Four repeats of the experiment gave consistent results. **d** Binding analysis of scFvF7. 0.5 µg of both recombinant LAG3 and glucose oxidase (GO) proteins were loaded in each replicate well on a 12% SDS-PAGE under reducing condition and transferred to filter paper. Strips from the filter were then incubated with the indicted primary antibodies. Anti-6 his mAb was used as a positive control for LAG3 recombinant protein (which has a 6-histidines tag at its C terminal end). As additional negative controls, two strips were incubated respectively with the anti-Flag antibody and the HRP-conjugated anti-mouse antibody, in order to monitor eventual non-specific signals due to the secondary antibodies used to detect scFv antibodies. The ELISA shown below (performed with the same secondary Abs) is a check for the reactivity of the scFv-containing supernatants (scFvGO and scFvF7) used in the assay. Arrows indicate relevant signals. Molecular markers in kilodaltons (kDa) are reported on the right. A representative of three independent assays is shown

ICBs are selected on the basis of their capacity to reverse the lymphocyte-dependent immune exhaustion, i.e., a typical signature affecting TILs infiltrating solid tumors, and particularly the CD8^{+} T sub-population [12]. In this context, the F7 Ab's potential as a novel ICB candidate for new antitumor combination therapies relies on the assessment of its activity on antigen-specific CD8^{+} T lymphocytes. To this end, we carried out functional assays on a couple of antigen-specific CD8^{+} T cell clones, i.e., an HLA-B7 restricted clone specific for HIV-1 Nef, and an HLA-A.02-restricted clone specific for the human melanoma antigen Mart-1.

Initially, we were interested in assessing the detection LAG3 expression with the reconstituted F7 Ab following the activation of the CD8^{+} T cell clones. The two clones were co-cultivated with HLA-matched B-lymphoblastoid cell lines (B-LCLs) previously treated with 100 ng/mL of each specific peptide. As a positive control, CD8^{+} T lymphocytes isolated from healthy donor peripheral blood mononuclear cells (PBMCs) were treated with PHA for three days. Stimulated cells were incubated with reconstituted F7 Ab and stained with a PE labeled anti-human IgG secondary Ab. As shown in Fig. 6, upon activation, both CD8^{+} T activated cell clones bound the reconstituted F7 Ab, a result which was consistent with the expected LAG3 upregulation.

These data validated the two CD8^{+} T cell clones as reliable substrates for functional assays devoted to assessing the activity of the divalent scFvF7.

Increased IFN-γ production in antigen-specific CD8^{+} T lymphocytes treated with Fc-reconstituted scFvF7 Ab

B-LCLs were treated with specific peptides for 3 h. The peptide loaded APCs were then co-cultured with the appropriate CD8^{+} T cell clones at a 1:1 cell ratio in the presence or absence of either an unspecific IgG control or

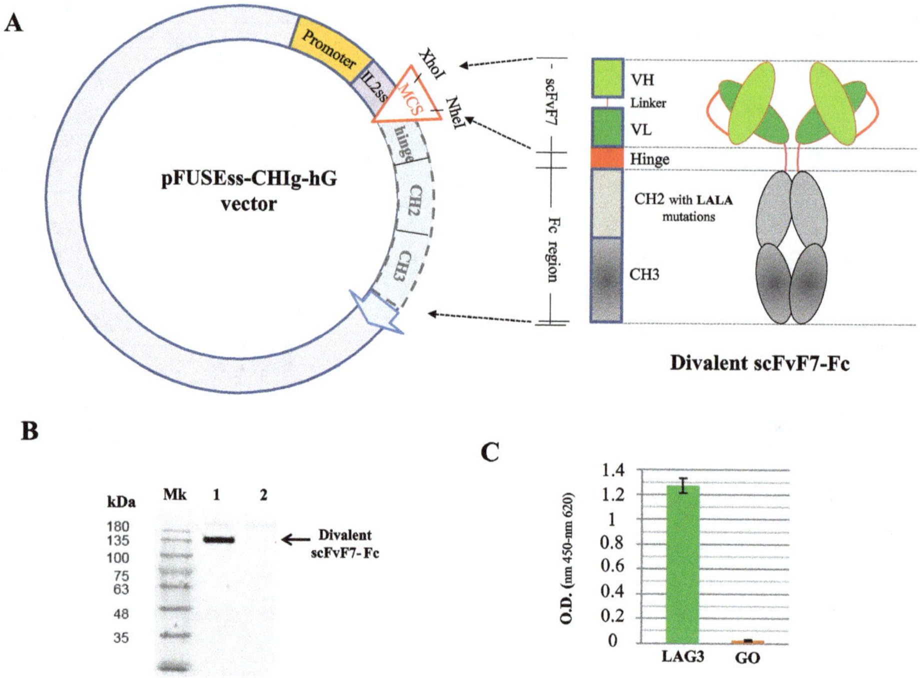

Fig. 4 Construction and characterization of the divalent scFvF7-Fc Ab. **a** Scheme of the vector expressing the scFvF7-Fc open reading frame. Indicated are the locations of the hEF1α/HTLVI promoter, IL-2 signal sequences (IL-2 SS) acting as signal peptide, multiple cloninig site (MCS), and hinge. Domain map of the resulting protein product is also illustrated and, on the right, the overall structure of the divalent Ab. **b** Anti 6 × histidine western blot analysis carried on supernatants of CHO cells transfected with either the scFvF7-Fc expressing vector (lane 1), or transfected with the empty vector (lane 2). Molecular weight markers (Mk) are reported on the left. The results are representative of four independent assays. **c** *ELISA* performed with supernatants of CHO cells transfected with the vector expressing scFvF7-Fc. Either recombinant LAG3 or glucose oxidase (GO) were coated in the microwells (0.5 μg). The mean values ±SEM are shown as calculated from results of three independent assays

anti-LAG3 Abs (either the reconstituted scFv or a murine mAb) in 100 μL of total volume. After an overnight (o.n.) incubation, supernatants were harvested, and IFN-γ production was evaluated by ELISA. The results reported in Fig. 7a (and Additional file 2: Figure S7A^{II}, and Additional file 3: Figure S7A^{III}) clearly demonstrate that the treatment with the Fc-reconstituted scFvF7 increased the IFN-γ release even more efficiently than the murine anti-LAG3 mAb validated for functional activity. The effect is more pronounced in the Nef-specific CD8$^+$ T cells, this finding being consistent with the apparently higher level of LAG3 detected by the divalent scFvF7 Fc compared to that of Mart-1 specific CD8$^+$ T lymphocytes (Fig. 6).

The specificity of the functional effect of scFvF7-Fc is supported by the data presented in Fig. 7b (and Additional file 4: Figure S7B^{II}), where the divalent Ab was incubated with cells in the presence or absence of increasing amounts of recombinant LAG3 protein. The F7 Ab induced IFN-γ production in a concentration dependent manner that was effectively inhibited with the addition of the soluble LAG3 antigen. The specific effect of the divalent F7 Ab on the CD8$^+$ T cell activation was further confirmed by IFN-γ ELISPOT assays (Fig. 7c and Additional file 5: Figure S7C^{II}).

Taken together, our functional data indicate that the human divalent scFvF7-Fc Ab efficiently binds LAG3 on the surface of antigen-specific CD8$^+$ T lymphocytes, thereby potently increasing their activation state. On the basis of its ability to counteract the LAG3 mediated inhibitory effect on antigen-activated CD8$^+$ T lymphocytes, scFvF7-Fc can be proposed as a novel candidate for the treatment of pathologies characterized by LAG3-driven lymphocyte exhaustion.

Discussion

Here we describe the experimental process aimed at the identification of novel anti-LAG3 human scFvs based on the previously described IORISS1 library [25]. This procedure allowed us to isolate the anti-LAG3 scFvF7. The combination of the results we obtained from western blot assay and ELISA are consistent with scFvF7 binding a conformational epitope of LAG3. Furthermore, functional assays carried out on antigen-specific CD8$^+$ T lymphocytes demonstrated that the divalent Fc-fused scFvF7 biological activity is comparable to the functionally characterized murine anti-LAG3 17B4 mAb. Considering the human origin of scFvF7, these results are of particular relevance in view of possible in vivo uses of

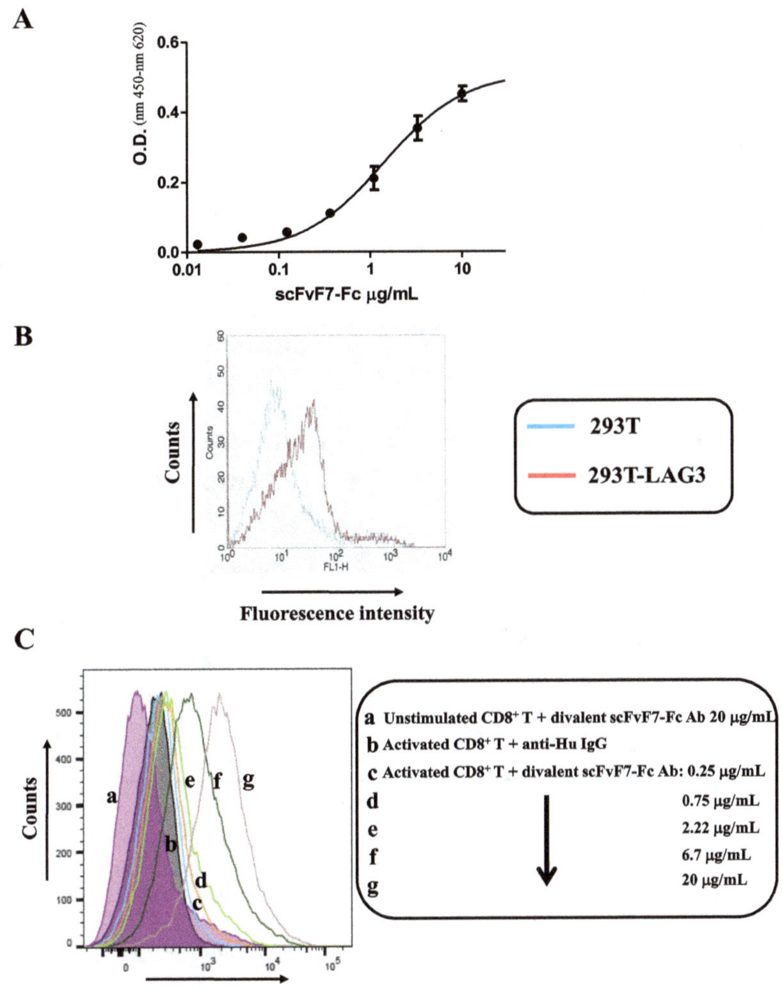

Fig. 5 Binding properties of the divalent scFvF7-Fc Ab. **a** Binding curve of purified divalent scFvF7-Fc on recombinant LAG3. 0.1 μg of recombinant LAG3 was coated on ELISA microwells, and then different amounts of the reconstituted F7 Ab were incubated for one hour. The Ab binding was finally revealed through the incubation with an HRP-conjugated anti-human antibody. The mean values ±SEM are shown as calculated from quadruplicate wells from a representative of two independent experiments. O.D.: optical density (450 nm–620 nm). **b** FACS analysis on LAG3 overexpressing 293 T cells. Both parental and LAG3 engineered 293 T cells were incubated with the divalent scFvF7-Fc Ab, and then with FITC-conjugated anti-human IgGs. Data are shown from a representative of six independent experiments. **c** Dose-response binding of the reconstituted F7 Ab on cell membrane of activated human $CD8^+$ T lymphocytes. PBMCs were isolated from the peripheral blood of healthy donors, and then the $CD8^+$ T fraction was isolated by immunomagnetic selection. The cells were activated with PHA and, after six days, labeled with the indicated amounts of the divalent scFvF7-Fc Ab. As control, activated cells were labeled with the secondary Abs alone. As an additional control, unstimulated $CD8^+$ T cells were labeled with the highest anti-LAG3 Ab concentration. The results shown are representative of two independent experiments

this scFv and derivatives thereof. In fact, monoclonal antibodies obtained with the classic hybridoma technology are extremely valuable instruments in many fields of application, but their use in the clinic is compromised by their xenogenic origin. Conversely, antibodies generated from human antibody libraries are entirely human, with the innate advantage of being only slightly or not at all immunogenic. Nevertheless, some drawbacks remain with the phage display-derived antibody fragments, such as their very short blood clearance. Therefore, the anti-LAG3 antibody here described was engineered in a recombinant 'scFv-Fc' format, in order to recover some important requirements for a potential drug compound, namely a structure with the typical avidity of natural antibodies (two binding sites), a prolonged half-life and a reduced ability to recall unwanted cell mediated responses [22, 23].

The interest in the identification and characterization of new ICBs is continually growing as witnessed by the varied clinical studies as well as by the quite recent publication of works describing the efforts to identify new anti-IR Abs. In particular, Everett et coll. Reported the isolation of

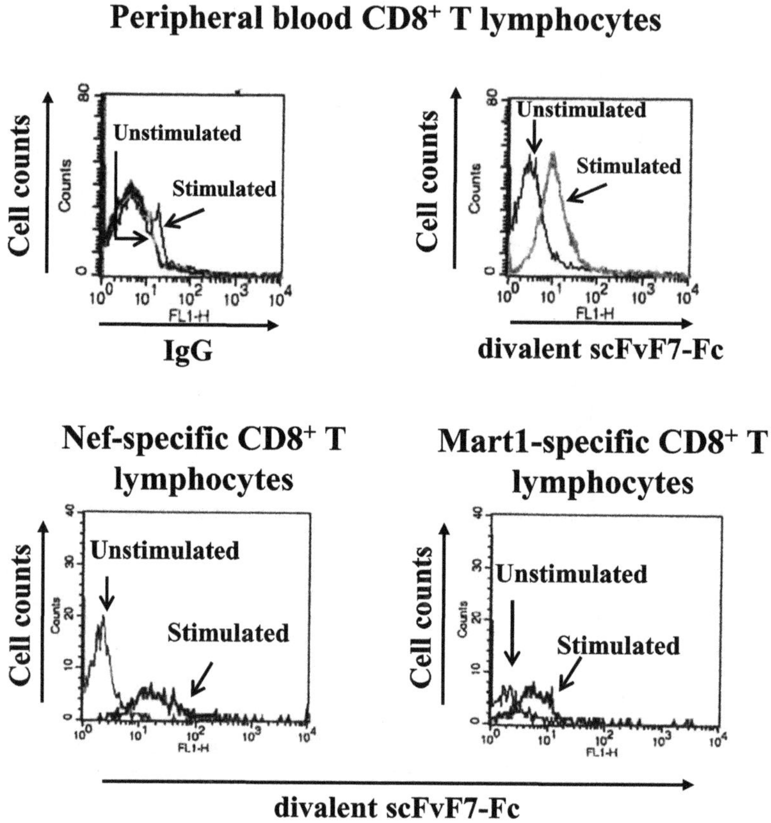

Fig. 6 Binding of the divalent scFvF7-Fc Ab on antigen-specific CD8$^+$ T lymphocytes. Both Nef-and Mart1-specific human CD8$^+$ T lymphocytes were stimulated through co-cultivation with HLA-matched B-LCLs previously treated with either specific (stimulated) or mismatched (unstimulated) peptides. After o.n. co-cultivation, the co-cultures were labelled with the reconstituted F7 Ab, followed by incubation with FITC-conjugated anti-human IgGs. Then, cells were labeled with PE-conjugated anti-CD8 mAb. As control, PBMC-derived CD8$^+$ T cells from either unstimulated or PHA-activated were labelled with either IgGs or the divalent scFvF7-Fc Ab (upper panels). In the bottom panels, histograms of FITC-related fluorescence profiles of PE-positive cells are shown. Results from a representative of four independent experiments are reported

novel anti-LAG3 Abs able to target both human and murine LAG3 by means of an additional engineered binding domain within the CH3 Fc region [27]. The in vitro activity of these Ab constructs resulted in an increased release of IL-2 from stimulated PBMCs. In addition, Sasso and colleagues described an Ab selection procedure based on binding phagemid particles on activated human lymphocytes to generate a collection of scFvs against different IRs including LAG3 [28]. Upon reconstitution into human IgG4 Fc, the respective biological activities were evaluated through cytometry-based lymphocyte proliferation assays and cytokine ELISA.

The biologic activity of scFvF7 was evaluated by means of a functional evaluation of CD8$^+$ T lymphocytes specific for either viral or tumor antigens. We focused our interest on CD8$^+$ T cells for the efficacy of scFvF7 as a new ICB candidate since CD8$^+$ T-driven immune reactivation is expected to have a significant therapeutic effect to counteract the exhaustion of this lymphocyte subpopulation in cancer [29] and even chronic infectious diseases [30].

Overall, it is now clear that the broad success of ICB-based immunotherapies will depend on the combination of different treatment options. In a rapidly evolving research landscape, the generation of a repertoire of different immune-modulatory Ab-based constructs could play an essential role in identifying novel therapeutic agents and LAG3 certainly represents one of the most appealing emerging targets. The great interest in LAG3 has been fueled by its sustained expression in dysfunctional T-cells associated with many human tumors [31–33]. In addition, there is strong evidence that LAG3 upregulation could mediate an escape mechanism from PD1 therapy, whose acquired resistance might be overcome by the combined treatment with anti-LAG3 ICBs [34–37]. Another more recent study, for its part, highlights how LAG-3 blockade could result in a superior T cell activation compared to other inhibitory pathways, including PD1/PDL-1 axis [38]. Furthermore, LAG3, together with TIGIT and Tim-3, has been predicted to have a better safety profile in the clinic compared to CTLA4 and PD1 combinations [1, 15]. Also, LAG3 blockade inhibits the suppressive activity of T_{reg} cells

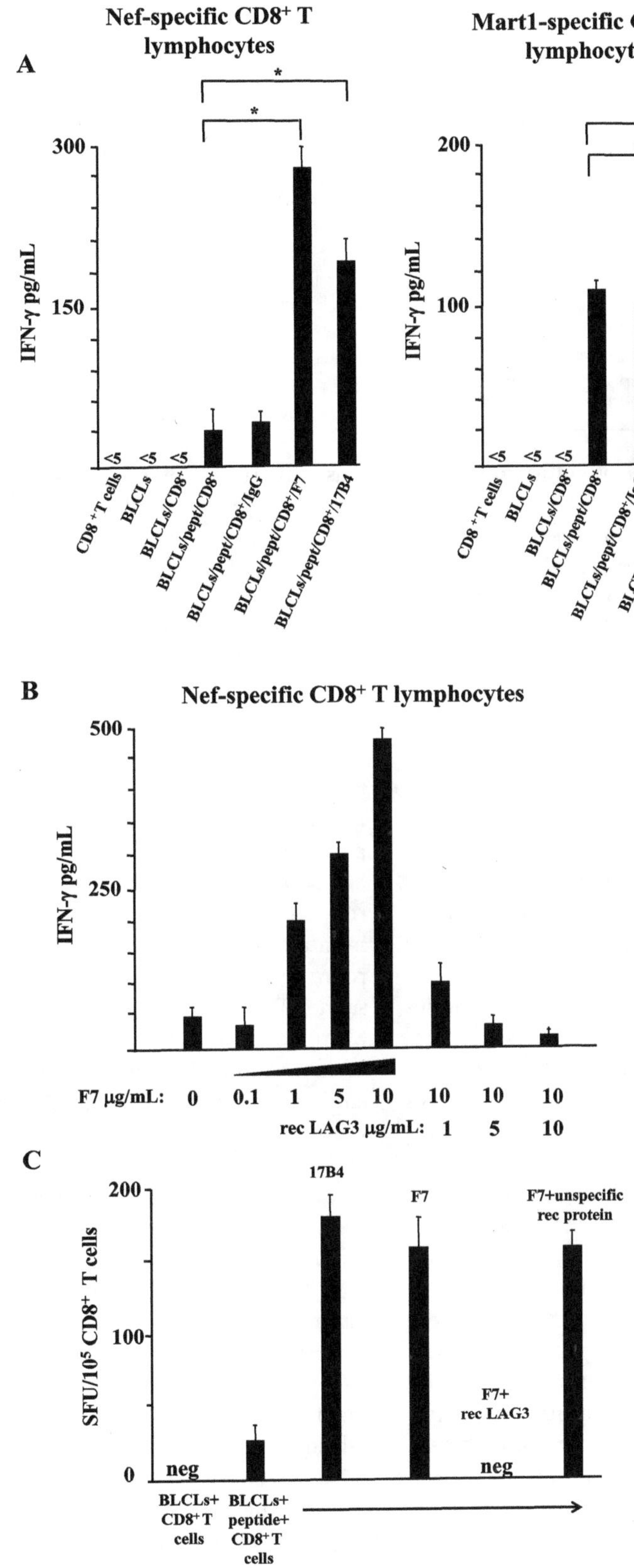

Fig. 7 (See legend on next page.)

(See figure on previous page.)

Fig. 7 The treatment with the divalent scFvF7-Fc Ab increases the activation of peptide-stimulated antigen-specific CD8+ T lymphocytes. **a** Both Nef- and Mart1-specific CD8+ T lymphocytes were put in co-culture with HLA-matched B-LCLs previously treated with appropriate peptides, and in the presence of either IgGs, divalent scFvF7-Fc Ab, or the murine anti-LAG3 17B4 mAb. As control, both B-LCLs (either untreated or treated with appropriate peptides) and CD8+ T cells were cultivated either alone or in co-culture. After o.n. incubation, supernatants were harvested, clarified, and tested for IFN-γ contents. The mean values ±SEM are shown as calculated from triplicate wells from a representative of three independent experiments. **b** Dose-response effect of the divalent scFvF7-Fc Ab and inhibition by recombinant LAG3. Co-cultures comprising B-LCLs and Nef-specific CD8+ T lymphocytes were set up as reported for panel A, albeit in the presence of different amounts of the reconstituted F7 Ab. Conditions comprising the highest concentration of the divalent scFvF7 and increasing doses of recombinant LAG3 were also run. After o.n. cultivation, supernatants were tested for IFN-γ contents. The mean values ±SEM are shown as calculated from triplicate wells from a representative of two independent experiments. **c** Effects of the divalent scFvF7-Fc as sensed by IFN-γ ELISPOT assay. IFN-γ-specific CD8+ T cell response as detected by ELISPOT assay after o.n. co-cultures of Nef-specific CD8+ T cells with B-LCLs pre-incubated or not with the specific peptide. The co-cultures were run in the presence or absence of either the divalent scFvF7-Fc Ab or the murine anti-LAG3 17B4 mAb. The former co-cultures were carried also out also in the presence of either recombinant LAG3, or the same amounts of an unrelated recombinant protein (i.e., HIV-1 gp120). Results are expressed as spot-forming units (SFU)/10^5 CD8+ T cells, and are representative of two experiments performed with triplicate wells

both in vitro and in vivo in a model of autoimmune pulmonary vasculitis [39], thus raising the intriguing hypothesis that a similar beneficial effect may be obtained in the tumor microenvironment. For all these reasons, LAG3-specific ICBs are currently being evaluated in phase I and phase II trials in a variety of solid and hematological malignancies either as monotherapy or in combination with anti-PD1 antibodies.

Conclusions

The increasing number of Abs against emerging regulatory molecules is expected to play a key role in promoting systematic analyses for the definition of robust biomarkers useful in predicting therapeutic activity and/or toxicity, as well as in designing patient-tailored precision medicine combination treatments. In the present study we describe the isolation and functional activity of a new anti-LAG3 recombinant Ab. Whereas further characterization studies are needed to support the potential of scFvF7 as a clinical candidate for the treatment of disease conditions in combination with other ICBs, data presented supports this agent as a valid scaffold for the development of novel anti-LAG3-based interventions.

Finally, in addition to the human nature of the scFvF7, making it a suitable reagent for clinical application, the flexibility of the described biotechnological platform allows an easy exploration of alternative therapeutic constructs, including bispecific or bifunctional Abs. These novel Abs could also have enhanced functional activity by substituting one antigen-binding portion with other ligands/cytokines and promoting the pairing of the heterodimers through insertion of specific mutations (e.g. Knob/Hole) [40] in constant regions.

Methods

Ab phage library

The IORISS1 naive human Ab library [24] consists in a large array (i.e., more than 10^9 Ab combinations) of scFv polypeptides displayed on the surface of M13 phage. It was constructed by recovering light and heavy Ab chains from a group of healthy donors, then randomly joining them by means of an appropriate peptide linker, and cloning the resulting scFvs into a phagemic vector as previously described [25]. It is noteworthy that the scFv inserts have been cloned upstream of two tag sequences, Flag tag and 6-his tag, respectively suitable to detect scFv in immunochemical assays and to purify them by nichel chromatography.

Biopanning strategy for selection LAG3 specific scFvs

LAG3-specific scFv Abs were selected from IORISS1 phage library as described elsewhere [25]. Briefly, a number of phages (~ 10^{13} colony forming units, *cfu*), proper to represent the complexity of the library, was placed in an immunotube (Nunc Maxisorp) previously coated o.n. at room temperature (r.t.) with 4 mL of phosphate buffered saline (PBS) containing the recombinant LAG3 antigen (Acrobiosystem) at a concentration of 5 μg/mL. After 2 h of incubation, the specific phages were detached by an elution carried out with a solution of 100 mM triethylamine (1 mL), ten minutes later neutralized with 0.5 mL of 1 M Tris–HCl pH 7.4. Afterwards, the phages were amplified in TG1 *E. coli* bacteria [supE hsd15 thi 1(lac-proAB) F′(traD36 proAB+ lacIqlacZ1M15)], as previously reported [25]; in detail, phages were used to infect TG1 bacteria in exponential growth phase, and the latter distributed on a plate with solid selective medium consisting of 2xTY (Sigma), 2% glucose and ampicillin (100 μg/mL). The day after, 50 mL of liquid selective medium was inoculated with the bacteria grown on the selective plate in order to obtain a 600 nm optical density (O.D.) of 0.05–0.1; when the culture reached the O.D. of 0.4–0.5, the bacteria were re-infected with the M13 K07 helper phage in a ratio phages/bacteria of about 20: 1, and incubated at 30 °C o.n.. Finally, the phages released into the supernatant were precipitated by means of PEG (Poly-Ethylene Glycol) 6000 before being used for a new cycle of selection on the LAG3 antigen.

An aliquot of TG1 bacteria infected with the phages from each panning cycle were also induced with isopropyl β-dithiogalactopyranoside (IPTG, Sigma) (final concentration > 1 mM) and used to perform a policlonal ELISA in order to monitor the enrichment of the LAG3-specific population (Table 1). In addition, bacteria infected with phages from the last panning was properly diluted and spread on selective plates in order to obtain single bacterial clones to be tested for their ability to secrete LAG3-specific scFvs.

For ELISA monoclonal screening, one hundred individual clones were inoculated in 96-well plates (Nunc) in a selective medium with 0.1% glucose; after 2 h of growth at 37 °C, clones were induced as above described and incubated for additional 12 h at 30 °C with shacking. Therefore, the scFv-containing supernatants were recovered, spun down (1800×*g* for 10 min) and used as primary antibodies against LAG3 protein.

DNA characterization and sequences

Plasmid DNAs from individual bacterial colonies were amplified and analyzed by restriction enzyme digestion using *Nco* I and *Not* I (i.e., the restriction sites used to clone the antibody fragments in IORISS1 library) to verify the presence of the scFv-encoding inserts. Afterwards, the scFv ORFs from the positive clones were sequenced with an automated DNA sequencer (Eurofins Genomics) using the following primers: scFv forward: 5′-atgaaacaaagcactattgcact-3′; scFv reverse: 5′-ttgatattcacaaacgaatgg-3′. The sequences were finally analyzed using the IMGT database.

ELISA on recombinant LAG3 protein

For screening procedure, ninety-six-well ELISA plates were coated with 50 μL/well of 10 μg/mL of recombinant LAG3 in PBS at 4 °C. The following day, a blocking solution consisting of 2% non-fat milk in PBS (MPBS) was added, and after 2 h the plates were washed with PBS containing 0.1% Tween 20 (TPBS). Plates were incubated for 2 h at r.t. with 50 μL of supernatants containing soluble scFv Abs, anti-flag M2 Ab (2.5 μg/mL, Sigma), and HRP- conjugated anti-mouse Ab (5 μg/mL, Dako). All Abs were resuspended in 2% MPBS. The reaction was visualized with 3,3′-5,5′-tetramethylbenzidine (BM blue, POD substrate, Roche), and stopped by adding 50 μL of 1 M sulfidric acid. The reaction was detected by an ELISA reader (Model 680 microplate reader, BioRad), and results expressed in terms of absorbance (A) = $A_{450\ nm}$-$A_{620\ nm}$.

The same ELISA procedure described above was performed to verify the reactivity of the scFvs-containing bacterial supernatants that were used for the Western blot assays.

The coating step for the assays to test the transfected CHO-derived supernatants was performed, as previously with 0.5 μg of recombinant LAG3/well, while for the dose-response binding analysis of the purified Fc-reconstituted scFv, the antigen was lowered to 0.1 μg /well. In both cases, the reconstituted Ab was incubated for one hour at r.t., followed by the incubation with a HRP-conjugated anti-human Ab (Thermo Fisher Scientific). The remaining procedures were performed as above described.

Soluble scFv purification

For the production of soluble scFvs, specifc phage-TG1 *E. coli* infected cells were cultured at 37 °C in 2 × TY containing 100 μg/mL ampicillin and 0.1% glucose up to 0.5 O.D. After induction of Ab expression by adding IPTG 1 mM, cells were incubated o.n. at 30 °C. Then, the bacterial cultures were centrifuged, and scFv containing supernatants collected. ScFv Abs were precipitated with ammonium sulfate, and then dialyzed in PBS. 6 × His-tagged scFv Abs were purified by immobilized metal affinity chromatography using Ni_{2+}-nitriloacetic acid agarose (Qiagen). ScFv fragments were eluted with 250 mM imidazole in PBS, dialyzed, aliquoted, and stored at − 80 °C.

Eukaryotic cell cultures

PBMCs were obtained by Ficoll–Hypaque density gradient centrifugation from heparinized blood samples of healthy donors. Purified T cells were obtained by PBMC negative selection using a kit for human T cell selection (Miltenyi) according to the manufacturer's protocol, and cultivated in RPMI medium supplemented with 10% of heat-inactivated fetal calf serum (FCS)(HyClone). Both CHO and HEK293T cells were cultivated in DMEM plus 10% FCS. Both HLA-A02 and HLA–B7 B-LCLs were grown in RPMI medium supplemented with 10% FCS. B-LCLs were generated by in vitro EBV infection of ex vivo B lymphocyte cultures. Both isolation and expansion of $CD8^+$ T cell clones specific for Mart (melanoma-associated recognized by T cells)-1 and HIV-1 Nef have been previously described [41, 42]. Mart1-specific $CD8^+$ T cells recognize the HLA-A.02-restricted $AAGIGILTV_{27-35}$ peptide sequence, whereas Nef-specific $CD8^+$ T cells recognize the HLA-B7-restricted $TPGPGVRYPL_{128-137}$ peptide. Both antigen-specific $CD8^+$ T-cell clones were cultivated in RPMI plus 10% AB human serum (Gibco), and regularly monitored for their specificity.

Flow cytometry assays on human T cells

T cells isolated from PBMCs were seeded at the concentration of 2×10^6/mL, and activated with 5 μg/mL of PHA. During the screening phase, scFvF7 was preliminarily tested on TCD4 cells at a one shot concentration

(50 μg/mL), just to have a 'yes/no' response about its ability to recognize the antigen on the cell surface. About 1×10^6 cells were resuspended in PBS containing primary scFv antibody and incubated for 1 h at r.t.. After washing, cells were resuspended for 30 min at 4 °C in PBS containing 25 μg/mL of an anti-FLAG M2 mouse antibody (Sigma) and then incubated for another 30 min at 4 °C with 6 μg/mL of an FITC-goat anti-mouse IgG (Pierce, IL, USA). An irrelevant antibody directed to an irrelevant antigen (glucose oxidase, GO) was used as negative control. After staining, the cell samples were washed, maintained at 4 °C and immediately analyzed by FACScan (Becton-Dickinson, NJ, USA) equipped with 15 mW, 488-nm, argon laser. Fluorescence compensation was determined using samples stained with anti-glucose oxidase scFv and goat FITC-conjugated anti-mouse secondary antibody.

For the cytofluorimetric analysis on HEK293 T cells (both LAG3-negative parental and the derivative LAG3 expressing cell lines, kindly provided by Pantaleo's team, Lausanne), 2.5 $\times 10^5$ cells were incubated with Fc-reconstituted scFvF7 Ab (10 μg/mL) for 1 h at r.t., and after washing with an goat FITC-conjugated anti-human IgG1 (Pierce) for 30 min at 4 °C.

Similarly, both PBMC-derived T cells and antigen-specific $CD8^+$ T lymphocytes (1×10^6 cells/test) were incubated with Fc-reconstituted anti-LAG3 scFv, at different concentrations as specified in each experimental assay (comprised between 0.1 and 20 μg/mL) for 1 h at r.t.; the murine 17B4 anti-LAG3 mAb (EnzoLab, 10 μg/mL) was used as control. Then, cells were washed and incubated with FITC-conjugated secondary either anti-human (for detecting Fc-reconstituted scFvF7) or anti-mouse (for detecting mouse 17B4 mAb) goat IgG1 (Pierce). In all cases, stained cells were analyzed on FACscan platform (BD Biosciences).

Western blotting

For the analysis of the scFv binding in Western blot assay, 0.5 μg of both recombinant LAG3 and, as control, glucose oxidase (GO), were loaded in multiple wells, run on 12% SDS-PAGE (under reducing or non-reducing conditions, as specified in each assay) and transferred to nitrocellulose membrane. Separate filter strips were then incubated for 2 h with bacterial supernatants from the anti-LAG3 clone scFvF7 and, as control, from a previously characterized anti-GO clone (scFvGO) [24], both of them previously tested in ELISA to verify their reactivities. Afterwards, the strips were washed and incubated again with 5 μg/mL anti-flag M2 mouse antibody (Sigma) in 2% MPBS. After an additional incubation for 1 h at r.t. in the presence of 5 μg/mL of a goat anti-mouse antibody HRP-conjugated (Dako), the reaction was developed and visualized using a chemiluminescence reagent (NEN Life Science Products). A mouse anti-6 × His mAb (Serotec, 1 μg/mL diluted in 2%MPBS) was used as positive control to detect the recombinant LAG3 protein (which is provided with a 6 × His-tag at its C-terminus).

For monitoring transfection efficiency of CHO cells with the vector encoding the Fc-reconstituted anti-LAG3 antibody, supernatants from transfected cells (50 μL/lane) were separated by 10% SDS-PAGE, and then transferred to nitrocellulose membrane. After blocking with 5% MPBS, the membranes were washed and incubated for 1 h at r.t. with HRP-conjugated, goat anti-human IgG (1:2000 dilution, (Thermo Fisher Scientific)) and developed as above described.

Genetic construction and production of the divalent anti-LAG3 scFvF7-Fc Ab

The anti-LAG3 scFv coding sequence was amplified from IORISS1 phagemide vector [24] by PCR with tailored primers. The amplified fragment was then digested with *Xho* I and *Eco* RI enzymes, gel purified, and ligated in frame to the *Xho* I/*Eco* RI double digeted pFUSEss-CHIg-HG1 vector (Invivogen) expressing the sequences of both human Fc CH2, including LALA mutations [26], and CH3 Fc domains. The resulting construct was transfected into CHO cells for pilot production of bivalent scFvF7-Fc Ab. Briefly, 10^6 cells were transfected with 1 μg of plasmid DNA per well in a six-well tissue culture plate using Lipofectamine Plus reagent (Life Technologies) according to the manufacturer's instructions. Supernatants were collected 48 h later to evaluate Ab production through both western blot analysis and ELISA.

Large-scale production of the divalent scFvF7-Fc Ab

Divalent scFvF7-Fc Ab was produced through transient transfection of CHO DG44 cells followed by incubation for 6 days in ProCHO5 serum free medium (Lonza). The scFvF7-Fc Ab was purified from the cell medium using a protein A column (Thermo Fisher) with Abs eluted with 100 mM glycine buffer pH 3.0 into a 1 M Tris-HCl eluate at pH 8.0. Abs were dialyzed twice against PBS, concentrated using a JumboSep centrifuge filter with 3 kDa molecular weight cut-off (Pall Laboratories), and sterile filtered with Millex GP 0.22 μm pore size (Millipore). Prior to evaluation of scFvF7-Fc in the in vitro functional recovery assays, Abs were tested with the FDA-licensed Endosafe-PTS kit (Charles River Laboratory) which showed < 5 EU of endotoxins per mg of protein.

IFN-γ ELISA and ELISPOT assays

$CD8^+$ T cells were pre-incubated with different concentrations of the reconstituted anti-LAG3 scFv for 2 h, and then co-cultured at a 2:1 ratio with HLA-matched B-LCLs previously pulsed with 100 ng/mL of the appropriate peptide. The co-cultures were incubated o.n., and finally the supernatants were clarified and assayed for IFN-γ content by ELISA (Immunological Sciences). For ELISPOT assays, the co-cultures were carried out for 16 h in ELISPOT microwells previously coated with a mAb against human IFN-γ (clone D1K, Mabtech). Afterwards, the cells were removed, and a biotinylated Ab against human IFN-γ added, followed by the addition of a streptavidin-alkaline phosphatase. The plate was then developed using BCIP/NBT substrate (Sigma). Spot-forming cells were analyzed and counted using an ELISPOT reader (Amplimedical Bioline A-EL-VIS GmbH).

Statistical analysis

Statistical analyses were performed by using GraphPad Prism version 5.01 software. Data are expressed as means ± standard error of the mean (SEM). The student's t test was used to determine significance. A p value of < 0.05 was considered statistically significant.

Supplementary information

Additional file 1: Figure S3D^{II}. Characterization of the anti-LAG3 scFvF7 (II). A. ELISA conducted in parallel on intact recombinant LAG3 protein and heat-denatured LAG3 protein (by boiling for 5 min). The plate was coated with 0.5 µg of antigen per microwell (GO, intact recombinant LAG3 and heat-stressed recombinant LAG3). After blocking step (2% MPBS for two hours at r.t.) wells were incubated for 2 h at r.t. with 50 µL of scFvF7 (25 µg/mL) together with anti-flag M2 Ab (2.5 µg/mL, Sigma) and HRP- conjugated anti-mouse Ab (5 µg/mL, Dako). A mouse anti-6 his mAb (which recognizes the 6-histidines tag at C terminal end of LAG3 protein)(Serotec) and commercial mouse anti-LAG3 mAb 17B4 (that recognizes the 30 aa extra-loop of the first N-terminal D1 domain of human LAG3)(EnzoLab) were used as positive controls. All Abs were resuspended in 2% MPBS. O.D.: optical density. B. Western blotting assay with LAG3 under reducing and non-reducing conditions. 0.5 µg of glucose oxidase (GO) and of recombinant LAG3 proteins (the latter under three different conditions, namely: -DTT (reducing agent)/–boiling, –DTT/+boiling 5 min, +DTT/+boiling 5 min) were loaded as specified in replicate wells on a 12% SDS-PAGE and transferred to filter paper. Portions from the filter were then incubated with the indicated primary antibodies. An Anti-6 his mAb was used as a positive control for LAG3 recombinant protein (which has a 6-histidines tag at its C terminal end). Arrows indicate relevant signals. The different molecular weights observed for LAG3 are obviously attributable to the impact of the different experimental conditions (non-reducing and reducing) on the SDS-PAGE separation. Molecular markers in kilodaltons (kDa) are reported on the right. The reactivity of the supernatants (scFvGO and scFvF7) used was previously checked in ELISA (bottom). (PPTX 125 kb)

Additional file 2: Figure S7A^{II}. The treatment with the divalent scFvF7-Fc Ab increases the activation of peptide-stimulated Nef-specific CD8^{+} T lymphocytes in terms of IFN-γ secretion (Exp. I and Exp. III). For details see Legend of Fig. 7a. (PPTX 40 kb)

Additional file 3: Figure S7A^{III}. The treatment with the divalent scFvF7-Fc Ab increases the activation of peptide-stimulated Mart1-specific CD8^{+} T lymphocytes in terms of IFN-γ secretion (Exp. II and Exp. III). For details see Legend of Fig. 7a. (PPTX 41 kb)

Additional file 4: Figure S7B^{II}. Dose-response effect of the divalent scFvF7-Fc Ab and inhibition by recombinant LAG3 (Exp. II). For details see Legend of Fig. 7b. (PPTX 36 kb)

Additional file 5: Figure S7C^{II}. Effects of the divalent scFvF7-Fc as sensed by IFN-γ ELISPOT assay (Exp. II). For details see Legend of Figure 7c. (PPTX 36 kb)

Abbreviations

B-LCLs: B-lymphoblastoid cell lines; DCs: Dendritic cells; ELISA: *Enzyme-linked immunosorbent assay; ELISPOT*: Enzyme-linked immunospot; Fc: Fragment crystallizable; FCS: Fetal calf serum; GO: Glucose oxidase; HLA: Human leukocyte antigen; ICB: Immune checkpoint blockers; IFN: Interferon; Ig: Immunoglobulin; IPTG: Isopropyl β-dithiogalactopyranoside; IR: Inhibitory receptor; kDa: Kilodaltons; LAG3: Lymphocyte activate gene-3; mAb: Monoclonal antibody; O.D.: Optical density; o.n.: Overnight; PBMCs: Peripheral blood mononuclear cells; PBS: Phosphate buffered saline; PHA: Phytohaemagglutinin; r.t.: Room temperature; scFv: Single-chain variable fragment; SE: Standard error; TILs: Tumor-infiltrating lymphocytes

Acknowledgments

We thank Pietro Arciero, Istituto Superiore di Sanità, for its technical support. We thank Martin Bennett for his help in revising the manuscript for grammar and style.

Authors' contributions

AA isolated and characterized the scFvF7, participated in the design of the research and drafted the manuscript; CA performed the experiments with CD8^{+} T cell clones; AM worked on the design and genetic construction of the diabody scFvF7-Fc; MFL carried out biochemical assays and binding studies with scFvs; MG performed western blot analyses; MA analyzed data from cytofluorimeter analyses and interpreted the results; CF and GP produced and characterized the divalent scFvF7; SV coordinated and supervised the study design on the basis of the scFv isolation and divalent antibody construction; MFE wrote the paper and supervised the cell immunology experiments. All authors have seen and agreed to the final version of the paper. All authors read and approved the final manuscript.

Funding

The research was supported by "in-kind" resources from the Istituto Superiore di Sanità (Rome, Italy) and by Grant PE-2016-02364927 Italian Health Minister.

Ethics approval and consent to participate

Human specimens used in some experiments described in our ms are PBMCs from healthy donors. Multiple stocks of human peripheral blood mononuclear cells (PBMCs) isolated from a large number of healthy donors ($n > 20$), have been stored in liquid nitrogen in our laboratory. The PBMCs have been purified from buffy coats obtained from the Transfusion Center of the University of Rome "La Sapienza" as waste material resulting from the procedures of plasma/platelet/red blood isolation from whole blood of healthy donors. They were volunteers and consented for their donated blood to be used for research. In line with the Italian law on this matter (Legislative Decree of the Italian Ministry of Health of January 25, 2001, published on the Official Gazette of April 3, 2001) donor ages ranged between 21 and 60 years for women and 21–65 years for men. There have been no restrictions for blood donors as to sex and ethnic background, except for children, mentally disabled individuals and pregnant women. All donors were healthy, non-medicated consulting adults, and were pre-screened for exposure to infectious agents. Blood was obtained by venipuncture following standard medical procedures using sterile technique and a compression bandage in the area of blood removal. A record of the donor and hematocrit result is stored on file at the Transfusion Center. Research investigators do not have any access to personal data or other information concerning the donors in agreement with the Italian law Number 675 of December 31, 1996.

Consent for publication

Not applicable.

Competing interests

Alessandro Ascione (AA), Michela Flego (MFL), Alessandra Mallano (AM), Stefano Vella (SV) and Maurizio Federico (MFE) are co-inventors of a patent recently filed, regarding the human monoclonal antibody against LAG3 described in the present manuscript (Date of patent filing: 12/07/2019. Title: Human recombinant antibody against LAG3 membrane receptor, medical and diagnostic uses thereof).

Apart from that the authors declare that they have no competing interests.

Author details

[1]National Center for Global Health, Istituto Superiore di Sanità (ISS), Viale Regina Elena 299, 00161 Rome, Italy. [2]Service of Immunology and Allergy, Lausanne University Hospital, University of Lausanne, Lausanne, Switzerland.

References

1. Andrews LP, Marciscano AE, Drake CG, Vignali DA. LAG3 (CD223) as a cancer immunotherapy target. Immunol Rev. 2017;276(1):80–96.
2. Triebel F, Jitsukawa S, Baixeras E, Roman-Roman S, Genevee C, Viegas-Pequignot E, et al. LAG-3, a novel lymphocyte activation gene closely related to CD4. J Exp Med. 1990;171(5):1393–405.
3. Huard B, Mastrangeli R, Prigent P, Bruniquel D, Donini S, El-Tayar N, et al. Characterization of the major histocompatibility complex class II binding site on LAG-3 protein. Proc Natl Acad Sci U S A. 1997;94(11):5744–9.
4. Baixeras E, Huard B, Miossec C, Jitsukawa S, Martin M, Hercend T, et al. Characterization of the lymphocyte activation gene 3-encoded protein. A new ligand for human leukocyte antigen class II antigens. J Exp Med. 1992; 176(2):327–37.
5. Kouo T, Huang L, Pucsek AB, Cao M, Solt S, Armstrong T, et al. Galectin-3 shapes antitumor immune responses by suppressing CD8+ T cells via LAG-3 and inhibiting expansion of Plasmacytoid dendritic cells. Cancer Immunol Res. 2015;3(4):412–23.
6. Liu W, Tang L, Zhang G, Wei H, Cui Y, Guo L, et al. Characterization of a novel C-type lectin-like gene, LSECtin: demonstration of carbohydrate binding and expression in sinusoidal endothelial cells of liver and lymph node. J Biol Chem. 2004;279(18):18748–58.
7. Mao X, Ou MT, Karuppagounder SS, Kam TI, Yin X, Xiong Y, et al. Pathological alpha-synuclein transmission initiated by binding lymphocyte-activation gene 3. Science. 2016;353(6307):aah3374.
8. Mastrangeli R, Micangeli E, Donini S. Cloning of murine LAG-3 by magnetic bead bound homologous probes and PCR (gene-capture PCR). Anal Biochem. 1996;241(1):93–102.
9. Workman CJ, Dugger KJ, Vignali DA. Cutting edge: molecular analysis of the negative regulatory function of lymphocyte activation gene-3. J Immunol. 2002;169(10):5392–5.
10. Liang B, Workman C, Lee J, Chew C, Dale BM, Colonna L, et al. Regulatory T cells inhibit dendritic cells by lymphocyte activation gene-3 engagement of MHC class II. J Immunol. 2008;180(9):5916–26.
11. Hemon P, Jean-Louis F, Ramgolam K, Brignone C, Viguier M, Bachelez H, et al. MHC class II engagement by its ligand LAG-3 (CD223) contributes to melanoma resistance to apoptosis. J Immunol. 2011;186(9):5173–83.
12. Demeure CE, Wolfers J, Martin-Garcia N, Gaulard P, Triebel F. T lymphocytes infiltrating various tumour types express the MHC class II ligand lymphocyte activation gene-3 (LAG-3): role of LAG-3/MHC class II interactions in cell-cell contacts. Eur J Cancer. 2001;37(13):1709–18.
13. Pardoll DM. The blockade of immune checkpoints in cancer immunotherapy. Nat Rev Cancer. 2012;12(4):252–64.
14. June CH, Warshauer JT, Bluestone JA. Is autoimmunity the Achilles' heel of cancer immunotherapy? Nat Med. 2017;23(5):540–7.
15. Anderson AC, Joller N, Kuchroo VK. Lag-3, Tim-3, and TIGIT: co-inhibitory receptors with specialized functions in immune regulation. Immunity. 2016; 44(5):989–1004.
16. Weiner LM, Surana R, Wang S. Monoclonal antibodies: versatile platforms for cancer immunotherapy. Nat Rev Immunol. 2010;10(5):317–27.
17. Le Mercier I, Lines JL, Noelle RJ. Beyond CTLA-4 and PD-1, the generation Z of negative checkpoint regulators. Front Immunol. 2015;6:418.
18. Sharma P, Allison JP. The future of immune checkpoint therapy. Science. 2015;348(6230):56–61.
19. Marshall HT, Djamgoz MBA. Immuno-oncology: emerging targets and combination therapies. Front Oncol. 2018;8:315.
20. Dempke WCM, Fenchel K, Uciechowski P, Dale SP. Second- and third-generation drugs for immuno-oncology treatment-the more the better? Eur J Cancer. 2017;74:55–72.
21. Popovic A, Jaffee EM, Zaidi N. Emerging strategies for combination checkpoint modulators in cancer immunotherapy. J Clin Invest. 2018; 128(8):3209–18.
22. Andersen DC, Reilly DE. Production technologies for monoclonal antibodies and their fragments. Curr Opin Biotechnol. 2004;15(5):456–62.
23. Weisser NE, Hall JC. Applications of single-chain variable fragment antibodies in therapeutics and diagnostics. Biotechnol Adv. 2009;27(4):502–20.
24. Ascione A, Capecchi B, Campitelli L, Imperiale V, Flego M, Zamboni S, et al. Human monoclonal antibodies in single chain fragment variable format with potent neutralization activity against influenza virus H5N1. Antivir Res. 2009;83(3):238–44.
25. Pasello M, Zamboni S, Mallano A, Flego M, Picci P, Cianfriglia M, et al. Design and construction of a new human naive single-chain fragment variable antibody library, IORISS1. J Biotechnol. 2016;224:1–11.
26. Schlothauer T, Herter S, Koller CF, Grau-Richards S, Steinhart V, Spick C, et al. Novel human IgG1 and IgG4 fc-engineered antibodies with completely abolished immune effector functions. Protein Eng Des Sel. 2016;29(10):457–66.
27. Everett KL, Kraman M, Wollerton FPG, Zimarino C, Kmiecik K, Gaspar M, et al. Generation of Fcabs targeting human and murine LAG-3 as building blocks for novel bispecific antibody therapeutics. Methods. 2019;154:60-69. https://doi.org/10.1016/j.ymeth.2018.09.003. Epub 2018 Sep 9
28. Sasso E, D'Avino C, Passariello M, D'Alise AM, Siciliano D, Esposito ML, et al. Massive parallel screening of phage libraries for the generation of repertoires of human immunomodulatory monoclonal antibodies. mAbs. 2018;10(7):1060–72.
29. Peggs KS, Quezada SA, Allison JP. Cancer immunotherapy: co-stimulatory agonists and co-inhibitory antagonists. Clin Exp Immunol. 2009;157(1):9–19.
30. Miller BC, Sen DR, Al Abosy R, Bi K, Virkud YV, LaFleur MW, et al. Subsets of exhausted $CD8^+$ T cells differentially mediate tumor control and respond to checkpoint blockade. Nat Immunol. 2019;20(3):326–36.
31. Nguyen LT, Ohashi PS. Clinical blockade of PD1 and LAG3--potential mechanisms of action. Nat Rev Immunol. 2015;15(1):45–56.
32. Baitsch L, Baumgaertner P, Devevre E, Raghav SK, Legat A, Barba L, et al. Exhaustion of tumor-specific CD8(+) T cells in metastases from melanoma patients. J Clin Invest. 2011;121(6):2350–60.
33. Matsuzaki J, Gnjatic S, Mhawech-Fauceglia P, Beck A, Miller A, Tsuji T, et al. Tumor-infiltrating NY-ESO-1-specific CD8+ T cells are negatively regulated by LAG-3 and PD-1 in human ovarian cancer. Proc Natl Acad Sci U S A. 2010;107(17):7875–80.
34. Woo SR, Turnis ME, Goldberg MV, Bankoti J, Selby M, Nirschl CJ, et al. Immune inhibitory molecules LAG-3 and PD-1 synergistically regulate T-cell function to promote tumoral immune escape. Cancer Res. 2012; 72(4):917–27.
35. Taube JM, Young GD, McMiller TL, Chen S, Salas JT, Pritchard TS, et al. Differential expression of immune-regulatory genes associated with PD-L1 display in melanoma: implications for PD-1 pathway blockade. Clin Cancer Res. 2015;21(17):3969–76.
36. Koyama S, Akbay EA, Li YY, Herter-Sprie GS, Buczkowski KA, Richards WG, et al. Adaptive resistance to therapeutic PD-1 blockade is associated with upregulation of alternative immune checkpoints. Nat Commun. 2016;7:10501.
37. Huang RY, Francois A, McGray AR, Miliotto A, Odunsi K. Compensatory upregulation of PD-1, LAG-3, and CTLA-4 limits the efficacy of single-agent checkpoint blockade in metastatic ovarian cancer. Oncoimmunology. 2017; 6(1):e1249561.
38. Lichtenegger FS, Rothe M, Schnorfeil FM, Deiser K, Krupka C, Augsberger C, et al. Targeting LAG-3 and PD-1 to enhance T cell activation by antigen-presenting cells. Front Immunol. 2018;27(9):385.
39. Huang CT, Workman CJ, Flies D, Pan X, Marson AL, Zhou G, et al. Role of LAG-3 in regulatory T cells. Immunity. 2004;21(4):503–13.
40. Atwell S, Ridgway JB, Wells JA, Carter P. Stable heterodimers from remodeling the domain interface of a homodimer using a phage display library. J Mol Biol. 1997;270(1):26–35.
41. Di Bonito P, Grasso F, Mochi S, Petrone L, Fanales-Belasio E, Mei A, et al. Anti-tumor CD8+ T cell immunity elicited by HIV-1-based virus-like particles incorporating HPV-16 E7 protein. Virology. 2009;395(1):45–55.
42. Rivoltini L, Kawakami Y, Sakaguchi K, Southwood S, Sette A, Robbins PF, et al. Induction of tumor-reactive CTL from peripheral blood and tumor-infiltrating lymphocytes of melanoma patients by in vitro stimulation with an immunodominant peptide of the human melanoma antigen MART-1. J Immunol. 1995;154(5):2257–65.

3

Quantitative assessment of LASSO probe assembly and long-read multiplexed cloning

Syukri Shukor[1†], Alfred Tamayo[1†], Lorenzo Tosi[1], H. Benjamin Larman[2] and Biju Parekkadan[1,3,4*]

Abstract

Background: Long Adapter Single-Stranded Oligonucleotide (LASSO) probes were developed as a novel tool for massively parallel cloning of kilobase-long genomic DNA sequences. LASSO dramatically improves the capture length limit of current DNA padlock probe technology from approximately 150 bps to several kbps. High-throughput LASSO capture involves the parallel assembly of thousands of probes. However, malformed probes are indiscernible from properly formed probes using gel electrophoretic techniques. Therefore, we used next-generation sequencing (NGS) to assess the efficiency of LASSO probe assembly and how it relates to the nature of DNA capture and amplification. Additionally, we introduce a simplified single target LASSO protocol using classic molecular biology techniques for qualitative and quantitative assessment of probe specificity.

Results: A LASSO probe library targeting 3164 unique *E. coli* ORFs was assembled using two different probe assembly reaction conditions with a 40-fold difference in DNA concentration. Unique probe sequences are located within the first 50 bps of the 5′ and 3′ ends, therefore we used paired-end NGS to assess probe library quality. Properly mapped read pairs, representing correctly formed probes, accounted for 10.81 and 0.65% of total reads, corresponding to ~ 80% and ~ 20% coverage of the total probe library for the lower and higher DNA concentration conditions, respectively. Subsequently, we used single-end NGS to correlate probe assembly efficiency and capture quality. Significant enrichment of LASSO targets over non-targets was only observed for captures done using probes assembled with a lower DNA concentration. Additionally, semi-quantitative polyacrylamide gel electrophoresis revealed a ~ 10-fold signal-to-noise ratio of LASSO capture in a simplified system.

Conclusions: These results suggest that LASSO probe coverage for target sequences is more predictive of successful capture than probe assembly depth-enrichment. Concomitantly, these results demonstrate that DNA concentration at a critical step in the probe assembly reaction significantly impacts probe formation. Additionally, we show that a simplified LASSO capture protocol coupled to PAGE (polyacrylamide gel electrophoresis) is highly specific and more amenable to small-scale LASSO approaches, such as screening novel probes and templates.

Keywords: Long adapter single-stranded oligonucleotides (LASSO), Genomic libraries, Multiplex PCR, Multiplex cloning

* Correspondence: biju_parekkadan@hms.harvard.edu
†Syukri Shukor and Alfred Tamayo contributed equally to this work.
[1]Center for Surgery, Innovation, and Bioengineering, Department of Surgery, Massachusetts General Hospital, Harvard Medical School and the Shriners Hospitals for Children, 51 Blossom Street, Boston, MA 02114, USA
[3]Harvard Stem Cell Institute, Cambridge, MA 02138, USA
Full list of author information is available at the end of the article

Background

Techniques that enable multiplexed amplification of specific DNA sequences from large and complex templates, namely genomes and transcriptomes, are invaluable tools for identifying mutations and functional genomic studies [1, 2]. PCR (polymerase chain reaction) amplification is the most common technique used for exponential enrichment of specific sequences from complex templates because it offers operational simplicity and a measure of specificity. Though more technically challenging, DNA padlock probe techniques are an alternative method for sequence amplification that offers high specificity in multiplex approaches by using two template complementary sequences on the same probe, commonly known as molecular inversion probes (MIPs). However, a major constraint of MIPs is the inability to capture sequences greater than 150 base pairs [3, 4]. In an effort to improve upon padlock probe technology, we have previously shown that Long Adapter Single-Stranded Oligonucleotide (LASSO) probes improve the target capture size limitation of MIPs, and demonstrated the massively multiplexed capture-by-circularization of kilobase long genomic regions (up to ~ 4 kb) from *E. coli* genomic DNA [5].

A critical element of LASSO is the creation of thousands of LASSO probes in a single reaction. Since the sequence identities of LASSO probes cannot be discerned by standard molecular biology techniques such as electrophoretic mobility, we did not have a clear understanding of LASSO probe assembly efficiency and how that relates to successful captures downstream. To this end, we set out to quantify LASSO probe assembly using next-generation sequencing techniques (NGS), and correlate those results to the success rate of LASSO capture. Our results confirm the critical nature of LASSO probe assembly and suggest how this step might be improved.

Results

Quality assessment of LASSO probe assembly by NGS

As previously described, a general schematic of LASSO probe assembly and capture is shown in Fig. 1a, b [5]. LASSO probes consist of a constant non-specific linker sequence (long-adapter) flanked by two ~ 50 bps sequences (probe arms) complementary to unique genomic targets. Massively parallel cloning by LASSO requires a multiplexed probe assembly process wherein pre-LASSO probes are fused to adapter sequences and amplified by a PCR-based approach (Fig. 1a). Since LASSO probes created this way are of a uniform size, electrophoretic mobility cannot fully assess the quality of probe assembly. To gain a better understanding of LASSO probe assembly, we focused our analysis on the unique probe arms by using paired-end next-generation sequencing (NGS). This approach allowed us to analyze the 5′ and 3′ sequences of individual LASSO probes within a large library.

In this study, we synthesized a LASSO probe library targeting 3164 unique *E. coli* ORF (open reading frame)-eome sequence as previously described. For comparison, we used two different reaction volumes during a critical step in the probe assembly process, decreasing the reaction volume to 50 µl from the previously used 2 ml [5], thereby dramatically increasing the concentration of DNA in this step. The resulting LASSO probes amplified by PCR were then prepared and submitted for paired-end NGS (2 × 75 bp Illumina MiSeq). Only read pairs from PCR amplicons corresponding to full-length LASSO probes (377 base pairs) were considered in our analysis, and therefore mapped to the probe library. Total raw sequencing reads for 50 µl and 2 ml were 572044 and 457694 reads respectively. Out of the total reads, 300332 (52.5%) and 297049 (64.9%) reads from 50 µl and 2 ml passed pairing and read filtering (Additional file 2: Table S1). Median read depths for all mapped pairs were 69 and 65 for 50 µl and 2 ml, respectively.

Our analysis sheds light on probe arm pair distribution, probe enrichment depth, and probe library coverage. For the purpose of our analysis, a properly assembly LASSO probe is a concordant probe. Concordant probes, as we define them, have read pairs mapped to both 5′ and 3′ ends of one unique reference sequence, confirming that both probe arms originated from the same sequence as intended. However, if read pairs contain probe arms that originate from different probes, then the probe in question has mismatched probe arms and is termed a discordant probe.

Concordant probes only accounted for 10.8 and 0.61% of read pairs when we compared probe assembly for the 2 ml and 50 µl and reaction conditions, respectively (Additional file 2: Table S1). This result suggested a very low rate of proper LASSO probe assembly because most probes were discordant. The distribution of LASSO probe enrichment depth, concordant or discordant, was not significantly different between the probe assembly sets and appeared random with regard to probe arm identity, indicating no biases in probe sequencing depth when probes were arranged according to expected target capture length (Fig. 2a). Despite full representation of the probe library in all mapped reads, we observed a dramatic difference in overall concordantly mapped probe coverage of 80.32% (2 ml) versus 20.73% (50 µl). As expected, we also did not detect a relationship between probe library coverage and expected target capture length, indicating that these results are not explained by a sequence bias between assembly protocols (Fig. 2b). These experiments suggest that while concordant probe assembly was highly inefficient in both cases, the 2 ml self-circularization reaction condition resulted in significantly higher library coverage for concordant probes. We

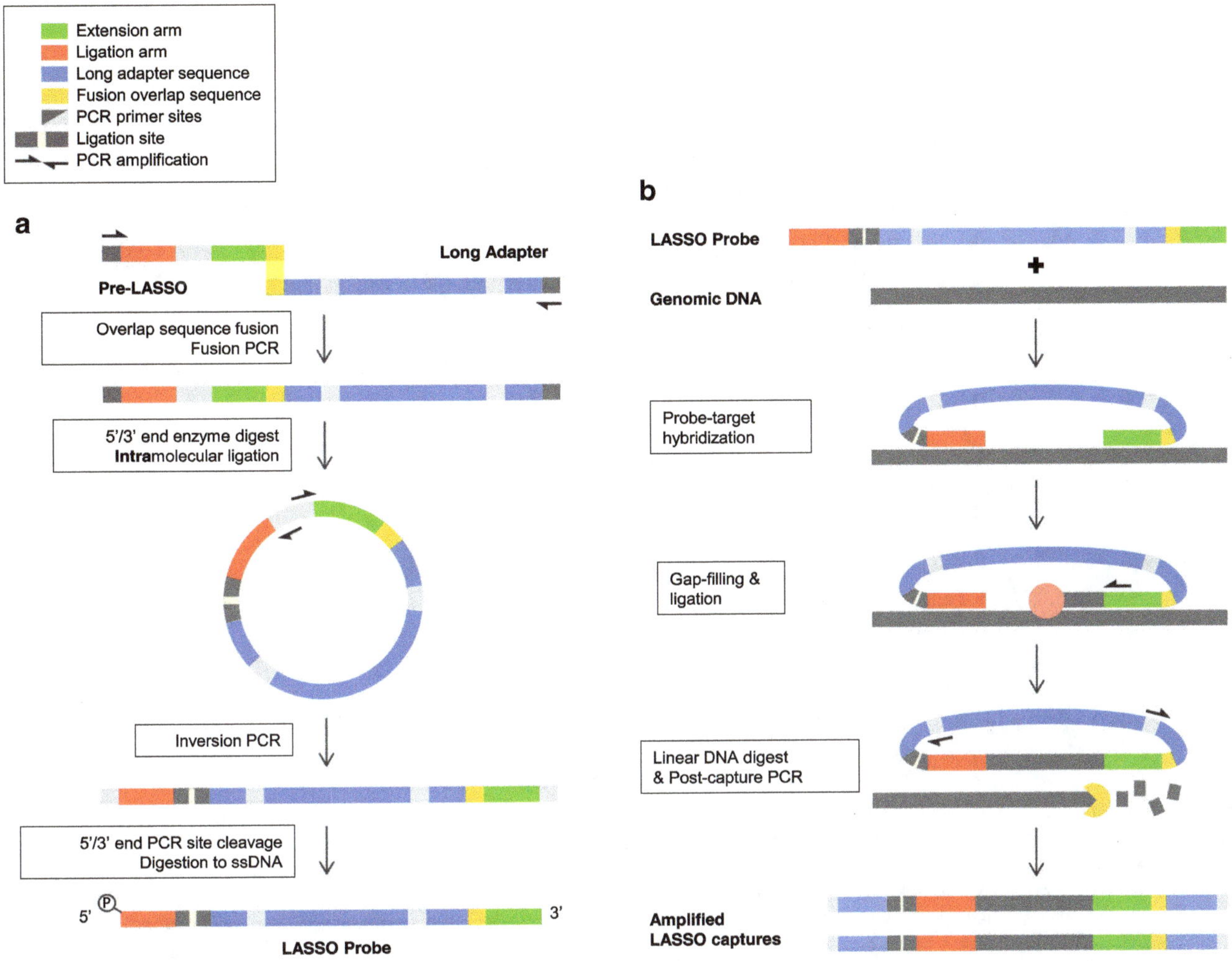

Fig. 1 LASSO probe de novo synthesis and capture. **a** Schematic of LASSO probe assembly protocol outline. **b** LASSO target capture outlines probe-template hybridization, target capture and dissociation in a single reaction followed by post-capture PCR

next sought to determine if these differences in probe library quality would affect downstream LASSO capture.

Effect of LASSO probe assembly on multiplexed capture

The LASSO capture and amplification protocol was previously established and is summarized in Fig. 1b [5]. In short, the extension and ligation probe arms hybridize to the start and end of genomic sequence targets. DNA polymerase and template-dependent ligase then fill in the genomic gap between both arms to produce circular LASSO probes containing captured genomic target sequences, which are then amplified during post-capture PCR with a primer set nested within the LASSO probe backbone.

To understand how LASSO capture is affected by the quality of probe assembly, we used NGS to assess multiplexed LASSO capture of an *E. coli* ORFeome, as previously described by Tosi et al. [5]. Additional file 1: Figure S1 plots capture targets ranging from 0.4 to 4.6 kb in size in increasing order, and approximately half of target ORFs are between 0.4 and 1 kb long. Resultant LASSO post-capture PCR products were sheared and sequenced using the Illumina HiSeq platform (50 bp reads). 6982080 and 6968823 total reads were sequenced for captures using the 50 μl or 2 ml probe assembly reaction mixture, respectively (Additional file 3: Table S2). These reads were then filtered for read duplicates and aligned to both *E. coli* gDNA and LASSO backbone sequence commonly shared across all probes to compare levels between genomic capture and probe backbone sequences amplified during post-capture PCR.

Using probes created by the 50 μl assembly reaction condition, 734097 (10.51%) of LASSO capture reads were aligned to the genome, with 5085185 (72.8%) of reads aligning to the LASSO probe linker sequence (Fig. 3a and Additional file 3: Table S2). Somewhat inversely, when probes were created

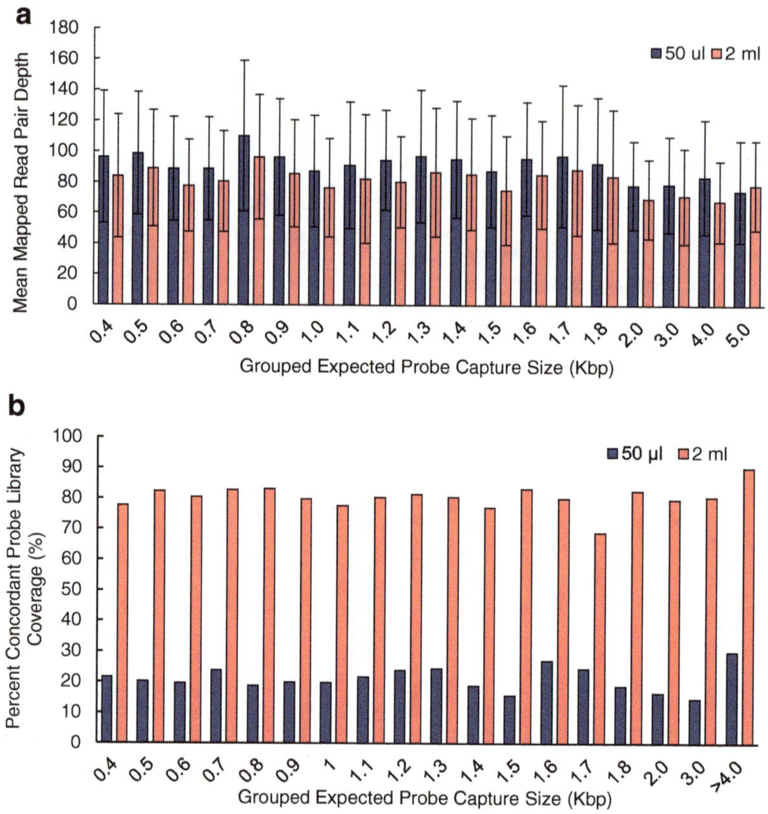

Fig. 2 Probe assembly NGS data analysis. **a** Mean read depth of all sequencing reads mapped to the LASSO probe library. The reference probe library sequences ($N = 3164$) were grouped according to ranges of expected capture size in increasing order to highlight biases in probe formation and predict downstream capture performance. Read depth is defined as the number of reads that map to a specific reference sequence. **b** On the horizontal axis, probe library sequences were grouped according to expected probe capture size ranges. The percentages of ORFs represented by concordant probes within these expected capture size ranges were plotted for both LASSO assembly methods. Concordant probes are properly formed probes with paired-end reads that map to a unique probe reference sequence

using the 2 ml reaction condition, 4556686 (65.4%) of reads aligned to the *E. coli* genome, while 1743643 (25.02%) of reads aligned to the LASSO linker sequence (Fig. 3a and Additional file 3: Table S2). We next asked if the genome aligned reads were specific capture targets using an analysis analogous with that of a previous report [5]. Targets are specific ORFs targeted by LASSO probes, while non-targets consist of ORFs unrepresented within the probe library and intronic regions. For all reads aligned to *E. coli* gDNA, median normalized read depths for targets/non-targets were 148.9/38.0 for captures performed with probes assembled using the 50 µl reaction condition, and 124.6/10.1 for those using the 2 ml reaction condition, indicating a 4-fold difference in target enrichment in favor of the 2 ml probe assembly reaction condition (Fig. 3b). Statistically significant enrichment of LASSO targets over non-targets was only observed for 2 ml (Student's *t*-test *p*-value of 2.753×10^{-76} and 0.082 for 50 µl and 2 ml reaction conditions respectively) (Fig. 3b and Additional file 3: Table S2). Using an analysis previously developed [5], we observed similar results on a per base enrichment plot of capture products longer than 1 kb in both captures done with 50 µl and 2 ml ligated probes (Fig. 3c).

While these results demonstrate differences in capture depth, we next asked if capture coverage was affected. Comparing high-quality captures of 50 µl to 2 ml ligated probes at a strict cutoff of ten-fold read coverage, Table 1 shows a two-fold higher LASSO target capture enrichment (590 versus 1282) when using a 2 ml probe assembly reaction condition, while a similar analysis revealed slightly less capture enrichment of non-targets (454 versus 388 non-targets. Plotting Table 1 according to groups of increasing capture target size, we observed that the frequency of high-quality captures for 50 µl to 2 ml ligated probes decreased from 0.4 kb to 1.4 kb and 2.0 kb, respectively (Additional file 1: Figure S1). These results suggest that the probe assembly conditions significantly impact the quality of LASSO capture, and that reducing the concentration of DNA in the probe self-circularization reaction significantly improves probe library quality, which in turn improves downstream capture depth and enrichment.

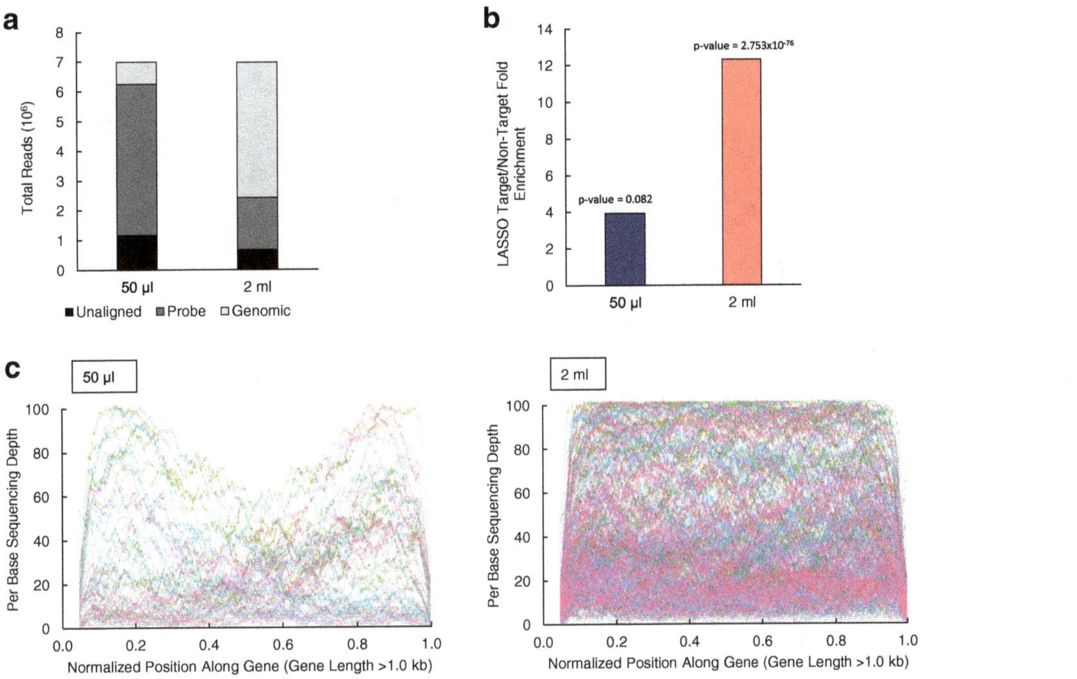

Fig. 3 Multiplexed LASSO capture NGS data analysis **a** Distribution of reads aligned to *E. coli* genome and LASSO probe backbone reference sequences. Unaligned reads were also included to reflect total raw sequencing data. **b** Target enrichment values were derived from the ratio of median normalized read depths of targets over median normalized read depths of non-targets. Statistical significance of target enrichment over non-targets was determined using Student's *t-test* with a 0.99 confidence interval (p-value < 0.01. **c** The lengths of LASSO capture targets longer than 1.0 kb with full sequencing coverage were normalized to 1 (Base-pair position normalization). The sequenced depth for each base in captures was then plotted for each normalized position along gene to visualize for target enrichment levels and capture efficiency

Simplified single target LASSO capture assessment by gel electrophoresis

Foreknowing the future importance of LASSO capture optimization for new libraries, we further developed a simplified single target LASSO protocol which can be used for small scale cloning or optimization experiments. While LASSO capture quality and efficiency is best evaluated using NGS approaches, initial efforts and optimization are more likely to be performed using inexpensive standard molecular biology techniques, such as gel electrophoresis. To better understand the limitations of this approach, we employed high-resolution polyacrylamide gel electrophoresis (PAGE) and highly sensitive dyes to quantify LASSO capture of individual targets under simplified conditions.

Table 1 High-quality LASSO captures for both target and non-target sequences

	Captured Sequences (> 10-fold Coverage)		Expected Sequences
	50 µl	2 ml	
Targets	590	1282	3164
Non-Targets	454	388	4434

Fold-coverage refers to the number of times that a certain sequence is fully sequenced. High-quality captures were strictly defined as sequences with more than ten-fold sequencing coverage, calculated as a function of sequencing read length (50 bp for HiSeq sequencing platform) multiplied by aligned read depth divided by sequence length

LASSO capture parameters and genomic target similar to that used previously were implemented with some modifications [5]. LASSO probes targeting a single 137 base pair target on the M13 bacteriophage genome were designed and synthesized as mature LASSO probes to eliminate LASSO assembly related errors and non-specific capture errors possibly associated with highly multiplexed reactions. We also performed successful LASSO captures of up to 4 kb using this approach, however, these results will be reported in a separate as yet unpublished study [6]. Post-capture PCR reactions were then resolved on both 4% agarose and 4–12% polyacrylamide gels.

A major band of 267 base pairs in size, consisting of 137 bp of capture target and two ~ 50 bp LASSO probe linker sequences, was resolved on agarose gels and PAGE (Fig. 4a). However, PAGE but not agarose gels resolved a series of minor bands ranging from < 0.1 to 0.2 kb, which are likely unreacted LASSO probes. A minor band laddering was also resolved by PAGE but not agarose throughout all capture lanes upwards of 0.3 to 1 kb. These bands were absent in all control lanes, suggesting non-specific LASSO capture or multimer formation during PCR amplification. Pixel density plots from imaged PAGE gels, assigning bands above 0.3 kb as noise, estimate the capture-to-noise ratio to be ~ 10 using this technique under simplified capture conditions

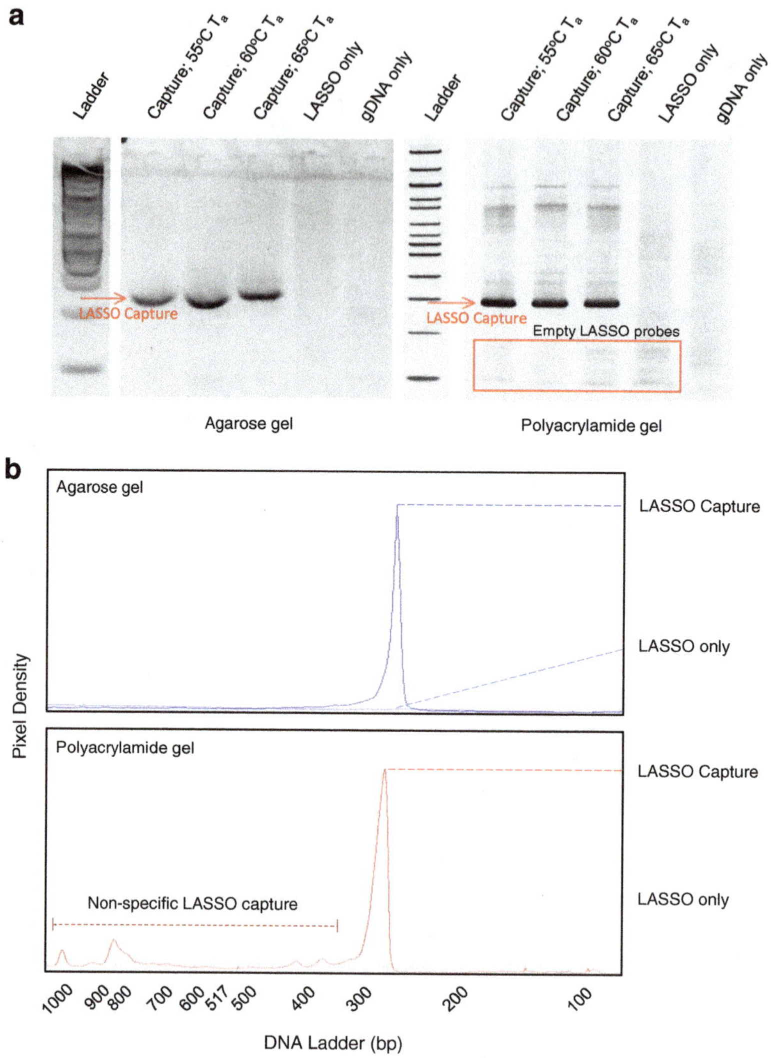

Fig. 4 Single-Target LASSO capture gel resolvability. **a** Single target captures resolved on agarose (left) and polyacrylamide gels (right). Both gels are scaled to PAGE DNA ladder, with the major band slightly underneath the 300 bp ladder band. For control lanes, GDNA only control only contains M13 genomic DNA while LASSO only controls only contain M13_137 probes. Annealing temperatures for captures are denoted by "T_a", while gDNA and LASSO probe only controls were done at 60 °C **b** Pixel density plots of single target LASSO captures on agarose and PAGE were obtained by ImageJ plot lane functions. In both plots, LASSO captures (60 °C T_a) are overlaid with a LASSO probe only control to compare non-specific capture with baseline pixel density

(Fig. 4b and Additional file 4: Table S3). These results suggest that PAGE coupled to high sensitivity dyes should be employed instead of agarose gels during LASSO optimization, and that the highest observable enrichment using this simplified system is 10-fold over background.

Discussion

We envision an array of future applications for LASSO, such as exploring long non-coding RNAs, producing chromosome-scale functional gene fragments, and preparing long-read sequencing libraries [7]. In order to realize the value of LASSO for these diverse applications, optimizing this powerful cloning tool for robustness is critical. This study introduced quantitative and highly sensitive quality control steps to the established LASSO probe assembly and target capture protocol. Here, we used NGS approaches to analyze LASSO probe assembly and correlated our results with a previously established capture and cloning NGS analytical approach.

A critical phase in the LASSO probe assembly protocol is the self-circularization step (Fig. 1a), wherein the commonly shared Eco-RI digested probe ends are intramolecularly ligated to each other. When attempting to generate thousands of probes in a single reaction by this manner, there exists a strong possibility that intermolecular ligations would manifest as mismatched probe arms on a mature LASSO probe (Fig. 5). Indeed, our analysis revealed that most LASSO probes assembled this way were mismatched (Additional file 2: Table S1).

From the outset, we reasoned that modulating the probe assembly reaction conditions at the critical step of self-circularization would impact LASSO probe assembly and possibly, in turn, capture. The two conditions we compared were not significantly different in the efficiency of probe assembly, given that more than 90 % of LASSO probes contained mismatched arms in either case. However, using a probe assembly protocol that decreases the DNA concentration during ligation by ~40 fold dramatically improved LASSO probe library coverage to ~80% (Fig. 2 and Additional file 2: Table S1) [5]. Importantly, only LASSO probes generated with our 2 ml self-circularization assembly protocol with ~80% probe library coverage produced significant successful captures (Fig. 3), suggesting that library coverage is critical for successful LASSO capture rather than concordant probe depth-enrichment. These experiments confirm that the self-circularization step is a critical area for further improvements to the LASSO protocol. In addition, we introduce an abbreviated LASSO protocol that bypasses de novo probe assembly altogether with limited scalability (Fig. 4), but whereby nearly all probes are properly assembled; providing a feasible standard upon which to base future probe assembly optimization efforts.

Conclusions

Our results demonstrate that self-circularization DNA concentration significantly impacts probe formation. NGS analysis suggests that the uniformity of concordant LASSO probe coverage is critical for efficient target capture. In addition, we introduce a single target LASSO capture protocol which does not require de novo LASSO probe assembly, making the technique more accessible to inexperienced users and more efficient for small scale approaches, such as screening novel LASSO probes or templates.

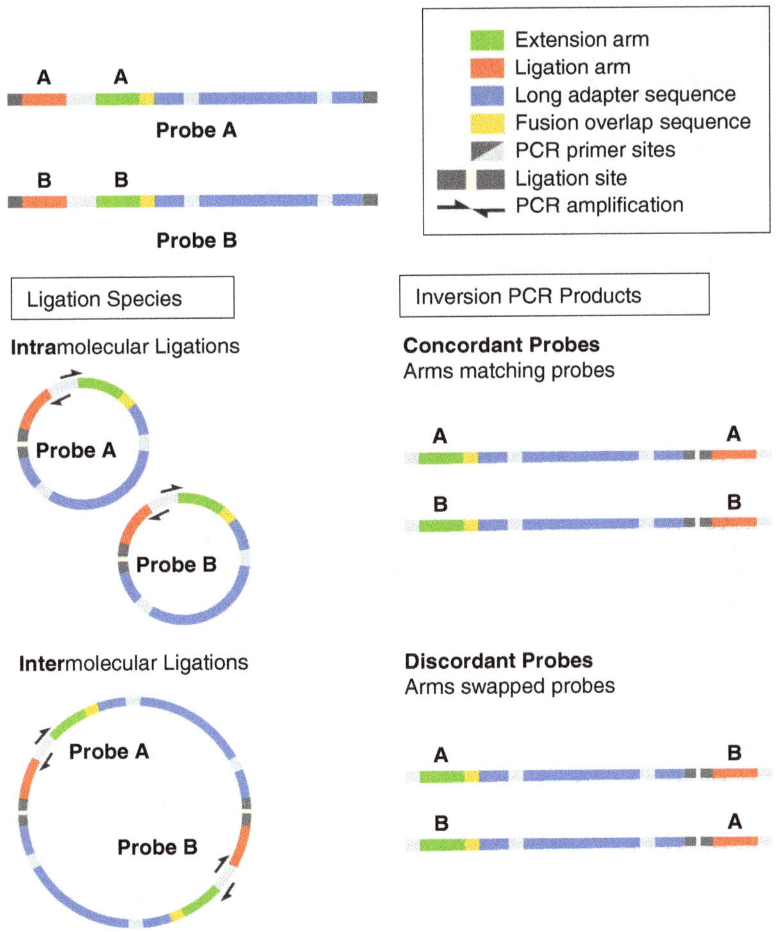

Fig. 5 Possible ligation events within probe assembly. Both Probe A and Probe B are formed after the Fusion PCR step prior to restriction enzyme digest and self-circularization. Examples of intermolecular ligation errors formed during self-circularization step (ligation reaction) of the LASSO probe assembly. Resultant discordant probes from intermolecular ligations shown here have mismatched arms from separate probe sequences

Methods

LASSO probe component design

Pre-LASSO probes

As previously reported [5], our Pre-LASSO design algorithm parsed 4140 cDNA sequences of *E. coli* K-12 MG1655 (Ensemble) to produce a pool of pre-LASSO probes targeting 3164 ORFs. Resulting Pre-LASSO probe sequences were synthesized as pooled single-stranded DNA oligonucleotides derived from programmable DNA microarray (Custom Array Inc.) approximately 160 bp long and had the design as 5′-GAGTATTACCGCGGCGAATTC-ligation arm (variable)-AACACTTCTTGCGGCGATGGTTCCTGGCTCT TCGATC-extension arm (variable)-AGAGAAGTCC-TAGCACGGTAACC-3′. The pre-LASSO algorithm only targeted *E. coli* ORFs larger than 400 bp and eliminated extension and ligation arm sequences containing EcoRI digestion sites.

Long adapter sequence

Table 2 shows the 242 bp Long Adapter sequence used to assemble LASSO probes for *E. coli* ORFeome captures. While the M13_137 sequence was fully synthesized LASSO probes for single target M13 Bacteriophage genomic capture. The long adaptor has a 5′-AGAGAAGTCCTAG-CACGGTAACC sequence similar to the 3′ end of pre-LASSO probes functioning as the overlap sequence during Fusion PCR (Fig. 1a).

LASSO probe assembly

The LASSO probe assembly protocol is the same as that previously described [5], with exceptions during the ligation protocol (self-circularization). The primers used during Fusion PCR were FusionBlaF and RFP200EcoR1 (Table 3). Approximately 45 μl of solution containing gel-purified fusion PCR product as described were digested by adding 5 μl of CutSmart 10X buffer (NEB) and 1 μl (20 units/μl) of EcoRI-HF (NEB) for 1 h at 37 °C followed by a denaturation step for 10 min at 80 °C. EcoRI digested DNA was purified with Agencourt AMPure XP beads (Beckman Coulter) and eluted in 40 μl ddH_2O.

Table 2 LASSO long adapter sequence

Sequence Name	Sequence (5′-3′)
Long Adapter for Multiplexed LASSO Probes	AGAGAAGTCCTAGCACGGTAACCTCCGAGGAT GTCATCAAAGAGTTTAAAGAGTTTATGAGATTT AAGGTCAAGATGGAGGGAAGCGTCAACGGAC ACGAGTTCGAGATTGAGGGAGAAGGAGAAGG CCGGCCTTACGAGGGCACACAAACCGCTAAG CTCAAGGTCACAAAAGGAGGACTAACTATAA CGGTCCTAAGGTAGCGAACCCTCCCCTTCTC CTGGGATATTCTGAGCCCTCAGTTCCAGTAC GGAAGCGAATTCCAGCTT

Table 3 Primer sequences for LASSO assembly and capture

Primer	Sequence (5′-3′)
BlaF	GAGTATTACCGCGGCGAATTC
RFPR200EcoRI	AAGCTGGAATTCGCTTCCGTACTG GAACTGAGGGC
SapINew	GGTTCCTGGCTCTTCGATC
TiolNew	A*T*C*GCCGCAAGAAGTGTU
PCR1kbCaptF400	GTGAAACTCAGAGGAAC CAACTTCC
ICeuI200CaptF	CTCCCCTTCTCCTGGGATATTCTG
M13_137_R(Tiol)	TTCAAAGCGAACCAGACCGG
M13_137_F(SapINew)	/5Phos/GCAATCCGCTT TGCTTCTGAC

"*" symbol denotes phosphorothioate bonds while /5Phos/ indicates a phosphorylated 5′ end

50 μl ligation protocol

For the 50 μl self-circularization method, the reaction was performed in a total volume of 50 μl 1X T4 Ligase Buffer (NEB) containing approximately 5 ng EcoRI-digested fusion PCR product (0.1 ng/μl) and 1 μl of T4 DNA ligase (400 units) which was lastly added into the reaction. The reaction was performed in a thermocycler (Eppendorf Mastercycler) for 1 h at 25 °C followed by a 10-min denaturation step at 65 °C.

2 ml ligation protocol

As for the 2 ml ligation protocol, self-circularization reaction was performed in a 15 ml conical tube (Corning). A reaction volume of 2 ml 1X T4 Ligase Buffer (NEB) containing approximately 5 ng of EcoRI digested fusion PCR product (0.1 ng/μl). Ten microliters of T4 DNA ligase (4000 units) was the final component added into the reaction. Ligation was performed in a 16 °C water bath for 16 h. The reaction was then vacuum spun for 3 h in a Savant SpeedVac concentrator (Thermo Scientific). Concentrated ligation was adjusted to 100 μl with ddH_2O, purified by Agencourt AMPure XP beads (Beckman Coulter), and finally eluted in 50 μl ddH_2O.

For ligations from 50 μl and 2 ml methods, 1 μl of λ-Exonuclease (5 U/μl) (NEB) and 1 μl of Exonuclease I (20 U/μl) (NEB) was added directly into the PCR tubes containing self-circularized DNA to digest any linear DNA. Digestion parameter was 30 min at 37 °C followed by 20 min at 80 °C.

Inversion PCR was performed in a 25 μl reaction. Ten microliters of circularized DNA from 50 μl or 2 ml ligation methods was aliquoted with 2.5 μl of 10X Klentaq Mutant Buffer (DNA Polymerase Technology), 0.5 μl of deoxynucleotides (dNTPs) (NEB), 10 μl ddH_2O, 1 μl of primers TiolNew and SapINew, and finally 0.1 μl of Omni-Klentaq LA (DNA Polymerase Technology). Both SapI and Tiol-New primers anneal with opposite orientations on the conserved sequence connecting the ligation and extension arms of pre-LASSO sequences (5′-AACACTTCTTGC GGCGATGGTTCCTGGCTCTTCGATC-3′). The SapI-New primer contains a SapI/BspQI restriction site. TiolNew primer has 3 phosphorothioate bonds on its 5′-end, indicated by "*" in between primer bases. On the 3′ end of TiolNew, a "U" indicates a deoxy-uracil moiety included for USER enzyme (NEB) cleavage (Table 3). The PCR thermal profile was as follows: - 5 min at 95 °C; 30 cycles of 15 s at 95 °C, 20 s at 55 °C, and 40 s at 72 °C; ending with an elongation step for 5 min at 72 °C.

The inversion PCR products were subsequently purified by Agencourt AMPure XP beads (Beckman Coulter) and finally eluted in 40 µl ddH_2O. The concentration of purified inversion PCR products was determined by Nanodrop. Twenty microliters of purified inversion PCR products were then brought up to 40 µl with 4 µl 10X CutSmart Buffer (NEB) and 16 µl of ddH_2O. The following enzymes were sequentially added (thermal profiles included): i) 1 µl of BspQI restriction enzyme (NEB) (1 h at 50 °C; 20 min at 80 °C), ii) 1 µl of Lambda Exonuclease (NEB) (30 min at 37 °C; 20 min at 80 °C), and iii) 2 µl USER (Uracil-Specific Excision Reagent) enzyme (NEB) (30 min at 37 °C).

The sequential enzyme digests above i) cleave SapI-New primer site, ii) digest DNA strands without 5′ phosphorothioate bonds, and iii) cleave TiolNew primer site to yield capture-ready mature LASSO probes.

Paired-end NGS for inversion PCR products

Non-sheared Inversion PCR products were sent for paired-end NGS as per manufacturer's protocol (2 × 75 bp MiSeq Reagent Nano Kit V3) (Illumina). Raw paired-end sequencing outputs (R1 and R2 files) from Illumina MiSeq were processed and analyzed using a computational pipeline to quantitate distribution of properly formed (concordant) probes over improperly formed (discordant) probes depending on how read pairs align to the probe library reference sequences (N = 3164). BBTools Repair parsed reads from R1 and R2 files and fixed read pairs into the correct order based on read identifiers assigned by the Illumina sequencer. Reads in one file without a pair in another file (singletons) were discarded [8]. Trimmomatic default parameters were then used to trim Illumina adapter sequences from paired read ends. TruSeq3 adapter sequences were used as reference for adapter trimming [9]. Read pairs were mapped to a probe library consisting of 3164 unique LASSO probe sequences that are 377 bp long by Bowtie2 ('--very-sensitive' preset) [10]. Post alignment, SAMtools selected only for probes with MAPQ alignment scores equals to or more than 30. These alignments are likely to be correctly mapped with a probability of 0.999. Reads were then ordered and indexed before sequencing duplicate removal by Picard (REMOVE_SEQUENCING_DUPLICATES = true). Our LASSO capture computational workflow includes a post-alignment duplicate removal step to account for the drawbacks of NGS sequencing platforms [11, 12]. Further SAMtools filtering steps separated mapped read pairs as concordant or discordant probes [13]. All mapped reads were grouped according to expected probe capture sizes in increasing order. Standard deviation read depths for probes in each group were then obtained (Excel).

Multiplexed kilobase-long targets LASSO captures

Overall, multiplexed LASSO ORF-eome captures were performed as previously described by Tosi et al. [5]. Both gap-filling and linear DNA digestion solution were prepared fresh for every capture experiment. Components of the gap-filling solution were 5 µl 10X Ampligase DNA buffer (Epicentre), 2 µl dNTP solution (1 mM), 1 µl Ampligase DNA Ligase (Epicentre), 0.2 µl Omni Klentaq LA, and 42 µl ddH_2O. Linear DNA digest solution components were 8 µl Exonuclease I enzyme (NEB), 8 µl Exonuclease III enzyme (NEB), 8 µl EcoRI enzyme and 8 µl ddH_2O.

One microliter of capture-ready LASSO probes (~ 10 ng), 2 µl of *E. coli* K-12 (Migula) gDNA (500 ng), 1.5 µl 10X Ampligase Buffer, and 10.5 µl ddH_2O for a total reaction volume of 15 µl. Probe-genomic target hybridization was done for 5 min at 95 °C followed by 60 min at 60 °C. Five microliters of gap-filling mix was subsequently added into the reaction and incubated for 30 min at 60 °C. After an incubation step for 3 min at 95 °C to dissociate probes from targets, 2 µl of linear DNA digestion solution was added with an incubation time of 1 h at 37 °C followed by a final denaturation step for 20 min at 80 °C.

Post-Capture PCR reaction consisted of 5 µl of LASSO captures, 2.5 µl 10X Klentaq Mutant Buffer (DNA Polymerase Technology), 0.5 µl dNTP solution (10 mM), 1 µl of each primers IceuI200CaptF200 and PCR1kbCaptF400, 15 µl of ddH_2O, and 0.1 µl Omni-Klentaq LA DNA polymerase (Table 3). PCR thermal profile were: 5 min at 95 °C; 30 cycles of 15 s at 95 °C, 20 s at 55 °C, and 2 min at 72 °C; and a final extension step of 5 min at 72 °C.

Single-end NGS for multiplexed LASSO captures

LASSO capture sequencing library preparation and computational analysis were performed as previously described with exceptions described below [5]. Trimmomatic was used for Illumina adapter sequence clipping and poor quality read filtering. Two separate Bowtie 2 read alignments were done against *E. coli* K-12 genome and LASSO backbone sequence (extension and ligation arm sequences excluded) [9]. SAMtools filtered only for high quality reads with MAPQ scores of at least 30. Another read filtering step with Picard and SAMtools removed duplicate reads [13].

Filtered alignment files were then analyzed with Bedtools *genomecov* and *coverage* function [14]. Reference *.bed* files were divided as target ORFs, untargeted ORFs, and intergenic sequences to account for the full *E. coli* genome sequence. Bedtools output files contained information on read depth per base pair, sequence fraction coverage, normalized read counts, and sequencing statistical summary, which were analyzed with Excel (Microsoft). Our analysis compared target ORFs and non-targets - untargeted ORFs and intergenic sequences combined. Statistical significance between target and non-target enrichment was performed as previously described [5].

Single target M13 Bacteriophage LASSO capture & semi-quantitative analysis

Probe design

Unlike LASSO probes used in the *E. coli* ORF-eome captures, single-target LASSO probes were synthesized de novo as GeneBlock Fragments (IDTDNA). These probes were designed with a ~ 400 bp Long Adapter (Table 4) flanked with ligation/extension arms targeting a 137 base pair sequence within the M13 Bacteriophage genome, a low complexity single-stranded circular genome. The ligation arm of this probe is the same as previously used while the extension arm was determined by our LASSO design algorithm [5]. The total length for this fully synthesized single target probe is 540 bp without TiolNew and SaplNew primer sites.

Mature LASSO probe generation

For preparative purposes, we used primers targeting the ligation and extension arms in the PCR reaction to amplify the full length of LASSO probes. PCR reaction consisted 1 μl of M13_137 probe (10 ng), 5 μl Multiplex PCR 5X Master Mix (NEB), 17 μl ddH_2O, and 1 μl each of primers M13_137_R and M13_137_F (Table 3). PCR thermal profile was 2 min at 95 °C; 30 cycles of 20 s at 95 °C, 35 s at 55 °C, and 1 min at 68 °C; with 5 min at 68 °C. PCR reaction was run in 1.0% agarose gel. Amplified band (540 bp) was excised and purified with QIAEX II Gel Extraction Kit (Qiagen); 10 μl of QXI silica-gel particles, and eluted in 25 μl ddH_2O. The gel extraction yields were quantitated with Nanodrop.

Mature probes were produced via digestion of gel extracted PCR products into single-stranded DNA by Lambda Exonuclease. Enzyme activity prefers 5′ phosphorylated end on the DNA strand amplified by the M13_137_F primer. The reactions were done with 13.5 μl of gel extracted LASSO probes, 2.5 μl 10X CutSmart Buffer, 9 μl of ddH_2O and 1 μl Lambda Exonuclease at 37 °C for 1 h.

Table 4 Single-target M13 LASSO probe design

Sequence Name	Sequence (5′-3′)
M13_137	GCAATCCGCTTTGCTTCTGACTATAATAGTCAG GGTAAAGACCTAGAGAAGTCCTAGCACGGTAA CCTCCGAGGATGTCATCAAAGAGTTTAAAGAG TTTATGAGATTTAAGGTCAAGATGGAGGGAAG CGTCAACGGACACGAGTTCGAGATTGAGGGA GAAGGAGAAGGCCGGCCTTACGAGGGCACA CAAACCGCTAAGCTCAAGGTCACAAAAGGAG GACTAACTATAACGGTCCTAAGGTAGCGAACC CTCCCCTTCTCCTGGGATATTCTGAGCCCTCAG TTCCAGTACGGAAGCAAAGCCTATGTTAAACA CCCTGCCGACATCCCTGACTATCTGAAGCTCT CCTTCCCTGAAGGCTTCAAGTGGGAGAGATTC ATGAACTTCGAGGACGGAGGCGTGGTGACAG TCACACAAGATAGCACCCTCCAGGACGGAGA GTTTATTTATAAGGTGAAACTCAGAGGAACCA ACTTCCCCTCCGATGGCCCTGTCATGAATTCTT GGAGTTTGCTTCCGGTCTGGTTCGCTTTGAA

Single target LASSO captures

Probe-target hybridization reaction was set-up in triplicates of 15 μl reactions in PCR tubes. Each reaction consists 1 μl of M13_137 probes (~ 5.0 fmol), 1 μl of M13mp18 Bacteriophage ssDNA (~ 0.5 fmol) (NEB), 11.5 μl of ddH_2O, and 1.5 μl of 10X Ampligase Buffer. In contrast to what was previously done, M13_137 captures were done with a ten-fold molar excess of probes to genomic DNA [5]. Hybridization step was initiated with a 5-min denaturation step at 95 °C followed by a hybridization step with hybridization temperatures (T_a) of 55 °C, 60 °C, and 65 °C for each reaction using our thermocycler's gradient temperature setting.

Five microliters of gap-filling solution was subsequently added – 3.85 μl ddH_2O, 0.5 μl Ampligase Buffer, 0.2 μl dNTP solution, 0.25 μl of Ampligase DNA ligase (Epicentre), and 0.2 μl Omni-Klentaq LA DNA Polymerase. Reactions were then incubated according to their respective T_a at 55 °C, 60 °C, and 65 °C for 30 min ended by a 3-min denaturation step at 95 °C.

Post-capture PCR

Five microliters of single-target LASSO captures was mixed with 5 μl Multiplex PCR 5X Master Mix, 13 μl ddH_2O, and 1 μl each of LASSO post-capture primers PCR1kbCaptF400 and ICeuI200CaptF (Table 3). PCR thermal profiles were as follows: 3 min at 95 °C; 30 cycles of 20 s at 95 °C, 35 s at 55 °C, and 45 s at 68 °C; with 3 min at 68 °C.

LASSO post-capture PCR visualization

A 4.0% agarose gel was cast and pre-stained with ethidium bromide. The whole post-capture PCR reaction was mixed with 5 μl of 6X Purple Gel Loading Dye (NEB). Novex 4–12% 12-well TBE Polyacrylamide Gels (Thermo Fisher) were used to resolve 2.5 μl of PCR amplified LASSO captures in 1X Purple Gel Loading Dye. For capture-to-noise ratio quantitation, four two-fold serial dilutions were made starting with a 1X PCR solution containing 8 μl of PCR product (60° C T_a), 12 μl of ddH_2O, and 4 μl Purple Loading Dye. Four two-fold serial dilutions of 1 kb Plus DNA Ladder (Thermo Fisher) were included for relative capture DNA mass estimation. PAGE for DNA quantitation was performed in duplicates. All PAGE gels were stained with SYBR Gold Nucleic Acid Gel Stain (Thermo Fisher) for 10 min before visualization via a gel imager (FastGene).

LASSO capture signal-to-noise ratio quantitation

Since captures with varying annealing temperature yielded no significant difference on PAGE, we randomly chose the capture done with 60 °C for signal-to-noise ratio quantitation. Stained PAGE gels were imaged with a gel imager and saved in high resolution *.tiff* format with similar exposure levels. Gel color profiles were inverted – black bands on white background - and analyzed with ImageJ (Fiji). Uniform vertical rectangular selections were done on every gel lane to plot for pixel density of resolved bands. Using ImageJ Plot Lanes function, standard curves correlating pixel density to expected DNA mass were plotted using DNA ladder bands; derived trendlines were used to estimate DNA mass. Signal-to-noise ratio was obtained only from samples in which its estimated DNA mass from band pixel density was within the linear portion of the trendline. Visible minor bands corresponding to non-specific LASSO capture (excluding bands from empty probes) were gated and classified as noise. Pixel density ratios between major bands versus total noise, or minor bands, were derived [13].

Additional files

Additional file 1: Figure S1. LASSO Probe Library Target Expected & Observed Frequency. Histogram groups the frequency of expected and observed high-quality LASSO ORF-eome target captures (More than ten-fold depth coverage) grouped in increasing capture sizes from 0.4 to > 4.0 kb. (PPTX 43 kb)

Additional file 2: Table S1. NGS Read Accounting of Sequenced Assembled LASSO Probes The table shows raw sequencing output (MiSeq 2 × 75 bp platform) of inversion PCR products before and after read pairing, sequencing duplicate removal, and alignment using our NGS pipeline. (XLSX 8 kb)

Additional file 3: Table S2. NGS Read Accounting of Sequenced LASSO *E. coli* ORF-eome Captures. The table shows raw sequencing output (HiSeq 1 × 50 bp platform) of PCR amplified multiplexed ORF-eome capture. Reads were first aligned to both *E. coli* genome and LASSO probe backbone sequence, then read mapped exclusively to genomic DNA was cleaned, filtered, and separated according to targets or non-targets (untargeted ORFs and intergenic regions). Also included are the median read depths of gDNA sequencing enrichment. (XLSX 9 kb)

Additional file 4: Table S3. Single Target Capture ImageJ Pixel Density Plot. Single target LASSO captures were visualized on PAGE gels and imaged. Major and minor bands were highlighted using ImageJ to obtain pixel density plots. (XLSX 8 kb)

Abbreviations

LASSO: Long adapter single stranded oligonucleotide; NGS: Next generation sequencing; ORF: Open reading frame; PAGE: Polyacrylamide gel electrophoresis; PCR: Polymerase chain reaction

Acknowledgments

Y.Y contributed in 50 µl and 2 ml de novo LASSO synthesis and capture of *E. coli* ORFeome. V.S provided the earlier versions of pre-LASSO design algorithm and NGS analysis pipeline.

Authors' contributions

BP, AT, LT, HBL and SS conceived this study. LT performed *E. coli* ORF-eome LASSO probe assembly, capture, and sequencing library preparation. SS and AT designed and performed the data analysis for NGS sequencing results and designed and performed the abbreviated single target LASSO capture experiments. SS and AT wrote the manuscript, LT and, HBL and BP edited the manuscript. All authors approved the final version of the manuscript.

Authors' information

Not applicable.

Funding

This research was conducted with support under Grant Nos. R01GM127353 (B.P., B.L.), R01EB012521 (B.P.) awarded by the National Institutes of Health. The funding body had no role in the design of the study nor the collection, analysis, and interpretation of data, nor in the writing of the manuscript.

Ethics approval and consent to participate

Not applicable

Consent for publication

Not applicable

Competing interests

A patent on the technology has been filed (PCT/US2016/035919). The authors declare no other competing financial interests.

Author details

[1] Center for Surgery, Innovation, and Bioengineering, Department of Surgery, Massachusetts General Hospital, Harvard Medical School and the Shriners Hospitals for Children, 51 Blossom Street, Boston, MA 02114, USA. [2]Division of Immunology, Department of Pathology, Johns Hopkins University, Baltimore, MD, USA. [3]Harvard Stem Cell Institute, Cambridge, MA 02138, USA. [4]Department of Biomedical Engineering, Rutgers University, Piscataway, NJ 08854, USA.

References

1. Mamanova L, Coffey AJ, Scott CE, Kozarewa I, Turner EH, Kumar A, et al. Target-enrichment strategies for next-generation sequencing. Nat Methods. 2010;7(2):111–8 Available from: http://www.ncbi.nlm.nih.gov/pubmed/20111037. Cited 23 Mar 2019.
2. Hardenbol P, Yu F, Belmont J, Mackenzie J, Bruckner C, Brundage T, et al. Highly multiplexed molecular inversion probe genotyping: over 10,000 targeted SNPs genotyped in a single tube assay. Genome Res. 2005;15(2):269–75 Available from: http://www.ncbi.nlm.nih.gov/pubmed/15687290. Cited 30 Aug 2018.
3. Niedzicka M, Fijarczyk A, Dudek K, Stuglik M, Babik W. Molecular Inversion Probes for targeted resequencing in non-model organisms. Sci Rep. 2016;6(1):24051 Available from: http://www.ncbi.nlm.nih.gov/pubmed/27046329. Cited 21 Mar 2019.
4. Akhras MS, Unemo M, Thiyagarajan S, Nyrén P, Davis RW, Fire AZ, et al. Connector inversion probe technology: a powerful one-primer multiplex DNA amplification system for numerous scientific applications. PLoS One. 2007;2(9):e915.
5. Tosi L, Sridhara V, Yang Y, Guan D, Shpilker P, Segata N, et al. Long-adapter single-strand oligonucleotide probes for the massively multiplexed cloning

of kilobase genome regions. 2017;1:92 Available from: https://doi.org/10.1038/s41551-017-0092.
6. Jingqian Liu, Syukri Shukor, Shuxiang Li, Alfred Tamayo, Lorenzo Tosi, Benjamin Larman, Vikas Nanda, Wilma K. Olson, Parekkadan B. Computational simulation of adaptor length-dependent LASSO probe capture efficiency. 2019. Unpublished.
7. Albert TJ, Molla MN, Muzny DM, Nazareth L, Wheeler D, Song X, et al. Direct selection of human genomic loci by microarray hybridization. Nat Methods. 2007;4(11):903–5.
8. BBTools - DOE Joint Genome Institute. Available from: https://jgi.doe.gov/data-and-tools/bbtools/. Cited 5 Sept 2018.
9. Bolger AM, Lohse M, Usadel B. Trimmomatic: a flexible trimmer for Illumina sequence data. Bioinformatics. 2014;30(15):2114–20 Available from: http://www.ncbi.nlm.nih.gov/pubmed/24695404. Cited 5 Sept 2018.
10. Langmead B, Salzberg SL. Fast gapped-read alignment with Bowtie 2. Nat Methods. 2012;9(4):357–9 Available from: http://www.nature.com/articles/nmeth.1923. Cited 5 Sept 2018.
11. Nakamura K, Oshima T, Morimoto T, Ikeda S, Yoshikawa H, Shiwa Y, et al. Sequence-specific error profile of Illumina sequencers. Nucleic Acids Res. 2011;39(13):e90 Available from: https://academic.oup.com/nar/article-lookup/doi/10.1093/nar/gkr344. Cited 5 Sept 2018.
12. Ebbert MTW, Wadsworth ME, Staley LA, Hoyt KL, Pickett B, Miller J, et al. Evaluating the necessity of PCR duplicate removal from next-generation sequencing data and a comparison of approaches. BMC Bioinformatics. 2016;17(Suppl 7):239 Available from: http://www.ncbi.nlm.nih.gov/pubmed/27454357. Cited 13 Aug 2018.
13. Li H, Handsaker B, Wysoker A, Fennell T, Ruan J, Homer N, et al. The sequence alignment/map format and SAMtools. Bioinformatics. 2009;25(16):2078–9 Available from: https://academic.oup.com/bioinformatics/article-lookup/doi/10.1093/bioinformatics/btp352. Cited 31 July 2018.
14. Quinlan AR, Hall IM. BEDTools: a flexible suite of utilities for comparing genomic features. Bioinformatics. 2010;26(6):841–2 Available from: https://academic.oup.com/bioinformatics/article-lookup/doi/10.1093/bioinformatics/btq033. Cited 5 Sept 2018.

4

A novel starch-binding laccase from the wheat pathogen *Zymoseptoria tritici* highlights the functional diversity of ascomycete laccases

Majid Haddad Momeni[1], Paolo Bollella[2,3], Roberto Ortiz[4], Esben Thormann[4], Lo Gorton[2] and Maher Abou Hachem[1*]

Abstract

Background: Laccases are multicopper oxidases, which are assigned into auxiliary activity family 1 (AA1) in the CAZy database. These enzymes, catalyzing the oxidation of phenolic and nonphenolic substrates coupled to reduction of O_2 to H_2O, are increasingly attractive as eco-friendly oxidation biocatalysts. Basidiomycota laccases are well characterized due to their potential in de-lignification of lignocellulose. By contrast, insight into the biochemical diversity of Ascomycota counterparts from saprophytes and plant pathogens is scarce.

Results: Here, we report the properties of the laccase from the major wheat pathogen *Zymoseptoria tritici* (*Ztr*Lac1A), distinguished from common plant fungal pathogens by an apoplastic infection strategy. We demonstrate that *Ztr*Lac1A is appended to a functional starch-binding module and displays an activity signature disfavoring relatively apolar phenolic redox mediators as compared to the related biochemically characterized laccases. By contrast, the redox potential of *Ztr*Lac1A (370 mV vs. SHE) is similar to ascomycetes counterparts. The atypical specificity is consistent with distinctive sequence substitutions and insertions in loops flanking the T1 site and the enzyme C-terminus compared to characterized laccases.

Conclusions: *Ztr*Lac1A is the first reported modular laccase appended to a functional starch-specific carbohydrate binding module of family 20 (CBM20). The distinct specificity profile of *Ztr*Lac1A correlates to structural differences in the active site region compared to previously described ascomycetes homologues. These differences are also highlighted by the clustering of the sequence of *Ztr*Lac1A in a distinct clade populated predominantly by plant pathogens in the phylogenetic tree of AA1 laccases. The possible role of these laccases in vivo merits further investigations. These findings expand our toolbox of laccases for green oxidation and highlight the binding functionality of CBM-appended laccases as versatile immobilization tags.

Keywords: Carbohydrate binding module family 20 (CBM20), Cyclic voltammograms, Laccase, Oxidoreductase, Plant, Pathogen, Starch, *Zymoseptoria tritici*

* Correspondence: maha@bio.dtu.dk
[1]Department of Biotechnology and Biomedicine, Technical University of Denmark, Søltofts Plads, 2800 Kgs, Lyngby, Denmark
Full list of author information is available at the end of the article

Background

Laccases (EC 1.10.3.2) are multicopper oxidases (MCOs), which catalyze monoelectric oxidation of a variety of substrates using molecular oxygen that is reduced to water [1–3]. A mononuclear copper binding site designated as a type-1 Cu (blue Cu) and a trinuclear copper site, involving one type-2 and two type-3 Cu (coupled binuclear Cu) binding sites are typically present in laccases. Substrates are oxidized in vicinity of the high redox potential mononuclear site (Cu1 or T1), which is responsible for the blue color and distinctive absorbance of these enzymes at about 600 nm [1, 4, 5]. Thus, the type-1 Cu possesses a higher redox potential as compared to its counterparts in the type 2 and type 3 sites [6]. The electrons from the oxidation of substrates at the T1 site are transferred through the protein via the Cys-His pathway to the trinuclear cupper cluster at the T2/T3 sites, where the electrons are transferred to oxygen (O_2) [5].

Laccases are considered as eco-friendly catalysts due to their ability to oxidize a broad range of organic and synthetic substrates using only oxygen as the electron acceptor with no toxic byproducts. Some potential substrates are large in size precluding their accommodation into the active sites of laccases. Small redox mediators, e.g. 2,2-azino-bis(3-ethylbenzothiazoline-6-sulfonic acid) diammonium salt (ABTS), 2,6 dimethoxyphenol, DMP and syringaldazine (Fig. 1), generally act as electron shuttles that mediate the oxidation of substrates too large for the direct oxidation.

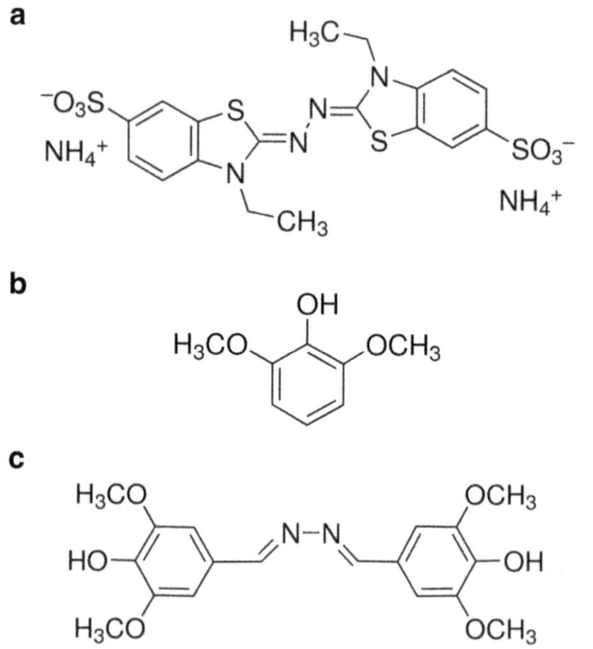

Fig. 1 Chemical structures of three synthetic redox mediators, **a** 2,2-azino-bis (3-ethylbenzothiazoline-6-sulfonic acid) diammonium salt, ABTS (**b**) 2,6 dimethoxyphenol, DMP (**c**) syringaldazine

Although, laccases are produced by diverse taxonomic groups including bacteria, insects, lichens, plants and fungi, the latter category mainly dominates in biotechnological applications. The wide utilization of fungal laccases in biotechnological applications is likely attributed to the stability, industrial efficient heterologous expression systems and high redox potential of Basidiomycota enzymes [3, 7]. Particularly, the abundance of laccases in almost all wood-rotting fungi has attracted interest in the context of de-lignification of lignocellulose, but laccases are also implicated in lignification, oxidative stress management in plants and are frequently active on an array of phenolic compounds (e.g. phenols, polyphenols, benzenothiols and anilines) [8].

Fungal laccases are produced by several groups ranging from yeasts, white and brown rot fungi (both from the Basidiomycota phylum), or Ascomycota as well as mycorrhizal species. Laccases are classified into auxiliary activity family 1 (AA1) in the Carbohydrate-Active Enzyme (CAZy) database [9]. Laccases from white rot basidiomycetes are distinguished by their high redox potentials ($E^0 \approx 0.8$ V for the Cu1 site) [10], as compared to ascomycete and/or bacterial laccases, which display redox potential $E^0 = 0.4–0.7$ V [11]. Notably, the genomes of phytopathogenic ascomycete fungi like e.g. *Botrytis cinera* [12], *Magnaporthe grisea* [13] and *Fusarium oxysporum* [14] each encode several putative laccases of AA1 subfamily 3 (AA1_3), which harbors most ascomycete laccases according to the CAZy classification, whereas basidiomycetes counterpart are assigned into AA1 subfamily 1 (AA_1). Functional insight into AA1_3 laccases from phytopathogens, which may contribute to the understanding of the properties and possible roles of these enzymes during pathogenesis, lags behind.

Zymoseptoria tritici (syn. *Mycosphaerella graminicola*, *Septoria tritici*) is responsible for the wheat disease *Septoria tritici* blotch (STB) [15, 16]. This plant disease has been recognized as the most devastating for wheat production in Europe [17, 18], which is also highlighted by the ranking of the causative fungus amongst the top ten fungal pathogens [19]. *Z. tritici* is distinguished by an infection within the apoplastic space (outside the plasma membrane but within the plant cell wall) that differentiates it from other globally recognized fungal phytopathogens e.g. *Fusarium* [20] and *Magnaporthe* spp. [21]. The genome of *M. graminicola* IPO323, reveals a significant reduction in the number of carbohydrate active enzymes (CAZymes) that target plant cell walls, especially those possessing a carbohydrate binding module (CBM) [22], consistent with this apoplastic infection strategy. Remarkably, starch-binding modules of CBM family 20 (CBM20) are encoded by four genes from *Z. tritici*: one encoding a hypothetical protein, two encoding putative α-glucosidases and a fourth encoding a laccase.

In this study, we produced and characterized the first modular laccase (hereafter referred to as *Ztr*Lac1A) comprising a catalytic module of AA1_3 and a starch-binding module of CBM20. Curiously, this enzyme has a preference to the negatively charged redox mediator 2, 2′-azino-bis(3-ethylbenzothiazoline-6-sulphonic acid) (ABTS) as compared to more apolar mediators. This activity profile is the converse of fungal laccases of AA1_3 characterized to date.

To provide a rationale for this different specificity, we carried out bioinformatic analyses, which revealed marked differences between *Ztr*Lac1A and characterized counterparts in the vicinity of the T1 site, highlighting this region as an important specificity determinant. Moreover, variations in sequence were also observed at the C-terminus, which has been shown to be an important activity signature in laccases. The modularity and atypical specificity profile of *Ztr*Lac1A offer an expansion of our enzymatic tool box of green oxidative enzymes and promote our understanding of the basic features that govern the specificity of AA1 laccases.

Methods

Materials and media

ABTS, syringaldazine and 2,6 DMP (2,6-dimethoxyphenol) and other chemicals were from Sigma-Aldrich (St. Louis, MO, USA). Restriction enzymes and molecular biology reagents were from New England Biolabs (Ipswich, MA, United States). All chemicals were of analytical grade. The *Myrothecium verrucaria* bilirubine oxidase (*Mv*BOD, 3.61 mg/mL) was from Novozymes (Bagsværd, Denmark). The osmium polymer ([Os (4,4′-dichloro-2,2′-bipyridine)2(poly-vinylimidazole)10Cl]·Cl, $E^{o\prime} = 0.350$ V vs. Ag|AgCl$_{sat}$) (Additional file 1: Figure S1) was kindly provided by Prof. Dónal Leech and Dr. Peter Ó Conghaile from Biomolecular Electronics Research Laboratory, National University of Ireland (Galway, Ireland) and prepared as previously reported [23].

The standard yeast selection medium YPDS (Yeast Extract Peptone Dextrose with Sorbitol) comprised 1% yeast extract, 2% peptone, 2% dextrose (glucose), 1 M sorbitol, 2% agar (all w/v, Invitrogen, Carlsbad, CA, USA). The standard yeast expression media BMGY/BMMY (Buffered Glycerol-complex Medium/Buffered Methanol-complex Medium, both from Invitrogen) comprised 1% yeast extract, 2% peptone, 100 mM potassium phosphate, pH 6.0, 1.34% YNB, 4×10^{-5}% biotin (all w/v), 1% glycerol or 0.5% methanol (both v/v).

Cloning, expression and purification of recombinant enzyme

The genomic DNA encoding the laccase of auxiliary family 1 [9] (http://www.cazy.org/) from *Zymoseptoria tritici* IPO323 (GenBank accession XP003852363) was synthesized by Invitrogen (Carlsbad, CA, USA). The forward GACAT**TTCGAA**ACGATGCGGTTAC and reverse primer GATTG**TCTAGA**TCAGACTCCGGAATC were used to amplify this gene fragment that encodes the mature peptide of the enzyme lacking the native signal peptide (amino acid residues 46–723, hereafter designated as *Ztr*Lac1A) using *Turbo* DNA polymerase and a PCR protocol involving preheating at 95 °C (30 s), followed by 16 cycles of denaturation at 95 °C (30 s), annealing at 55 °C (1 min) and extension at 68 °C (1 min). The PCR amplicon (2197 bp) was cloned within the BstBI and XbaI restriction sites of *p*PICZαA vector (Invitrogen, Carlsbad, CA, USA) using standard molecular biology protocols. The resulting recombinant *p*PICZαA-*Ztr*lac1A plasmid was transformed into *Escherichia coli* DH5α and clones were selected on low salt LB agar plates supplemented with 25 μg mL^{-1} Zeocin. Sequencing was performed using the 5′-AOX promoter 3′-AOX terminator universal primers and the internal primers 5′-GGGTTTGAATTATGAGGATCCG-3′ and 3′-TTGACTTGCCAGTAGAGGGTG-5′).

The recombinant plasmid was linearized using PmeI and transformed by electroporation into *Pichia pastoris* X-33 cells according to the manufacturer's protocol. Transformants were selected on YPDS plates supplemented with 100 μg mL^{-1} Zeocin according to the manufacturer's recommendation (EasySelect™ Pichia Expression Kit, Invitrogen) after incubation at 30 °C for 3–5 days. Single transformants were re-streaked on new plates and used to inoculate 5 ml buffered complex medium BMGY. The best secreting transformants were selected based on the SDS-PAGE analysis and activity measurements using ABTS as substrate.

An overnight culture (25 mL, BMGY medium, 30 °C) was grown to $OD_{600} = 5$ prior to inoculation into 0.9 L of the same medium and propagated overnight to $OD_{600} \approx 7$. Thereafter, cells were harvested by centrifugation (7000 ***g***, 30 min, 4 °C), re-suspended in BMMY to $OD_{600} \approx 1$, and re-incubated at 30 °C. The culture was supplemented with methanol (0.5% v/v) every 24 h prior to harvesting after 72 h of induction. Culture supernatants were recovered by centrifugation (7500 ***g***, 30 min, 4 °C), pooled, supplemented with $(NH_4)_2SO_4$ to 0.5 M and incubated for 18–20 h, followed by centrifugation (8000 ***g***, 20 min, 4 °C) and filtration (0.5 μm) prior loading onto a 40 ml β-cyclodextrin-Sepharose affinity column and purification as previously described [24]. The elution fractions containing *Ztr*Lac1A were analyzed using SDS-PAGE and activity measurements according to the protocol described below. Fractions displaying high purity and activity were pooled and concentrated using Amicon filters (MWCO, 30 kDa, Millipore). The concentrated protein loaded onto a HiLoad 36/60 Superdex gel filtration column (GE Healthcare,

Uppsala, Sweden) and eluted with 20 mM Na acetate, 150 mM NaCl, pH 5.5 at 0.9 ml min^{-1}. The fractions with highest purity and activity were pooled, concentrated and the protein concentration was determined by measuring the absorbance at 280 nm (A_{280}) using the theoretical extinction coefficient ε_{280} = 79074 $M^{-1}.cm^{-1}$ Expasy (http://web.expasy.org/protparam). All Purification steps were performed using an ÄKTA purifier chromatograph (GE Healthcare) at 7 °C. *N*-Glycosylation and *O*-glycosylation sites were predicted using NetNGyc 1.0 (http://www.cbs.dtu.dk/services/NetNGlyc/), and NetOGlyc (http://www.cbs.dtu.dk/services/NetNGlyc/), respectively. To examine *N*-glycosylation, a 9 µl of supernatant was initially denatured at 100 °C (10 min) and furthermore was used in EndoH treatment at 37 °C for 90 min according to the manufacturer's instructions (New England Biolabs) and finally analyzed using SDS-PAGE.

Isothermal titration calorimetry

The isothermal titration calorimetry (ITC) experiments were performed using an iTC200 instrument (GE healthcare, Northampton, MA, USA). *Ztr*Lac1A (145 µM) dialyzed against 20 mM NaOAc, pH 5.5 was titrated with 2.1 mM β-cyclodextrin (β-CD) dissolved in the same buffer at 25 °C with an initial injection of 0.4 µL followed by 15 injections of 2 µL. A control titration into the dialysis buffer was used to compensate the heat of dilution. A one-binding site model was fit to the ITC data to determine the equilibrium association constant (K_a), the molar binding enthalpy (ΔH) and the stoichiometry of binding (N_o) using the ITC analysis plug in ORIGIN software provided with the instrument.

Biochemical characterization

The kinetic parameters of *Ztr*Lac1A (K_m and k_{cat}) were determined by monitoring the increase in absorbance of oxidized ABTS (A_{405}, ε_{405} = 36.8 $mM^{-1}\,cm^{-1}$) [25] in 96-well microtiter plates using a PowerWave XS plate reader (BioTeK, Winooski, VT, USA) using eight different concentrations (0.032–4 mM) in 40 mM sodium acetate buffer pH 4.2 at 25 °C. The *Ztr*Lac1A activity as function of pH was measured for 10 min at 25 °C in the pH range 2.2–8.0, using ABTS, as substrate in McIlvaine buffer.

The laccase activity was also determined using 2,6-DMP (A_{469}, ε_{469} = 19.6 $mM^{-1}\,cm^{-1}$) and syringaldazine (A_{525}, ε_{525} = 65.0 $mM^{-1}.cm^{-1}$) in the range of 0.4–8 mM and 0.006–0.11 mM, respectively. The reactions were initiated by the addition of the enzyme to a final concentration of 0.26 µM to pre-temperated mixtures containing substrate and buffer to a final volume of 250 µL. The experiments were carried out in triplicates and the kinetic parameters were determined by fitting the Michaelis-Menten equation to the initial rate data at each substrate concentration using ORIGIN 8.0 (OriginLab, Northampton, MA). An additional activity assay was also carried out using the phenolic substrates coniferyl aldehyde, *p*-coumaric, caffeic acids, vanillin, 3,5 dimethoxybenzoic acid and 3,4 dihydroxybenzoic acid in 2 mM concentrations. All the reactions were performed in 50 mM NaAc pH 5.2 in 0.5 ml reaction volume for 30 and 90 min. The structures of these substrates and their analysis are described in Additional file 1: Figure S2a–e.

Electrochemical measurements and electrode modification

The redox potential of the T1 site for *Ztr*Lac1A was determined using modified graphite electrode (GE) in the presence and absence of a redox mediator. Electrodes were prepared as previously described for bioelectrochemical studies on multicopper oxidase (MCO) [26]. The spectrographic low-density graphite rods (Ø 3.05 mm, Sigma) were polished by fine emery paper (Turfbak Durite, P1200), thoroughly rinsed with Milli-Q water and dried at room temperature. The GE/Os/ *Ztr*Lac1A (0.16 mM) modified electrodes were prepared by transferring Os-polymer (5 µL, 5 mg mL^{-1}) and freshly prepared poly(ethylene glycol) diglycidyl ether solution (PEGDGE, 2 µL, 10 mg/mL water) to the top of the GE and subsequent incubation for 10 min, followed by adding the enzyme solution (12.6 mg/mL). Thereafter, the modified electrodes were dried at room temperature prior to incubation overnight at 4 °C to complete the cross-linking. Furthermore, in the direct electron transfer approach (in absence of a redox mediator), the same immobilization procedure was also applied but in the absence of the Os-polymer and cross-linking agent. To begin with, the enzyme-modified electrodes were press-fitted into a Teflon holder and rinsed thoroughly with acetate buffer to remove any weakly bound material. All electrode modifications were prepared in triplicate.

A three-electrode cell was used with an Ag|AgCl KCl_{sat} (E = 0.199 V vs. SHE, all redox potentials are reported vs. SHE) as reference electrode and a Pt plate as counter electrode. The laccase modified electrodes were used as working electrode. Cyclic voltammetry measurements were performed in an 0.1 M acetate buffer at pH 4.0 also containing 0.1 M $NaClO_4$ at room temperature. Prior to the measurements in the absence of O_2, N_2 gas was bubbled through the measuring solution at least for 20 min to maintain O_2-free conditions. Oxygen saturated conditions were achieved by purging pure O_2 in the buffer solution for at least 20 min. All cyclic voltammetry measurements were conducted using a potentiostat Metrohm Autolab Potentiostats/Galvanostats (Model PGSTAT128N, Metrohm Autolab B.V., Utrecht, The Netherlands) equipped with Nova 2.1 as software.

Phylogenetic analysis and homology modelling

Most of the ascomycete laccases (about 90%) assigned into AA1_3 in the CAZy database were retrieved. In addition, the catalytic module of *Ztr*Lac1A was also used as a query in a BLASTp search (https://www.ncbi.nlm.nih.gov/BLAST/) against the non-redundant protein sequence database [27] to retrieve additional basidiomycete sequences within an E-value $< 5 \times 10^{-71}$, sequence identity > 31% and a minimum sequence coverage of 88%. In total 221 sequences from ascomycete and basidiomycetes laccase catalytic modules (AA1) were aligned using MAFFT software standard settings (mafft.cbrc.jp/alignment/). The phylogenic tree, calculated using neighbor joining method with bootstrap value of 1000 in MAFFT server and it was visualized by DENDROSCOPE [28].

A homology model for *Ztr*Lac1A was generated using the laccase structure from *Melanocarpus albomyces* (PDB code 1GW0, 44% sequence identity) as a template and the Schrödinger package. The template enzyme represents the closest structurally and biochemically characterized (Table 1) homologue of *Ztr*Lac1A.

Results

Heterologous expression of the modular laccase from *Zymoseptoria tritici* (*Ztr*Lac1A)

The non-codon optimized synthetic gene encoding the modular laccase from *Zymoseptoria tritici* (*Ztr*Lac1A) (Additional file 1: Figure S3a), was cloned into *Pichia pastoris*. The recombinant protein was expressed and purified using β-cyclodextrin-Sepharose affinity chromatography, confirming the binding functionality of the CBM20. The purification was polished with an additional gel filtration step and the yield of pure protein was about 1.5 mg L^{-1} culture and a specific activity of 45.5 U mg^{-1}. *Ztr*Lac1A migrated as smeary bands with an apparent molecular mass of 93 kDa compared to the theoretically calculated value (79 kDa). The larger size is consistent with *O*- and/or *N*-glycosylation, which are typically observed in recombinant proteins produced in *P. pastoris* [30]. Among seven putative *N*-glycosylation sites in the catalytic module, only three exhibit high score (> 0.6), Asn 209, Asn252 and Asn366. The putative *O*-glycosylation sites were predicted in serine/threonine rich linker regions (Additional file 1: Figure S3a) [31]. Endo-H treatment of the recombinant enzyme resulted in a reduction of about 15 kDa in molecular mass estimated from the migration shift in SDS-PAGE analysis (Additional file 1: Figure S3b) suggesting the protein is mainly decorated with *N*-glycans.

*Ztr*Lac1A is a functional laccase

The activity of *Ztr*Lac1A was examined using the synthetic mediator ABTS at pH 4.2, whereas activity on syringaldazine and 2,6-DMP was assayed at pH 6.0. Notably, *Ztr*Lac1A is mainly active on ABTS, compared to a negligible activity towards 2,6-DMP and no activity towards syringaldazine, which is different from the closest related ascomycete laccases from *Melanocarpus albomyces* laccase (*Mal*Lac1A) (Table 1).

Isothermal titration calorimetry (ITC)

The binding of *Ztr*Lac1A to the starch mimics β-CD, a typical model substrate used to demonstrate the functionality of starch-binding proteins [32], was quantified using ITC (Fig. 2). The data revealed that *Ztr*Lac1A binds to β-CD with an association constant of $K_a = (3.07 \pm 0.24) \times 10^4\ M^{-1}$ ($K_d = 32.6 \pm 2.4\ \mu M$, $\Delta G = -6.17$ kcal mol^{-1}). The binding was largely driven by a favorable enthalpy ($\Delta H = -7.15 \pm 0.19$ kcal mol^{-1}), which is compensated by an unfavorable change in entropy ($-T\Delta S =$ 1.03 kcal mol^{-1}). The fitted stoichiometry of $N_o = 1.5 \pm 0.1$ is lower, but consistent with the presence of two β-CD binding sites in canonical CBM20 modules [32], which is likely to be due to residual partial occupancy from the β-CD elution during purification.

Electrochemical characterization

In order to study the electrocatalytic behavior of *Ztr*Lac1A, modified graphite electrode (GE) cyclic voltammograms (CVs) were collected in absence (black curve) and presence (red curve) of O_2 in 0.1 M acetate buffer at pH 4.0, 0.1 M $NaClO_4$ (Fig. 3a). It was not possible to define the onset potential for the electroreduction of O_2 from these CVs, while this was clearly displayed in the subtracted curve at 0.569 V vs. SHE (Fig. 3). A second electroreduction process was observed at approximately 0.2 V and attributed to direct non-enzymatic oxygen reduction at the GE surface. Unmodified GE studied at the same conditions showed only electro-

Table 1 Kinetic parameters for *Ztr*Lac1A were determined in triplicates, using 40 mM McIlvaine buffer at pH 4.2 compared to the *Mal*Lac1A using the same substrates

	ABTS	2,6-DMP	Syringaldazine
*Mal*Lac1A[a]			
K_m (mM)	0.4	0.011	0.037
k_{cat} (s^{-1})	28.1	10.2	40.2
k_{cat}/K_m (s^{-1} mM^{-1})	70.3	927.3	1085.6
*Ztr*Lac1A			
K_m (mM)	0.25 ± 0.01	1.00 ± 0.05	ND[b]
k_{cat} (s^{-1})	14.0 ± 0.3	0.14 ± 0.00	ND[b]
k_{cat}/K_m (s^{-1} mM^{-1})	53.3 ± 3.3	0.14 ± 0.01	ND[b]

[a]The kinetic parameters for *Mal*Lac1A were recalculated from values measured previously by Andberg et al., using ABTS in 25 mM succinate buffer at pH 4.5, and 2,6-DMP and syringaldazine activities were performed in 40 mM MES buffer at pH 6.0 both at 25 °C. The error in all measurements was estimated to ±15% [29]. [b]ND: Not determined due to lack of activity

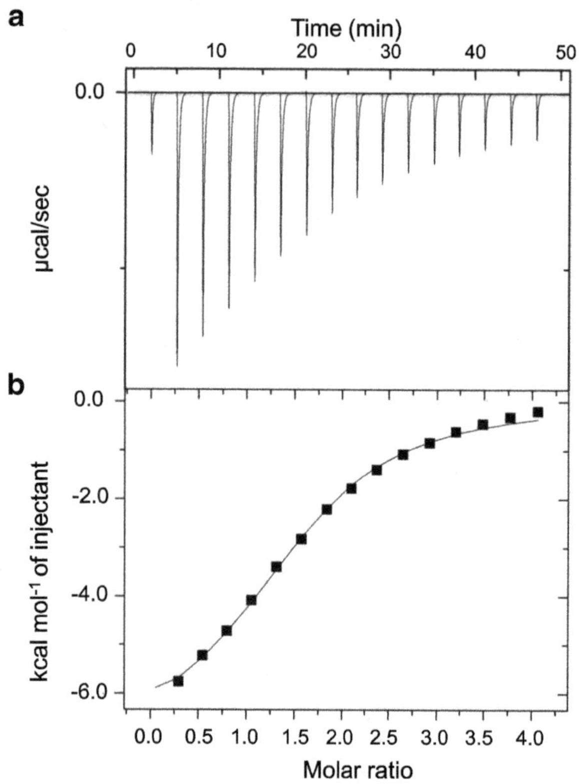

Fig. 2 Isothermal titration calorimetry of the binding of *Ztr*Lac1A to β-CD, carried out at 25 °C at pH 5.5. **a** Binding thermogram and **b** The normalized integrated heat response (black squares) plotted against the molar ratio of injected ligand to the protein concentration in the cell. The fit of a one-site binding model (black line) to the binding isotherm

reduction at 0.2 V as shown in (Additional file 1: Figure S4). The peaks observed with $E^{\circ\prime}$ 0.408 V vs. SHE (Fig. 3a) are attributed to quinone moieties at the GE as previously reported [33, 34]. The *Ztr*Lac1A modified GE electrodes do not show O_2 electroreduction at this potential implying that there is no electrical communication between these quinone moieties and the enzyme.

Furthermore, to confirm the results, *Ztr*Lac1A was co-immobilized with an Os-redox polymer and polyethylene glycol diglycidyl ether (PEGDGE) (GE/Os/*Ztr*Lac1A) using previously reported methods [35, 36]. An Os-polymer with ($E^{\circ\prime}$ = 0.549 V vs. SHE) was used as a surface immobilized redox mediator [37] to facilitates electron transfer between electrode surface and the T1 site of the enzyme. This mediated electron transfer is thermodynamically favorable if the $E^{\circ\prime}$of the polymer is more negative than $E^{\circ\prime}$ of the T_1 site. A polymer with $E^{\circ\prime}$ close to that of the T1 site is desirable. This together with the highly flexible backbone and highly cationic nature of the polymer facilitate enzyme immobilization and electron transfer between the $Os^{2+/3+}$ containing complexes, the enzyme and electrode. The polypyrrole side chains present reactive $-NH_2$ groups, which are used in combination with PEGDGE to stabilize the immobilization through covalently links between the enzyme and polymer. This results in a 3D-hydrogel incorporating very high amounts of both immobilized enzyme molecules as well as mediating functionalities onto the electrode surface with easy access for substrates and products to diffuse to/from the enzyme active site. The electrochemical characterization of *Ztr*Lac1A entailed the determination of the onset potential for the oxygen electro-reduction (E_{O2}), which is close to the formal potential of the T_1 site (E_{T1}) [38]. The onset potential was estimated by physically adsorbing the enzyme onto graphite electrodes (GE) with or without an Os-polymer of $E^{\circ\prime}$ 0.549 V vs. Standard hydrogen electrode (SHE) previously characterized [37].

The Os-polymer facilitates transfer of electrons between the electrode and enzyme in a process known as mediated electron transfer (MET). CVs corresponding to GE/Os/*Ztr*Lac1A are shown in Fig. 3b with the subtracted curves, where the peaks observed at 0.532 V vs. SHE are attributed to the Os-polymer. Two electrocatalytic processes were observed in the presence of O_2 with the first having its onset potential at 0.660 V vs. SHE, which is attributed to the MET between the Os-polymer and *Ztr*Lac1A. The second process starting from 0.4 V vs. SHE is attributed to the direct electroreduction of O_2 in the surface of the GE as previously discussed.

Phylogenetic and sequence analysis

The aligned sequences were from the ascomycetes laccases retrieved from the CAZy database and the basidiomycetes sequences from a BLASTp search of the non-redundant protein sequences database using the catalytic module of *Ztr*Lac1A (556 residues) as a query. This analysis revealed that the closest orthologues to *Ztr*Lac1A are laccases from the plant pathogens *Zymoseptoria brevis* (*Zbr*Lac1A) and *Sphaerulina musiva* (*Smu*Lac1A) sharing 98 and 69% sequence identity, respectively. In total, 221 sequences (490–650 amino acid residues, 55–75 kDa), including ascomycete and basidiomycetes laccases, were included in the alignment and phylogenetic tree.

*Ztr*Lac1A together with eight closest homologues segregate into a single branch, which is hereafter designated as the *Septoria* cluster in the phylogenetic analysis. Remarkably, six of these putative laccases share the same modular organization as *Ztr*Lac1A (Additional file 1: Figure S3a) with variations in the length of linker (12–17 amino acid residues) separating the binding and catalytic modules (Additional file 1: Table S1). The *Septoria* cluster (bold blue in Fig. 4) is adjacent to the two closest structurally characterized ascomycete laccases, *Melanocarpus albomyces*

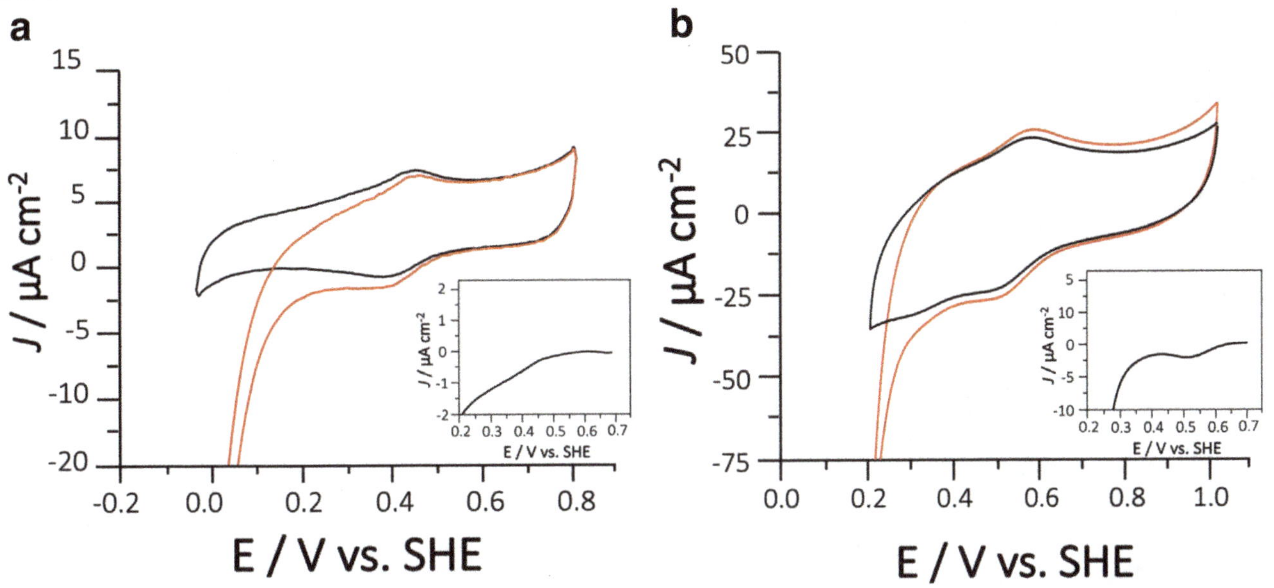

Fig. 3 **a** Cyclic Voltammograms (CVs) recorded with *Ztr*Lac1A modified low-density GE in N_2 (black line) and in O_2 (red line); (inset Fig. 3a) subtracted curve in order to determine the onset of O_2 reduction. **b** CVs recorded with GE/Os-polymer/*Ztr*Lac1A under N_2 (black line) and in O_2 (red line) conditions. (Inset Fig. **b**) subtracted curve in order to determine the onset of O_2 reduction. Conditions: 0.1 M acetate buffer at pH 4.0 in 0.1 M $NaClO_4$; scan rate 10 mVs^{-1}

(*Mal*Lac1A) and *Thielavia arenaria* (*Tar*Lac1A), displaying 44% sequence identity to *Ztr*Lac1A (in brown). Two other structurally characterized ascomycete laccases from *Botyris aclada* (*Bac*Lac1A) and *Aspergillus niger* (*Ani*Lac1A) [39] that share 41 and 36% sequence identity to *Ztr*Lac1A, respectively, segregate in two distant clades. The basidiomycetes orthologues populate a distinct clade, which includes some intermediate ascomycete sequences from *A. niger* and *Fusarium oxysporum* (Fig. 4).

Distinctive structural features in ZtrLac1A compared to other fungal laccases

To analyse the structural elements responsible for segregation of the *Ztr*Lac1A in a distinct cluster, we compared these enzymes to related sequences of AA1_3 in CAZy. To assess the functional relevance of these differences, we generated homology models of *Ztr*Lac1A based on the closest structurally characterized ascomycete laccase from *M. albomyces* (*Mal*Lac1A) as a template (44% identity to *Ztr*Lac1A). A good quality model was obtained as judged by LGscore and MaxSub qualitative values of 6.18 and 0.24 calculated via the ProQ online server [40], respectively. Overall, the model was similar to *Mal*Lac1A as reflected by the root mean square deviation (RMSD) of 0.78 Å for the superimposition of 528 C_α backbone atoms out of 556 between these enzymes, which also shared three disulphide bridges that stabilize their structures. In *Ztr*Lac1A, the disulfide bridges between Cys303-Cys340, Cys118-Cys544, and Cys10-Cys19 at the N-terminus stabilize the fold analogous to counterparts in the *Tar*Lac1A and *Ani*-Lac1A structures.

The sequence alignment between *Ztr*Lac1A and the characterized laccases revealed differences in three functionally relevant regions including loops surrounding the active site, C-terminus and copper 1 (T1) active center.

The substrate-binding site in fungal laccases is defined by four loops that flank the T1 site [39]. Ascomycete laccases have elongated loops in this region compared to basidiomycetes laccases. Notably, Loop A (N373 – L384) in *Ztr*Lac1A is further elongated with two, four and six residues compared to the corresponding loops in the known structural homologues *Ani*Lac1A, *Mal*Lac1A, *Tar*Lac1A, and *Bac*Lac1A, respectively (Fig. 5a). Significant sequence substitutions are also observed in this loop, e.g. the T368, N363 and V387 in *Mal*Lac1A (and *Tar*Lac1A), *Bac*Lac1A and *Ani*Lac1A laccases are substituted with the K381 in *Ztr*Lac1A (Fig. 5a). Similarly, Loop B comprising T435 – N444 in *Ztr*Lac1A, is also elongated with up to six residues compared to *Tar*-Lac1A, *Mal*Lac1A, *Bac*Lac1A as well as *Ani*Lac1A possessing the shortest B loop. Notably, a proline (P423) in *Mal*Lac1A and *Tar*Lac1A, is substituted with glutamine (Q437) in *Ztr*Lac1A (Additional file 1: Figure S5).

*Ztr*Lac1A also displays important difference at the C-terminus, with three proline residues in the structurally characterized laccases, being substituted to leucine, glutamine or lysine (Fig. 5b). Mutation of the leucine at this

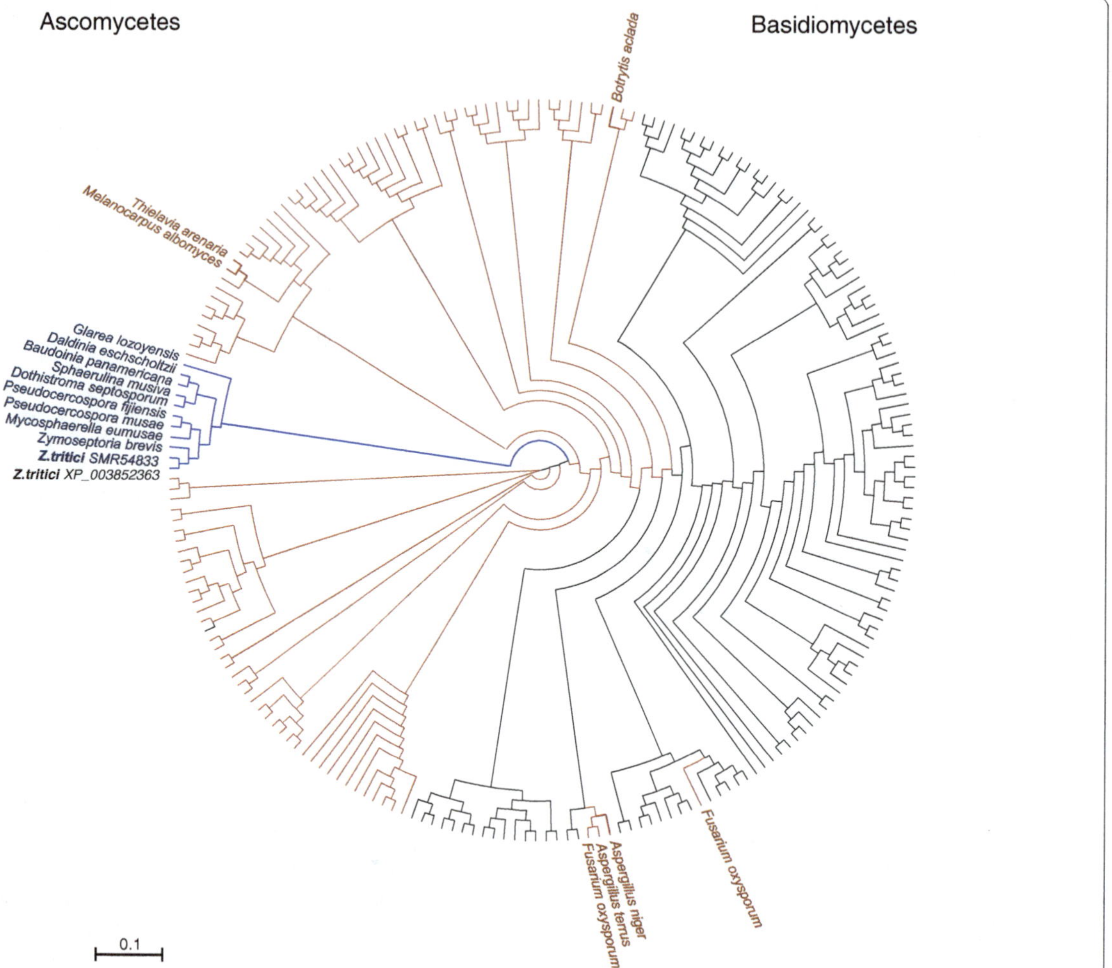

Fig. 4 Phylogenetic analysis of *Ztr*Lac1A and related fungal laccases. The phylogram shows 221 orthologues to the catalytic module of *Ztr*Lac1A, all assigned into AA1 according to the CAZy classification. The *Septoria* cluster is highlighted in blue and the two sequences from the *Zymoseptoria tritici* species are in bold, whereas basidiomycetes sequences are in black. The only four structurally characterized ascomycete laccases are in brown. The tree illustrates the segregation of *Septoria* in a distinct clade of the phylogenetic tree consistent with the unique biochemical profile and the prevalent occurrence with a CBM20 module as opposed to other enzymes

position to methionine was shown to reduce the activity of *Bal*Lac1A highlighting the functional importance of C-terminus [41].

Discussion

The first starch binding modular laccase from fungal pathogens that use apoplastic infection strategy

Carbohydrate active enzymes (CAZymes) are frequently modular featuring one or more catalytic modules (CMs) appended to non-catalytic auxiliary modules. The most common non-catalytic modules are ancillary CBMs, which potentiate the deconstruction of insoluble polysaccharides by mediating substrate targeting and prolonged contact of cognate enzymes [42]. Presence of CBMs is also observed in oxidoreductases targeting complex polysaccharides [43]. In this study, we report the properties of the first modular laccase possessing a starch-binding CBM20. This enzyme stems from the major wheat pathogen *Z. tritici* causing massive crops yield losses globally [17, 18]. *Z. tritici* has a "stealth" pathogenesis, with an apoplastic and symptomless initial growth within the plant cell wall. During later stages, loss of plant cell integrity, leakage of nutrients to the intercellular space concomitant with acceleration of fungal growth and finally necrotic lesions on leaves are typically observed [19]. Despite the significant reduction in CAZymes and CBMs [44], the genome of *Z. tritici* encodes at least 24 putative starch degrading enzymes including α-glucosidases (8 of GH31 and 1 of GH133), α-amylases (14 sequences from GH13) and a glucoamylase (GH15) (http://www.cazy.org) [45]. This genomic expansion with putative starch targeting enzymes is also reflected by the presence of seven encoded proteins possessing putative starch binding domains including four

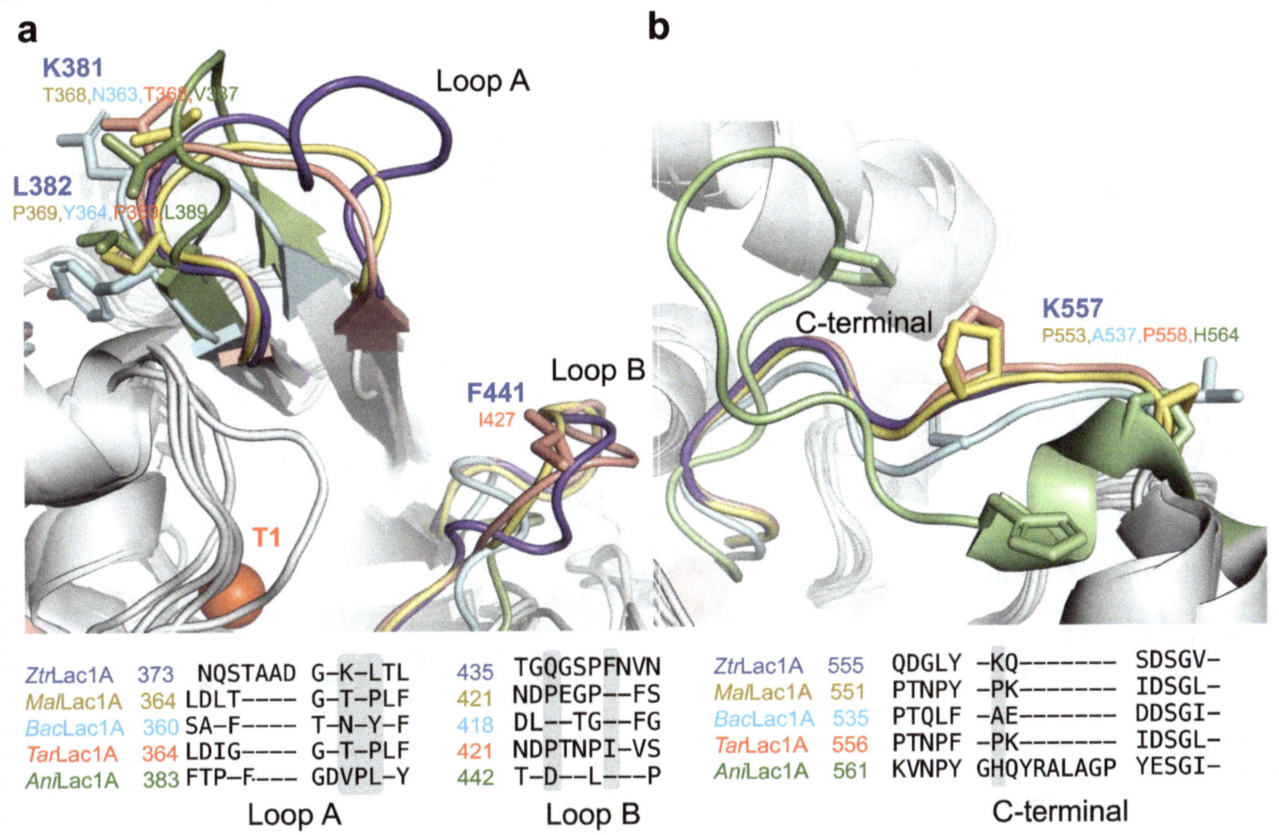

Fig. 5 Superimposition of modeled *Ztr*Lac1A with the structures from the ascomycetes *M. albomyces* (*Mal*Lac1A, 1GW0), *T. arenaria* (*Tar*Lac1A, 3PPS), *Botyris aclada* (*Bac*Lac1A, 3SQR) and *A. niger* (*Ani*Lac1A, 5LM8). **a** The loop A flanking the T1 substrate binding site in *Ztr*Lac1A (dark blue) is variable with respect to size and sequence from the structurally characterized *Mal*Lac1A (light brown), *Tar*Lac1A (red), *Bac*Lac1A (Cyan) as well as *Ani*Lac1A (green). **b** Superposition of C-termini of *Ztr*Lac1A, *Mal*Lac1A, *Tar*Lac1A, *Bac*Lac1A and *Ani*Lac1A. The Cu 1 and Cu 2–3 are shown as red spheres. Structures were rendered using PYMOL v1.8 software (Schrödinger, LLC, Palo Alto, CA). The sequence alignment of these functionally relevant structural elements is shown and divergent segments and amino acid substitutions that involve large changes in chemistry or length of sidechains are highlighted in grey shades

assigned into CBM20. Notably, large changes in chloroplasts are observed during infection, especially during the later stages, which revealed an expansion of chloroplasts before cell collapse [46]. The chloroplasts are the sites of starch synthesis in the leaves [46], and interestingly a recent transcriptional analysis revealed a substantial upregulation of α-amylases and oxidoreductases in the taxonomically related pine pathogen *Dothistroma septosporum* also during the later stage of infection [47]. This fungus possesses a homologue of the *Z. tritici* laccase characterized in the present study (Additional file 1: Figure S3, Additional file 1: Table S1). In fact, all the seven modular laccases possessing a CBM20 stem from taxonomically related pathogens from the order Capnodiales. Taken altogether, the evolution of starch binding laccases in this group of pathogens and the histological and transcriptional response seems to highlight an important role of starch targeting enzymes in the necrotic stage of infection. Our data provide compelling evidence for the functionality of the CBM20, which possesses moderate affinity (K_d 33 μM) typical of counterparts from amylolytic enzymes. Therefore, *Ztr*Lac1A is likely to be efficiently targeted to starch granules, but the role of the laccase and the substrates it oxidizes at the starch granule surface in vivo remain intriguing and merits further studies.

ZtrLac1A displays a unique activity profile and sequence signatures compared to ascomycete characterized laccases

The kinetic data of *Ztr*Lac1A on the negatively charged redox mediator ABTS verified its laccases functionality (Table 1). Interestingly, the K_m of this enzyme on ABTS was 60% lower than that measured for the closest related ascomycete laccases *Mal*Lac1A (Table 1). While, *Mal*-Lac1A displays an increasing affinity and efficiency on the more apolar phenolic mediators DMP and syringaldazine compared to ABTS [29], *ZtrLac1A* was almost inactive on these substrates (Table 1). This sequence signature is also consistent with changes in the sequence and length of loops flanking the T1 binding site, where the substrate binding and oxidation occurs as compared to *Mal*Lac1A and other currently characterized ascomycete laccases (Fig. 5, Additional file 1: Figure S5). Notably, the loops flanking the active site of *Ztr*Lac1A, exhibit a lower content of negatively charged and apolar residues compared to structurally characterized laccases. These changes, together with the elongation of loop A with four amino acid residues including a lysine (Fig. 5a), are consistent with a more positive electrostatics and occlusion of the active site in *Ztr*Lac1A. These substitutions may provide a rationale for better affinity of the negatively charged ABTS and the lack of activity on the two other less polar mediators. The C-terminal loop of laccases acts as a plug that occludes the T2 and T3 sites, where the second substrate O_2 binds [29]. A single mutation of the C-terminal residue (L559A) and the deletion of the four C-terminal residues of *Mal*Lac1A caused a severe activity drop of the enzyme [29]. We have also observed important substitutions in this C-terminal segment making it more flexible compared to characterized counterpart due to the loss of three closely located proline residues (Fig. 5b).

Another interesting difference is a conserved residue close to the Cu ligands at the T1 site (Additional file 1: Figure S6). This residue is either a phenylalanine as in *Ztr*Lac1A, or a leucine as in characterized ascomycete laccases (Additional file 1: Figure S6), but methionines are also observed in a few non-characterized ascomycete laccases. Site-directed mutagenesis of Cu ligands have shown large effects on the redox potential [39, 48]. According to the current classification, laccases are arranged in three groups: low (340–490 mV), middle (470–710 mV) and high potential (730–780 mV) [49]. The difference in the T_1 potentials depends strongly on the axial Cu-ligands of the T1 site being methionine for low potential, leucine for middle and phenylalanine for high potentials [49].

The bioelectrochemical O_2 reduction by *Ztr*Lac1A immobilized on GE electrodes gives an estimate of the potential of T1, being 0.569 V vs. SHE when the electrons are transferred directly from the electrode to the T1 site of the enzyme. The successfully mediated electroreduction observed for GE/Os/*Ztr*Lac1A is an indication of the closeness of the $E^{0\prime}$ of T_1 to the one of the Os-polymer used (0.55 V). The assignment of the onset to be from the T_1 site is based on previously published results reporting the electro-reduction of O_2 at carbon modified electrodes with "blue" MCO [50–52]. Subsequently, electron transfer proceeds internally through the T_2/T_3 multicopper site, where the electroreduction of O_2 takes place.

*Ztr*Lac1A is thus classified as a middle redox potential accordingly, which is in agreement with previous values reported for the few other ascomycete characterized laccases [48].

Conclusions

*Ztr*Lac1A displays an atypical substrate preference, but shows a typical redox potential range compared to ascomycetes characterized counterparts. This first demonstration of activity of a modular laccase with a starch binding module offers an attractive possibility for reversible immobilization of laccases on cheap material like starch or cellulose, and allows targeting of these enzymes without using classical covalent immobilization techniques that typically reduce enzymatic activity. Finally, further studies are needed to get insight into the role of this enzyme in the pathogenic life cycle of fungi from the order Capnodiales and to identify physiologically relevant substrates.

Additional file

Additional file 1: Table S1. Septoria cluster laccases. **Figure S1.** Chemical structure of the osmium polymer. **Figure S2.** Activity screening on aromatic (mainly phenolics) susbtrates. **Figure S3.** ZtrLac1A primary structure and purification. **Figure S4.** Cyclic voltammograms. **Figure S5.** Sequence alignment of ZtrLac1A and homologues from AA1_3. **Figure S6.** Architecture of Cu1 copper site in characterized ascomycetes laccases. (PDF 747 kb)

Abbreviations

2,6-DMP: 2,6-dimethoxyphenol; ABTS: 2,2′-azino-bis(3-ethylbenzothiazoline-6-sulfonic acid); *Ani*Lac1A: *Aspergillus niger* laccase; *Bac*Lac1A: *Botyris aclada* laccase; CBM: Carbohydrate binding module; CMs: Catalytic modules; CV: Cyclic voltammograms; GE: Graphite electrode; *Mal*Lac1A: *Melanocarpus albomyces* laccase; MCOs: Multicopper oxidases; SHE: Standard hydrogen electrode; STB: *Septoria tritici* blotch; *Tar*Lac1A: *Thielavia arenaria* laccase; *Ztr*Lac1A: *Zymoseptoria tritici* laccase; β-CD: β-cyclodextrin

Acknowledgments

We are grateful to Prof. Dónal Leech and Dr. Peter Ó Conghaile from Biomolecular Electronics Research Laboratory; National University of Ireland Galway (Galway, Ireland) for providing us with osmium polymer.

Authors' contributions

MAH and MHM conceived the study and designed the experiments. MHM performed all bioinformatics, molecular biology and biochemical characterization. LG along PB, RO and ET have contributed to the bio-electrochemical study. MAH and MHM wrote the first draft of the manuscript and all authors participated in writing and approved the final version.

Funding

This study was supported by a Novo Nordisk foundation grant (NNF12OC0000769) within the "Biotechnology-based Synthesis and Production" programme to MAH, and a H.C. Ørsted-Marie Curie Actions COFUND Post-doc grant and Novo Nordisk foundation (NNF17OC0025642) within Biotechnology-based Synthesis and Production Research program to MHM. Carlsberg Foundation is acknowledged for an instrument grant to acquire the ITC instrument. None of the funders took part in any aspect of the scientific content of the study.

Ethics approval and consent to participate
Not applicable.

Consent for publication
Not applicable.

Competing interests
The authors declare that they have no competing interests.

Author details
[1]Department of Biotechnology and Biomedicine, Technical University of Denmark, Søltofts Plads, 2800 Kgs, Lyngby, Denmark. [2]Department of Biochemistry and Structural Biology, Lund University, P.O. Box 124, 221 00 Lund, Sweden. [3]Department of Chemistry and Drug Technologies, Sapienza University of Rome, Piazzale Aldo Moro 5, 00185 Rome, Italy. [4]Department of Chemistry, Technical University of Denmark, Kemitorvet 207, 2800 Kgs, Lyngby, Denmark.

References

1. Solomon EL, Sundaram UM, Machonkin TE. Multicopper oxidases and oxygenases. Chem Rev. 1996;96:2563–606.
2. Giardina P, Faraco V, Pezzella C, Piscitelli A, Vanhulle S, Sannia G. Laccases: a never-ending story. Cell Mol Life Sci. 2010;67:369–85.
3. Solomon EI, Heppner DE, Johnston EM, Ginsbach JW, Cirera J, Qayyum M, Kieber-Emmons MT, Kjaergaard CH, Hadt RG, Tian L. Copper active sites in biology. Chem Rev. 2014;114:3659–853.
4. Solomon EI, Chen P, Metz M, Lee SK, Palmer AE. Oxygen binding, activation, and reduction to water by copper enzymes. Angew Chem Int Ed Engl. 2001;40:4570–90.
5. Kamitaka Y, Tsujimura S, Kataoka K, Sakurai T, Ikeda T, Kano K. Effects of axial ligand mutation of the type I copper site in bilirubin oxidase on direct electron transfer-type bioelectrocatalytic reduction of dioxygen. J Electroanalytical Chem. 2007;601:119–24.
6. Reinhammar BRM. Oxidation-reduction potentials of the electron acceptors in laccases and stellacyanin. Biochim Biophys Acta. 1972;275:245–59.
7. Rodjers CJ, Blandford CF, Giddens SR, Skamnioti P, Armstrong FA, Gurr SJ. Designer laccases: a vogue for high-potential fungal enzymes? Trends Biothechnol. 2010;28:63–72.
8. Claus H. Laccases: structure, reactions, distribution. Micron. 2004;35:93–6.
9. Levasseur A, Drula E, Lombard V, Coutinho PM, Henrissat B. Expansion of the enzymatic repertoire of the CAZy database to integrate auxiliary redox enzymes. Biotechnol Biofuels. 2013;6:41.
10. Kiiskinen LL, Viikari L, Kruus L. Purification and characterisation of a novel laccase from the ascomycete *Melanocarpus albomyces*. Appl Microbiol Biotechnol. 2002;59:198–204.
11. Baldrian P. Fungal laccases - occurrence and properties. FEMS Microbiol. 2006;30:215–42.
12. Slomczynski D, Nakas JP, Tanenbaum SW. Production and characterization of laccase from *Botrytis cinerea*. Appl Environ Microbial. 1995;61:907–12.
13. Iyer G, Chattoo BB. Purification and characterization of laccase from the rice blast fungus, *Magnaporthe grisea*. FEMS Microbiol Lett. 2003;227:121–6.
14. Canero Cordoba D, Roncero MIG. Functional analyses of laccase genes from *Fusarium oxysporum*. Amercian phytopathol. Soci. 2007;98:509–18.
15. O'Driscoll A, Kildea S, Doohan F, Spink J, Mullins E. The wheat-Septoria conflict: a new front opening up? Trends Plant Sci. 2014;19:602–10.
16. Torriani SFF, Melichar JPE, Mills C, Pain N, Sierotzki H, Courbot M. *Zymoseptoria tritici:* a major threat to wheat production, integrated approaches to control. Fungal Genet Biol. 2015;79:8–12.
17. Fones H, Gurr S. The impact of *Septoria tritici* blotch disease on wheat: an EU perspective. Fungal Genetic Biol. 2015;79:3–7.
18. Jørgensen LN, Hovmøller MS, Hansen JG, Lassen P, Clark B, Bayles R, B. Rodemmann, Flath K, Jahn M, Goral T, et al. IPM strategies and their dilemmas including an introduction to www.eurowheat.org. J Integr Agric 2014;13:265–281.
19. Dean R, Van Kan JA, Pretorious ZA, Hammond-Kosack KE, Di Pietro A, Spanu PD, Rud JJ, Dickman M, Kahmann R, Elis J, Foster GD. The top 10 fungal pathogens in molecular plant pathology. Mol Plant Pathol. 2012;13:414–30.
20. Windels CE. Economic and social impacts of *Fusarium* head blight: changing farms and rural communities in the northern Great Plains. Phytopathology. 2000;90:17–21.
21. Howard RJ, Valent B. Breaking and entering: host penetration by the fungal rice blast pathogen *Magnaporthe grisea*. Annu Rev Microbiol. 1996;50:491–512.
22. Goodvin SB, MBarek SB, Dhillon B, et al. Finished genome of the fungal wheat pathogen *Mycosphaerella graminicola* reveals Dispensome structure, chromosome plasticity, and stealth pathogenesis. PLoS Genet. 2011. https://doi.org/10.1371/journal.pgen.1002070.
23. Mano N, Kim HH, ZhangY, Heller A. An oxygen cathode operating in a physiological solution. J Am Chem Soc. 2002;124:6480–6.
24. Nekiunaite L, Isaksen T, Vaaje-Kolstad G, Abou Hachem M. Fungal lytic polysaccharide monooxygenases bind starch and β-cyclodextrin similarly to amylolytic hydrolases. FEBS Lett. 2016;590:2737–47.
25. Niku-Paavola ML, Karhunen E, Salola P, Raunio V. Ligninolytic enzymes of the white-rot fungus *Phlebia radiate*. Biochem J. 1988;254:877–84.
26. Shleev S, Jarosz-Wilkolazka A, Khalunina A, Morozova O, Yaropolov A, Ruzgas T, Gorton L. Direct heterogeneous electron transfer reactions of laccases from different origins on carbon electrodes. Bioelectrochemistry. 2005;67:115–24.
27. Altschul SF, Madden TL, Schäffer AA, Zhang J, Zhang Z, Miller W, Lipman DJ. Gapped BLAST and PSI-BLAST: a new generation of protein database search programs. Nucleic Acids. 1997;25:3389–402.
28. Huson DH, Scornavacca C. Dendroscope 3: an interactive tool for rooted phylogenetic trees and networks. Syst Biol. 2012;61:1061–7.
29. Andberg M, Hakulinen N, Auer S, Saloheimo M, Koivula A, Rouvinen J, Kruus K. Essential role of the C-terminus in *Melanocarpus albomyces* laccase for enzyme production catalytic properties and structure. FEBS J. 2009;276:6285–300.
30. Borodina I, Jensen BM, Wagner T, Abou Hachem M, Sondergaard I, Poulsen LK. Expression of enzymatically inactive wasp venom phospholipase A1 in *Pichia pastoris*. PLoS One. 2011;6:e21267.
31. Jeoh T, Michener W, Himmel ME, Decker SR, Adney WS. Implications of cellobiohydrolase glycosylation for use in biomass conversion. Biotechnol Biofuels. 2008;1:10.
32. Christiansen C, Abou Hachem M, Janecek S, Viksø-Nielsen A, Blennow A, Svensson B. The carbohydrate-binding module family 20-diversity, structure, and function. FEBS J. 2009;276:5006–29.
33. Swain GM. Handbook of electrochemistry, ed. C G Zoski Elsevier Amsterdam. 2007;111-V.
34. McCreery RL. Advanced carbon electrode materials for molecular electrochemistry. Chem Rev. 2008;108:2646–87.
35. Heller A. Electrical wiring of redox enzymes. Acc Chem. 1990;23:128–34.
36. Zafar MN, Tasca F, Boland S, Kujawa M, Patel I, Peterbauer CK, Leech D, Gorton L. Wiring of pyranose dehydrogenase with osmium polymers of different redox potentials. Bioelectrochemistry. 2010;8038–42.
37. Kavanagh P, Jenkins P, Leech D. Electroreduction of O_2 at a mediated *Melanocarpus albomyces* laccase cathode in a physiological buffer. Electrochem Commun. 2008;10(7):970–2.
38. Dos Santos L, Climent V, Blandford CF, Armstrong FA. Mechanistic studies of the 'blue' cu enzyme, bilirubin oxidase, as a highly efficient electrocatalyst for the oxygen reduction reaction. Phys Chem Chem Phys. 2010;12:13962–74.
39. Ferraroni M, Westphal AH, Borsari M, Tamayo-Ramos JA, Briganti F, de Graaff LH, van Berkel WJH. Structure and function of *Aspergillus niger* laccase McoG. Biocatalysis. 2017;3(1):16.
40. Wallner B, Elofsson A. Protein Sci. 2003;12:1073–86.
41. Osipov E, Polyakov K, Kittl R, Shleev S, Dorovatovsky TT, Hann S, Ludwig R, Popov V. Effect of the L499M mutation of the ascomycetous *Botrytis aclada* laccase on redox potential and catalytic properties. Acta Crystallogr D Biol Crystallogr. 2014;70:2913–23.
42. Gilbert HJ, Knox JP, Boraston AB. Advances in understanding the molecular basis of plant cell wall polysaccharide recognition by carbohydrate-binding modules. Curr Opin Struct Biol. 2013;23:669–77.
43. Nekiunaite L, Arntzen M, Svensson B, Vaaje-Kolstad G, Abou Hachem M. Lytic polysaccharide monooxygenases and other oxidative enzymes are abundantly secreted by *Aspergillus nidulans* grown on different starches. Biotechnol Biofuel. 2016;9:187.
44. do Amaral AM, Antoniw J, Rudd H-KKE. Defining the predicted protein Secretome of the fungal wheat leaf pathogen *Mycosphaerella graminicola*. PLoS One. 2012;7:e49904.

45. Kema GHJ, Yu DZ, Rijkenberg FHJ, Shaw MW, Baayen RP. Histology of the pathogenesis of *Mycosphaerella graminicola* in wheat. Phytopathology. 1996; 86:777–86.
46. Zeeman SC, Delatte T, Messerli G, Umhang M, Stettler M, Mettler T, Streb S, Reinhold H, Kötting O. Starch breakdown: recent discoveries suggest distinct pathways and novel mechanisms. Funct Plant Biol. 2007;34:465–73.
47. Bradshaw RE, Guo YD, Sim AD, Kabir MS, Chettri P, Ozturk IK, Hunziker L, Ganley RJ' Cox MP. Genome-wide gene expression dynamics of the fungal pathogen *Dothistroma septosporum* throughout its infection cycle of the gymnosperm host *Pinus radiate*. Mol Plant Pathol. 2016;17:210–24.
48. Durão P, Chen Z, Silva CS, Soares CM, Pereira MM, Todorovic S, Hildebrandt P, Bento I, Lindley PF, Martins LO. Proximal mutations at the type 1 copper site of CotA laccase: spectroscopic, redox, kinetic and structural characterization of I494A and L386A mutants. Biochem J. 2008;412:339–46.
49. Shleev S, Tkac J, Christenson A, Ruzgas T, Yaropolov AI, Whittaker JW, Gorton L. Direct electron transfer between copper-containing proteins and electrodes. Biosens Bioelectron. 2005;20:2517–54.
50. Berezin IV, Bogdanovskaya VA, Varfolomeev SD, Tarasevich MR, Yaropolov AI. Bioelectrocatalysis. Equilibrium oxygen potential in the presence of laccase. Dokl Akad Nauk SSSR. 1978;240:615–8.
51. Yaropolov AI, Kharybin AN, Emnéus J, Marko-Varga G, Gorton L. Electrochemical properties of some copper-containing oxidases. Bioelectrochem Bioenerg. 1996;40:49–57.
52. Christenson A, Dimcheva N, Ferapontova EE, Gorton L, Ruzgas T, Stoica L, Shleev S, Yaropolov AI, Haltrich D, Thorneley RNF, Aust SD. Direct electron transfer between ligninolytic redox enzymes and electrodes. Electroanalysis. 2004;16:1074–92.

Identification and characterization of an Endo-glucanase secreted from cellulolytic *Escherichia coli* ZH-4

Jian Pang[1,3], Junshu Wang[2*], Zhanying Liu[1,3*], Qiancheng Zhang[1] and Qingsheng Qi[2]

Abstract

Background: In the previous study, the cellulolytic *Escherichia coli* ZH-4 isolated from bovine rumen was found to show extracellular cellulase activity and could degrade cellulose in the culture. The goal of this work was to identify and characterize the secreted cellulase of *E. coli* ZH-4. It will be helpful to re-understand *E. coli* and extend its application in industry.

Results: A secreted cellulase was confirmed to be endo-glucanase BcsZ which was encoded by *bcsZ* gene and located in the cellulose synthase operon *bcsABZC* in cellulolytic *E. coli* ZH-4 by western blotting. Characterization of BcsZ indicated that a broad range of pH and temperature tolerance with optima at pH 6.0 and 50 °C, respectively. The apparent Michaelis–Menten constant (K_m) and maximal reaction rate (V_{max}) for BcsZ were 8.86 mg/mL and 0.3 μM/min·mg, respectively. Enzyme activity of BcsZ was enhanced by Mg^{2+} and inhibited by Zn^{2+}, Cu^{2+} and Fe^{3+}. BcsZ could hydrolyze carboxymethylcellulose (CMC) to produce cello-oligosaccharides, cellotriose, cellobiose and glucose.

Conclusions: It is confirmed that extracellular cellulolytic capability of *E. coli* ZH-4 was attributed to BcsZ, which explained why *E. coli* ZH-4 can grow on cellulose. The endo-glucanase BcsZ from *E. coli*-ZH4 has some new characteristics which will extend the understanding of endo-glucanase. Analysis of the secretion characteristics of BcsZ provided a great reference for applying *E. coli* in multiple industrial fields.

Keywords: Cellulolytic *Escherichia coli* ZH-4, Secretory endo-glucanase, BcsZ, Enzyme characterization

Background

Cellulose biomass is the most abundant carbohydrate on the earth. It can be hydrolyzed to reducing sugars for production of biofuels and chemicals, and thus has a great economic and commercial potential [1]. Cellulose as the main component of plant cell wall consists of linear long chains of β-1, 4 glucose units. Hydrolyzing cellulose by cellulase is ideal and promising for its utilization in environmentally friendly and high efficiency manner [2]. However, the cooperative action of three kinds of cellulolytic enzymes (endo-glucanase, exo-glucanase, and β-glucosidase) is essential in hydrolysis of cellulose to glucose [3]. Among three kinds of cellulolytic enzymes, endoglucanases plays an important role in the process of cellulose hydrolysis because it hydrolyzes the glycosidic bond randomly and shorten the cellulose chains in the initial stage of cellulose breakdown [4]. Cellulase is produced by various cellulolytic bacteria and fungi which have been isolated from different environment [5]. Isolating cellulolytic microorganisms from various environment and characterizating their cellulase are crucial for understanding the evolution mechanism of cellulolytic microorganisms and the hydrolysis mechanism of cellulase, which will promote their application in industry. In recent years, cellulase from bacteria was focused again because the glycoside hydrolases of cellulolytic bacteria are very diverse [6].

The previous study revealed that a cellulase (Cel-CD) from *Bacillus sp.* can be secreted into culture medium when Cel-CD was overexpressed in *E.coli* with or without its signal peptide, which indicated that *E. coli* has the capacity of secreting cellulase [7]. In addition, cellulolytic *E. coli* will be got when expressed this cellulase in

* Correspondence: wangjunshusdu@163.com; zyliu1979@163.com
[2]State Key Laboratory of Microbial Technology, Shandong University, Qingdao 266237, China
[1]School of Chemical Engineering, Inner Mongolia University of Technology, Hohhot 010051, Inner Mongolia, China
Full list of author information is available at the end of the article

E. coli, and the *E. coli* has a potential application to produce enzymes and chemicals directly from lignocellulose biomass [8].

In our previous study, a cellulolytic *E. coli* ZH-4 was isolated from the rumen [9]. *E. coli* ZH-4 is capable of converting corn straw to ethanol and hydrogen anaerobically. Extracellular endo-glucanase and β-glucosidase activity were detected. The results indicated that such enzymes were expressed and secreted in cellulolytic *E. coli* ZH-4. Genome sequence analysis of *E. coli* ZH-4 revealed an endoglucanase gene (Genbank accession number KY965823) encoding a BcsZ homolog.

From another point of view, cellulose is a major structural component in bacteria, which provides cell-surface and cell–cell interaction in various of biofilm models, and protects cells against chlorine treatment [10–12]. The previous study showed that inactivation of BcsZ altered the cellulose-associated phenotypes in *Salmonella enterica* serovar Typhimurium, such as rdar biofilm morphotype, cell clumping, biofilm formation, pellicle formation and flagella-dependent motility [10]. The hydrolase activity of BcsZ is hypothesized to mediate alignment of each β-1, 4 -glucan for proper cellulose microfibril formation [13]. *BcsZ* is a conserved component of the cellulose synthase operon *bcsABCZ*, which encodes the cellulose synthase BcsAB and the outer membrane porin for cellulose translocation and secretion [10, 14]. BcsZ belongs to Glycoside Hydrolase family 8 with endo-1,4-D-glucanase activity. BcsZ hydrolyzes glycosidic bonds by a pair of acidic residues inverting the anomeric configuration at the new reducing end [15]. The crystal structure analysis of BcsZ from *E. coli* showed an $(\alpha/\alpha)_6$-barrel fold. BcsZ binds 4 glucan moieties of cellopentaose via highly conserved residues exclusively on the non-reducing side of its catalytic center [13]. However, whether BcsZ is responsible for cellulolytic ability of *E. coli* ZH-4 is uncertain. Little is known about the characteristic of BcsZ-ZH-4. Enzymatic hydrolysate of BcsZ-ZH-4 from cellulose is unknow.

In this study, the cellulolytic ability of *E. coli* ZH-4 was verified from BcsZ. The endoglucanase was assessed through transcription, expression and secretion. BcsZ from *E. coli* ZH-4 was functionally expressed in *E. coli* BL21 (DE3), and the recombinant protein was purified and characterized.

Results

Identification and verification of secreted cellulase in cellulolytic *E. coli* ZH-4

The mature protein in the culture medium was analyzed through western blotting to identify the extracellular endoglucanase in ZH-4. The coding gene was located on the operon of *bcsABZC*. Meanwhile, the extracellular protein was verified by western blotting using BcsZ antiserum (Fig. 1). The transcription level of *bcsZ* in cellulolytic *E. coli* ZH-4 was also found to be 2.6 ± 0.25 and 6.0 ± 0.26 fold higher than that of *E. coli* MG1655 under aerobic and anaerobic condition respectively (Fig. 2).

A signal peptide sequence of 1–21 amino acid residues was found in *E. coli* ZH-4 by the analysis of the protein sequence. In compared to *E. coli* MG1655, there were two amino acids difference in BcsZ of *E. coli* ZH-4: Ser63 to Phe (Ser-Phe) and Ala71to Val (Ala-Val). The *bcsZ* located in the downstream of *bcsB* and was supposed under the control of the *bcsB* promoter. DNA-binding transcriptional dual regulator FNR (Fnr) regulates *bcsBZ* operon expression under anaerobiosis, and the putative FNR-binding site was identified in upstream of this operon. Genetic analysis of the operon indicated that there was no difference with that of MG1655 in the regulation and transcription region.

Expression and purification of endo-glucanase BcsZ

The *bcsZ* gene amplified from *E. coli* ZH-4 was cloned in pET-28a vector, and then overexpressed in *E. coli* BL21 (DE3). BcsZ was detected in culture medium (Fig. 3a, line 1) with the recombinant cell by Sodium dodecyl sulfate polyacrylamide gel

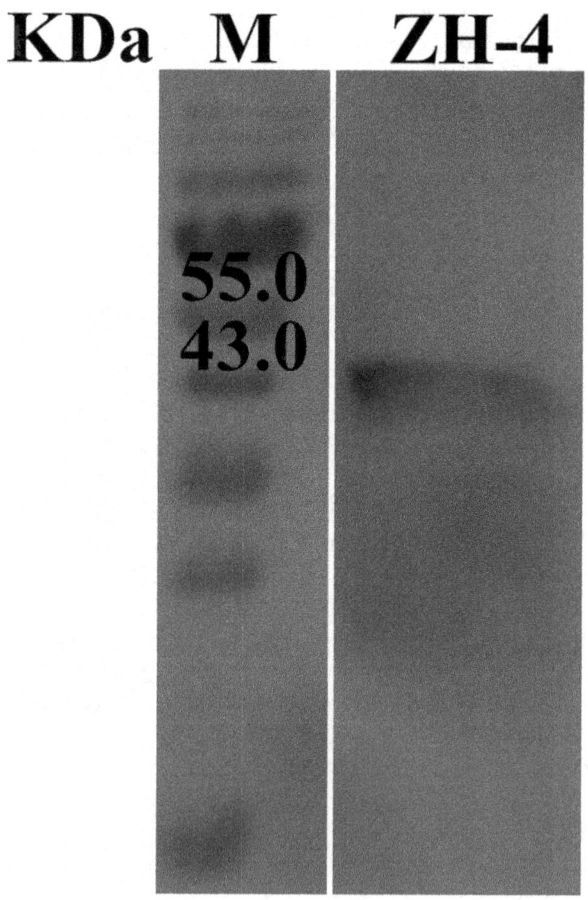

Fig. 1 Western Blotting analysis of BcsZ in culture medium

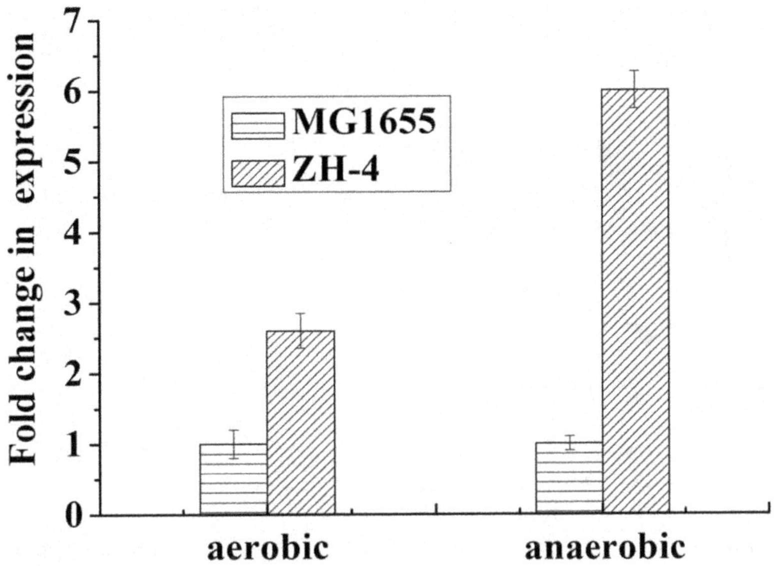

Fig. 2 The fold change in gene expression of *bcsZ* in *E. coli* ZH-4 and MG1655

electrophoresis (SDS-PAGE). The result indicated that endo-glucanase, BcsZ, can be secreted to the outside of cell. The crude protein of BcsZ was purified, and the purified protein appeared as a single protein band on SDS–PAGE gel with a molecular mass of 41.7 KD, which was consistent with prediction (Fig. 3a, line 3). The *E. coli* BL21 (DE3) carrying the pET-28a vector (empty) was used as control (Fig. 3b).

Characterization of the endo-glucanase BcsZ

The optimal temperature and pH of purified BcsZ were determined. As shown in Fig. 4a, the optimum temperature was 50 °C. With the rising of the temperature, the enzyme activity began to decrease, but retained over 70% at the temperature of till 65 °C. The enzyme activity sharply decreased when the temperature is over 65 °C, and was remained by 40% when the temperature was 80 °C. From Fig. 4b, the BcsZ displayed

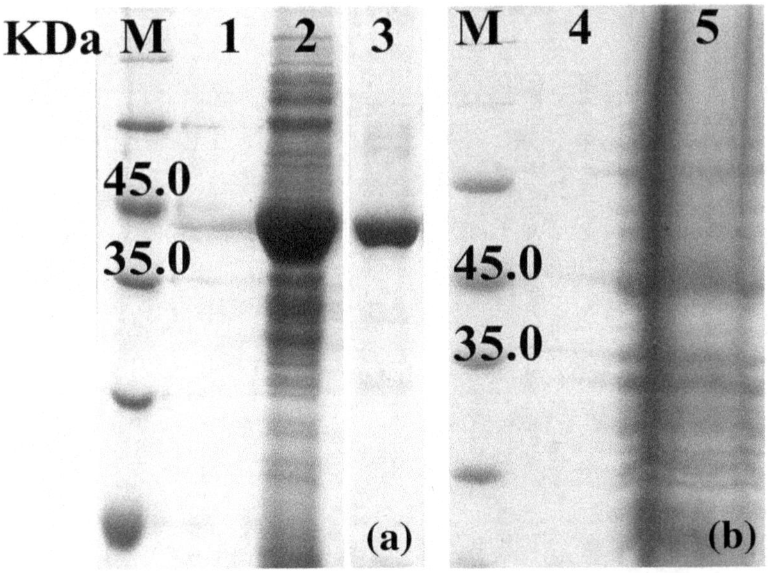

Fig. 3 SDS–PAGE analysis of recombinant BcsZ protein stained with coomassie blue (**a**). M: Protein molecular weight marker; Lane 1: BcsZ in culture medium; Lane 2: BcsZ in cells; Line 3: The purified BcsZ. The *E. coli* BL21 (DE3) carrying the empty plasmid was used as control (**b**). Line 4: culture medium; Line 5: cells

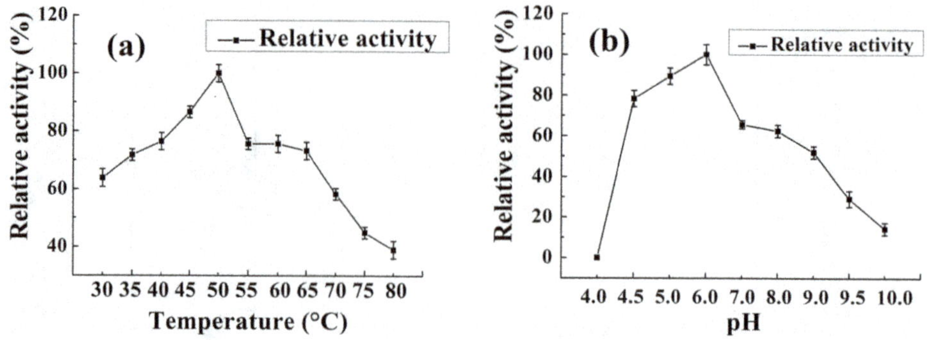

Fig. 4 The optimum temperature (**a**) and pH (**b**) of BcsZ

optimal activity at pH 6.0. It had relatively high enzyme activity at the broad pH range of 4.5–9.0. BcsZ was sensitive to the acidic condition. It remained about 80% activity when pH at 4.5, and enzyme activity was not detected when the pH was 4.0. The BcsZ protein has a tolerance to alkali solution because it retained over 15% of the maximum enzyme activity when the pH was 10.0.

The thermal stability result was shown in Fig. 5, and showed BcsZ displayed high thermal stability at 50 °C and 60 °C with about 90 and 70% of its enzyme activity remained, respectively. The enzyme activity wasn't detected at 70 °C after 0.5 h incubation.

The substrate specificity results were shown in Table 1. As shown in Table 1, BcsZ displayed enzyme activity with CMC, RAC and glucan (data not shown) as substrate. The enzyme showed a weak catalytic activity on Avicel, xylan, cellobiose, laminarin and chitin. The K_m value of BcsZ for CMC as substrate was 8.86 mg/mL, and the calculated V_{max} was 0.3 μM/min·mg.

Effect of metal ion on BcsZ activity

The effect of metal ions on the enzyme activity of BcsZ was examined (Fig. 6). The enzyme activity was improved 165% by Mg^{2+}. Fe^{3+} inhibited BcsZ, and 74% enzyme activity was lost with10 mM Fe^{3+}. Zn^{2+} also inhibited the enzyme of BcsZ, and there was 66% of enzyme activity was left with 10 mM Zn^{2+}. The enzyme was strongly inhibited by Cu^{2+}, there is no enzyme activity detected when Cu^{2+} concentration reached 5 mM in the reaction system.

Analysis of hydrolysis products by TLC

The hydrolysis products of CMC and RAC were analyzed by TLC. As shown in Fig. 7, glucose was released when BcsZ hydrolyzed RAC (G9). The hydrolysis

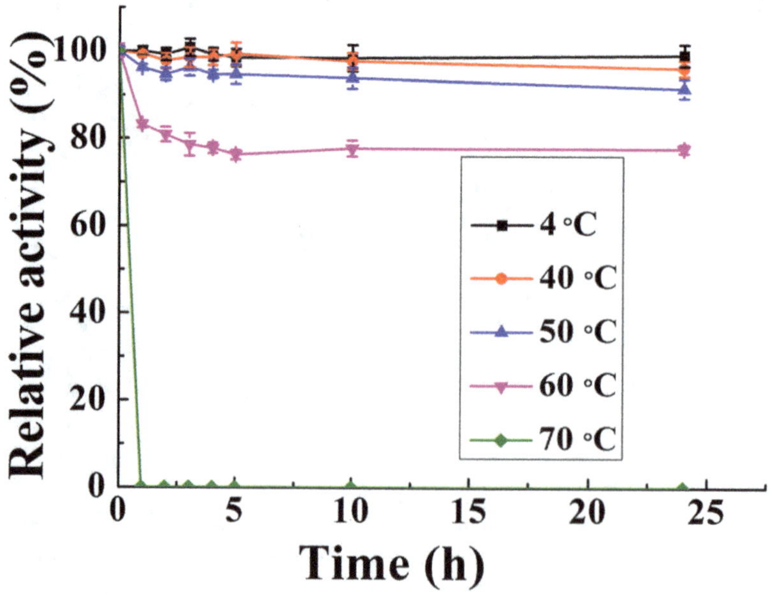

Fig. 5 The thermostability of BcsZ

Table 1 Substrate specificity of BcsZ

Substrate	Specific activity (IU/mg)
CMC	0.84 ± 0.05
RAC	0.52 ± 0.08
Avicel	< 0.01
xylan	< 0.01
cellobiose	< 0.01
laminarin	< 0.01
chitin	< 0.01

products were glucose, a small amount of cellobiose and cellotriose, and unknown cello oligosaccharides from CMC (G7) as substrate.

Discussion

In our previous study, a cellulolytic *E. coli* ZH-4 was isolated from bovine rumen, this strain could produce extracellular cellulases. In this study, secretion of BcsZ to culture medium was confirmed by western blotting (Fig. 1). These results explained the reason ZH-4 formed a clear zone in anaerobic Hungate roll tubes (containing cellulose and Congo red) [9]. While *E. coli* strain MG1655 and W3110 didn't generate the above phenotype at the same condition. Secretion of cellulase (BcsZ) and its contribution to the cellulolytic capacity of *E. coli* ZH-4 were confirmed, which were consistent with our previous supposition and the published [9]. The previous studies showed overexpression of BcsZ in *E. coli* leads to its secretion to extracellular space and formation of clear zone on congo red plates [16, 17]. Further analysis confirmed that the transcription level of *bcsZ* in ZH-4 was higher than that of MG1655, especially under anaerobic condition. The increased transcription level of *bcsZ* is critical to the cellulolytic activity of *E. coli* ZH-4. Sequence analysis showed that the promoter, Fnr and FNR-binding site of *bcsZ* in *E. coli* ZH-4 is consistent with MG1655 and W3110. The genome of *E. coli* ZH-4 (5.3 Mb) is larger than *E. coli* MG1655 (4.6 Mb). It is supposed that some regulatory factors and elements regulate *bcsZ* transcription and expression in *E. coli* ZH-4. This may lead to the increased transcription level of *bcsZ* in *E. coli* ZH-4. Further elevated transcription level investigation of *bcsZ* may reveal the difference in BcsZ secretion between *E. coli* ZH-4 and MG1655.

The endo-glucanase BcsZ had the broad pH range and strong alkali tolerance. These characteristics of BcsZ might be affected by the rumen habitat and consistent with the previous reports. For instance, Gong et al. cloned and identified some novel hydrolase genes from a dairy cow rumen. The purified recombinant enzyme displayed optimal activity at pH 6.0 and 50 °C. It was stable over a broad pH range, from pH 4.0 to 10.0 [18]. Chang et al. reported an endoglucanase from yak rumen microorganisms and the optimal conditions for enzyme activity were 50 °C and pH 5.0 [19]. BcsZ is mesophilic enzyme which is in line with the other endoglucanse in some microorganisms such as *Bacillus sp.* HSH-810 [20], *Komagataeibacter xylinus* [21].

Metal ions significantly affected the enzymatic catalytic activity based on the above results. Endoglucanase of GH8 family has two conserved glutamate residues at the active site and the cellulase activity might be restrained when metal ions is bound to the radical of these residuals [22]. Mg^{2+} might enhance cellulase activity through altering the dimensional structure of BcsZ or stabilizing the enzyme structural conformation. Co^{2+}, Mn^{2+}, and

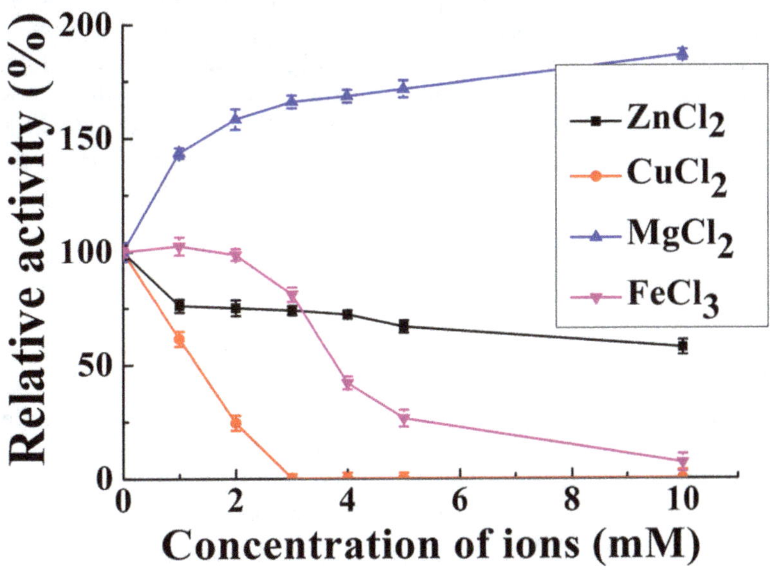

Fig. 6 The effect of metal ions on the activity of BcsZ

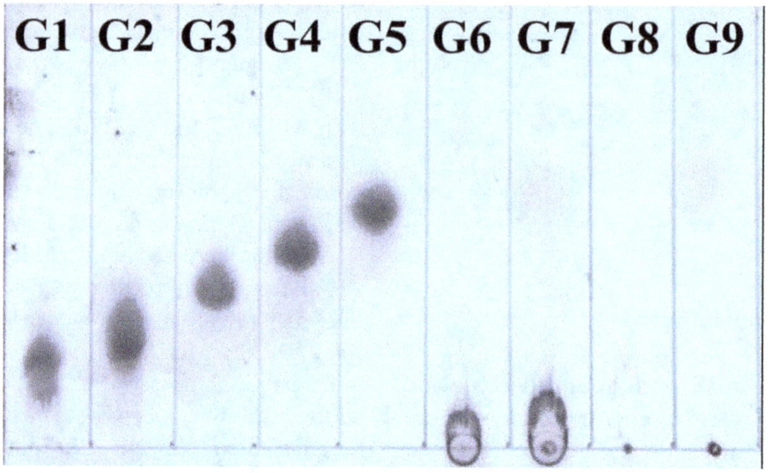

Fig. 7 TLC analysis of hydrolysates of CMC and RAC catalyzed by BcsZ. Standards of Cellopentose (G1), Cellotetraose (G2), Cellotriose (G3), Cellobiose (G4) and Glucose (G5); CMC incubated with inactivated BcsZ as the control (G6), CMC incubated with BcsZ (G7); RAC incubated with inactivated BcsZ as the control (G8), RAC incubated with BcsZ (G9)

Fe^{2+} enhanced the enzyme activity of GHF9 endoglucanase from *Reticulitermes speratus*, but Pb^{2+} and Cu^{2+} inhibited its enzyme activity [23]. Meleiro et al cloned and overexpressed an endoglucanase (Egst) from *Scytalidium thermophilum* that showed high catalytic activity at harsh condition. NH^{4+}, Na^{+}, K^{+}, $Ba^{2}+$, Ca^{2+} and Mg^{2+} had little effect on Egst. Cu^{2+} presented a slight inhibitory effect and Hg^{2+} had a strong inhibitory effect on Egst [24].

BcsZ could hydrolyze CMC-Na, RAC and glucan. BcsZ had a higher affinity for CMC than Thcel9A (12.02 mg/mL) from *Thermobifida halotolerans* YIM 90462 T [25], the purified endoglucanase (21.01 mg/mL) from *Aspergillus niger B03* [26] and C67–1 (37 mg/mL) from metagenomes of buffalo rumens [27]. It had a weaker affinity for CMC than the endoglucanase from *Bacillus sp.* (0.8 mg/mL) [28]. The V_{max} value (0.3 μM/min·mg) of BcsZ was lower than Cell-1 (0.84 μmol/min·mg) [29] and an endoglucanse (1000 μM/min) [28]. The V_{max} value of BcsZ was low indicating this enzyme had a weak catalytic efficiency for CMC.

The BcsZ from *E. coli* ZH-4 could be classified as an endo-β-1, 4 glucanase according to its product pattern. Some other endo-β-1, 4 glucanases also have the above properties. An endoglucanase was expressed and characterized from *Serratia proteamaculans* CDBB-1961 by Cano-Ramírez, and then was applied to hydrolyze CMC to glucose and cello oligosaccharides [30]. The endo-β-1, 4 glucanase (E4–90) from *Termomonospora fusca* hydrolyzed CMC to produce cellobiose, cellotriose, cellotetraose and glucose [31]. As the previous study, a purified β-1, 4 endoglucanase (LbGH5) could hydrolyze CMC and phosphoric acid swollen cellulose (PASC) with the enzymatic hydrolysate including cellobiose, cellotriose, cellotetraose and a small amount of cellopentaose and glucose [32]. In contrast, a thermophilic endo-1, 4-β-glucanase from *Sulfolobus shibatae* hydrolyzed cellotetraose and cellopentose not cellobiose or cellotriose. The products of CMC hydrolysis were cellobiose, cellotriose, cellotetraose and cellopentose [33]. Cellobiose and cellotriose was produced from PASC and Avicel respectively, and cellobiose, cellotriose and cellotetraose were produced from CMC with Endo-glucanase EG5C-1 [34]. The release of significant amount of glucose also explained why *E.coli* ZH-4 can grow on cellulose. Our experiments also showed that expression of BcsZ in *E.coli* BL21 resulted in the growth of *E.coli* BL21 on cellulose. These identification and characterization of BcsZ may improve understanding the adaptation of *E. coli* to different environment.

Conclusions

BcsZ was shown to be responsible for extracellular endoglucanase activity and cellulolytic capability of *E. coli* ZH-4. Expression, purification and characterization of the endoglucanase BcsZ showed that it was thermotolerant and pH tolerant. This enzyme could hydrolyze CMC to produce glucose, cellobiose, cellotriose and unknown cello oligosaccharides, which explained why *E. coli* ZH-4 can grow on cellulose.

Methods

Bacterial strains and culture conditions

The cellulolytic *E. coli* ZH-4 was isolated from the bovine rumen by our laboratory and it was preserved in China General Microbiological Culture Collection Center (CGMCC) (Preservation No. 12427). *E. coli* DH5α was used for plasmid amplification, and *E. coli* BL21 (DE3) was used as the host for recombinant protein expression with pET28a as vector. *E. coli* strains were routinely cultured in Luria–Bertani (LB) medium at 37 °C with shaking at 220 rpm. Kanamycin was added when needed at a final concentration of 50 μg/mL.

Quantitative RT-PCR analysis

Total RNA was extracted using the RNAprep pure Cell/ Bacteria Kit (TIANGEN) following manufacturer's instruction. The concentration and quality of RNA was measured using the Nano-Drop spectrophotometry (NanoDrop Technologies, Wilmington, DE, USA). The cDNA was synthesized according to the First Stand cDNA Synthesis Kit (TOYOBO) following the manufacturer's protocol. The primers were displayed in Table 2. The PCR reaction included 12.5 uL SYBR Green Realtime PCR Master Mix (TOYOBO), 2.5 uL diluted cDNA (500 mM) reaction mixture, 1uL each forward and reverse primer (10 umol) and 8.0 uL ddH_2O. The RT-PCR assays were performed on a 7900HT Fast Real-Time PCR System (Applied Biosystems, Carlsbad, California, USA). The relative gene expression calculated by the equation of $2^{(-\Delta\Delta Ct)}$ method (Applied Biosystems Research Bulletin No. 2 P/N 4303859).

Western blot analysis

Western blotting was performed as described by Sambrook [35]. Proteins separated by SDS-PAGE were transferred onto polyvinylidene difluoride (PVDF) membranes (Millipore, Billerica, MA, USA) using the Bio-Rad semi-dry apparatus. The blots were incubated with primary anti-BcsZ serum in 1: 2000 dilution in 2% (wt/vol) skimmed milk for 1 h with agitation at room temperature followed by washes with PBST buffer (PBS buffer with 1% Tween) for 3 times. The secondary antibody with horseradish peroxidase (HRP) conjugated goat anti-rabbit (1:2000 dilution) was performed as above. Blots were developed with the ECL Plus Kit (Thermo Scientific, US) following the manufacturer's directions. The in-gel identification of secreted protein by mass spectrometry analysis was performed by Shenzhen BGI gene co., LTD.

Sequence analysis

bcsZ gene sequences from *E. coli* ZH-4 and *E. coli* MG1655 was compared using the BLASTN program (https://blast.ncbi.nlm.nih.gov/Blast.cgi).

Plasmid construction

To overexpress BcsZ, *bscZ* gene was amplified from *E. coli* ZH-4 using primers BcsZ-F (5′- CGCGGATCCG GGTGTGAATTTGCGCATTCCT-3′) and BcsZ-R (5′- CATGCCATGGGCAATGTGTTGCGTAGTGGAAT-3′). The PCR products were then digested and cloned into *Bam*HI and *Nco*I sites of pET-28a plasmid. His-tag was located in C-terminus of *bcsZ*. The constructed expression vector pET-28a-BcsZ was verified by DNA sequencing.

Table 2 The primers of qRT-PCR

Primers	Sequences (5′-3′)
bcsZ-F	GAGAACAGTAAGTGGGAAGTGC
bcsZ-R	AACGCTGCTCTTTCCACAAACG
16S rRNA-F	GCTCAACCTGGGAACTGC
16S rRNA-R	CCACGCTTTCGCACCTGA

Protein expression and purification

Overnight culture of *E. coli* BL21 (DE3) harboring plasmid pET-28a-BcsZ was inoculated to fresh LB medium supplemented with 50 μg/mL kanamycin in 1:100 dilution and cultured at 37 °C with shaking. The *E. coli* BL21 (DE3) carrying the empty plasmid was used as control. To induce *bscZ* gene expression, isopropyl-β- d-thiogalactoside (IPTG) was added to the culture at a final concentration of 0.25 mM when the optical density at 600 nm reached 0.4~0.6. Cells were harvested by centrifugation (8000×g, 10 min) after 24 h cultivation at 25 °C with shaking at 220 rpm. The resulting cells were suspended in 10 mM phosphate buffer saline (PBS) with PMSF and DNase I. The cells were lysed using JNBIO JN-3000 PLUS high-pressure cell press. Then, the crude cell lysate was prepared by centrifugation (12,000×g, 30 min) to remove the cell debris.

The BcsZ was further purified by affinity chromatography using HisPur Cobalt Resin (Thermo Fisher Scientific Inc) according to the manufacturers' instruction. The purity and homogeneity of the purified enzyme was detected by SDS–PAGE.

Enzyme assay and protein determination

The enzyme activity of BcsZ was determined by the standard DNS method [36]. The mixture of 1.5 mL 1% (w/v) CMC solubilized in 50 mM phosphate buffer (pH 6.0) and 0.5 mL purified BcsZ was incubated at 50 °C for 30 min. The the reaction was stopped by the addition of 3 mL of DNS. The release of reducing sugars was determined by the absorbance at 540 nm. One unit of activity is defined as the amount of enzyme that released 1 μmol of reducing sugars/min from CMC. Protein concentration was measured by Bradford method using bovine serum albumin as standard [37].

Basic biochemical character of BcsZ

The optimum pH was determined by measuring BcsZ activity in different buffers (50 mM) of pH ranging from 4.0 to 10.0: citric acid buffer for pH 4.0–6.0; phosphate buffer for pH 6.0–8.0; Glycine-NaOH buffer for pH 8.5–10.0. The optimum temperature was identified by incubating the enzyme in phosphate buffer (pH 6.0) at different temperatures (30–80 °C). For thermostability determination of BcsZ, it was incubated at different temperatures 4, 40, 50, 60 and 70 °C for 0.5, 1, 2, 3, 4, 5, 10 and 24 h. The residual endoglucanase activity was determined, respectively. The influence of metal ions on enzyme activity of BcsZ was measured in presence of tested metal ions at indicated concentration. All enzyme activity was measured as previously described. The highest enzymatic activity was used as benchmark of 100% activity when calculating relative activity.

Kinetic constants

The kinetic parameter values Michaelis-Menten constants (K_m) and maximum velocity (V_{max}) were determined to calculate the K_m value of BcsZ on CMC hydrolysis according to double-reciprocal Lineweaver–Burk plots (Eq. 1). The activity assay was performed with CMC at different concentrations at pH 6.0 and 50 °C for 10 min.

$$\frac{1}{V} = \frac{K_m}{V_{max}} \frac{1}{[S]} + \frac{1}{V_{max}} \tag{1}$$

Substrate specificity

The hydrolytic ability of BcsZ on 1% (w/v) of various substrates were determined by DNS method under optimal conditions. The substrate included Avicel, CMC-Na, regenerated amorphous cellulose (RAC), xylan, cellobiose, laminarin and chitin.

Thin layer chromatographic (TLC)

The products were analyzed by Thin Layer Chromatographic (TLC) as previously described when BcsZ hydrolyzed CMC and RAC [38].

Statistical analysis

All experiments were conducted in triplicate and the data were presented as mean values ± standard deviation.

Abbreviations

CGMCC: China General Microbiological Culture Collection Center; CMC: carboxymethylcellulose; DNS: dinitrosalicylic; GH: Glycoside hydrolase; HRP: horseradish peroxidase; IPTG: isopropyl-β- d-thiogalactoside; LB: Luria–Bertani; PBS: phosphate buffer saline; PVDF: polyvinylidene difluoride; RAC: Regenerated amorphous cellulose; SDS-PAGE: Sodium dodecyl sulfate-polyacrylamide gel electrophoresis; TLC: Thin layer chromatographic

Acknowledgments

Not applicable.

Authors' contributions

JP did the experimental work and wrote the draft manuscript. JSW revised and designed the draft manuscript. ZYL participated in the design of the study. QCZ took part in checking the results. QSQ participated in the design of the study and commented on the manuscript. All authors read and approved the final manuscript.

Funding

This work was supported by the National Natural Science Foundation of China [Grant number 31730003]; the National Natural Science Foundation of China (NSFC) [Grant number 61361016]; the Foundation of Talent Development of Inner Mongolia and the "Prairie talent" project of Inner Mongolia [Grant number CYYC20130034]; the National Natural Science Foundation of Inner Mongolia [Grant number 2018MS02019]; the Science and Technology Planning Project of Inner Mongolia; and the Graduate student education innovation planning project of Inner Mongolia [Grant number B2018111925]. These funding bodies had no role in the design of the study and collection, analysis, and interpretation of data and in writing the manuscript.

Ethics approval and consent to participate

Not applicable.

Consent for publication

Not applicable.

Competing interests

The authors declare that they have no competing interests.

Author details

[1]School of Chemical Engineering, Inner Mongolia University of Technology, Hohhot 010051, Inner Mongolia, China. [2]State Key Laboratory of Microbial Technology, Shandong University, Qingdao 266237, China. [3]Inner Mongolia Energy Conservation and Emission Reduction Engineering Research Center in Fermentation Industry, Hohhot 010051, China.

References

1. Ecem Oner B, Akyol C, Bozan M, Ince O, Aydin S, Ince B. Bioaugmentation with *Clostridium thermocellum* to enhance the anaerobic biodegradation of lignocellulosic agricultural residues. Bioresour Technol. 2018;249:620–5.
2. Zeng R, Hu Q, Yin XY, Huang H, Yan JB, Gong ZW, Yang ZH. Cloning a novel endo-1,4-β- d -glucanase gene from *Trichoderma virens* and heterologous expression in *E. coli*. AMB Express. 2016;6:108.
3. Seneesrisakul K, Guralp SA, Gulari E, Chavadej S. *Escherichia coli* expressing endoglucanase gene from Thai higher termite bacteria for enzymatic and microbial hydrolysis of cellulosic materials. Electron J Biotechnol. 2017;27:70–9.
4. Yennamalli RM, Rader AJ, Wolt JD, Sen TZ. Thermostability in endoglucanases is fold-specific. BMC Struct Biol. 2011;11:10.
5. Sadhu S, Maiti TK. Cellulase production by bacteria: a review. British Microbiol Res J. 2013;3:235.
6. Maki M, Leung KT, Qin W. The prospects of cellulase-producing bacteria for the bioconversion of lignocellulosic biomass. Int J Biol Sci. 2009;5:500.
7. Gao D, Wang S, Li H, Yu H, Qi Q. Identification of a heterologous cellulase and its N-terminus that can guide recombinant proteins out of *Escherichia coli*. Microb Cell Factories. 2015;14:49.
8. Gao D, Luan Y, Wang Q, Liang Q, Qi Q. Construction of cellulose-utilizing *Escherichia coli* based on a secretable cellulase. Microb Cell Factories. 2015;14:159.
9. Pang J, Liu ZY, Hao M, Zhang YF, Qi QS. An isolated cellulolytic *Escherichia coli* from bovine rumen produces ethanol and hydrogen from corn straw. Biotechnol Biofuels. 2017;10:165.
10. Ahmad I, Rouf SF, Sun L, Cimdins A, Shafeeq S, Le Guyon S, et al. BcsZ inhibits biofilm phenotypes and promotes virulence by blocking cellulose production in *Salmonella enterica serovar Typhimurium*. Microb Cell Factories. 2016;15:177.
11. Bokranz W, Wang X, Tschape H, Romling U. Expression of cellulose and curli fimbriae by *Escherichia coli* isolated from the gastrointestinal tract. J Med Microbiol. 2005;54:1171–82.
12. Grantcharova N, Peters V, Monteiro C, Zakikhany K, Romling U. Bistable expression of CsgD in biofilm development of *Salmonella enterica serovar typhimurium*. J Bacteriol. 2010;192:456–66.
13. Mazur O, Zimmer J. Apo- and cellopentaose-bound structures of the bacterial cellulose synthase subunit BcsZ. J Biol Chem. 2011;286:17601–6.
14. Standal R, Iversen TG, Coucheron DH, Fjaervik E, Blatny JM, Valla S. A new gene required for cellulose production and a gene encoding cellulolytic activity in *Acetobacter xylinum* are colocalized with the bcs operon. J Bacteriol. 1994;176:665–72.
15. Rabinovich ML, Melnick MS, Bolobova AV. The structure and mechanism of action of cellulolytic enzymes. Biochem Mosc. 2002;67:850–71.
16. Park Y, Yun H. Cloning of the *Escherichia coli* endo-1,4-D-glucanase gene and identification of its product. Mol Gen Genet. 1999;261:236–41.

17. Lin L, Fu C, Huang W. Improving the activity of the endoglucanase, Cel8M from *Escherichia coli* by error-prone PCR. Enzym Microb Technol. 2016;86:52–8.
18. Gong X, Gruninger RJ, Qi M, Paterson L, Forster RJ, Teather RM, et al. Cloning and identification of novel hydrolase genes from a dairy cow rumen metagenomic library and characterization of a cellulase gene. BMC Res Notes. 2012;5:566.
19. Chang L, Ding M, Bao L, Chen Y, Zhou J, Lu H. Characterization of a bifunctional xylanase/endoglucanase from yak rumen microorganisms. Appl Microbiol Biotechnol. 2011;90:1933–42.
20. Kim JY, Hur SH, Hong JH. Purification and characterization of an alkaline cellulase from a newly isolated alkalophilic *Bacillus sp.* HSH-810. Biotechnol Lett. 2005;27:313–6.
21. KOO HM. Expression and characterization of CMCax having β-1, 4-endoglucanase activity from *Acetobacter xylinum*. BMB Rep. 1998;31:53–7.
22. Demain AL, Newcomb M, Wu JH. Cellulase, clostridia, and ethanol. Microbiol Mol Biol Rev. 2005;69:124–54.
23. Zhang P, Yuan X, Du Y, Li JJ. Heterologous expression and biochemical characterization of a GHF9 endoglucanase from the termite *Reticulitermes speratus* in Pichia pastoris. BMC Biotechnol. 2018;18:35.
24. Meleiro LP, Carli S, Fonseca-Maldonado R, da Silva TM, Zimbardi A, Ward RJ, et al. Overexpression of a Cellobiose-glucose-halotolerant endoglucanase from *Scytalidium thermophilum*. Appl Biochem Biotechnol. 2018;185:316–33.
25. Zhang F, Chen JJ, Ren WZ, Nie GX, Ming H, Tang SK, Li WJ. Cloning, expression and characterization of an alkaline thermostable GH9 endoglucanase from *Thermobifida halotolerans* YIM 90462 T. Bioresour Technol. 2011;102:10143–6.
26. Dobrev GT, Zhekova BY. Biosynthesis, purification and characterization of endoglucanase from a xylanase producing strain *Aspergillus niger* B03. Braz J Microbiol. 2012;43:70–7.
27. Duan C, Xian L, Zhao GC, Feng Y, Pang PH, Bai XL, et al. Isolation and partial characterization of novel genes encoding acidic cellulases from metagenomes of buffalo rumens. J Appl Microbiol. 2009;107:246–56.
28. Sriariyanun M, Tantayotai P, Yasurin P, Pornwongthong P, Cheenkachorn K. Production, purification and characterization of an ionic liquid tolerant cellulase from *Bacillus sp.* isolated from rice paddy field soil. Electron J Biotechnol. 2016;19:23–8.
29. Zhou X, Kovaleva ES, Wu-Scharf D, Campbell JH, Buchman GW, Boucias DG, et al. Production and characterization of a recombinant beta-1,4-endoglucanase (glycohydrolase family 9) from the termite *Reticulitermes flavipes*. Arch Insect Biochem Physiol. 2010;74:147–62.
30. Cano-Ramirez C, Santiago-Hernandez A, Rivera-Orduna FN, Garcia-Huante Y, Zuniga G, Hidalgo-Lara ME. Expression, purification and characterization of an endoglucanase from *Serratia proteamaculans* CDBB-1961, isolated from the gut of Dendroctonus adjunctus (Coleoptera: Scolytinae). AMB Express. 2016;6:63.
31. Irwin D, Shin D-H, Zhang S, Barr BK, Sakon J, Karplus PA, et al. Roles of the catalytic domain and two cellulose binding domains of *Thermomonospora fusca* E4 in cellulose hydrolysis. J Bacteriol. 1998;180:1709–14.
32. Zhang F, Anasontzis GE, Labourel A, Champion C, Haon M, Kêmppainen M, et al. The ectomycorrhizal basidiomycete Laccaria bicolor releases a secreted beta-1,4 endoglucanase that plays a key role in symbiosis development. New Phytol. 2018;220:1309–21.
33. Boyce A, Walsh G. Expression and characterisation of a thermophilic endo-1,4-beta-glucanase from *Sulfolobus shibatae* of potential industrial application. Mol Biol Rep. 2018;45:2201–11.
34. Wu B, Zheng S, Pedroso MM, Guddat LW, Chang S, He B, et al. Processivity and enzymatic mechanism of a multifunctional family 5 endoglucanase from *Bacillus subtilis* BS-5 with potential applications in the saccharification of cellulosic substrates. Biotechnol Biofuels. 2018;11:20.
35. Sambrook J, Fritsch EF, Maniatis T. Molecular cloning: a laboratory manual. CSH. 1989.
36. Miller GL. Use of dinitrosalicylic acid reagent for determination of reducing sugar. Anal Chem. 1959;31:426–8.
37. Bradford MM. A rapid and sensitive method for the quantitation of microgram quantities of protein utilizing the principle of protein-dye binding. Anal Biochem. 1976;72:248–54.
38. Zhang C, Wang Y, Li Z, Zhou X, Zhang W, Zhao Y, et al. Characterization of a multi-function processive endoglucanase CHU_2103 from *Cytophaga hutchinsonii*. Appl Microbiol Biotechnol. 2014;98:6679–87.

6

Syneresis investigations of lacto-fermented sodium caseinate in a mixed model system

Soumaya El Bouchikhi[1,2], Philippe Pagès[3], Yassir El Alaoui[1], Azeddine Ibrahimi[2] and Yahya Bensouda[1*]

Abstract

Background: The textural characteristics of fermented dairy products are important quality parameters that play a major role in their stability and consumer's acceptance. The aim of this study was to investigate the influence of sodium caseinate, starch, lactose and lactic acid bacteria as ferment on the syneresis in a mixed model system, and to evaluate their impact on the acid gel formation throughout pH and zeta potential monitoring. Accordingly, a protocol was designed to perform an experimental design by using a mixture of the selected factors.

Results: A significant decrease of syneresis was detected in all mixtures at 8% of sodium caseinate, ranging between a minimum of 1.8% and a maximum of 20.6% compared to the mixtures at 3% of sodium caseinate in which the syneresis decrease had ranged between a minimum of 22.2% and a maximum of 47.8%. The addition of starch had a significant impact on the acidification profile and on the syneresis of the fermented mixed model. Moreover, the monitoring of pH and zeta potential during the lacto-fermentation process has also led to a better understanding of the acid gelation and the syneresis variations.

Conclusion: Syneresis varies very closely with sodium caseinate concentration, starch concentration and also with their association, regardless of the concentrations of lactose and ferment. In fact syneresis could be reduced to an optimum level if a sodium caseinate-starch mixed system is employed: Less syneresis gels could be obtained at a sodium caseinate concentration above 5% if starch is used above 1%.

Keywords: Syneresis, Sodium caseinate, Lacto-fermentation, Starch, Acid gel, Non-dairy product

Background

Fermented food products have a great nutritional value and beneficial effect on host gut micro-biome due to their probiotics bacteria, which have a positive impact on human health [1]. In fact Gut micro-biome can directly influence host wellbeing by providing nutritional, metabolic and immunological benefits [2]. In the market, there is a growth in the development of fermented functional dairy products reflecting great interest in studying dairy products that demonstrate health benefits. This market expansion is highly attributed to the food technologist's research, in particular, the development of new ingredients as flavors, fat substitutes, and protein ingredients to be added in the product in order to improve different qualitative and technical aspects. However more innovative efforts are needed to develop new internal matrix from a variety of substrates, and evaluate its fermentation. On the basis of recent developments, it is anticipated that fermented functional products will continue to be a significant component within the functional food market [3].

Milk, is a natural source of human nutrition and functional food materials that serves as the raw material for the production of dairy products, but also used as foodstuffs in their own right and as ingredients in the manufacture of formulated and processed foods: proteins, lactose, water-soluble vitamins and minerals. Milk proteins can be divided into two species: casein and whey proteins. The beneficial health properties resulting from the consumption of fermented milks have been known for many years. Fermented dairy products are usually made from standardized milk with fat content of 0–3.5% and non fat milk solid content of 11–13% through acidification with lactic acid bacteria. Among the different process parameters such as heat treatment, fermentation, storage conditions and starter culture, milk base

* Correspondence: yahya.bensouda@um5.ac.ma
[1]Laboratory of Pharmaceutics, Faculty of Medicine and Pharmacy, Mohammed V University, Rabat, Morocco
Full list of author information is available at the end of the article

composition has a big impact on the gel structure formation.

To improve and optimize the properties and the nutritional qualities of fermented dairy and non-dairy products, a variety of ingredients are used on both the process and the composition. It was also reported that casein-based products tended to produce firmer gels with less syneresis than yogurts fortified with whey protein [4]. Damin et al. (2009) showed that yogurt made with milk supplemented with sodium caseinate resulted in significant increase in firmness. The increase in milk solids may be achieved by the addition of skimmed milk powder or through the concentration of a milk portion by ultrafiltration. In a previous study the influence of milk protein-based ingredients on the textural characteristics, the sensory properties, and the microstructure of probiotic yogurt during a refrigerated storage period of 28 days was studied; and it was reported that probiotic yogurt from milk fortified with sodium caseinate showed much higher values of firmness and viscosities than yogurt from skimmed milk powder or whey protein concentrate fortified milk [5]. Heat treatment also promotes chemical changes, which improve the milk solids as a fermentation substrate by inducing whey protein-casein interaction through β-lacto-globulin-κ-casein interactions, which improve the texture of the product and reduce syneresis [6]. Conventionally, syneresis is considered as a major defect in products that are based on acid-induced gels [7]. Consumers perceive the separation of whey (syneresis) from yogurt during shelf life as a defect, which is a major subject of concern in the dairy industry [8]. Therefore, The textural characteristics of fermented dairy products are considered as important quality parameters that play an important role in their stability and their consumer's acceptance. Moreover, additives such as pectin, guar gum, cornstarch and gelatin may be used to stabilize the product against syneresis, which also contributes to texture and viscosity.

The matrix or the gel formation is a key property in the acid induced gel formulation, in which casein plays a major role due to its structure properties [5]. It is also known that these gels consist of a particulate network of casein particles linked together to trap water and hold it in place [9, 10]. The main factors governing the formation of acid casein gels are casein concentration, pH, temperature and ionic strength [11, 12]. Before acidification the individual molecules of casein interact with each other and form associate structures called micelles, which are stabilized by electrostatic and steric interactions [4]. During fermentation, lactic acid bacteria use lactose to produce lactic acid that neutralizes gradually the negative charges on the surface of the casein micelles. This causes a decrease of the repulsive interactions and the onset of big aggregates, leading to a destabilization of the dispersion as soon as the pH drops below 5 [13]. The reduction in the negative charge on caseins occurs with the approach of the isoelectric point, which facilitates aggregation via electrostatic attraction between oppositely charged segments on the caseins as well as by enhanced hydrophobic interactions [14]. Electrostatic repulsion on the caseins decreases with the approach of the isoelectric point; in unheated milk this causes a collapse of the k-casein hairs (brush) onto the micelles surface (in heated milk the k-casein hairs are associated with denatured whey proteins and so do not collapse onto the micelle surface). The importance of both electrostatic and hydrophobic interactions for acid gelation was elegantly demonstrated by the cold acidification method developed by Roefs and van Vliet (1990) [15, 16]. At the isoelectric point of casein (pH 4.65), micelles lose completely their structure because of the whole dissolution of calcium's micelles and the entire denaturation of casein, forming an acid gel protein-network [17, 18]. It was also reported that starch addition increases the value of the rheological and physical properties of yoghurts by improving consistency, increasing viscosity and reducing syneresis [19], which contributes to the formation of more stable dispersed acidified milk gelled systems [20, 21]. The nature of interaction between milk proteins and polysaccharides used as stabilizers depends on many factors including the specific types of intermolecular forces between them, their concentrations and environmental factors such as pH, ionic strength and calcium content [11]. For exemple the interactions of modified starches and casein included electrostatic adhesion, steric stabilization and hydrogen bond [22]. Therefore, a comprehensive study integrating convenient combinations of the main factors involved in the acid gel formation is essential for a closer understanding of syneresis mechanism through the proposed mixed model system.

The purpose of this study was to use sodium-caseinate as the main substrate to create physical building blocks within the continuous phase that may substitute the functionality of the removed milk ingredients and to produce an acid gel-network capable of holding the serum phase with an optimal stability. Accordingly, an experimental design was proposed by using a mixture design of the following selected factors: sodium caseinate, starch, lactose and lactic acid bacteria to determine the effect of these formulations on the syneresis of the formed acid gel protein network. The effects of these formulation variables on syneresis were evaluated through pH and zeta potential monitoring. Fermentation was performed with lactic acid bacteria *Lactobacillus delbrueckii* ssp. *Bulgaricus* combined to *streptococcus thermophilus* [23, 24]. This work provided more insight on how to reduce syneresis in fermented non-milk based products, which could be a promising venue in their development.

Conclusion

This study demonstrated that sodium caseinate considerably decreased syneresis in the fermented model system compared to starch while ferment and lactose did not affect significantly syneresis. Moreover, the effect of starch on the acidification profile was revealed as an important element to consider in the understanding of the impact of reduced fermentation time on syneresis. It could be concluded from our trials that the optimum syneresis was achieved when the model bases were sodium caseinate and starch. Based on our results, we also showed that syneresis in the proposed model gel system could be reduced to an optimum level if a sodium caseinate-starch mixed system was employed: Less syneresis gels could be obtained at a sodium caseinate concentration above 5% if starch was used above 1%. Therefore, it will be worthwhile to make an exhaustive study integrating the most convenient combination of the selected factors and different steps of yogurt like manufacturing.

Materials and methods

Ingredients

The main ingredients used in this study were: Sodium caseinate (provided by Trade Bio-Industries Morocco); Native starch (Maïzena trade name, purchased from the local market); Lactose (LOBAL Chemie Laboratory) and Lactic acid bacteria: commercial yogurt starter culture freeze-dried containing *Strep. Thermophilus* and *Lactobacillus delbrueckii ssp. Bulgaricus* (provided by Trade Bio-Industries Morocco). Freshly distilled water was used for preparation of all samples.

Equipment

The main equipment used in this study was: Water bath set at 42 °C (Memmert WNE14); Fine test sieve (mesh width 0.25 mm); pH meter (Inlab sensor dairy Electrode. Mettler Toledo); Nanosizer ZS analyzer (Malvern version 6.2 serial number MAL1043134) and Design Expert 10 software edited by Stat-Ease inc.

Syneresis monitoring

The ranges of the studied factors proposed for the determination of functional design space, showed in Table 1, were determined according to the concentrations range of sodium caseinate, starch, lactose and lactic acid bacteria used in fermented dairy products manufacturers [17] and designated as factors A, B, C and D respectively.

Table 1 Lower and upper limits of components used to make the experimental design

Factors	Lower limit	Upper limit
Sodium caseinate (A)	3%	8%
Starch (B)	0%	2%
Lactose (C)	4%	9%
Ferment (D)	0.2 g/l	0.5 g/l

Main experimental design

To define the formulation space for sodium-caseinate mixtures, a Complete Factorial Plan performed with Design Expert was selected to evaluate and to model the effects of sodium caseinate, Lactose, ferment and starch on decreasing the syneresis of the fermented product. This provides maximum information from a limited number of experiments as indicated in Table 2.

Preparation of samples

Each of the nineteen formulas has been prepared as follows: Lactose was dissolved in pure distilled water; sodium-caseinate was added slightly under agitation at 250 rpm, and all heated in a bath at 70 °C to obtain a homogeneous solution. When starch was required, it was added once the glass beaker was removed from the water bath and cooled at the ambient temperature. The final mixture was water bath heated at 82–85 °C for 30 min to pasteurize and to allow hydration and swelling of starch granules. Samples were then cooled to 42 °C, which is the incubation temperature that is recommended for the starter culture, and inoculated with different doses of ferments. All these mixtures were poured into glass-jars and placed in a water-bath heated at 42 °C for 5–6 h. All samples were then stored at 4–5 °C for 24 h before proceeding to the evaluation of their syneresis. The analysis was carried out for 24 h. The volumes of the whey released by the fermented samples were measured by inverting each sample on fine test sieve (mesh width 0.25 mm) that was placed on top of a beaker. The volume of whey drained was collected in a graduated cylinder for each sampling time within 24 h weighed and recorded as index of syneresis (%) [25, 26].

Experimental design set to a fixed dry matter

In this experiment, the design was adapted to a fixed dry matter. The concentration of the ferment was fixed to 0.35 g/l and the dry matter content of all samples was set at 12% with respect to the same proportions used in the experimental design 1 as showed in Table 3. The syneresis was measured as indicated bellow.

pH and potential zeta monitoring

The purpose of this experiment was to evaluate the impact of starch and lactose on the acidification profile of

Table 2 The main experimental design for the selected factors. The completed plan is a "Complete Factorial Plan" with 4 Factors in 2 levels and 16 trials plus 3 points in the center. This plan made it possible to calculate the following and perform a Curvature test: The 4 Main Effects, the 6 Interactions of order 2, the 5 Interactions of order 3, the only Interaction of order 4

Sample	Space type	Sodium caseinate %(A)	Starch %(B)	Lactose %(C)	Ferment lg/l(D)	Dry matter %
1	Factorial	3	0	4	0.2	7
2	Factorial	3	0	4	0.5	7
3	Factorial	3	0	9	0.2	12
4	Factorial	3	0	9	0.5	12
5	Factorial	3	2	4	0.2	9
6	Factorial	3	2	4	0.5	9
7	Factorial	3	2	9	0.2	14
8	Factorial	3	2	9	0.5	14
9	Factorial	8	0	4	0.2	12
10	Factorial	8	0	4	0.5	12
11	Factorial	8	0	9	0.2	17
12	Factorial	8	0	9	0.5	17
13	Factorial	8	2	4	0.2	14
14	Factorial	8	2	4	0.5	14
15	Factorial	8	2	9	0.2	19
16	Factorial	8	2	9	0.5	19
17	Center	5.5	1	6.5	0.35	13
18	Center	5.5	1	6.5	0.35	13
19	Center	5.5	1	6.5	0.35	13

the fermented mixed model system. According to the syneresis results of the second experimental design (Table 3), four samples were designed to be pH monitored and syneresis measured as presented in Table 4. Sodium caseinate was set at 5% and ferment at 0.35 g/l in all sample. The pH was continuously measured along the fermentation process. by immersing the glass electrode of the pH meter on samples.

The selection of samples for zeta potential monitoring was based on the variation of the acidification profile observed in the four samples during their pH monitoring. The Samples that exhibited the shortest (sample i) and the longest (sample iv) time of gelation were selected. Zeta potential measurements were determined on dilute dispersions of each sample, the type of diluent greatly influenced the obtained results. Samples were diluted 100-fold with ultra pure distilled water before conducting the measurement while dilution with buffer solutions was unsuccessful.

Data analysis

All statistical analyses were performed using the Design Expert version 10, a statistical software package from Stat-Ease Inc. that is specifically dedicated to performing

Table 3 Experimental design with a fixed rates of ferment and dry matter

Corresponding sample in Table 2	Sample	Sodium caseinate %(A)	Starch %(B)	Lactose %(C)	Ferment lg/l(D)	%Dry matter
1–2	I	5.142	0	6.857	0.35	12%
3–4	II	3	0	9	0.35	12%
5–6	III	4	2.66	5.33	0.35	12%
7–8	IV	2.57	1.714	7.714	0.35	12%
9–10	V	8	0	4	0.35	12%
11–12	VI	5.647	0	6.35	0.35	12%
13–14	VII	6.857	1.714	3.428	0.35	12%
15–16	VIII	5.05	1.26	5.68	0.35	12%

Table 4 Formulas for pH and Zeta Monitoring

Sample	Sodium caseinate %	Starch %	Lactose %	Ferment g/l
i	5	2	9	0.35
ii	5	0	9	0.35
iii	5	2	4	0.35
iv	5	0	4	0.35

design of experiments (DOE). In order to reduce scattering effects and to compare the samples, all physical results had been normalised. The Residual analysis, the coefficient of determination (adjusted R^2), the significance of the models and the lack of fit were used to check the quality of the model. The robustness of the models was evaluated by determining the squared correlation coefficient (R^2) for predicted versus measured values in cross-validation and the ratio of standard deviation (SD) to root mean square error of calibration

Table 5 Syneresis results

Sample	Sodium Caseinate %(A)	Starch %(B)	Lactose %(C)	Ferment %(D)	Syneresis %	Dry Matter %
Influence of the selected factors on syneresis						
1	3	0	4	0.2	47.8	7
2	3	0	4	0.5	35.6	7
3	3	0	9	0.2	41.4	12
4	3	0	9	0.5	41	12
5	3	2	4	0.2	31.4	9
6	3	2	4	0.5	25.4	9
7	3	2	9	0.2	21.2	14
8	3	2	9	0.5	15.8	14
9	8	0	4	0.2	20.6	12
10	8	0	4	0.5	19	12
11	8	0	9	0.2	17.6	17
12	8	0	9	0.5	17.2	17
13	8	2	4	0.2	10.2	14
14	8	2	4	0.5	8.4	14
15	8	2	9	0.2	9.2	19
16	8	2	9	0.5	1.8	19
17	5.5	1	6.5	0.35	27.9	13
18	5.5	1	6.5	0.35	27.8	13
19	5.5	1	6.5	0.35	27.4	13
Syneresis after fixed rates of ferment and dry matter						
I	5.142	0	6.857	0.35	33	12
II	3	0	9	0.35	63	12
III	4	2.66	5.33	0.35	24	12
IV	2.57	1.714	7.714	0.35	48	12
V	8	0	4	0.35	5	12
VI	5.647	0	6.35	0.355	36	12
VII	6.857	1.714	3.428	0.35	12	12
VIII	5.05	1.26	5.68	0.35	29	12
Syneresis on formulas of pH and zeta potential monitoring						
i	5	2	9	0.35	28	16
ii	5	0	9	0.35	46	14
iii	5	2	4	0.35	20	11
iv	5	0	4	0.35	45	9

(RMSEC) of the data sets. The ratio of the SD to the RMSEP, called the ratio of prediction to deviation (RPD), is the factor, by which the prediction accuracy has been increased compared to using the mean composition for all samples. This ratio is desired to be larger than 2 for a good calibration. An RPD ratio less than 1.5 indicates poor predictions and the model cannot be used for further prediction.

Results and discussion

Syneresis evaluation

Casein gels are dynamic by nature [10]. The excessive rearrangements of casein particles gel network during gelation and storage have been implicated as being responsible for whey separation and several rheological conditions of acid-induced gel network [8]. However this spontaneous process that is subsequent to the gel contraction with no external forces applied remain far from being completely controlled [27].

The first global experimentation with Design Expert 10 software showed a large variations ranging between a minimum of 1.8% when all factors were at their high level and a maximum of 47.8% when they were at their low level as showed in Table 5, with a "very normal distribution" of the residues (Fig. 1). The significance tests were valid and it was not necessary to transform the response to improve its modeling even if the three centered points showed a very little curvature and lack of adjustment. The Pareto Chart clearly indicated the two factors (A: Sodium caseinate and B: starch) above the Bonferroni limit (Fig. 2), showing their dominant influence on the syneresis.

The separate influence of the four parameters also showed the dominant influence of A & B factors (Fig. 3). The Syneresis decreased highly (19.44 and 14.60%) when the factors A (Sodium caseinate) and B (Starch) varied from their low (3 and 0) to their high level (8 and 2) while the Syneresis decreased smoothly (only 4.16 and 4.40%) when the factors C (Lactose) and D (Ferments) varied from their low level (4 and 0.2) to their high level (9 and 0.5) respectively. Furthermore, the action of the two factors whose effects were the most important (A: Sodium caseinate & B: Starch) showed a positive interaction (+ 1.70); Moreover the results revealed that when sodium caseinate and starch were combined together, sodium caseinate had more significant effect on syneresis than starch (Fig. 4). This was also confirmed by the second experimental design in which the dry matter content of all samples was fixed to 12% and syneresis measurements ranged between a minimum of 5% for the sample at 8% sodium caseinate and a maximum of 63% for the sample with 3% sodium caseinate regardless of their lactose concentrations (Fig. 5). This variation could be linked to the protein content in relation to the total solids level. Accordingly, the casein content played the major role of all solids contents in reducing syneresis, which was also observed in other previous works [28]. These results may be explained in terms of a complete denaturation of the protein at its isoelectric point showing more hydrophilic parts, which improved the water holding capacity of the resulting caseinate gel and therefore exhibited less syneresis [17]. In addition, it was reported that higher concentrations of casein originate a protein matrix less prone to whey expulsion, probably due to larger particles volume that leads to an increase of the osmotic pressure in the gel system by inter-

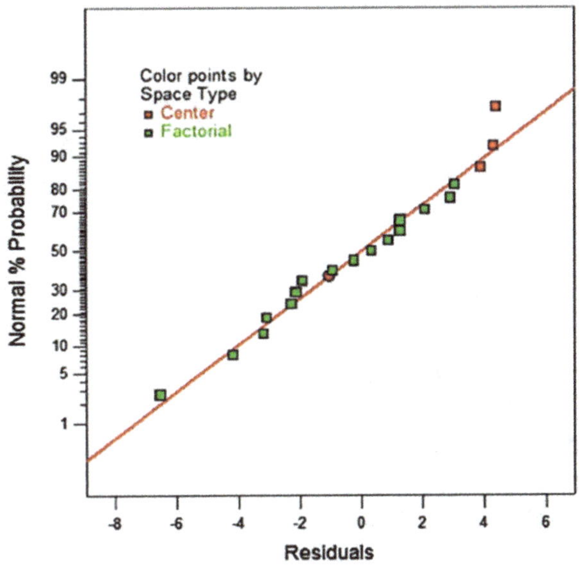

Fig. 1 Normal Plot of Residuals

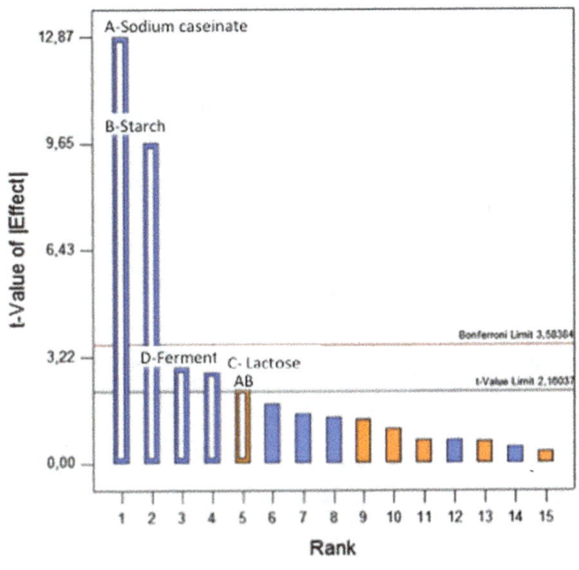

Fig. 2 Pareto Chart

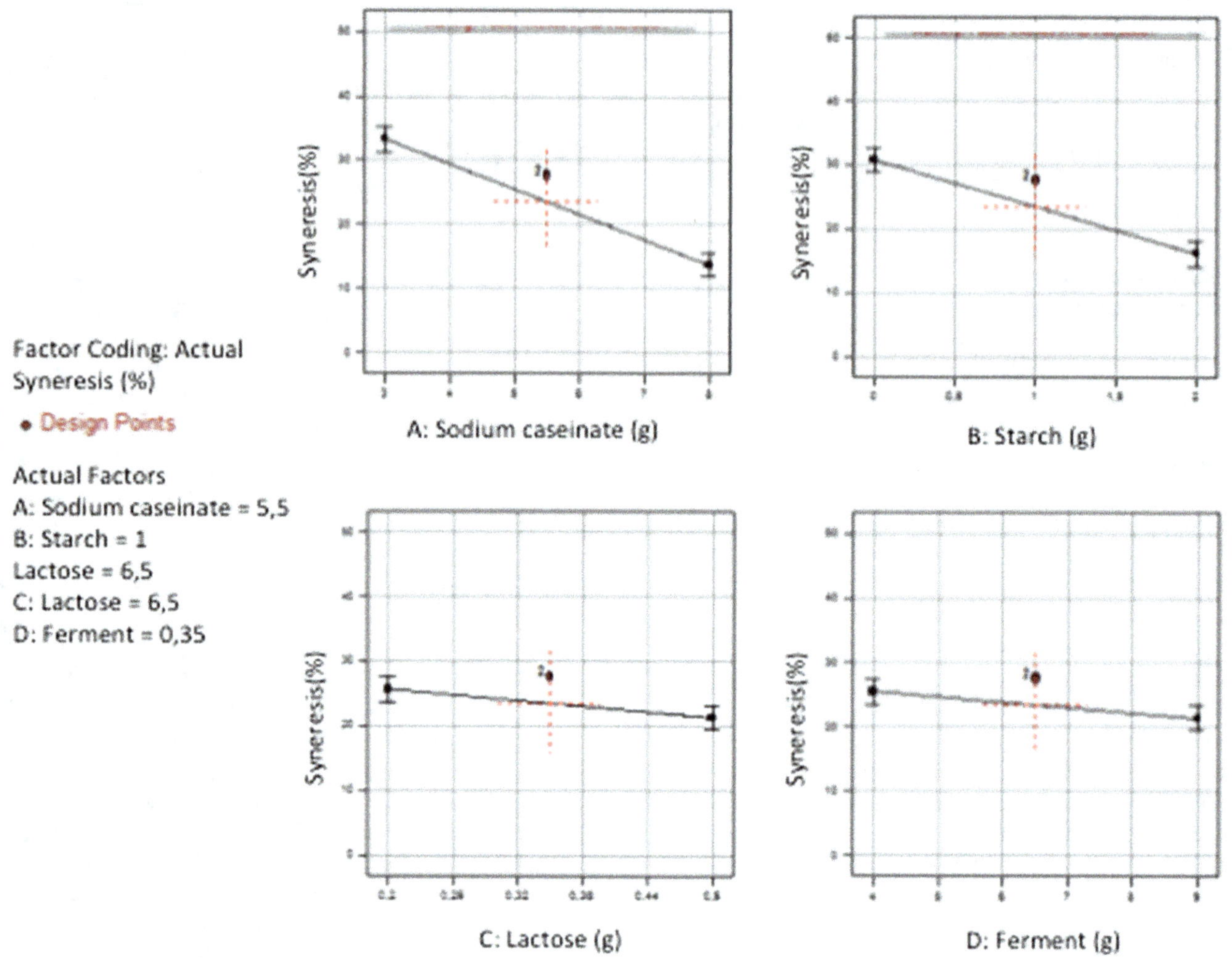

Fig. 3 Visualization of the average negative influence of the four factors

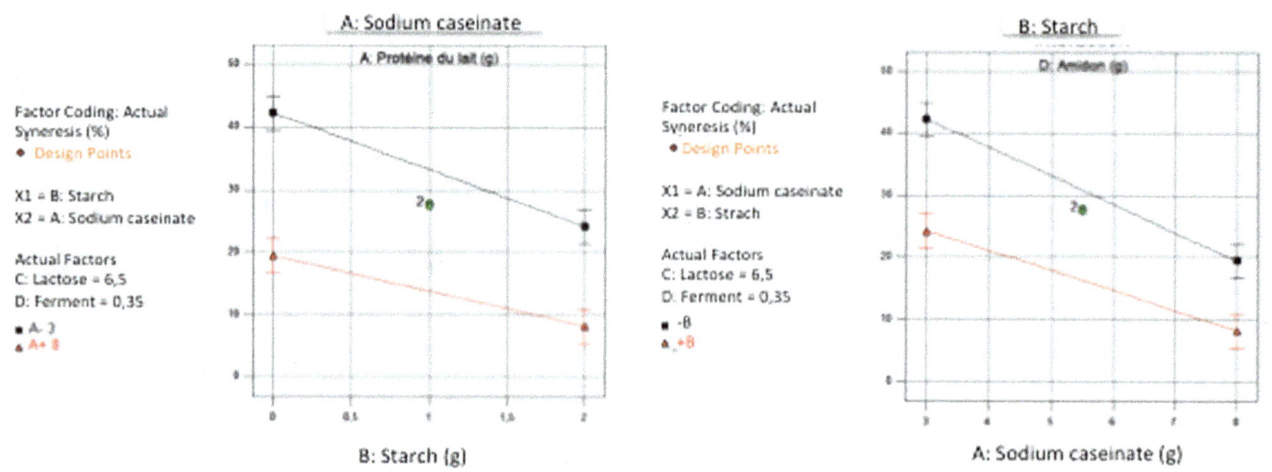

Fig. 4 Visualization of the combined influence of factor **a** (sodium caseinate) and factor **b** (Starch) under two symmetrical angles

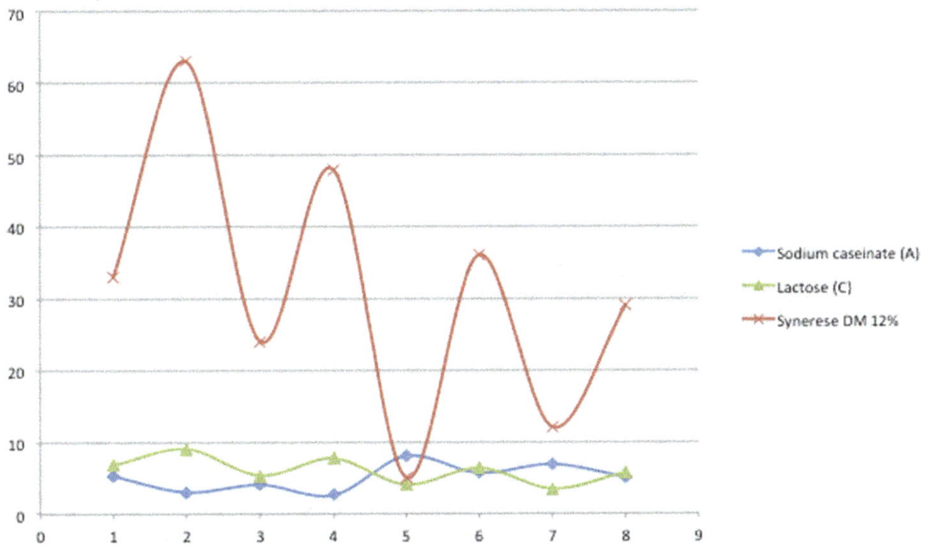

Fig. 5 Impact of lactose and sodium caseinate on syneresis in a model system set at 12% of Dry Matter (DM)

particles repulsion, thus more resistance to gel shrinkage. This kind of stabilization can be alternatively viewed as an osmotic effect [7]. Furthermore, Kunitz in 1928 also reported that the two important factors that must play a major role in the phenomenon of syneresis, were namely the osmotic pressure and the elasticity of the polymeric gel matrix [29].

The significant impact of starch on syneresis confirms the fact that starches along with other hydrocolloids have been evaluated in yoghurt formulations as ways to achieve a desired viscosity and reduce syneresis [30]. The range of starch and milk ingredients interactions is large and the effect on the physico-chemical properties of the mixed system depends on their relative concentrations, the physico-chemical properties of starch and milk ingredients as well as the composition of milk ingredients. In these systems, Zuo et al.(2008) suggested that the main effect of starch granules is to increase the milk protein concentration during swelling by absorbing water from the continuous phase [31]. Therefore, the starch granules have to be destabilized by disrupting the native crystalline structure into an amorphous one that swells by absorbing a considerable amount of water. In fact, upon heat treatment, Hydroxyl groups in glucose monomers of amylose and amylopectin bind water and thus form a viscous solution in which the casein micelles are trapped and reduce syneresis [21]. Accordingly, the functionality of the native starch is related to the gelatinization process rather to its interaction with the caseinate aggregates [7, 22]. This results is in accordance with ours findings. Therefore, the decrease of syneresis among the studied samples could be chiefly attributed to sodium caseinate and then to starch.

pH & zeta monitoring

To understand the effect of these formulation variables on syneresis, and their impact on the quality characteristics of the formed acid gel, a pH and zeta potential monitoring of four designed samples were conducted for controlling the fermentation process and predicting the physical properties of the final product in terms of acidification and ionic strength. Based on our results, syneresis was reduced when sodium caseinate concentration was used above 5%. Therefore, to evaluate starch and lactose impact on the acidification process all samples were designed to be set at 5% of sodium caseinate, while lactose and starch concentrations ranged between (4–9%) and (0–2%) respectively as indicated in Table 4. During fermentation process, acidification monitoring was carried out in an incubator by maintaining the temperature at 42 °C until the fermentation was complete and the pH reached 4.5–4.6. The fermentation time can be considered as the time required for the pH to decrease till end point, presupposing that the required quality properties have been developed [6]. The pH values measured at different time intervals showed that the pH decrease is slightly faster in both samples with 2% starch compared to the acid gels samples prepared without starch (Fig. 6). The difference in the incubation time was found to be significant in samples with 2% starch, 285 min comparing to 315 min in samples without starch. It can be concluded that lactose enhancement did not have any significant impact on the incubation time of samples, while the addition of starch affected slightly the acidification profile of the fermented model and therefore its fermentation time (average increase in tpH4.5 of 30 min). A similar observation was

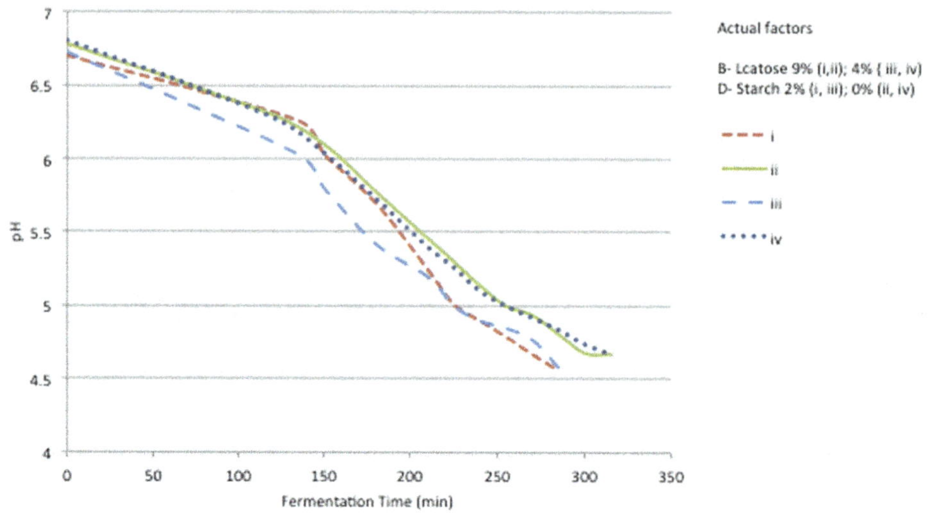

Fig. 6 Effect of starch addition and lactose on the acidification profiles of samples set at 5% of A-sodium caseinate and 0,35 g/l of C-Ferment

reported by Williams et al. (2003) upon the addition of starch to milk before acidification to make yogurt [32].

Furthermore, the two samples showing the highest and the shortest acidification profile were zeta potential monitored. This later had been used as an indicator of the electrical charge of caseinates' particles to measure the strength of repulsion or attraction between particles in both samples along the fermentation process. It was observed that the zeta potential decreased in both samples during acidification as showed in Fig. 7. However, a faster decrease of zeta potential was noticed in sample (i), which is probably responsible for its short fermentation time comparing to sample (iv).

In fact as the pH decreased, the proteins charges got neutralized, which minimized their electrostatic and steric forces inducing a new arrangement of micelles. Consequently, the system became unstable earlier in the sample with starch and tended to flocculation because of the less repulsive forces with the DLVO classical theory [33]. Therefore, an increase in the aggregation state of sodium caseinate and the formation of an acid gel protein network [34]. According to our results, the addition of starch shortened the incubation time, and had a detrimental effect on reducing syneresis in the fermented model, probably because of the reduction of free water and the increase of the solids contents proportion.

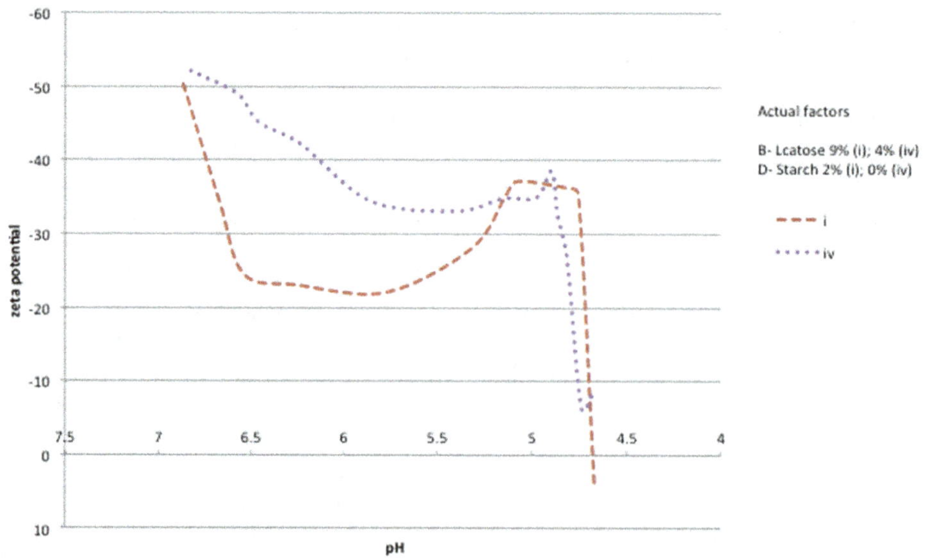

Fig. 7 Zeta potential monitoring in samples set at 5% of A-sodium caseinate and 0,35 g/l of C-Ferment

Acknowledgements
Not applicable.

Authors' contributions
S. El B participated in the design of the study, carried out the experimental procedures, analyzed the results and wrote the manuscript; Ph. P contributed to the main part experimental design and its statistical interpretation; Y. El A participated in Nanosizer ZS analyzer measurements; A. I oversaw the study and revised the document; Y. B conceived the study, provided guidance for experimental work, read and formatted the document. All authors approved the final manuscript.

Funding
Not applicable.

Ethics approval and consent to participate
Not applicable.

Consent for publication
Not applicable.

Competing interests
The authors declare that they have no competing interests.

Author details
[1]Laboratory of Pharmaceutics, Faculty of Medicine and Pharmacy, Mohammed V University, Rabat, Morocco. [2]Laboratory of Medical Biotechnology, Faculty of Medicine and Pharmacy, Mohammed the Vth University in Rabat, Rabat, Morocco. [3]PhP Stats, Création et analyse d'information, Conseil, études et formations en statistique, 19, rue Pasteur, 94170 Le Perreux, France.

References

1. Pang G, Xie J, Chen Q, Hu Z. How functional foods play critical roles in human health. Food Sci Human Wellness. 2012;1(1):26–60.
2. Valentini L, Pinto A, Bourdel-Marchasson I, Ostan R, Brigidi P, Turroni S, Hrelia S, Hrelia P, Bereswill S, Fischer A, et al. Impact of personalized diet and probiotic supplementation on inflammation, nutritional parameters and intestinal microbiota - the "RISTOMED project": randomized controlled trial in healthy older people. Clin Nutr. 2015;34(4):593–602.
3. Marsh AJ, Hill C, Ross RP, Cotter PD. Fermented beverages with health-promoting potential: past and future perspectives. Trends Food Sci Technol. 2014;38(2):113–24.
4. Early R. Dairy products and milk-based food ingredients; 2012. p. 417–45.
5. Akalın AS, Unal G, Dinkci N, Hayaloglu AA. Microstructural, textural, and sensory characteristics of probiotic yogurts fortified with sodium calcium caseinate or whey protein concentrate. J Dairy Sci. 2012;95(7):3617–28.
6. Soukoulis C, Panagiotidis P, Koureli R, Tzia C. Industrial yogurt manufacture: monitoring of fermentation process and improvement of final product quality. J Dairy Sci. 2007;90(6):2641–54.
7. Mizrahi S. Syneresis in food gels and its implications for food quality; 2010. p. 324–48.
8. Lucey* JA. The relationship between rheological parameters and whey separation in milk gels. Food Hydrocoll. 2001;15:603–8.
9. Heertje I, Visser J, Smits P. Structure Formation in Acid Milk Gels, vol. 4; 1985. p. 267–77.
10. Luceyub* JA, TvVb KG, Geurtsb T, Walstrab P. Properties of acid casein gels made by acidification with Glucono-Nactone. 2. Syneresis, permeability and microstructural properties. Int Dairy J. 1997;7(6–7):389–97.
11. O'Kennedy BT, Mounsey JS, Murphy F, Duggan E, Kelly PM. Factors affecting the acid gelation of sodium caseinate. Int Dairy J. 2006;16(10):1132–41.
12. Manab* A. Casein polysaccaharides interaction – a review. Int J ChemTech Res. 2017;10:01–9.
13. Sadeghi M, Madadlou A, Khosrowshahi A, Mohammadifar M. Acid-induced gelation behavior of casein/whey protein solutions assessed by oscillatory rheology. J Food Sci Technol. 2014;51(9):2113–9.
14. Horne DS. Casein interactions: casting light on theBlackBoxes, the structure in dairy products. Int Dairy J. 1998;8(3):171–7.
15. Lucey jA. Perspectives on casein interactions. Int Dairy J:56–65.
16. S.P.F.M Roefs TW: Structure of acid casein gels 2. Dynamic measurements and type of interaction forces colloids and surfaces 50:161–175.
17. Lapointe-Vignola C. Fondation de technologie laitière du Q: Science et technologie du lait: transformation du lait: Presses internationales Polytechnique; 2002.
18. Guo C, Campbell BE, Chen K, Lenhoff AM, Velev OD. Casein precipitation equilibria in the presence of calcium ions and phosphates. Colloids Surf B: Biointerfaces. 2003;29(4):297–307.
19. Considine T, Noisuwan A, Hemar Y, Wilkinson B, Bronlund J, Kasapis S. Rheological investigations of the interactions between starch and milk proteins in model dairy systems: a review. Food Hydrocoll. 2011;25(8):2008–17.
20. Lobato-Calleros C, Ramírez-Santiago C, Vernon-Carter EJ, Alvarez-Ramirez J. Impact of native and chemically modified starches addition as fat replacers in the viscoelasticity of reduced-fat stirred yogurt. J Food Eng. 2014;131:110–5.
21. Gyawali R, Ibrahim SA. Effects of hydrocolloids and processing conditions on acid whey production with reference to Greek yogurt. Trends Food Sci Technol. 2016;56:61–76.
22. N-x S, Liang Y, Yu B, C-p T, Cui B. Interaction of starch and casein. Food Hydrocoll. 2016;60:572–9.
23. Bourlioux P, Braesco V, Mater DDG. Yaourts et autres laits fermentés. Cahiers de Nutrition et de Diététique. 2011;46(6):305–14.
24. Tamime AY, Robinson RK. 7 - biochemistry of fermentation. In: Tamime and Robinson's yoghurt. Third ed: Woodhead Publishing; 2007. p. 535–607.
25. Kneifel W, Paquin P, Abert T, Richard JP. Water-holding capacity of proteins with special regard to Milk proteins and methodological aspects—a review. J Dairy Sci. 1991;74(7):2027–41.
26. Cueva O, Aryana KJ. Quality attributes of a heart healthy yogurt. LWT Food Sci Technol. 2008;41(3):537–44.
27. Lucey JA. Formation and physical properties of Milk protein gels. J Dairy Sci. 2002;85(2):281–94.
28. Fiszman SM, Lluch MA, Salvador A. Effect of addition of gelatin on microstructure of acidic milk gels and yoghurt and on their rheological properties. Int Dairy J. 1999;9(12):895–901.
29. Kunitz M. Syneresis and swelling of gelatin. The Journal of General Physiology. 1928;12(2):289–312.
30. Keogh MK, O'Kennedy BT. Rheology of stirred yogurt as affected by added Milk fat, protein and hydrocolloids. J Food Sci. 1998;63(1):108–12.
31. Zuo JY, Hemar Y, Hewitt S, Saunders A. Effect of the extent of pasting on the dynamic rheological properties of acidified skim milk gels containing normal rice starch. Food Hydrocoll. 2008;22(8):1567–73.
32. Williams R, Glagovskaia O, Augustin MA. Properties of stirred yogurts with added starch: effects of alterations in fermentation conditions. Aust J Dairy Technol. 2003;58(3):228–32.
33. Tholstrup Sejersen M, Salomonsen T, Ipsen R, Clark R, Rolin C, Balling Engelsen S. Zeta potential of pectin-stabilised casein aggregates in acidified milk drinks. Int Dairy J. 2007;17(4):302–7.
34. Lucey WJLJA. Structure and physical properties of yogurt gels: effect of inoculation rate and incubation temperature. American Dairy Science Association. 2004.

7

Soybean plants expressing the *Bacillus thuringiensis cry8*-like gene show resistance to *Holotrichia parallela*

Di Qin[1,2], Xiao-Yi Liu[3], Cristina Miceli[2], Qi Zhang[1] and Pi-wu Wang[1*]

Abstract

Background: *Cry8*-like from *Bacillus thuringiensis* (*Bt*) encodes an insecticidal crystal (Cry) protein. *Holotrichia parallela* (Coleoptera: Scarabaeoidae), commonly known as the dark black chafer, is a troublesome pest of soybean (*Glycine max*). To test whether *cry8*-like can confer resistance against *H. parallela* to soybean, we introduced *cry8*-like from the *Bt* strain HBF-18 into soybean cultivar Jinong 28.

Results: Quantitative reverse transcription-PCR analysis demonstrated that *cry8*-like was expressed most highly in soybean leaves. In addition, Southern blot assays revealed that one copy of the integrated fragment was present in the transformed plants. Eight independent *cry8*-like transgenic lines were subsequently fed on by *H. parallela*. Under *H. parallela* feeding stress, the survival rates of the non-transgenic plants were 92% lower than those of the transgenic plants. The mortality rate of *H. parallela* increased when the larvae fed on the roots of T_1 transgenic soybean plants. Moreover, the surviving larvae were deformed, and their growth was inhibited.

Conclusions: Collectively, our data suggest that transgenic soybean plants expressing the *cry8*-like gene are more resistant to *H. parallela* than non-transgenic plants and that transgenic expression of the *cry8*-like gene may represent a promising strategy for engineering pest tolerance. The events generated in this study could thus be utilized in soybean breeding programs.

Keywords: *Holotrichia parallela*, *Cry8*-like, *Bacillus thuringiensis* endotoxin, Transgenic soybean

Background

Soybean (*Glycine max* L. Merr.) is a globally important crop species that is grown for its oil and protein; moreover, its seeds are used for human consumption, animal feed, and industrial raw materials [1]. Soybean is self-pollinated, and soybean seeds are rich in protein (38–40%), in addition, soybean crops are utilized in crop rotations because of the ability of the plants to increase the amount of nitrate in the soil via symbiotic nitrogen fixation. However, soybean yields are severely affected by insects [2].

Holotrichia parallela (Coleoptera: Scarabaeoidea), commonly known as the dark black chafer, is one of the most important pests in agriculture and forestry in China [3]. *H. parallela* has a three-stage life cycle: the egg, larva and adult (Fig. 1). The larvae of *H. parallela* live in the soil and prefer to feed on plant roots during their two-year larval stage, and adult *H. parallela* insects feed on fresh leaves. *H. parallela* can cause considerable damage to soybean—from 10 to 30% yield losses in a typical year and up to 50% yield losses during severe infestations—and are responsible for $1–2 billion in losses per year worldwide [4]. Currently, farmers apply insecticidal sprays for the control of these insect pests, however, because these larvae are soil dwelling, it is difficult to find a specific chemical product that can control these insects.

Conventional breeding approaches for insect resistance have limited success due to a lack of sources of resistance. As an alternative approach, genetic engineering using *cry* genes from the soil bacterium *Bacillus thuringiensis* (*Bt*) offers great potential for improving crop species [5, 6].

Bt genes encode a protoxin that is transformed to an active toxin via enzyme cleavage in the insect midgut [7]. Different types of *cry* genes have been used successfully to develop insect-resistant crop species. According to

* Correspondence: wangpiwu189@163.com
[1]Biotechnology Center of Jilin Agricultural University, Jilin Agricultural University, Changchun 130118, People's Republic of China
Full list of author information is available at the end of the article

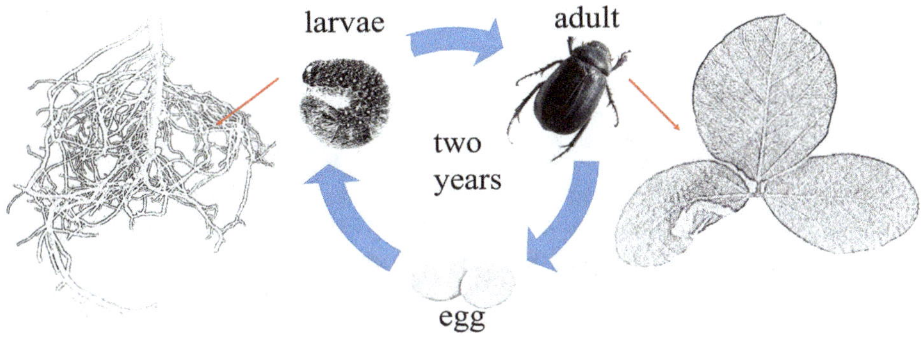

Fig. 1 Three-stage life cycle of *Holotrichia parallela*

insecticidal specificity, *cry* genes are mainly divided into 4 types including anti-Lepidoptera, anti-Diptera, anti-Coleoptera and anti-nematodes types. The *Bt* genes that provide resistance to Coleoptera include *cry3, cry4, cry7, cry8,* and *cry23* [8]. Development of insect resistance via transgenesis with fused *cry* genes has been previously applied to protect cotton from damage to cotton bollworm insects [9]. A novel *cry8* gene exhibiting activity against the larvae of *Holotrichia oblita* and *H. parallela* was identified [10]. Furthermore, a chimeric *cry8Ea1* gene flanked by MARs in transgenic peanut plants effectively controls *H. parallela* [11], but no studies have investigated transgenic soybean plants expressing *cry8*-like genes.

Therefore, in the present study, we used *Agrobacterium*-mediated transformation to transform callus of the soybean cultivar Jinong 28. Jinong 28 has good agronomic characteristics but no resistance to *H. parallela.* We produced eight independent transgenic soybean lines expressing *cry8*-like under the control of the CaMV 35S promoter. Insect feeding assays performed for a period of 96 h indicated effective protection of the soybean plants against *H. parallela* compared to the Jinong 28 control plants.

Results

Production of transgenic soybean plants expressing the cry8-like gene

The recombinant plasmid designated pCAMBIA3300-cry8-like was introduced into *Agrobacterium tumefaciens* strain (Fig. 2a). The callus tissue used was originally derived from cotyledon-nodes of soybean (Fig. 2b). About 300 soybean cotyledon calluses were subjected to transformation. The putative transgenic embryos were grown with the addition.

of $50\,mg{\cdot}L^{-1}$ kanamycin to select the transgenic plants. Approximately 3% of the stems (8 plants) developed on the kanamycin-selected cotyledons (Fig. 2c, d). 8 individual T_0 plants (primary transformants) established in the green house that grew normally, flowered and set seeds (Fig. 2e). Since the transformants developed by the transformation strategy used could be chimeras in the T_0 generation, screening of T_1 generation plants for the identification of putative transformants was essential.

Detection of the cry8-like gene in the T_0/T_1 generations

Southern blot assay was used to detect the presence and determine the copy number of the *cry8*-like gene in the selected eight putative transgenic lines. One copy of *cry8*-like was detected in T_0 generation of each line (Fig. 3a). T_1 generation of transgenic lines were obtained by selfing and inheritance of *cry8*-like investigated by PCR and Southern blot assay. PCR analysis of 971 plants revealed a 3:1 segregation ratio indicating that the transgene was stably integrated in the soybean genome. Plants of the T_1 generation retained the single copy number of *cry8*-like insertion (Fig. 3b) and both in the T_0 and T_1 generations, each part of the transgene (*cry8*-like coding region, 35S promoter and *Nos* terminator) was detected in each selected line (Additional file 1: Fig. S1).

Detection of the cry8-like expression in the transgenic lines

Expression of *cry8*-like was studied by quantitative real-time PCR (qRT-PCR). The qRT-PCR results indicated that *cry8*-like was expressed in all organs (leaves, stems, and roots) of the transgenic soybean plants, with the highest expression level detected in the leaves (Fig. 4a). Two-way ANOVA (including all interactions and simple and main effects) was used to test for significant differences among the different organs and lines. The interaction between soybean lines and organs had a significant effect on the expression of *cry8*-like (Fig. 4b red), but the F-value of the interaction was small compared to that of the main effects. The results indicated that the simple and main effects of the lines and organs on *cry8*-like expression were more significant than the interaction effects. Two-way ANOVA revealed that there was a difference among the different *cry8*-like transgenic lines ($df = 7$, $F = 435$, $P < 0.001$) and the different organs ($df = 2$, $F = 8033$, $P < 0.001$) (Fig. 4b). The expression

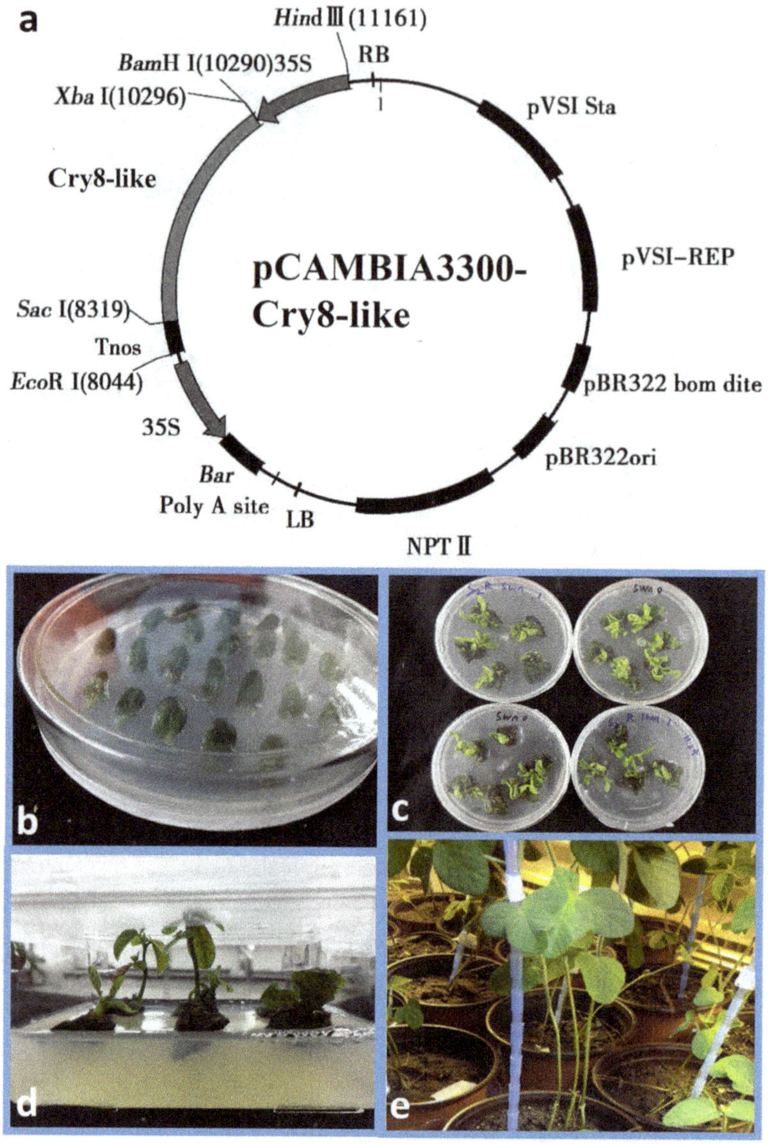

Fig. 2 *Agrobacterium*-mediated transformation of soybean. (**a**) Diagram of pCAMBIA3300-cry8-like used for transformation. (**b**) Soybean cotyledon nodes cultured on shoot-induction medium (**c**) and (**d**) Shoot development. (**e**) 3-month-old T_0 transgenic soybean plants grown under greenhouse conditions in soil

level of *cry8*-like in the eight lines followed the order Jinong 28-*cry*-4>Jinong 28-*cry*-5>Jinong 28-*cry*-3>Jinong 28-*cry*-2 ≥ Jinong 28-*cry*-1 ≥ Jinong 28-*cry*-7 ≥ Jinong 28-*cry*-6>Jinong 28-*cry*-8 (Fig. 4c).

Quantifying the level of cry endotoxin in the different organs

An enzyme-linked immunosorbent assay (ELISA) was used to quantify the level of Cry endotoxin in the different organs (leaves, stems, and roots) of the transgenic soybean plants (T_0/T_1 generation). Variable amounts of toxin accumulated in all transgenic plants and in all organs, whereas the nontransformed control plants did not show any accumulation of Cry endotoxin (Table 1). The Jinong 28-cry-4 transgenic plants showed the highest level of Cry protein accumulation (see Additional file 2: Table S1). The ELISA also revealed that high levels of endotoxin accumulated in the leaves and roots and that the lowest level of endotoxin accumulated in the stems (see Additional files 3: Table S2). The interaction between soybean lines and organs had a significant effect on the accumulation of BT protein (Table 2). Compared with that of the main effects, the F-value of the interaction is small, and the results indicate that the simple effects of the lines and organs on Bt toxin accumulation were more significant than were the main effects.

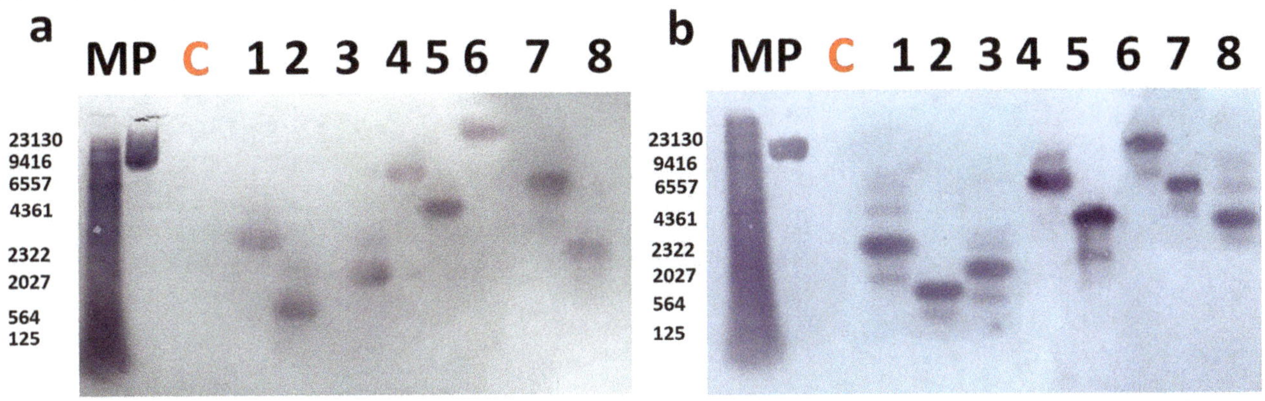

Fig. 3 Presence of the *cry8*-like gene in different generations. (**a**) Southern blot analysis of the copy number of the *cry8*-like expression cassette in T_0 plants. (**b**) Copy number of the *cry8*-like expression cassette in T_1 plants. M: DNA marker; c: Jinong 28 controls

Tolerance of the cry8-like transgenic plants to *H. parallela*

After 4, 6 and 8 days of third-instar *H. parallela* larval feeding, extensive damage was incurred on the non-transgenic control plants (Fig. 5a). The eight independent *cry8*-like transgenic lines showed a high degree of plant survival after 8 days of feeding, and the maximum number of larvae were paralyzed after 4 days of feeding on these lines. The fresh weight of the transgenic lines was also greater than that of the controls, with the weight of the Jinong 28 controls being significantly lower (78%) than that of the eight transgenic lines (Fig. 5b). We also investigated the effects of *H. parallela* feeding

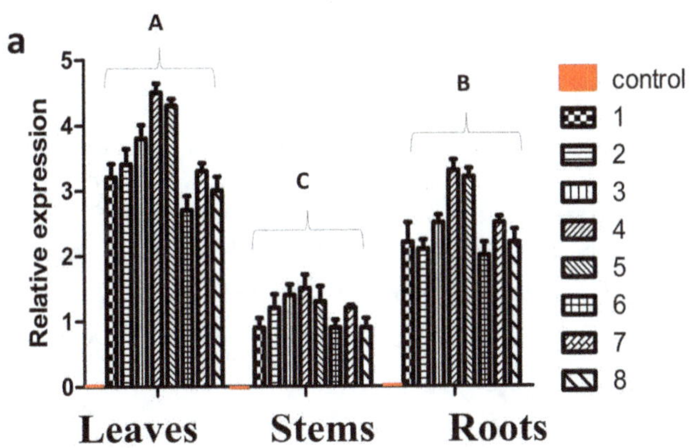

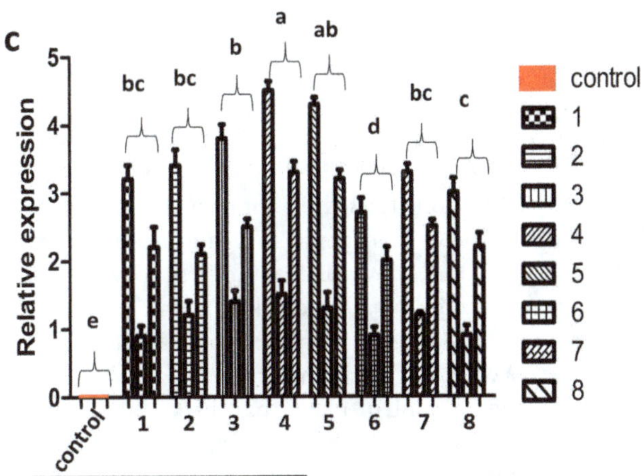

b

Source	Type III Sum of Squares	df	Mean Square	F	Sig.
corrected Model	81.129[a]	23	3.527	857.608	0.000
Intercept	417.191	1	417.191	101432.256	0.000
Organs	66.087	2	33.043	8038.884	0.000
Soyean lines	12.54	7	1.791	435.55	0.000
Organs*lines	2.502	14	0.179	43.455	0.000
Error	0.197	48	0.004		
Total	498.517	72			
Corrected Total	81.326	71			

Fig. 4 Detection of *cry8*-like expression by quantitative real-time PCR (qRT-PCR). (**a**) *cry8*-like expression in different organs of the eight transgenic lines (1–8) and the non-transformed control (red). (**b**) Test of between-subject effects. F: F-value; Sig: significant. (**c**) Comparison of the *cry8*-like expression levels of the eight transgenic lines. Different letters indicate significant differences at $P < 0.05$

Table 1 Bt toxin accumulation in the transgenic plants

Plant line (T_0)	Bt toxin expression ($ng{\cdot}mg^{-1}$) in leaves	Bt toxin expression ($ng{\cdot}mg^{-1}$) in stems	Bt toxin expression ($ng{\cdot}mg^{-1}$) in roots	Plant line (T_1)	Bt toxin expression ($ng{\cdot}mg^{-1}$) in leaves	Bt toxin expression ($ng{\cdot}mg^{-1}$) in stems	Bt toxin expression ($ng{\cdot}mg^{-1}$) in roots
T_0/ Jinong28	–	–	–	T_1/ Jinong28	–	–	–
T_0/ Jinong28-cry-1	12 ± 1	9 ± 1	10 ± 2	T_1/ Jinong28-cry-1	13 ± 2	11 ± 1	12 ± 1
T_0/ Jinong28-cry-2	15 ± 2	11 ± 2	14 ± 2	T_1/ Jinong28-cry-2	11 ± 1	8 ± 1	10 ± 2
T_0/ Jinong28-cry-3	14 ± 2	11 ± 1	13 ± 1	T_1/ Jinong28-cry-3	18 ± 1	15 ± 1	16 ± 2
T_0/ Jinong28-cry-4	17 ± 2	12 ± 1	16 ± 2	T_1/ Jinong28-cry-4	19 ± 1	17 ± 2	18 ± 1
T_0/ Jinong28-cry-5	15 ± 2	10 ± 2	14 ± 1	T_1/ Jinong28-cry-5	17 ± 2	15 ± 2	17 ± 1
T_0/ Jinong28-cry-6	8 ± 2	5 ± 2	6 ± 1	T_1/ Jinong28-cry-6	10 ± 2	6 ± 1	8 ± 2
T_0/ Jinong28-cry-7	11 ± 1	7 ± 2	9 ± 3	T_1/ Jinong28-cry-7	12 ± 2	9 ± 2	10 ± 2
T_0/ Jinong28-cry-8	6 ± 2	5 ± 1	5 ± 1	T_1/ Jinong28-cry-8	8 ± 1	5 ± 2	6 ± 1

Note: Expression of Bt toxin in T_0/T_1 transgenic plants by ELISA in leaves, stems and roots. Results are means ± SE

stress on the growth of the transgenic plants. After 2 days of *H. parallela* larval feeding on V_2-stage plants, the leaves of the non-transgenic controls started to wilt, whereas the *cry8*-like transgenic soybean lines continued to grow well. After 4 days of feeding, the leaves of the controls were dry, yellow, and dead, whereas most leaves of the transgenic soybean plants remained green for 1 week. After 2 weeks of third-instar *H. parallela* feeding, all non-transgenic control plants had been eaten by the larvae, and the eight *cry8*-like transgenic lines displayed enhanced tolerance to *H. parallela* (Fig. 5c). We detected higher seedling survival rates on the transgenic plants than on the non-transgenic controls subjected to larval feeding. The survival rate of the non-transgenic controls was 92% lower than the survival rate of the eight transgenic lines (Fig. 5d). These results further demonstrated that, compared with the controls the *cry8*-like transgenic lines displayed increased tolerance to *H. parallela*.

Insect bioassays

After *H. parallela* fed on leaves or roots of the eight transgenic plants for 8 days, the relative growth rate (RGR), relative consumption rate (RCR), and survival rate were determined. The RGR (Fig. 6a) and the RCR (Fig. 6b) of the larvae feeding on the roots of the eight

Table 2 Test results of between-subject effects

Source	Type III Sum of Squares	df	Mean Square	F	Sig.
corrected Model	1355.259a	26	52.125	16.855	0.000
Intercept	5622.241	1	5622.341	1817.97	0.000
Organs	66.704	2	33.252	10.784	0.000
lines	1273.295	8	159.241	51.491	0.000
Organs*lines	14.672	16	1.194	1.282	0.003
Error	83.576	27	3.093		
Total	7061	54			
Corrected Total	1438.759	53			

Note: F: F value, Sig: significance

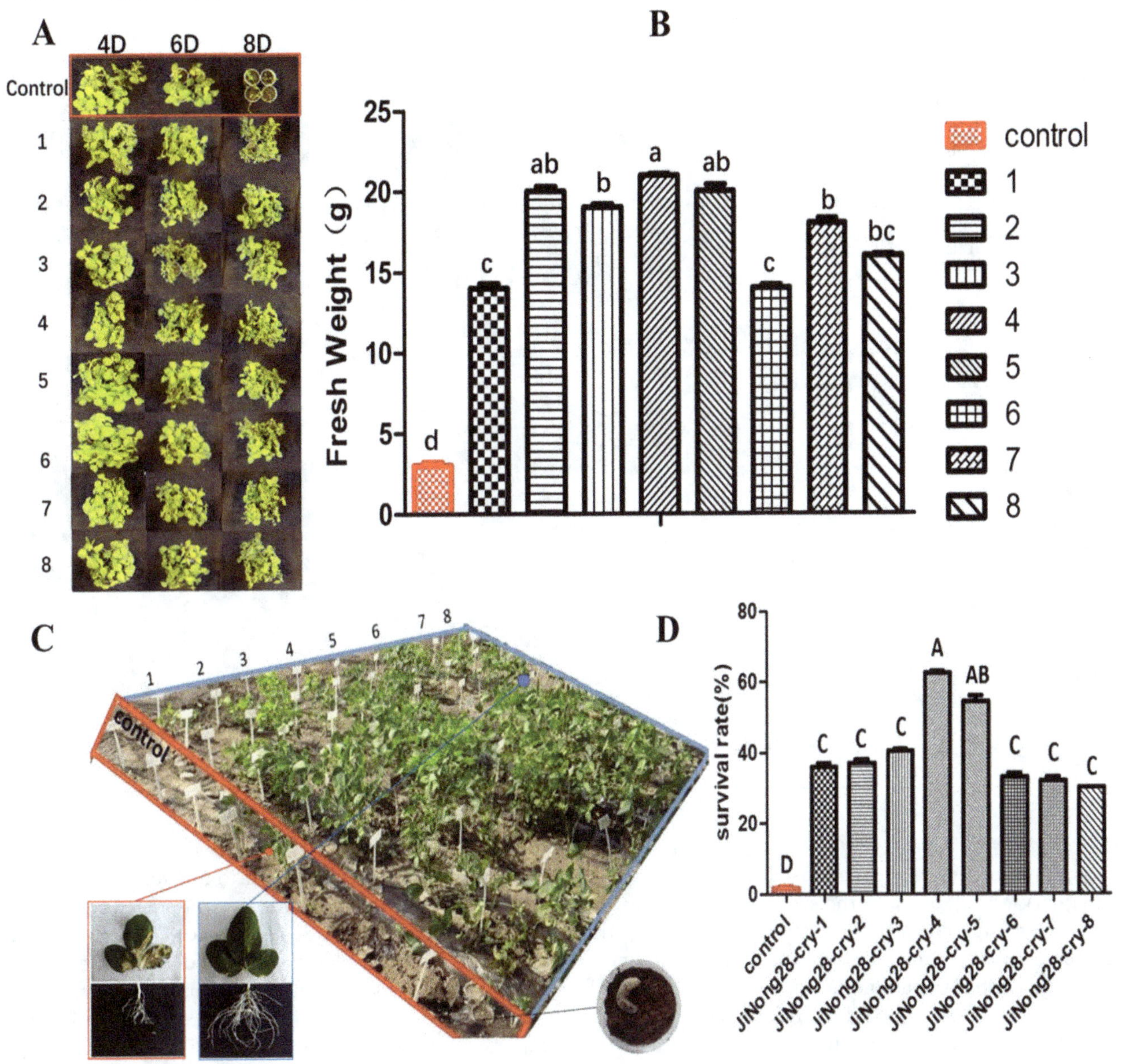

Fig. 5 Tolerance of the *cry8*-like transgenic plants to *H. parallela*. (**a**) Representative images of the insect bioassay performed on transgenic soybean plants with *H. parallela* larvae after 4, 6, and 8 days of feeding. (**b**) Fresh weight of soybean plants after 8 days of *H. parallela* larval feeding. The bars represent the means ± standard errors; the mean values followed by the different letters are significantly different ($P < 0.05$), as determined by Duncan's multiple-range test. (**c**) Phenotypes of 25-day-old plants treated with *H. parallela* for 14 days. Control: Jinong 28; 1–8: Jinong 28-cry-1, Jinong 28-cry-2, Jinong 28-cry-3, Jinong 28-cry-4, Jinong 28-cry-5, Jinong 28-cry-6, Jinong 28-cry-7, and Jinong 28-cry-8; (**d**) Analysis of the survival rates of seedlings after 14 days of *H. parallela* feeding. The different uppercase letters indicate significant differences at $P < 0.01$, as determined by Duncan's multiple-range test

independent *cry8*-like transgenic lines were significantly lower than those feeding on the roots of the non-transgenic controls. The survival rates of the larvae consuming the roots of the transgenic lines declined, whereas the survival rates of the larvae consuming the roots of the non-transgenic controls were significantly higher (100% survival) (Fig. 6c). The RGR, RCR and survival rate of *H. parallela* adults feeding on the leaves were also investigated, the RGR and RCR values of the *H. parallela* adults from the *cry8*-like transgenic soybean treatments were significantly lower than those of *H. parallela* adults consuming non-transgenic leaves (Fig. 6d and e, respectively). A significant difference in the survival rate of *H. parallela* adults between the *cry8*-like transgenic lines and the non-transgenic Jinong 28 controls was found (Fig. 6f). After *H. parallela* larvae fed on

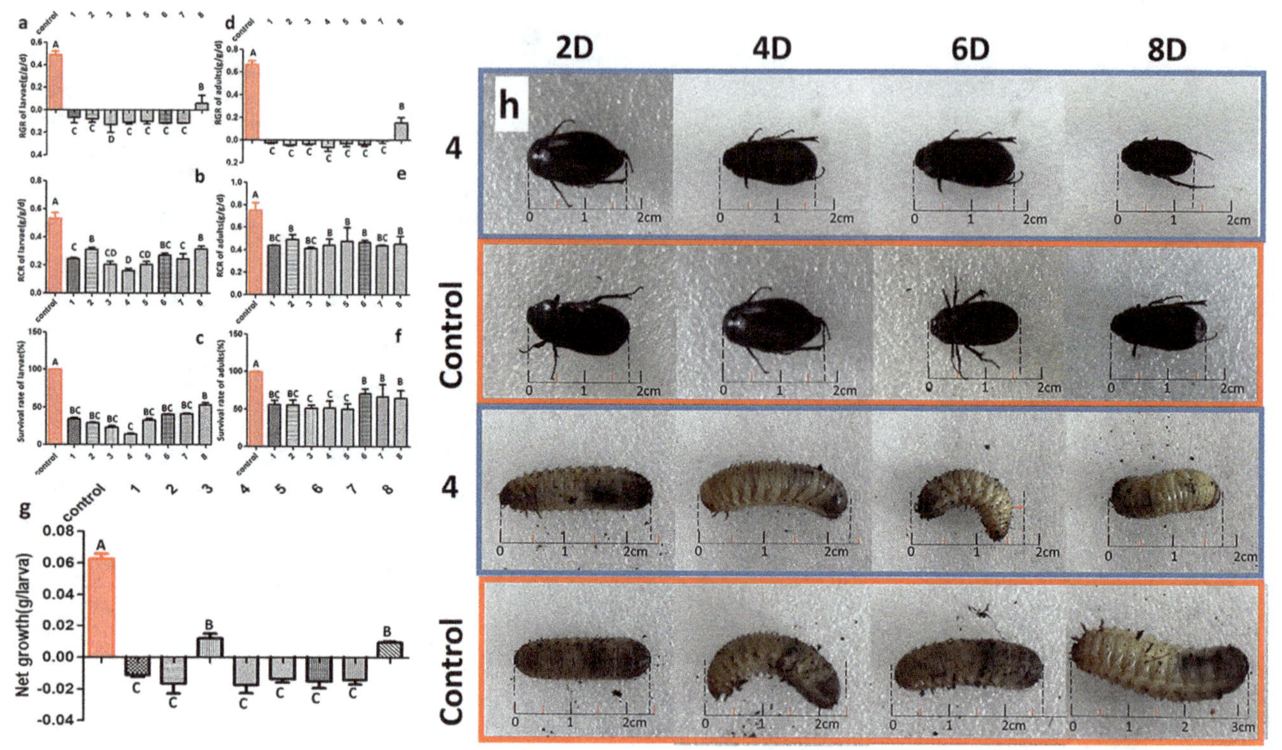

Fig. 6 Insect bioassays. RGR(**a**), RCR (**b**) and survival rate (**c**) of *H. parallela* larvae consuming the transgenic lines after 8 days. The RGR (**d**), RCR (**e**), and survival rate (**f**) of *H. parallela* adults were also investigated. The red zone represents the Jinong 28 controls. 1–8 indicate the eight independent *cry8*-like transgenic soybean plants. (**g**) Net growth of larvae after 8 days of feeding. (**h**) The size of *H. parallela* larvae and adults after 8 days. 4: The size of *H. parallela* consuming Jinong 28-cry-4 transgenic lines; The different uppercase letters indicate significant differences at $P < 0.01$

the roots of the transgenic lines for 8 days, the net growth rate of the larvae decreased from 0.017 to −0.018 g/larvae (Fig. 6g), and the size of larvae decreased. However, compared with the larvae consuming the transgenic plants, the larvae consuming the non-transgenic controls had significantly higher net growth (0.06 g/larvae) (Fig. 6g), and their size also increased (Fig. 6h). Compared with that reared on the non-transgenic controls, the size of adult *H. parallela* insects reared on the transgenic plants expressing the *cry8*-like gene decreased (Fig. 6 h).

Agronomic performance of cry8-like transgenic lines in the field

The agronomic characteristics such as pods per plant, seeds per plant, hundred-seed weight (HSW), seed weight per plant (SWPP), plant height and plot yield of the T_2 *cry8*-like transgenic soybean plants grown in the field were compared with those of the non-transgenic plants grown in the field. The seeds of both transgenic and control plants presented 100% germination under normal growth conditions. The growth and fertility of the transgenic plants were compared with those of the control plants. All eight independent transgenic plants showed an exhibited phenotype and were fertile (Table 3). One-way ANOVA was used to test for significant differences among the different lines. Significant differences ($P < 0.01$) were detected in the growth and fertility status of the transgenic plants compared with the control plants. The average height of the eight transgenic lines under *H. parallela* stress was twice that of the non-transgenic Jinong 28 controls. The yield of the eight transgenic lines was significantly higher (4 times) than that of the non-transgenic controls, and the seed number per plant was four times that of the Jinong 28 controls.

Discussion

Insect pests are a major limitation to soybean production [12]. *H. parallela* is a beetle belonging to the Scarabaeoidea (subfamily Melolonthidae) and causes considerable damage in soybean production worldwide [13]. *H. parallela* larvae are difficult to control with chemical insecticides because the larvae reside in the soil.

Cry toxins (such as *cry1F*, *cry2A* and *cry8*) have been widely used in genetically modified organisms for pest control because the toxic effects are based on the

Table 3 Yield of *cry8*-like transgenic soybean plants in the field

Line	Pods per plant	Seed number per plant	hundred-seed weight (g)	seed weight per plant (g)	Average plant height (cm)	Plot yield (g)
Jinong28	30A	100A	10.91A	16.4A	40.292A	324.84A
Jinong28-cry-1	78B	222B	19.71B	35.4B	106.021B	1319.80B
Jinong28-cry-2	70B	204B	19.43B	34.9B	106.893B	1307.28B
Jinong28-cry-3	72B	188B	19.82B	34.0B	102.397B	1364.52B
Jinong28-cry-4	70B	235B	19.29B	38.2B	105.899B	1342.20B
Jinong28-cry-5	71B	215B	19.459B	32.5B	101.180B	1314.4B
Jinong28-cry-6	69B	210B	28.79B	32.5B	104.010B	1322.24B
Jinong28-cry-7	70B	218B	20.51B	32.2B	100.887B	1375.24B
Jinong28-cry-8	68B	200B	21.43B	32.6B	104.897B	1326.24B
F values	80.534	8.483	6.413	5.312	250.451	253.187
P values	0.000	0.002	0.006	0.005	0.000	0.000

Note: Different upper letters indicate significant differences at ($P < 0.01$)

disruption of midgut cells within insect pests [14, 15]. Transgenic soybeans that carry multiple *Bt* genes that provide resistance to important insect pests in agricultural crops have been produced [16, 17]. Numerous studies have indicated that Bt proteins can suppress major pests thus reducing insecticide use [18, 19]. In 1994, a paper described the engineering of *cry* gene into soybean plants and compared the insecticidal activity of soybean plants [20]. In 2000, A transgenic lineage of the soybean 'Jack' expressing a synthetic *cry1Ac* gene (Jack-Bt) was evaluated for resistance to lepidopteran pests, compared with untransformed soybean, Jack-Bt showed three to five times less defoliation from corn earworm [21]. In 2013, a study showed that the *Cry1Ac*-expressing *Bt* soybean provided good protection against *Helicoverpa armigera,* while, limited resistance was also found for transgenic soybean against pest [22]. Most studies of *Bt* transgenic soybean have shown promising insecticidal activity against *lepidopteran* pests, not against *Scarabaeidae* pest [22]. The *Bt* strain HBF-18 (CGMCC2070), which carries two novel *cry* genes (*cry8*-like and *cry8Ga*) [23], has previously been reported to exhibit toxicity to *H. parallela* [24]. Compared with *cry8Ga*, *Cry8*-like exhibits 3-fold greater toxicity [23], confirming that *cry8*-like would be the best choice for expression in transgenic soybeans. These findings suggest that the insertion of an exogenous *cry8*-like gene into soybean is a feasible approach to produce insect resistance.

In this study, we successfully inserted the *cry8*-like gene into soybean to confer resistance against *H. parallela* to transgenic plants. We obtained eight independent transgenic soybean lines, each containing a single copy of the *cry8*-like gene. In the case of soybean, leaves and roots are the major plant parts damaged by *H. parallela*. Therefore, a high level of expression of Bt protein in soybean leaves and roots is important to obtain effective pest control. The present ELISA results showed that the Cry8-like levels in the two *Bt* soybean lines were relatively high, ranging from 16 to 19 $ng{\cdot}mg^{-1}$. Although levels did not reach those previously reported [25], by comparison with untransformed Jinong28, the transgenic lines showed five times greater resistance than Jinong28 to artificially infestations of *H. parallela*. In conventional field plots at one location in 2018, transgenic soybean plants also showed four times greater resistance than untransformed Jinong28 to natural infestations of *H. parallela*. According to Walker et al. [21], a transgenic line Jack-Bt also showed four times greater resistance than untransformed Jack to natural infestations of lesser cornstalk borer. The results of the bioassay revealed a significant difference in the survival of *H. parallela* reared on the eight independent transgenic soybean lines. The survival rate of *H. parallela* consuming the Jinong 28-cry-4 transgenic line was the lowest (14%), while, the survival rate of *H. parallela* consuming the Jinong 28-cry-8 transgenic line was the highest (57%). Accordingly, the Jinong 28-cry-4 transgenic line had the highest accumulation of Bt protein, and the Jinong 28-cry-8 transgenic line had the lowest accumulation of Bt protein. These results suggest that the efficacy of insect protection was based on the level of protein accumulation in the transgenic plants. The variable insect mortality levels was mostly in agreement with the protein expression levels in the transgenic plants, and the protein expression levels in the transgenic plants was related to the intensity of expression of the *cry8*-like gene. The variation in transgene expression has been attributed to several factors, including differences with respect to chromosome location, copy number, and transgene construct fidelity. We investigated the transcriptional expression of the *cry8*-like gene in the eight independent transgenic soybean lines. All transgenic plants showed variable levels of *cry8*-like expression. The *cry8*-like construct was probably inserted at different locations in the plant genome, thus producing differences in Cry toxin concentration, as previously suggested in [22].

Both the cumulative and relative food consumption rates of the *H. parallela* larvae and adults feeding on the transgenic soybean plants were significantly reduced compared to those of the larvae feeding on the control plants. The mortality of *H. parallela* larvae mostly reached 100% after 2 weeks of feeding, and that of adults was 85%. The results showed that the transgenic plants exerted differential effects on the larvae and adults and that the larvae seemed to be more sensitive to the Bt toxin than the adults. Some researchers also reported that the first and second instar are the most sensitive to Cry protein [26]. The high toxicity of *Cry8*-like to *H. parallela* larvae can be related to the natural tolerance of insects to the Bt proteins.

The evolution of insect resistance directly threatens Bt toxins efficacy in the transgenic plants [27]. The transgenic soybean plants expressing a *Cry1Ac* gene that showed a poor control of *Spodoptera* species is one example [28], as well as the western corn rootworm showing resistance to *Cry3Bb1* maize in the USA [29]. Therefore, the transgenic crops including two or more Bt toxins, and novel Bt toxins are considered better options [30]. In our experiment, the novel *cry8*-type gene was used in the transgenic soybeans for the control of *Scarabaeidae* beetles. In general, our results showed that transgenic soybean plants expressing the *cry8*-like gene are resistant to *H. parallela*, but a limitation of the study is that only one *cry* gene was used. Therefore, as already suggested in [22], future research should focus on the multiple-gene Bt transgenic soybeans to delay the fast development of resistance in the pest.

Conclusions

In summary, our study demonstrated that *cry8*-like expression is an effective technique for improving resistance to *H. parallela* in soybean. The eight lines developed in this study showed enhanced resistance to *H. parallela* adults and larvae. In addition, these lines did not have any negative impact on the agronomic traits studied.

Methods

Biological materials

The 1971-bp cDNA of the *cry8*-like gene from the *Bacillus thuringiensis* strain HBF-18 (from the Jilin Academy of Agricultural Sciences) was ligated into the *Bam*HI-*Sac*I site of pCAMBIA3300 [31] to place the coding region under the regulatory control of the 35S promoter and the *Nos* terminator. The recombinant plasmid was named pCAMBIA3300-cry8-like and introduced into *Agrobacterium tumefaciens* strain LBA4404 via the freeze-thaw method [32]. We introduced *cry8*-like from the *Bt* strain HBF-18 (from the Jilin Academy of Agricultural Sciences) into the soybean cultivar Jinong 28 (Approval number 2011010), the seeds of Jinong 28 were provided by the Biotechnology Center of Jilin Agricultural University. Jinong 28 has good agronomic characteristics but no resistance to *H. parallela.* We used *Agrobacterium*-mediated transformation [33] to transform callus of the soybean cultivar Jinong 28. The transformation process was divided into five sequential steps: bacterial inoculation, cocultivation, resting, selection and plant regeneration. Eight independent transgenic lines were harvested and propagated in the greenhouse: Jinong 28-cry-1, Jinong 28-cry-2, Jinong 28-cry-3, Jinong 28-cry-4, Jinong 28-cry-5, Jinong 28-cry-6, Jinong 28-cry-7, and Jinong 28-cry-8. These transformed plants were used as sources of plant materials. Untransformed plants of cultivar Jinong 28 were used as controls. T_0 seeds from the primary transformants were harvested and analyzed for the identification of transformants.

Polymerase chain reaction analysis

Genomic DNA from T_0 and T_1 plants and from the controls was extracted from the leaf tissues following the method of Kim and Hamada [34]. Quantification of DNA was performed using a Nanodrop spectrophotometer (Thermo Fisher Scientific, Waltham, Masachusetts, USA). The genomic DNA from the T_0 and T_1 transgenic plants was used to perform PCR to detect the presence of the *cry8*-like gene. The specific primers for PCR were designed based on the highly conserved sequence of the *cry8*-like gene via Primer 5.0 software (Additional file 4: Table S3). The PCR conditions were as follows: 94 °C for 5 min; 40 cycles of 95 °C for 30 s, 58 °C for 30 s, and 72 °C for 30 s; and a final extension of 72 °C for 10 min. Genomic DNA from Jinong 28 was used as a negative control, and the vector was used as a positive control. The PCR products were subsequently analyzed by agarose gel electrophoresis.

Quantitative reverse transcription-PCR

qRT-PCR analysis was performed to determine the transcript abundance of *cry8*-like. The gene-specific primer pairs P3 (5′-TTTGGATCCAAGCTTTCTAGACCCGG GCCTAT-3′) and P4 (5′-TTTGAGCTCTCAAAGTTC ATCCTTCTCGGAGT-3′) were used to amplify *cry8*-like. A *lectin* gene (GenBank: A5547–127) was used as a reference gene, which was amplified with the primer pairs P5 (5′-GCACTTAAGATACTCTAGGTAC-3′) and P6 (5′-CCACCTCCCTACTATCCATT-3′). At the three-leaf stage (V_3), 0.3 g of leaves, 0.3 g of stems and 0.3 g of roots from each plant were washed thoroughly with ddH_2O, and then immediately frozen in liquid nitrogen for qRT-PCR analysis. Total RNA was extracted using an RNA isolation kit (Omega Bio-tek, Norcross, GA, USA). cDNA synthesis was then performed using a reverse transcription kit (Omega Bio-tek, Norcross, GA, USA). Three biological replicates with three technical replicates of each qRT-PCR were performed. The qRT-PCR analysis was performed using a Bio-Rad CFX system (Amersham Biosciences, Little Chalfont, Buckinghamshire, UK).

Southern blot analysis

Genomic DNA was isolated from leaf tissues as described previously. Approximately 2 µg of DNA was digested with *Bam*HI, separated on 0.8% agarose gel and then blotted to Hybond™-N+ nylon membrane (Amersham Biosciences, Little Chalfont, Buckinghamshire, UK). Purified *cry8*-like gene was used as a probe, and a DIG DNA Labeling Detection Kit (Roche Company, Basel, Swiss) was used for Southern blots according to the manufacturer's instructions.

Enzyme-linked immunosorbent assay (ELISA) for the quantitative estimation of the Bt protein

We analyzed 27 samples (three replicates of each of the 8 transgenic plants plus a control plant) for the presence of the Cry8-like protein with an enzyme-linked immunosorbent assay (ELISA). Non-transgenic soybean cultivar Jinong 28 was used as a control. We essentially followed a previously published protocol [11]. The BCA protein assay kit (CWBIO, Beijing, China) was used to estimate the total soluble protein from different tissues, the anti-Cry8-like antibody was prepared in a rabbit. The protein concentration was calculated according to a standard method [35]. The optical density (OD) values was measured at 450 nm wavelength using Micro-plate reader (Biocompare, South San Francisco, USA).

Insect collecting and feeding

The *H. parallela* insects were collected from Jilin Agricultural University, Changchun, China, located at 43°48′ N, 125°24′ E. The *H. parallela* larvae were collected from soybean fields in late May and reared in a glasshouse on fresh roots of non-transgenic soybean plants. The *H. parallela* beetles were maintained in a ventilated incubator with a layer of soil (18–20% moisture) and fresh leaves as a food source. Healthy individuals were selected and starved for 24 h before use in bioassays.

Insect bioassays

For the bioassays, *H. parallela* adults were fed leaves from the T_1 transgenic plants, which had been confirmed to be transgenic by PCR, and the larvae were fed roots. One hundred insects were released per Petri dish containing 30 g leaves from one independent line. Each experimental unit consisted of three Petri dishes, which represent the three replicates used for each soybean line. The plates were incubated in an acclimatized room (24–26 °C, 18-h light/6-h dark cycle). The light wavelength ranged from 480 to 780 nm. The leaves and roots were replaced with fresh material every 3 days. Observations of the behavior, mortality, and other parameters were recorded daily for 15 days. The larval survival rate was calculated as $Nn/N_0 \times 100$, where N*n* is the number of larvae on day *n* of the experiment and N_0 is the number of *H. parallela* larvae at the beginning of the experiment. The relative consumption rate (g/g/d) = weight of food eaten/(duration of feeding (d) × mean weight of the larvae). The weight of food consumed (g) = weight of diet after feeding (g)- weight of diet before feeding (g), and the relative growth rate (g/g/d) = weight gain of the larvae/(duration of feeding (d) × mean weight of the larvae).

Field trial evaluations

The *cry8*-like transgenic soybean plants and Jinong 28 control plants were grown at Jilin Agricultural University, Changchun, China, located at 43°48′ N, 123°24′ E. A randomized complete block design was used. The field was divided into three blocks of 300 m^2 (15 × 20 m), each of which was subdivided into nine subsections. Seeds were planted in nine random subsections of each block. The plants were grown under natural temperature, light and humidity conditions during the season. The climate of the region is considered semiarid, and the region is considered to be a representative region for typical soil in Northeast China. Two months after sowing, the *H. parallela* adults and third-instar larvae were released into the field (100/300 m^2). Pesticides were not applied prophylactically at the seeding stage. The following characteristics were measured: plant height, pods per plant, seed number per plant, grain yield and hundred-seed weight. The pods of each plant were counted for 100 plants per line, and the average number was calculated. The hundred-seed weight was determined from 100 randomly chosen soybean samples, and the yield was obtained by weighing the grain mass.

Data analysis

All data were analyzed via SPSS version 22.0 software (SPSS Inc., Chicago, IL, USA) [36], each with three replicates. Two-way ANOVA was used to test whether the soybean lines or organs had a significant effect on the accumulation of the Cry8-like protein and the *cry8*-like gene in the transgenic lines. These analyses of variance were performed in two steps. Two-way ANOVAs (including all interactions and simple and main effects) were used to test for significant differences among the transgenic lines. The means were subsequently tested a posteriori to the ANOVAs via Duncan's test at the 0.05 probability level. One-way ANOVA in conjunction with Student's t test was used to determine whether the differences in the agronomic characteristics between the eight different transgenic lines were significant. $P < 0.01$ was considered statistically significant.

Supplementary information

Additional file 1: Fig. S1. Detection of the different *cry8*-like transgene regions in the eight transgenic lines by PCR.

Additional file 2: Table S1. Significant differences among transgenic soybean lines.

Additional file 3: Table S2. Significant differences among plant organs.

Additional file 4: Table S3. Primers used for the experiments in this study.

Abbreviations

d: Day; ELISA: Enzyme-linked immunosorbent assay; H: Hour; HSW: Hundred-seed weight; qRT-PCR: Quantitative reverse transcription-polymerase chain reaction; RCR: Relative consumption rate; RGR: Relative growth rate; SWPP: Seed weight per plant

Acknowledgements

We are grateful for the generous grant from the Biotechnology Center of Jilin Agricultural University and the Jilin Academy of Agricultural Sciences that made this work possible.

Authors' contributions

WP and QD designed the experiments. QD, XL and ZQ planned and performed the experiments. CM and QD edited the manuscript. All authors discussed the results and commented on the manuscript. All authors have read and approved the manuscript.

Funding

The project was funded by National Genetically Modified New Varieties of Major Projects of China (2016X8004–004). The funding bodies had no role in the design of the study and collection, analysis, and interpretation of data and in writing the manuscript.

Ethics approval and consent to participate

Not applicable.

Consent for publication

Not applicable.

Competing interests

The authors declare that they have no competing interests.

Author details

[1]Biotechnology Center of Jilin Agricultural University, Jilin Agricultural University, Changchun 130118, People's Republic of China. [2]School of Biosciences and Veterinary Medicine, University of Camerino, MC, Camerino, Italy. [3]College of Life Sciences, Kyung Hee University, 1732, Deogyeong daero, Giheung-gu, Yongin-si, Gyeonggi-do 17104, Republic of Korea.

References

1. Hill AM, Katcher HI, Flickinger BD, Kris-Etherton PM. Human nutrition value of soybean oil and soy protein. Champaign, Illinois: AOCS Press; 2008.
2. Wang C-L. Soybean insects occurring at podding stage in Taichung. J Agric Res China. 1980;29:283–6.
3. Shi S, Cui J, Song P, Zang D, Li W, Wu T. Occurring regularity of *Holotrichia parallela* in summer soybean field. Soybean Sci. 2014;33:91–4.
4. Bravo A, Gómez I, Porta H, García-Gómez BI, Rodriguez-Almazan C, Pardo L, et al. Evolution of *Bacillus thuringiensis* cry toxins insecticidal activity. Microb Biotechnol. 2013;6:17–26.
5. Kumar S, Chandra A, Pandey KC. *Bacillus thuringiensis* (Bt) transgenic crop: an environment friendly insect-pest management strategy. J Environ Biol. 2008;29:641–53.
6. Ostlie K, Hutchison WD, Hellmich R. Bt corn and European corn borer. University of Nebraska: Lincoln; 1997.
7. Xue J, Liang GM, Crickmore N, Li H, He K, Song FP, et al. Cloning and characterization of a novel cry1A toxin from *Bacillus thuringiensis* with high toxicity to the Asian corn borer and other lepidopteran insects. FEMS Microbiol Lett. 2008;280:95–101.
8. Burlet A, Chapleur-Chateau M, Haumont-Pellegri B, Jansen F, Menzaghi F, Fernette B, et al. Characterization of cry genes in a Mexican *Bacillus thuringiensis* strain collection. Appl Environ Microbiol. 1998;64:4965.
9. Anilkumar KJ, Rodrigo-Simon A, Ferre J, Pusztai-Carey M, Sivasupramaniam S, Moar WJ. Production and characterization of *Bacillus thuringiensis* cry1Ac-resistant cotton bollworm *Helicoverpa zea* (Boddie). Appl Environ Microbiol. 2008;74:462–9.
10. Shu C, Yan G, Wang R, Zhang J, Feng S, Huang D, et al. Characterization of a novel cry8 gene specific to *Melolonthidae* pests: *Holotrichia oblita* and *Holotrichia parallela*. Appl Microbiol Biotechnol. 2009;84:701–7.
11. Geng L, Chi J, Shu C, Gresshoff PM, Song F, Huang D, et al. A chimeric cry8Ea1 gene flanked by MARs efficiently controls *Holotrichia parallela*. Plant Cell Rep. 2013;32:1211–8.
12. Rupe J, Luttrell RG. Effect of pests and diseases on soybean quality. In: Johnson LA, White PJ, Galloway R, editors. Soybeans: chemistry, production processing, and utilization. Champaign, Illinois: AOCS Press; 2008. p. 93–116.
13. Kogan M, Irwin M, Sinclair J, Slife F. Major world soybean diseases, weeds and insect pests: a diagnostic pictorial atlas. Urbana, Illinois: National Soybean Research Laboratory; 1997.
14. Bandyopadhyay D. Genetically modified crops, agriculture and biosafety. In: Securing our natural wealth. Singapore: Springer; 2018. p. 81–91.
15. Romeis J, Meissle M, Bigler F. Transgenic crops expressing *Bacillus thuringiensis* toxins and biological control. Nat Biotechnol. 2006;24:63–71.
16. Roh JY, Choi JY, Li MS, Jin BR, Je YH. *Bacillus thuringiensis* as a specific, safe, and effective tool for insect pest control. J Microbiol Biotechnol. 2007;17: 547–59.
17. Datta K, Vasquez A, Tu J, Torrizo L, Alam MF, Oliva N, et al. Constitutive and tissue-specific differential expression of the *crylA(b)* gene in transgenic rice plants conferring resistance to rice insect pest. Theor Appl Genet. 1998;97: 20–30.
18. Lu Y, Wu K, Jiang Y, Guo Y, Desneux N. Widespread adoption of Bt cotton and insecticide decrease promotes biocontrol services. Nature. 2012;487: 362–5.
19. Bravo A, Likitvivatanavong S, Gill SS, Soberon M. *Bacillus thuringiensis*: a story of a successful bioinsecticide. Insect Biochem Mol Biol. 2011;41:423–31.
20. Parrott WA, All JN, Adang MJ, et al. Recovery and evaluation of soybean plants transgenic for a *Bacillus thuringiensis* var. Kurstaki insecticidal gene. In Vitro Cell Dev Biol Plant. 1994;30(3):144–9.

21. Walker DR, All JN, McPherson RM, et al. Field evaluation of soybean engineered with a synthetic *cry1Ac* transgene for resistance to corn earworm, soybean looper, velvetbean caterpillar (Lepidoptera: Noctuidae), and lesser cornstalk borer (Lepidoptera: Pyralidae). J Econ Entomol. 2000; 93(3):613–22.
22. Yu H, Li Y, Li X, et al. Expression of *Cry1Ac* in transgenic Bt soybean lines and their efficiency in controlling lepidopteran pests. Pest Manag Sci. 2013; 69(12):1326–33.
23. Jiang J, Huang Y, Shu C, Soberon M, Bravo A, Liu C, et al. *Holotrichia oblita* midgut proteins that bind to *Bacillus thuringiensis* cry8-like toxin and assembly of the *H. oblita* midgut tissue transcriptome. Appl Environ Microbiol. 2017;83:e00541–17.
24. Bi Y, Zhang Y, Shu C, Crickmore N, Wang Q, Du L, et al. Genomic sequencing identifies novel *Bacillus thuringiensis* Vip1/Vip2 binary and cry8 toxins that have high toxicity to *Scarabaeoidea* larvae. Appl Microbiol Biotechnol. 2015;99:753–60.
25. Macrae TC, Baur ME, Boethel DJ, et al. Laboratory and field evaluations of transgenic soybean exhibiting high-dose expression of a synthetic *Bacillus thuringiensis cry1A* gene for control of Lepidoptera. J Econ Entomol. 2005; 98(2):577–87.
26. Shanmugam PS, Balagurunathan R, Sathiah N. Susceptibility of *Helicoverpa armigera* Hübner instars to *Bacillus thuringiensis* insecticidal crystal proteins. Pestic Res J. 2006;18(2):186–9.
27. Soberón M, Pardo-López L, López I, et al. Engineering modified Bt toxins to counter insect resistance. Science. 2007;318(5856):1640–2.
28. Bernardi O, Sorgatto RJ, Barbosa AD, et al. Low susceptibility of *Spodoptera cosmioides, Spodoptera eridania* and *Spodoptera frugiperda* (Lepidoptera: Noctuidae) to genetically-modified soybean expressing Cry1Ac protein. Crop Prot. 2014;58:33–40.
29. Gassmann AJ, Petzold-Maxwell JL, Keweshan RS, et al. Field-evolved resistance to Bt maize by western corn rootworm. PLoS One. 2011;6(7): e22629.
30. Sheikh AA, Wani MA, Bano P, et al. An overview on resistance of insect pests against Bt crops. J Entomol Zool Stud. 2017;5:941–8.
31. Wu N, Wang P, Lin N, Lu S, Feng Y, Rong J, et al. Construction of a chalcone reductase expression vector and transformation of soybean plants. Mol Med Rep. 2017;16:6178–83.
32. Wang X-Q, Shen X, He Y-M, Ren T-N, Wu W-T, Xi T. An optimized freeze-thaw method for transformation of *Agrobacterium tumefaciens* EHA 105 and LBA 4404. Pharm Biotechnol. 2011;18:382–6.
33. Donaldson PA, Simmonds DH. Susceptibility to *Agrobacterium tumefaciens* and cotyledonary node transformation in short-season soybean. Plant Cell Rep. 2000;19:478–84.
34. Kim SH, Hamada T. Rapid and reliable method of extracting DNA and RNA from sweetpotato, Ipomoea batatas (L). Lam Biotechnol Lett. 2005;27:1841–5.
35. Palmer HM. Using antibodies: a laboratory manual. J Antimicrob Chemother. 2000;45:413.
36. Verma JP. Data analysis in management with SPSS software. Springer Science & Business Media: Berlin, Germany; 2013.

8

Structure-based engineering of heparinase I with improved specific activity for degrading heparin

Chuan Zhang[1,2], Bao-Cheng Yang[1,2], Wen-Ting Liu[1,2], Zhong-Yuan Li[1,2], Ya-Jian Song[1,2], Tong-Cun Zhang[1,2] and Xue-Gang Luo[1,2*]

Abstract

Background: Heparinase I from *Pedobacter heparinus* (Ph-HepI), which specifically cleaves heparin and heparan sulfate, is one of the most extensively studied glycosaminoglycan lyases. Enzymatic degradation of heparin by heparin lyases not only largely facilitates heparin structural analysis but also showed great potential to produce low-molecular-weight heparin (LMWH) in an environmentally friendly way. However, industrial applications of Ph-HepI have been limited by their poor yield and enzyme activity. In this work, we improve the specific enzyme activity of Ph-HepI based on homology modeling, multiple sequence alignment, molecular docking and site-directed mutagenesis.

Results: Three mutations (S169D, A259D, S169D/A259D) exhibited a 50.18, 40.43, and 122.05% increase in the specific enzyme activity and a 91.67, 108.33, and 75% increase in the yield, respectively. The catalytic efficiencies (k_{cat}/K_m) of the mutanted enzymes S169D, A259D, and S169D/A259D were higher than those of the wild-type enzyme by 275, 164, and 406%, respectively. Mass spectrometry and activity detection showed the enzyme degradation products were in line with the standards of the European Pharmacopoeia. Protein structure analysis showed that hydrogen bonds and ionic bonds were important factors for improving specific enzyme activity and yield.

Conclusions: We found that the mutant S169D/A259D had more industrial application value than the wild-type enzyme due to molecular modifications. Our results provide a new strategy to increase the catalytic efficiency of other heparinases.

Keywords: Heparinase I, Specific activity, Site-directed mutagenesis, Homology modeling, Molecular docking

Background

Heparin is a mixed sulfated polysaccharide with specific properties and dispersions. It was originally found in the liver, so it was named heparin. Heparin levels are relatively high in the lungs, ileum and skin of most species analyzed [1]. Heparinase acts directly upon heparin, yielding 52% of a trisulfated disaccharide (O-(α-L-ido-4-enepyranosyluronic acid 2-sulfate)-(1–4)-2-sulfoamino-2-deoxy-D-glucose 6-sulfate) and 40% a tetrasaccharide, in addition to small amounts of hexa- and disaccharides [2]. Although heparin is widely used in medicine, its actual biological role remains unknown [3]. The polysaccharide compounds have good activity on immune regulation, anti-tumor, anti-coagulation, and anti-inflammatory properties and have been applied to the pharmaceutical industry [4, 5]. Low-molecular-heparin (LMWH) is a small fraction of heparin produced by physical, chemical or enzymatic hydrolysis of heparin. Compared with normal-molecular-weight heparin, low-molecular-weight heparin has less anti-factor IIa activity and a lower risk of bleeding [6]. Currently, there are physical, chemical, biological, and synthetic methods for preparing low-molecular-weight heparin. Among these methods, the bioenzymatic method has become an emerging method due to its advantages of mild conditions, strong selectivity and low pollution.

In general, heparinase is an enzyme that degrade heparin by lysis. At present, most research focuses on microorganism-derived heparinases [7]. There are three

* Correspondence: luoxuegang@hotmail.com
[1]Key Laboratory of Industrial Fermentation Microbiology of the Ministry of Education & Tianjin Key Lab of Industrial Microbiology, College of Biotechnology, Tianjin University of Science and Technology, Tianjin 300457, China
[2]State Key Laboratory of Food Nutrition and Safety, Tianjin 300457, China

variants of heparinase (HepI, –II, –III), which differ in the catalytic site and enzymatic activity. Heparinase I (HepI), first isolated from *Pedobacter heparinus.* is the polysaccharide lyase that has been most extensively used to study the physiological roles of heparin-like complex polysaccharides. Heparinase I (HepI) is a polysaccharide lyase that depolymerizes heparin and heparan sulfate to produce unsaturated disaccharides and oligosaccharides by the β-elimination mechanism [8, 9].

Moreover, heparinase can be widely used to produce low-molecular-heparins [10, 11]. LMWH has more stable and predictable anticoagulation activity than unfractionated heparin, and is widely used in the prevention and treatment of venous thromboembolism and in the treatment of myocardial infarction [8]. In addition, HepI is the most important tool for analyzing heparin structure [12] and eliminating heparin in human blood [13, 14]. However, whether in laboratory-scale heparin analysis or in industrial-scale low-molecular-weight heparin production, the original manufacture of *P. heparinus* HepI is costly, which largely hinders the application of HepI [15]. Although HepI has been heterologously expressed in *Escherichia coli*, the yield was very low due to the presence of inclusion bodies [16]. The high price, low yield and low activity of HepI greatly hinder its industrial application.

Using homology modeling, multiple sequence alignment, and site-directed mutagenesis, many enzymes have been modified to improve the activity, specificity and thermostability [17–20]. However, there are still no reports on the structure-based molecular evolution of HepI. Therefore, we proposed a rational design method based on the structure of Ph-HepI from *P. heparinus* to improve its catalytic activity. Through docking a substrate and a calcium ion into the Ph-HepI model structure, analyzing related residues and multiple sequence alignment with Bt-HepI, two residues that might increase catalytic activity were selected. Mutation analysis of key amino acid residues (Ala259 and Ser169) in the loops (251–280 and 165–172) that constitute the substrate cover site and calcium binding site of the Ph-HepI was performed. To the best of our knowledge, this is the first report on improving the specific enzyme activity of HepI by protein structural engineering.

Methods

Materials

The cloning strain and the expression strain were *E. coli* DH5α and *E. coli* Rosetta (DE3), respectively. Plasmid pE-SUMO was employed as an expression vector with a 6 × His tag and a SUMO tag upstream of the multiple cloning sites. Heparin sodium salt (185 U/mg) was purchased from Solarbio company (Beijing, China). DNA polymerases, restriction enzymes, and T4 ligase were purchased from ThermoFisher (USA). The His Gravi-Trap TALON was purchased from GE Health (Uppsala, Sweden).

Construction of HepI expression plasmid

The coding gene sequence of *P. heparinus* HepI (GenBank accession no. AAA24920.1) was optimized according to the codon usage table of *E. coli* (GenBank accession no. MK229006) and was synthesized by Genewiz (Beijing). The hepI gene (1089 bp) flanked by *Bsa* I and *Bam* HI sites was amplified by PCR using the forward primer 5′-GCGGTCTCGGTGGCGGTGGCAGTGATGATGATAA ACAGCA-3′ and a reverse primer 5′-TCGGATCCTTAA CGAGCAGTTTCGGAGTAAC-3′. A flexible peptide linker (G4S) sequence and a bovine enterokinase site (D4K) sequence were also inserted at the C-terminus of Ph-HepI. The PCR fragment was inserted into the *Bsa* I and *Bam* HI sites of the pE-SUMO vector.

Sequence alignment and homology modeling of HepI

An initial sequence alignment with *P. heparinus* HepI and *Bacteroides thetaiotaomicron* HepI (GenBank accession no. AAO79780.1) (Bt-HepI) was performed by ClustalW [21], and the resulting figure was prepared with ESPript 3.0 [22]. A phylogenetic tree of Ph-HepI and other characterized polysaccharide lyases from the PL13 family (http://www.cazy.org) was constructed by MEGA 6.0 using the neighbor-joining method, and the boot-strap value was set as 1000. A 3D model structure of *P. heparinus* HepI was built using SWISS-MODEL (https://www.swissmodel.expasy.org/) based on the known crystal structure of *B. thetaiotaomicron* HepI (Bt-HepI, PDB code: 3IKW) [23].

Site-directed mutagenesis

The recombinant plasmid was PCR-amplified with mutagenic oligonucleotides using a Mut Express II Fast Mutagenesis Kit (Vazyme, Nanjing). The mutant primer sequences are shown in Table 1. PCRs were performed using *Phanta Max Super-Fidelity DNA polymerase* for 33 cycles. Each 50 μL reaction mixture consisted of 2 μL of plasmid carrying the HepI, 25 μL of 2× Max buffer, 1 μL of dNTP Mix (10 mM each), 2 μL of 10 μM forward primer, 2 μL of 10 μM reverse primer, 1 μL of *Phanta Max Super-Fidelity DNA polymerase,* and ddH_2O up to 50 μL. Thermal cycling conditions were 95 °C for 15 s followed by 30 s of 95 °C for 15 s, 62 °C for 15 s, and 72 °C for 2 min, and a final step at 72 °C for 5 min. The PCR product was digested using *Dpn* I for 2 h at 37 °C. The reaction mixture consisted of 50 μL of amplification products and 1 μL of *Dpn* I. All gel-purified reaction products were recycled using the Gel Extraction Kit. The *Dpn* I digestion product was ligated into circular plasmids using Exnase II for 30 min at 37 °C. The

Table 1 Oligonucleotide primer pairs used for the construction of mutant enzymes

S. no.	Amino acid change	Primer sequences[a]
1	S169D	5'-CAGTGGCACGGTGCTCCGGAT(GCT)CGTACTCTTGTTGCTACT – 3' 5'-AGTAGCAACAAGAGTACGATC(AGC)CGGAGCACCGTGCCACTG – 3'
2	A259D	5'-GTGGCTTACTGATAAAGAT(GAA)GATCGTAACAACGCTA-3' 5'-TAGCGTTGTTACGATCATC(TTC)TTTATCAGTAAGCCAC-3'

[a]Base changes in the relevant triplets are shown in bold and underlined. The original triplets are given in the parentheses

reaction mixture consisted of 200 ng *Dpn* I digestion product, 4 μL of 5× CE II Buffer, 2 μL of Exnase II, and ddH_2O up to 20 μL. These plasmids were sequenced correctly (GENEWIZ, Suzhou) and transformed into expressed strains (*E. coli* Rosetta (DE3)).

Enzyme expression and purification

The recombinant strain carried plasmid pE-SUMO-HepI and was cultured for 12–15 h at 37 °C on LB agar plates containing 34 μg/mL chloramphenicol and 50 μg/mL kanamycin. The seed culture (1%) was then inoculated into 100 mL of LB medium with antibiotics and cultivated at 37 °C and 220 rpm until the optical density at 600 nm (OD_{600}) reached 0.6. The cells were induced with 0.6 mM isopropyl-β-D-thiogalactopyranoside (IPTG) at 30 °C for 9 h. Cells were harvested by centrifugation at 8000 rpm for 10 min at 4 °C. Cells were washed twice with Tris–HCl buffer (20 mM Tris, 200 mM NaCl, 5 mM $CaCl_2$, pH 7.4) and resuspended in 30 mL of the same buffer, followed by sonication (sonication 3 s, pause 4 s, 300 W) on ice for 10 min to disrupt the cells. Cell debris was separated by centrifugation at 12,000 rpm for 20 min at 4 °C. The expressed HepI was the soluble protein in the supernatant [24] and analyzed by resolving 12% SDS-PAGE. The target proteins stained with Coomassie Brilliant Blue R-250 were visualized by an Odyssey Infrared Imaging System (Gene Company, Beijing, China). The supernatant containing the target protein was filtered using a 0.45 mm filter membrane and then loaded onto a Hi-Trap metal-chelating affinity column. The column was washed with 10 mL of buffer (50 mM Tris, 300 mM NaCl, 5 mM $CaCl_2$, 50 mM imidazole, pH 7.4) to wash away protein. Finally, 3–5 mL of elution buffer (50 mM Tris, 300 mM NaCl, 10 mM$CaCl_2$, 150 mM imidazole, pH 7.4) was applied to obtain the purified SUMO-HepI. The expression yield of HepI was analyzed by SDS-PAGE, and measured by Bradford's method [25].

Determination of enzyme activity and kinetic parameters

The heparinase I activity was measured by the 232 nm method [26]. The reaction was implemented at 30 °C using heparin as the substrate in reaction buffer (containing 25 g/L heparin, 50 mM CH_3COONa, 5 mM $Ca(CH_3COO)_2$ pH 7.4). Heparin degradation was detected by the UV absorbance at 232 nm on a UV-3200 spectrophotometer (Mapada, Shanghai) and the activity was calculated using a molar extinction coefficient of 3800 $M^{-1} cm^{-1}$. One international unit (IU) was defined as the amount of protein that could form 1 μmol unsaturated uronic acid per minute at 30 °C.

The kinetic parameters (K_m, $Vmax$, k_{cat} and k_{cat}/K_m) of wild-type and mutated Ph-HepI were determined at 37 °C and pH 7.4. The *K*m and *V*max were calculated with Eadie-Hofstee plots.

Effects of temperature, pH and concentration of Ca^{2+} on enzyme activity

The optimal temperature of the pure enzyme was measured by assaying the enzyme activity at various temperatures (25 °C, 29 °C, 33 °C, 37 °C, 41 °C, 45 °C). Wild-type enzyme activity at 25 °C was defined as 100%, and the relative enzyme activity of the mutants was calculated based on wild-type enzyme activity. The optimal pH of the pure enzyme was measured by assaying the enzyme activity at various pH values within the range of 4–10. Wild-type enzyme activity under pH 4 was defined as 100%, and relative enzyme activity of the mutants was calculated based on wild-type enzyme activity. The optimal concentrations of Ca^{2+} for the pure enzyme was measured by assaying the enzyme activity at various concentrations of Ca^{2+} (0 mM, 5 mM, 10 mM, 50 mM, 100 mM, 200 mM). Wild-type enzyme activity under concentrations of 0 mM Ca^{2+} was defined as 100%, and relative enzyme activity of the mutants was calculated based on wild-type enzyme activity.

Mass spectrometry analysis of the product

Analysis of degraded low-molecular-weight heparin used oligosaccharide time-of-flight mass spectrometry (MALDI-TOF-MS) [27]. First, a low-molecular-weight heparin was irradiated with a laser to form a film cocrystallized with the matrix, and then the matrix was absorbed into the energy by the laser and then transmitted to the biomolecule. The main principle of time-of-flight mass spectrometry is that the ions are accelerated through the flight path by an electric field, and the mass-to-charge ratio (M/Z) of the ions ionized by the sample is determined from their time of arrival at the detector. Target plate: ground steel target. Matrix solution: 2,5-DHB (20 mg/L, T30 dissolved). Matrix additive: 1 mM NaCl. The sample was dissolved using 30%

acetonitrile (1 mg/mL). Sample Preparation: The solution included 1 μL of sample solution and 1 μL of matrix solution and was naturally dried on the sample target.

Detection of anti-Xa and anti-IIa activity of the product

LMWH anti-Xa and anti-IIa factor activity detection referred to the European Pharmacopoeia [28]. This method compares the anti-thrombin activity of a sample with the LMWH standard in vitro to determine the ability of the test sample to accelerate the inhibition of factor Xa and factor IIa. The reagents used were purchased from Shanghai Boatman Biotech.

REAGENT 1: Human antithrombin (AT): Each vial was reconstituted with exactly 1 mL of distilled water. It was shaken thoroughly until complete dissolution of the contents (vortex). The contents homogenized for 30 min at room temperature (18–25 °C) while shaking the vial from time to time. Just before use, the sample was diluted 1:5 with R4 buffer. REAGENT 2: Factor Xa/factor IIa: Each vial was reconstituted with exactly 1 mL of distilled water and shaken thoroughly until complete dissolution of the contents (vortex). The contents homogenized for 30 min at room temperature (18–25 °C) while shaking the vial from time to time. Just before use, each sample was diluted 1:5 with R4 buffer. REAGENT 3: Factor Xa/ factor IIa specific chromogenic substrate: Each vial was reconstituted with exactly 5 mL of distilled water and shaken thoroughly until complete dissolution of the contents (vortex). The contents homogenized for 30 min at room temperature (18–25 °C) while shaking the vial from time to time (vortex). REAGENT 4: Assay Reaction Buffer at pH 8.40: Ready-to-use vial of 25 ml. It was shaken before use.

The reaction was carried out in a microplate. Manual method: We added 40 μL of heparinized sample (at 1:10 dilution) and 40 μL of R1, mixed, and incubated at 37 °C for 2 min. We then introduced 40 μL of R2, mixed, and incubated at 37 °C for exactly 2 min (stage 1). Next, we introduced 40 μL of R3, mixed, and incubated at 37 °C for exactly 2 min (stage 2). The reaction was stopped by introducing 80 μL of citric acid (20 g/L). We mixed and measured the absorbance at 405 nm against the corresponding blank.

Statistical analysis

All experiments were performed independently at least three times, and the results are expressed as the mean ± standard deviation. The T-test was used for statistical analysis.

Results

Sequence analysis and homology modeling

Protein sequence analysis showed that Ph-HepI shared 62.5% identity with the *B. thetaiotaomicron* HepI and multiple sequence alignment (MSA) of the two heparinase I enzymes is shown in (Fig. 1). To explore the key sites of Ph-HepI, a homology model of the protein was built using SWISS-MODEL (Fig. 2) according to the known crystal structure of the *B. thetaiotaomicron* HepI. Additional file 1: Figure S1 shows the phylogenetic tree of Ph-HepI and the characterized polysaccharide lyases from the PL13 family. It can be seen that most characterized PL13 family polysaccharide lyases are from bacteria. Additional file 2: Figure S2 shows that the 3D model structure of *P. heparinus* HepI was reasonable. Residues in most favored regions were over 90% conserved. The stereochemistry of the structure was analyzed using structural analysis and verification server of the University of California, Los Angeles (http://services.mbi.ucla.edu/SAVES/). Finally, all structural figures were generated with PyMOL (www.pymol.org) [29].

Enzymatic interactions related residues and selection of mutational sites

It has been reported that residues D155, E222, W248, N345, and D346 in Bt-HepI are involved in calcium ion binding [23]. Based on the alignment assay, these residues in Bt-HepI were corresponded to S169, E228, W254, N351, and D352 in Ph-HepI (Fig. 3a). In the template, several amino acid residues formed a perfect lid that covered the substrate, including E73, Y75, K185, and D253, which correspond to E87, Y89, K199, and A259 in Ph-HepI (Fig. 3b). The amino acid residues 169 and 259 in Ph-HepI were different from the corresponding residues in Bt-HepI (Fig. 3c). Other residues in Ph-HepI, which also interacted with the substrate, shared no obvious counterparts in the template. The presence of a Ca^{2+} ion is crucial for heparinase I activity [30, 31]. From molecular docking, a Ca^{2+} binding site was also found in the active pocket of Ph-HepI. In general, the Ca^{2+} ion plays an important role in facilitating bridging between the enzyme and substrate in the complex.

The model structure of Ph-HepI revealed an unusually deep and narrow shape of the substrate tunnel. We performed docking of the structures of Ph-HepI and heparin to further investigate substrate binding (Fig. 4a).. The results showed that there was a perfect lid to cover the substrate, consisting of four amino acid residues (E87, Y89, K199, and A259). Compared with the two enzymes, the residue A259 is different in Ph-HepI. Bipyramidal coordination of Ca^{2+} ion in the Ph-HepI structure among the residues (S169, E228, W254, N351, and D352) involved in the substrate-binding pocket was also discovered (Fig. 4b). Ph-HepI had a similar structure to Bt-hepI, but the amino acid residue S169 was different from the D155 in Bt-HepI. Therefore, to improve the specific enzyme activity of Ph-HepI, three mutants, S169D, A259D, and S169D/A259D, were designed.

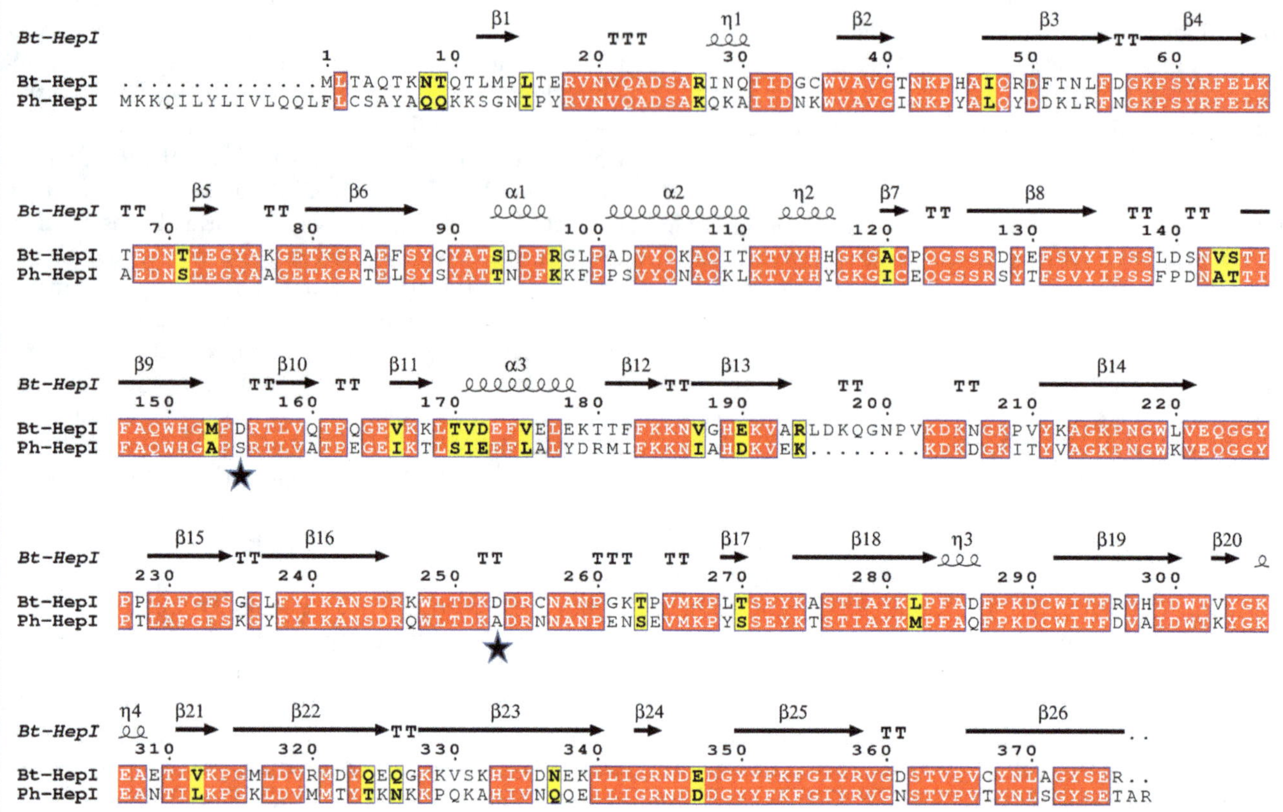

Fig. 1 Structural sequence alignment of Bt-HepI and Ph-HepI. Residues forming the secondary structures are highlighted in the Bt-HepI sequence. Identical and similar amino acid residues are shown by white letters on a red background and black letters on a yellow background, respectively. Two different amino acids that are not conserved in Ph-HepI and Bt-HepI are indicated by black stars

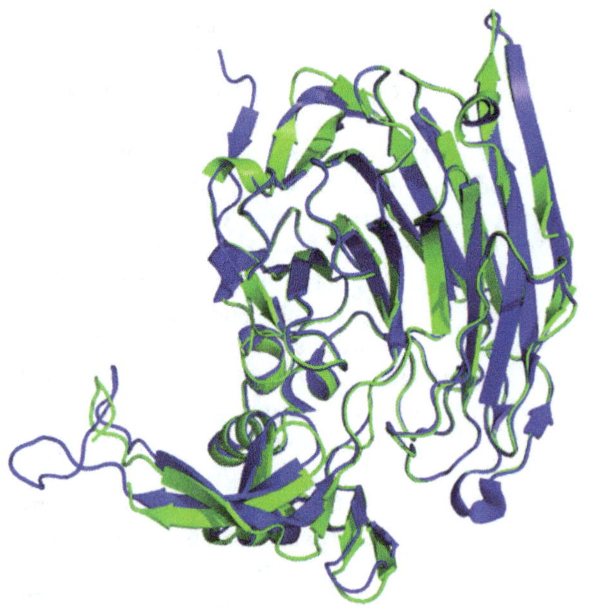

Fig. 2 The overall structure of Ph-HepI. Structure comparison of Ph-HepI (green) is superimposed with the template from *Bacteroides thetaiotaomicron* (blue; PDB ID: 3ikw)

Expression and purification of Ph-HepI mutants

The wild-type and mutant recombinant expression strains were successfully constructed. By analyzing the expression of wild-type and mutant Ph-HepI, it was found that there were different expression levels. Three mutants of Ph-HepI showed soluble expression, which was used to determine enzymatic properties. The purification of Ph-HepI wild-type and mutants was analyzed by 12% SDS-PAGE (Fig. 5). Obviously, the amount of protein expressed was significantly increased after the mutation. The protein concentrations of Ph-HepI wild-type, S169D, A259D and S169D/A259D were 0.12, 0.23, 0.25, and 0.21 mg/mL by quantitative protein and gray-level analysis, respectively.

Enzymatic characterization and kinetic parameters of Ph-HepI mutants

To investigate the effects of on the bioactivity of the mutation, enzymatic characteristics of wild-type and mutated enzymes were measured. As shown in Fig. 6, amino acid mutations did not alter the optimal reaction temperature or pH of enzymes, and they all had an optimal temperature and pH of 37 °C and 7, respectively (Fig. 6a-b). However, the optimal concentration of Ca^{2+}

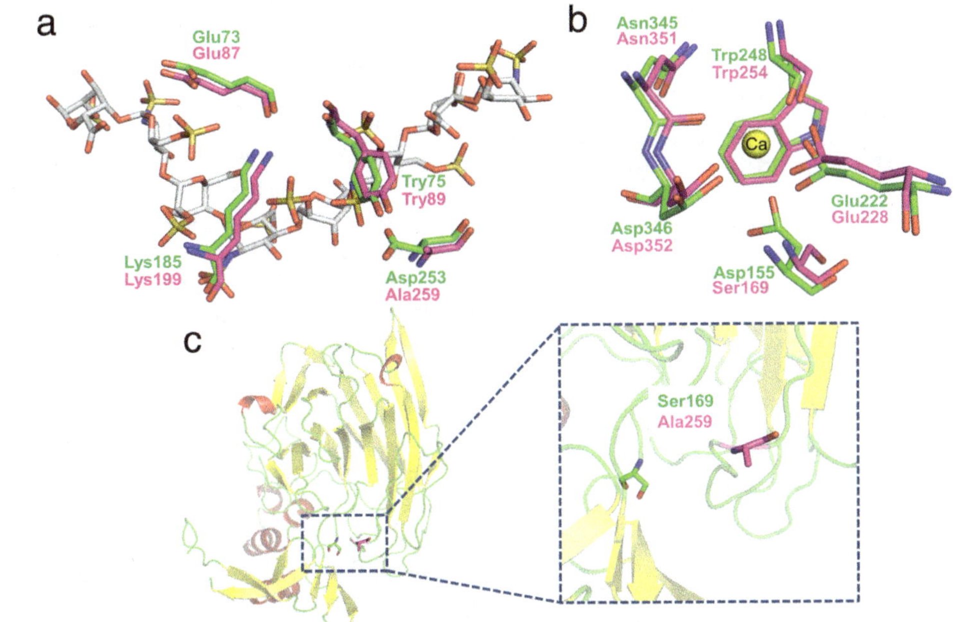

Fig. 3 Substrate recognition and calcium ion binding interaction sites. (**a**) The substrate is shown as sticks in gray. Residues from template and Ph-HepI are shown in green and magenta, respectively. (**b**) The calcium ion is shown as a yellow sphere. Residues from template and Ph-HepI are shown in green and magenta, respectively. (**c**) Cartoon representation of the structure model of Ph-HepI. The two residues (Ser169 and Ala259) in the Ph-HepI are indicated by markers in green and magenta, respectively. Two amino acids in Ph-HepI are different from template (Ser169, Ala259)

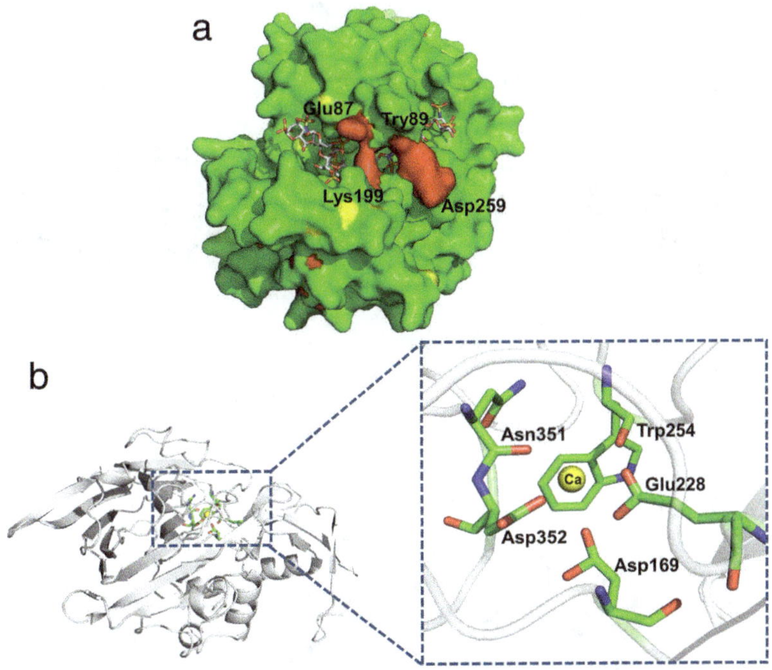

Fig. 4 Structures of Ph-HepI bound with heparin and calcium ion. (**a**) The binding of heparin in the positively charged canyon of Ph-HepI-A259D is shown as a surface charge presentation. The similar cover, including four amino acids (Glu87, Tyr89, Lys199, Asp259) on the surface, is shown in red. (**b**) The interaction between Ph-HepI and calcium ion (Ca^{2+}). The Ca^{2+} is shown as a yellow sphere. Amino acid residues Asp169, Glu228, Trp254, Asn351, and Asp352, which are essential for the interaction, are shown in sticks

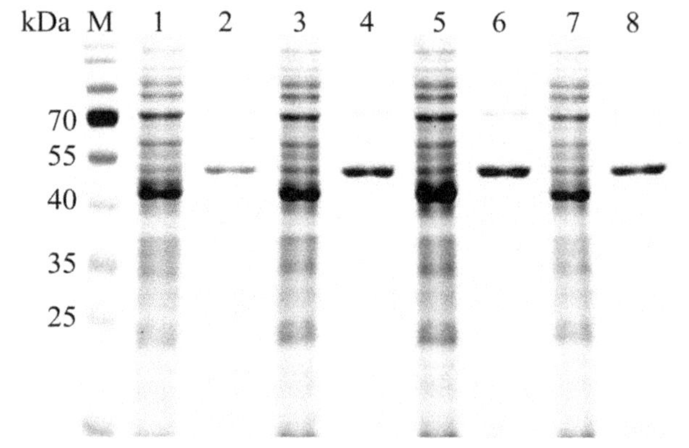

Fig. 5 The purified protein was analyzed by 12% SDS-PAGE and stained with Coomassie Brilliant Blue R-250. Lane M—protein marker; lanes 1, 3, 5, 7—wild-type Ph-HepI and S169D, A259D and S169D/A259D mutants are produced by inducing the recombinant bacteria; lanes 2, 4, 6, 8—purified wild-type Ph-HepI and S169D, A259D and S169D/A259D mutants. The band of Ph-HepI protein (53 kDa) is indicated by an arrow

was changed, and the optimal concentration of Ca^{2+} for wild-type and A259D mutations was 10 mM, but the optimal concentration of S169D and S169D/A259D mutations was 50 mM (Fig. 6c). The reason for this phenomenon might be that the amino acid at position 169 is in the binding region of Ca^{2+}, the Asp has a longer side chain than Ser, and it is an acidic amino acid with a stronger electron cloud density. When the Ser was mutated to Asp, the binding of the enzyme to Ca^{2+} was enhanced.

As shown in Table 2, compared with the wild-type enzyme, the K_m values of the mutant enzymes S169D, A259D, and S169D/A259D were lower than the wild-type enzyme by 68, 58 and 73%, respectively. Furthermore, compared with the wild-type enzyme, the catalytic constants (k_{cat}) of the mutated enzymes S169D, A259D, and S169D/A259D were higher than those of the wild-type enzyme by 20, 12, and 31%, respectively. Consequently, the catalytic efficiencies (k_{cat}/K_m) of the mutated enzymes S169D, A259D, and S169D/A259D were higher than those of the wild-type enzyme by 275, 164, and 406%, respectively.

Enzymatic activity and structural analysis of mutants

When the purified proteins were obtained, the enzyme activity of the wild-type and three mutants of Ph-HepI proteins were examined. As shown in Fig. 7, the S169D mutation exhibited an obvious improvement of hydrolysis activity on heparin, demonstrating that S169 is crucial for Ph-HepI activity. Another active-site mutation A259D, resulted in obvious

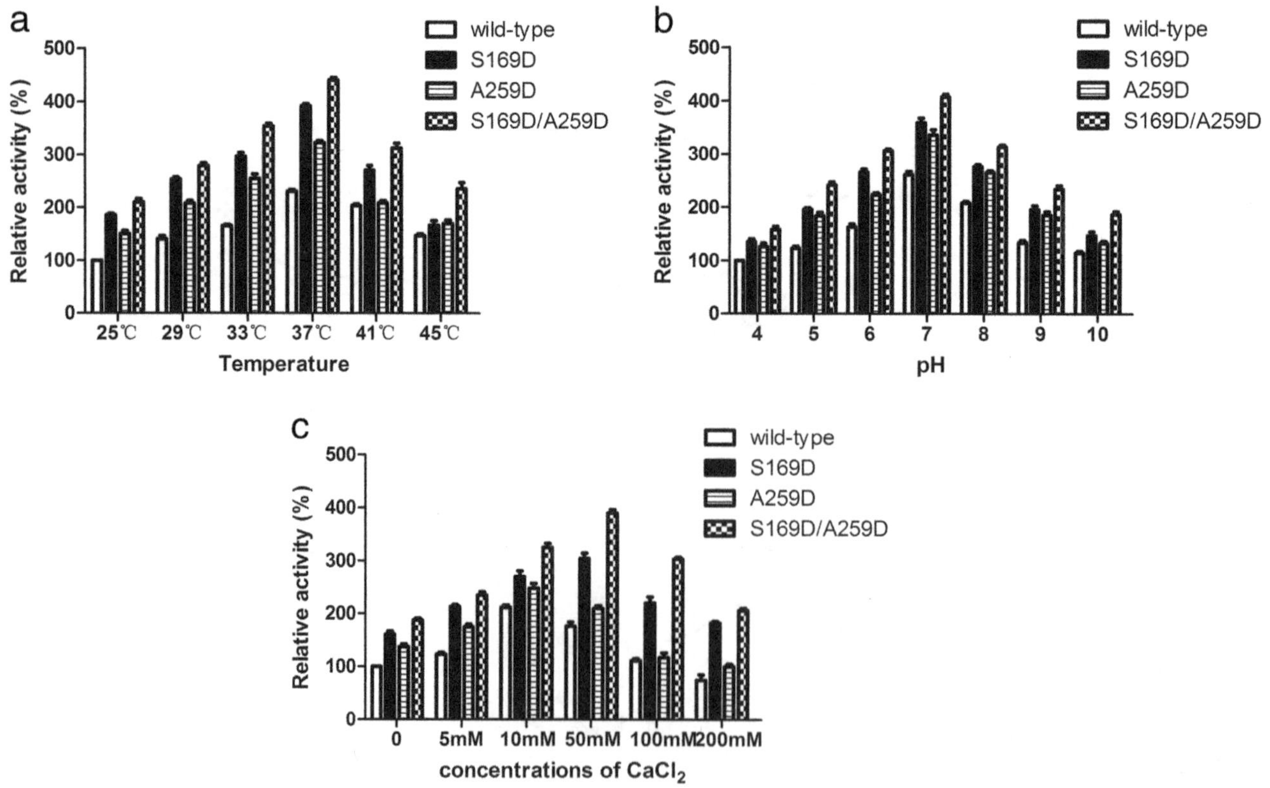

Fig. 6 Effects of temperature (**a**), pH (**b**) and concentration of Ca^{2+} (**c**) on the activities of mutated Ph-HepI

Table 2 Kinetic parameters of wild-type, mutated and combined mutations Ph-HepI

Enzyme	K_m (mM)	k_{cat} (s^{-1})	k_{cat}/K_m ($s^{-1}\ M^{-1} \times 10^{-3}$)
Wild type	0.78 ± 0.05	19.31 ± 0.17	24.76
S169D	0.25 ± 0.07	23.19 ± 0.09	92.76
A259D	0.33 ± 0.09	21.56 ± 0.15	65.33
S169D/A259D	0.21 ± 0.06	26.32 ± 0.03	125.33

improvement of enzymatic activity, showing that this mutation is essential for catalysis. The combined mutation S169D/A259D showed the highest enzyme activity toward heparin. Compared with the wild-type enzyme activity, the specific enzyme activity of mutations S169D, A259D and S169D/A259D increased by 50.18, 40.43, and 122.05% (Fig. 7).

To understand how mutations affect the catalytic efficiency of enzymes, we analyzed the structure of a series of mutants (S169D, A259D, and S169D/A259D). Asp is an acidic amino acid with a strong negative charge, and its side chain is longer in space than Ser. Therefore, when the amino acid at position 169 was mutated from Ser to Asp, and an ionic bond formed with the calcium ion. The amino acid Asp not only forms an ionic bond with a calcium ion, but

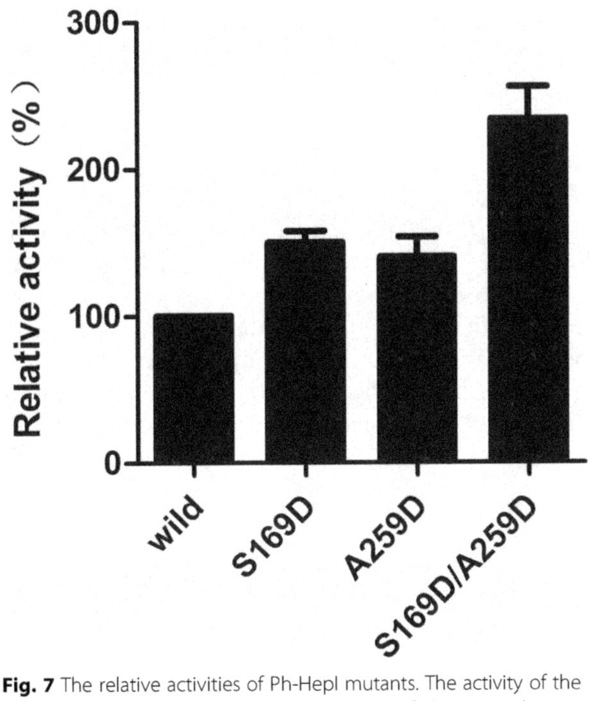

Fig. 7 The relative activities of Ph-HepI mutants. The activity of the Ph-HepI WT was taken as 100%. The mutants of Ph-HepI with considerable enzymatic activity are represented as bars. The error bars indicate the standard deviation obtained from three independent experiments

also forms a hydrogen bond with the adjacent Asp353. Finally, the Ph-HepI and substrate were more tightly bound, increasing the enzyme's specific enzyme activity (Fig. 8a-b). Additionally, when the amino acid at position 259 was mutated from Ala to Asp, a hydrogen bond formed between Asp 259 and Try 87. Furthermore, there were four amino acids (Glu87, Lys199, Tyr89, Asp259) above the substrate that formed a similar lid to cover the substrate by forming a hydrogen bond between them. A similar lid allowed the substrate to be better placed in the active pocket, thereby increasing the catalytic efficiency of the enzyme. The mutated amino acid Asp259 still formed hydrogen bonds with the surrounding amino acids Asp257 and Arg 261, as in the wild-type (Fig. 8b-c).

Mass spectrometry and activity determination of LMWH

To study the degradation of heparin by the enzyme after mutation, we performed mass-spectrometric detection and activity measurement on the degradation products. As shown in Additional file 3: Figure S3, we measured the mass spectra of the enzymatic hydrolysates of wild-type, S169D, A259D and S169D/A259D. Compared with the standards of LMWH, the enzymatically hydrolyzed products we obtained met the standard requirements. The molecular masses of the products were 4000–6000 Da. The anti-FXa and anti-FIIa activities of the enzymatic hydrolysate were determined by the chromogenic substrate method, and the results are shown in Table 3. The anti-FXa and anti-FIIa activity of the enzymatic hydrolysate met the requirements of the pharmacopoeia standard at 1.5–2.5 [28].

Discussion

Heparinase I is used for the production of low-molecular-weight heparin (LMWH) from heparin, and the LMWH can retain the anticoagulant properties [32]. There are many sources of heparin, and the porcine intestinal mucosa is the most important source [33], but the low production of LMWHs limits its development. The major degradation methods of heparin include chemical lysis and enzymatic lysis [34]. Although chemical lysis is simpler than enzymatic lysis, it has many disadvantages. The use of nitrite in the lysis process has caused serious environmental pollution, and the drug activity of prepared LMWHs is low. However, LMWH production still requires the use of chemical degradation instead of purely enzymatic degradation due to the low production and activity of enzymes [35]. Natural HepI is mainly produced by *Pedobacter heparinus* [36]. Since the yield of HepI in *P. heparinus* is low, recombinant expression of HepI using genetic engineering strategies has received

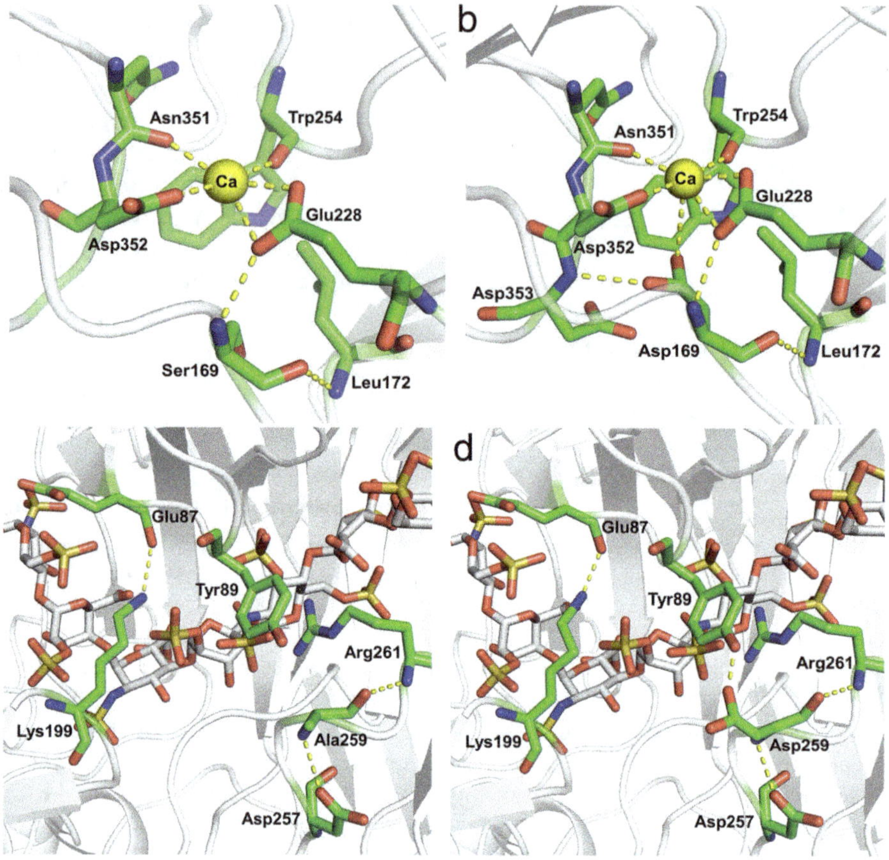

Fig. 8 Structural analysis of Ph-HepI wild-type and mutant enzymes. (**a**) In the wild-type, Ser169 forms two hydrogen bonds with Leu172 and Glu228. (**b**) In the mutant S169D, Asp169 forms three hydrogen bonds with adjacent amino acids (Leu172, Glu228, and Asp353) and forms an ionic bond with the calcium ion. (**c**) In the wild-type, Ala259 forms two hydrogen bonds with Asp257 and Arg261. (d) In the mutant A259D, Asp259 forms three hydrogen bonds with Tyr89, Asp257, and Arg261

widespread attention [16]. However, recombinantly expressed HepI in *E. coli* easily aggregates into insoluble inclusion bodies [10]. To solve this problem, fusion expression with a tag that has the ability to improve the formation of the correct conformation is a useful approach for the soluble expression of this protein [37–39]. In our previous research, to establish a soluble expression system of Ph-HepI, the *hepI* gene (coding for HepI) was fused with a hexahistidine-tag (6 × His) and genes encoding a small ubiquitin-like modifier (SUMO), a flexible peptide linker (G4S) and a bovine enterokinase site (D4K) [24].

By fusion to maltose binding protein (MBP), the first soluble expression of HepI was achieved in *E. coli* [39]. The soluble expression of HepI has been greatly improved by fusioing to a SUMO-tag, but the enzyme activity is still very low [40]. Therefore, this study improved the specific enzyme activity by using homology modeling, multiple sequence alignment, molecular docking, and site-directed mutagenesis.

Table 3 Anticoagulant activity and molecular weight of LMWH

Sample	Anti-FXa(IU/mg)	Anti-FIIa(IU/mg)	Anti-FXa/Anti-IIa	Relative molecular mass
1	121.72 ± 4.18	48.37 ± 5.02	2.5164 ± 0.15	4326.984
2	112.18 ± 9.32	59.01 ± 3.21	1.9010 ± 0.06	5168.031
3	113.24 ± 7.51	56.35 ± 6.08	2.0096 ± 0.13	4944.875
4	115.72 ± 3.29	58.92 ± 3.72	1.9640 ± 0.09	5209.882
5	119.89 ± 2.87	57.05 ± 3.02	2.1015 ± 0.12	5032.487

1:Low molecular weight heparin, 2:Heparin degraded by wild-type Ph-hepI,3:Heparin degraded by mutant S169D,4:Heparin degraded by mutant A259D, 5:Heparin degraded by mutant S169D/A259D

The molecular modification of heparinase is important to improve the performance of the enzyme. Substitution of Cys297 to serine in MBP-HepI offered a 30.6% increase in recovered total enzyme activity due to a mutation-induced thermostabilizing effect [31]. The clone *E. coli*-heparinase-I133/P316 with two amino acid substitutions has been screened and identified by error-prone PCR. Compared with the wild-type, the mutant HepI has a 57.8% increased enzyme activity [41]. In these studies, the HepI enzyme activity was improved by directed evolution and fusion expression, but these mutations were not well explained in the HepI structure.

To date, there is no crystal structure of Ph-HepI for reference. Therefore, our focus was to explore the structure and function of HepI through multisequence alignment and 3D homologous modeling (Fig. 2). Since Ph-HepI shared 62.5% identity with structurally known Bt-HepI from *Bacteroides thetaiotaomicron* (Fig. 1), a series of active residues of Ph-HepI were predicted (E87, Y89, D169, K199, E248, W254, D259, N351, and D352) through multiple-sequence alignment, homology modeling and molecular docking. Based on sequence alignment of Ph-HepI and Bt-HepI, the residues S169 and A259 in the Ph-HepI were found to be different from in the Bt-HepI, in which they were instead of D155 and D253 (Fig. 3). Three mutants (S169D, A259D, and S169D/A259D) were constructed by site-directed mutagenesis and analyzed for their enzymatic properties. Compared with the wild-type, the specific enzyme activity of the three mutants (S169D, A259D, and S169D/A259D) was increased by 50.18, 40.43, and 122.05%, respectively (Fig. 6). The optimum reaction temperature and pH of the three mutants were 37 °C and 7, which was consistent with the wild-type. However, the optimal concentration of Ca^{2+} for wild-type and A259D was 10 mM, while the optimal concentration of Ca^{2+} for S169D and S169D/A259D was 50 mM (Fig. 7).

Our study found that the exchange of several amino acids could produce more active enzymes, which had basically identical interresidue interactions to that observed in the Bt-HepI crystal structure. The exchange of S169 with aspartic acid resulted in an enzyme with higher activity, which promoted D169 to become the nucleophilic amino acid of Ph-HepI and resulted in its combination with calcium. This is consistent with the structure of heparinase I binding to calcium, which can increase the initial activity of Ph-HepI [31, 42]. When Ser was mutated to Asp in Ph-HepI, the enzyme activity was significantly improved. The underlying cause of increased enzyme activity may be the binding and activation of D169 with the calcium ion and the formation of new hydrogen bonds with nearby amino acid residues.

However, the amino acid at position 259 was Ala instead of Asp. Alanine is hydrophobic and did not form a hydrogen bond with Tyr89, but aspartic acid has a side-chain radical containing a carboxyl, which allowed it to produce hydrogen bond with Tyr89. Following the construction of the combined mutation (S169D/A259D), the specific enzyme activity had obvious improvements. The mutants only specifically increased the specific enzyme activity of the enzyme and did not change the enzyme degradation substrate. Altogether, our results provide structural information to reveal Ph-HepI's catalytic mechanisms and guide rational engineering of Ph-HepI for industrial applications.

Conclusions

In summary, the enzymatic activity of Ph-HepI was significantly improved through site-directed mutagenesis with the help of a rational structure-based design method. Compared with the wild-type enzyme, the enzyme activity and catalytic efficiency ($k$cat/Km) of the mutant S169D/A259D increased by 122.05 and 406.18%, respectively. We provided an effective method for the modification of heparinase I and laid the foundation for the industrial application of heparinase I.

Additional files

Additional file 1: Figure S1. Phylogenetic tree of characterized heparinase I of PL13 family (PDF 113 kb)

Additional file 2: Figure S2. The stereochemical spatial arrangement of amino acid residues in the modelled 3D structure of Ph-HepI in the favored region of the Ramachandran plot (PDF 106 kb)

Additional file 3: Figure S3. Mass spectrometry analysis of degradation products (PDF 92 kb)

Abbreviations

Bt-HepI: Heparinase I from *Bacteroides thetaiotaomicron*; HepI: Heparinase I; Ala: Alanine; Asp: Aspartic; *E. coli*: *Escherichia coli*; HepI: Heparinase I; IPTG: Isopropyl β-D-thiogalactoside; LB: Luria-Bertani; LMWH: Low molecular heparin; Ph-HepI: Heparinase I from *Pedobacter heparinus*; PL13: Polysaccharide lyases family 13; Ser: Serine

Acknowledgements

Not applicable.

Authors' contributions

CZ and BY performed experiments and analyzed data; WL, ZL, YS contributedmaterials; XL and TZ analyzed data and designed experiments; CZ and XL wrotethe paper. All authors read and approved the final manuscript.

Funding

This work was supported by the National Key R&D Program of China (2017YFD0400304), the National 863 project of China (No. 2012AA021505), and the project for the public service platform of strain breeding and fermentation technology of industrial microorganisms (17PTGCCX00190).

Ethics approval and consent to participate
Not applicable.

Consent for publication
Not applicable.

Competing interests
The authors declare that they have no competing interests.

References

1. Helena B. Nader, Helio K. Takahashi AHS and, Dietrich CP. Selective distribution of the heparin in mammals conspicuous presence of heparin in lymphoid tissues. Biochim Biophys Acta 1980;627:40–48.
2. Silva M, Dietrich C. Structure of heparin. Biol Chem. 1975;250:6841–6 http://www.ncbi.nlm.nih.gov/pubmed/13619777.
3. Yates EA, Gallagher JT, Guerrini M. Introduction to the molecules special edition entitled "Heparan sulfate and heparin: challenges and controversies": some outstanding questions in Heparan sulfate and heparin research. Molecules. 2019;24:1399. https://doi.org/10.3390/molecules24071399.
4. Folkman J, Langer R, Linhardt R, Haudenschild C, Taylor S. Angiogenesis inhibition and tumor regression caused by heparin or a heparin fragment in the presence of cortisone. Science (80-). 1983;221:719–25. doi:https://doi.org/10.1126/science.6192498.
5. Yaping Chen, Terry Maguire, Ronald E. Hileman, Jonathan R. Fromm, Jeffrey D. Esko RJL& RMM. Dengue virus infectivity depends on envelope protein binding to target cell heparan sulfate. Nat Med. 1997;3:866–871.
6. Quader MA, Stump LS, Sumpio BE. D P. specialty section : vascular surgery low molecular weight heparins : current use and indications. J Am Coll Surg. 1998;187:641–58.
7. Carl P.Dietrich, Maria E.silva YMM. Sequential Degradation of Heparin Fla . vobacterium. TRE J C Biol Chem 1973;248:6408–6415.
8. Ernst S, Venkataraman G, Winkler S, Godavarti R, Langer R, Cooney CLSR. Expression in Escherichia coli, purification and characterization of heparinase I from Flavobacterium heparinum. Biochem J. 1996;315:589–97.
9. Yang VC, Linhardt RJ, Bernstein H, Cooney CL, Langer R. Purification and characterization of heparinase from Flavobacterium heparinum. J Biol Chem. 1985;260:1849–57.
10. Pervin A, Gallo C, Jandik KA, Han XJLR. Preparation and structural characterization of large heparin-derived oligosaccharides. Glycobiology. 1995;5:83–95.
11. Ye F, Kuang Y, Chen S, Zhang C, Chen Y, Xing X-H. Characteristics of low molecular weight heparin production by an ultrafiltration membrane bioreactor using maltose binding protein fused heparinase I. Biochem Eng J. 2009;46:193–8. https://doi.org/10.1016/j.bej.2009.05.007.
12. Guerrini M, Beccati D, Shriver Z, Naggi A, Viswanathan K, Bisio A, et al. Oversulfated chondroitin sulfate is a contaminant in heparin associated with adverse clinical events. Nat Biotechnol. 2008;26:669–75.
13. Ameer GA, Harmon W, Sasisekharan R, Langer R. Investigation of a whole blood fluidized bed Taylor-Couette flow device for enzymatic heparin neutralization. Biotechnol Bioeng. 1999;62:602–8. https://doi.org/10.1002/(SICI)1097-0290(19990305)62:5<602::AID-BIT12>3.0.CO;2-M.
14. Linhardt RJ, Cooney CL, Tapper D, Zannetos CA, Larsen AK, Langer R. An immobilized microbial heparinase for blood deheparinization. Appl Biochem Biotechnol. 1984;9:41–55.
15. Ernst S, Garro OA, Winkler S, Venkataraman G, Langer R, Cooney CL, et al. Process simulation for recombinant protein production: cost estimation and sensitivity analysis for heparinase I expressed in Escherichia coli. Biotechnol Bioeng. 1997;53:575–82.
16. Sasisekharan R, Bulmer M, Moremen KW, Cooney CL, Langer R. Cloning and expression of heparinase I gene from Flavobacterium heparinum. Proc Natl Acad Sci. 1993;90:3660–4. https://doi.org/10.1073/pnas.90.8.3660.
17. Vasu V, Kumaresan J, Babu MG, Meenakshisundaram S. Active site analysis of cis-epoxysuccinate hydrolase from Nocardia tartaricans using homology modeling and site-directed mutagenesis. Appl Microbiol Biotechnol. 2012; 93:2377–86.
18. Deng Z, Yang H, Shin HD, Li J, Liu L. Structure-based rational design and introduction of arginines on the surface of an alkaline α-amylase from Alkalimonas amylolytica for improved thermostability. Appl Microbiol Biotechnol. 2014;98:8937–45.
19. Choi H, Kyeong HH, Choi JM, Kim HS. Rational design of ornithine decarboxylase with high catalytic activity for the production of putrescine. Appl Microbiol Biotechnol. 2014;98:7483–90.
20. Panigrahi P, Chand D, Mukherji R, Ramasamy S, Suresh CG. Sequence and structure-based comparative analysis to assess, identify and improve the thermostability of penicillin G acylases. J Ind Microbiol Biotechnol. 2015;42:1493–506.
21. Chenna R, Sugawara H, Koike T, Lopez R, Gibson TJ, Higgins DG, et al. Multiple sequence alignment with the Clustal series of programs. Nucleic Acids Res. 2003;31:3497–500.
22. Gouet P, Courcelle E, Stuart DI, Métoz F. ESPript: analysis of multiple sequence alignments in PostScript. Bioinformatics. 1999;15:305–8.
23. Han Y-H, Garron M-L, Kim H-Y, Kim W-S, Zhang Z, Ryu K-S, et al. Structural snapshots of heparin Depolymerization by heparin Lyase I. J Biol Chem. 2009;284:34019–27. https://doi.org/10.1074/jbc.M109.025338.
24. Yang BC, Zhang C, Wang C, Zhou H, Li ZY, Song YJ, et al. Soluble expression and purification of heparinase I in Escherichia coli using a hexahistidine-tagged small ubiquitin-like modifier as a fusion partner. Biotechnol Biotechnol Equip. 2017;31:1040–5.
25. Bradford MM. A rapid and sensitive method for the quantitation of microgram quantities of protein utilizing the principle of protein-dye binding. Anal Biochem. 1976;72:248–54. https://doi.org/10.1016/0003-2697(76)90527-3.
26. Bernstein H, VC Y, CL C, R L. Immobilized heparin lyase system for blood deheparinization Methods Enzymol 1988;137:515–529.
27. Marie A, Fournier F, Tabet JC. Characterization of synthetic polymers by MALDI-TOF/MS: investigation into new methods of sample target preparation and consequence on mass spectrum finger print. Anal Chem. 2000;72:5106–14.
28. Europe C of. European pharmacopoeia,Heparins,Low-molecular-mass (5th edition). Renouf Pub CoLtd. 2005;:1717–9.
29. Bramucci E, Paiardini A, Bossa F, Pascarella S. PyMod: Sequence similarity searches, multiple sequence-structure alignments, and homology modeling within PyMOL. BMC Bioinformatics. 2012;13 SUPPL.4:S2.
30. Chen S, Ye F, Chen Y, Chen Y, Zhao H, Yatsunami R, et al. Biochemical analysis and kinetic modeling of the thermal inactivation of MBP-fused heparinase I: implications for a comprehensive thermostabilization strategy. Biotechnol Bioeng. 2011;108:1841–51. https://doi.org/10.1002/bit.23144.
31. Chen S, Huang Z, Wu J, Chen Y, Ye F, Zhang C, Yatsunami R, Nakamura S XX. Combination of site-directed mutagenesis and calcium ion addition for enhanced production of thermostable MBP-fused heparinase I in recombinant *Escherichia coli*. Appl Microbiol Biotechnol. 2013;97(7) c:2907–16.
32. Laurent P, Dussarat G-V, Bonal J, Jego C, Talard P, Bouchiat C, et al. Low molecular weight heparins: a guide to their optimum use in pregnancy. Drugs. 2002;62:463–77.
33. Griffin CC, Linhardt RJ, Van Gorp CL, Toida T, Hileman RE, Schubert RL, et al. Isolation and characterization of heparan sulfate from crude porcine intestinal mucosal peptidoglycan heparin. Carbohydr Res. 1995;276:183–97.
34. Linhardt RJ, Gunay NS. Production and chemical processing of low molecular weight heparins. Semin Thromb Hemost. 1999;25(Suppl 3):5–16.
35. Liu X, Stange K, Fareed J, Hoppensteadt D, Jeske W, Kouta A, et al. Comparison of low-molecular-weight heparins prepared from bovine heparins with enoxaparin. Clin Appl Thromb. 2017;23:542–53.
36. D L Lohse, R J Linhardt. Purification and characterization of heparin lyases from Flavobacterium heparinum. J Theor Biol 1992;159:377–379. doi:https://doi.org/10.1016/S0022-5193(05)80731-3.
37. Huang J, Cao L, Guo W, Yuan R, Jia Z, Huang K. Enhanced soluble expression of recombinant Flavobacterium heparinum heparinase I in Escherichia coli by fusing it with various soluble partners. Protein Expr Purif. 2012;83:169–76. https://doi.org/10.1016/j.pep.2012.03.016.
38. Luo Y, Huang X, McKeehan WL. High yield, purity and activity of soluble recombinant Bacteroides thetaiotaomicron GST-heparinase I from Escherichia coli. Arch Biochem Biophys. 2007;460:17–24. https://doi.org/10.1016/j.abb.2007.01.029.
39. Chen Y, Xing X-H, Ye F, Kuang Y, Luo M. Production of MBP–HepA fusion protein in recombinant Escherichia coli by optimization of culture medium. Biochem Eng J. 2007;34:114–21. https://doi.org/10.1016/j.bej.2006.11.020.
40. Xu S, Qiu M, Zhang X, Chen J. Expression and characterization of an enhanced recombinant heparinase I with chitin binding domain. Int J Biol Macromol. 2017;105:1250–8.

41. Yu P, Jia T, Chen Y, Wu Y, Zhang Y. Improving the activity of heparinase I by the directed evolution, its enzymatic properties and optimal conditions for heparin degrading by recombinant cells. Biochem Eng J. 2016;114:237–43. https://doi.org/10.1016/j.bej.2016.07.011.
42. Ma X, Wang ZLS. Effect of CaCl2 as activity stabilizer on purification of heparinase I from Flavobacterium heparinum. J Chromatogr B Anal Technol Biomed Life Sci. 2006;843(2):209–15.

gRNA validation for wheat genome editing with the CRISPR-Cas9 system

Taj Arndell[1*], Niharika Sharma[2], Peter Langridge[3], Ute Baumann[3], Nathan S. Watson-Haigh[4] and Ryan Whitford[3]

Abstract

Background: The CRISPR-Cas9 system is a powerful and versatile tool for crop genome editing. However, achieving highly efficient and specific editing in polyploid species can be a challenge. The efficiency and specificity of the CRISPR-Cas9 system depends critically on the gRNA used. Here, we assessed the activities and specificities of seven gRNAs targeting 5-enolpyruvylshikimate-3-phosphate synthase (*EPSPS*) in hexaploid wheat protoplasts. EPSPS is the biological target of the widely used herbicide glyphosate.

Results: The seven gRNAs differed substantially in their on-target activities, with mean indel frequencies ranging from 0% to approximately 20%. There was no obvious correlation between experimentally determined and in silico predicted on-target gRNA activity. The presence of a single mismatch within the seed region of the guide sequence greatly reduced but did not abolish gRNA activity, whereas the presence of an additional mismatch, or the absence of a PAM, all but abolished gRNA activity. Large insertions (≥20 bp) of DNA vector-derived sequence were detected at frequencies up to 8.5% of total indels. One of the gRNAs exhibited several properties that make it potentially suitable for the development of non-transgenic glyphosate resistant wheat.

Conclusions: We have established a rapid and reliable method for gRNA validation in hexaploid wheat protoplasts. The method can be used to identify gRNAs that have favourable properties. Our approach is particularly suited to polyploid species, but should be applicable to any plant species amenable to protoplast transformation.

Keywords: Genome editing, CRISPR-Cas9, gRNA, Wheat, *Triticum aestivum*, Protoplasts, EPSPS

Background

Genome editing technologies enable the targeted and precise modification of plant genomes via the creation and subsequent repair of site-specific DNA double-strand breaks (DSBs) [1]. Over the last few years, the field of genome editing has been revolutionised by the introduction of the CRISPR (clustered regularly-interspaced short palindromic repeats)-Cas9 (CRISPR associated protein) system [2–7]. This system consists of the Cas9 endonuclease in complex with a small guide RNA (gRNA) that is engineered to target a specific site in the genome. The target site is defined by a 20 nucleotide guide sequence at the 5′ end of the gRNA, making programming of the system relatively straightforward. For the system to function, the target site must be located immediately 5′ to a protospacer adjacent motif (PAM) whose canonical form is 5′-NGG-3′ (for SpCas9 from *Streptococcus pyogenes*). The PAM is on the strand opposite to the strand bound by the gRNA. Site-specific DSBs generated by Cas9 are typically repaired through either of two competing pathways: non-homologous end joining (NHEJ) or homology directed repair (HDR). NHEJ, which is the predominant repair pathway in somatic plant cells [8], is error-prone and often produces small insertions/deletions (indels) that result in gene knockout (e.g. through frame shift or the creation of a premature stop codon) [9–12]. Alternatively, if an exogenous DNA donor template with homologous ends is delivered to the cell, then precise modifications (sequence insertion or replacement) can be made through HDR [9, 13–18]. If the homologous ends of the donor template are short (5–25 bp), then the DSB may be repaired through microhomology-mediated end joining (MMEJ) [19]. Due to its simplicity, flexibility, and high specificity, the CRISPR-Cas9 system has been and continues to be rapidly adopted by the plant research community for basic research and crop improvement.

* Correspondence: taj.arndell@csiro.au
[1]Present address: CSIRO, Agriculture and Food, Canberra, ACT, Australia

Although the CRISPR-Cas9 system has been successfully applied to many model and crop plants, editing efficiencies have varied greatly and in many cases have been lower than one would desire. In particular, gene knockout via NHEJ tends to be relatively inefficient in polyploidy species due to genetic redundancy [20–22]. Furthermore, the inherent low frequency of HDR remains a major challenge in plant genome editing [23]. In addition, despite the high specificity of the CRISPR-Cas9 system, off-target mutations can occur at sites that have sequence similarity to the target site [24–26], especially when there are no mismatches in the PAM-proximal 8–12 nucleotide 'seed region' of the guide sequence [27–30]. Such off-target sites may be present in non-target genes or non-target alleles. Off-target mutations are undesirable as they may confound results and/or produce impaired phenotypes, in which case they must be removed by backcrossing. Consequently, much effort has been directed toward improving the efficiency and specificity of the CRISPR-Cas9 system in plants and other organisms.

It is well-established that the efficiency and specificity of the CRISPR-Cas9 system depends critically on target site selection, as well as certain sequence features of the gRNA. Thus, an effective strategy for achieving efficient and specific editing is to use gRNAs that exhibit high activity and specificity. A number of bioinformatics tools have been developed for the in silico prediction of on-target and/or off-target gRNA activity [31–36]. Some of these tools can provide reliable predictions for potential off-target sites in a limited number of species, and predictions for on-target gRNA activity can reduce the time spent on gRNA screening [37]. However, the predictions are not always accurate, and the development and independent validation of these tools has been based on data obtained from non-plant species. Therefore, it is prudent to carry out experimental validation of gRNAs prior to commencing plant transformation experiments that require substantial investment of time and resources.

Here, we propose and test a strategy for assessing gRNA activity and specificity, using seven gRNAs targeting 5-enolpyruvylshikimate-3-phosphate synthase (*EPSPS*) in hexaploid wheat (*Triticum aestivum*). *EPSPS* is an ideal target for editing via HDR, as several well-characterised amino acid substitutions in this gene are known to confer resistance to the widely used herbicide glyphosate [38]. Furthermore, in wheat there is an opportunity to take advantage of its hexaploid nature by performing homoeoallele-specific editing, thereby potentially avoiding the severe yield penalty associated with homozygous amino acid substitutions and loss-of-function mutations in *EPSPS* [39–41]. Therefore, one of our aims was to identify a highly active and homoeoallele-specific gRNA targeting *ESPSP*. Three of our gRNAs were designed to be homoeoallele-specific based on the presence of single nucleotide polymorphisms (SNPs) between the three homoeologous copies of *EPSPS*. This allowed us to determine the frequency of off-target mutagenesis. The seven gRNAs were rigorously evaluated through transient expression of CRISPR-Cas9 reagents in wheat mesophyll protoplasts, followed by TIDE (Tracking of Indels by DEcomposition) [42] analysis of Sanger sequence reads, and CRISPResso [43] analysis of amplicon reads. On-target activities varied substantially between gRNAs, and there was no obvious correlation between experimentally determined and in silico predicted on-target gRNA activity. Off-target mutations in homoeoalleles of *EPSPS* were detected at low frequencies, whereas large insertions (≥20 bp) of DNA vector-derived sequence were detected at surprisingly high frequencies. One of the gRNAs exhibited several properties that make it potentially suitable for the development of non-transgenic glyphosate resistant wheat.

Results

Cloning and sequencing of *EPSPS* in wheat cv. Fielder

A vast number of intervarietal SNPs are known to exist in hexaploid wheat [44]. Therefore, for the purpose of designing effective gRNAs, we first obtained sequence information for the three homoeoalleles of *EPSPS* in our target wheat cv. Fielder. For each homoeoallele, two independent partial genomic clones of *EPSPS* were Sanger sequenced. In each case, the sequences of the two independent clones were identical. Thus, the following consensus sequences, covering a region from the 5′ end of intron 1 to the middle of exon 5, were obtained: TaEPSPS-FL-7AS [GenBank MG460802], TaEPSPS-FL-4AL [GenBank MG460801], and TaEPSPS-FL-7DS [GenBank MG460803]. These consensus sequences mapped unambiguously to chromosomes 7AS, 4AL (translocated from 7BS) and 7DS, as expected [45]. We identified three synonymous, homoeologous SNPs located at the 3′ end of exon 2, in close proximity to a universal mutation hotspot for resistance to glyphosate [46] (Fig. 1). We exploited these SNPs for the design of homoeoallele-specific gRNAs (see below).

Protoplast transformation

We designed seven gRNAs targeting a region of *EPSPS* that contains the universal mutation hotspot for resistance to glyphosate (Fig. 1). gRNA1, gRNA2 and gRNA7 were designed to target only one or two of the three *EPSPS* homoeoalleles, whereas the other four gRNAs were designed to target all three homoeoalleles. We transiently co-expressed Cas9 and each gRNA in wheat mesophyll protoplasts. To gauge transient transformation efficiencies, we used a positive control in which YFP was substituted for the gRNA. The proportion of fluorescent (YFP expressing) protoplasts in the positive control ranged from 64 to 72% (mean = 68%) (Additional file 1). We

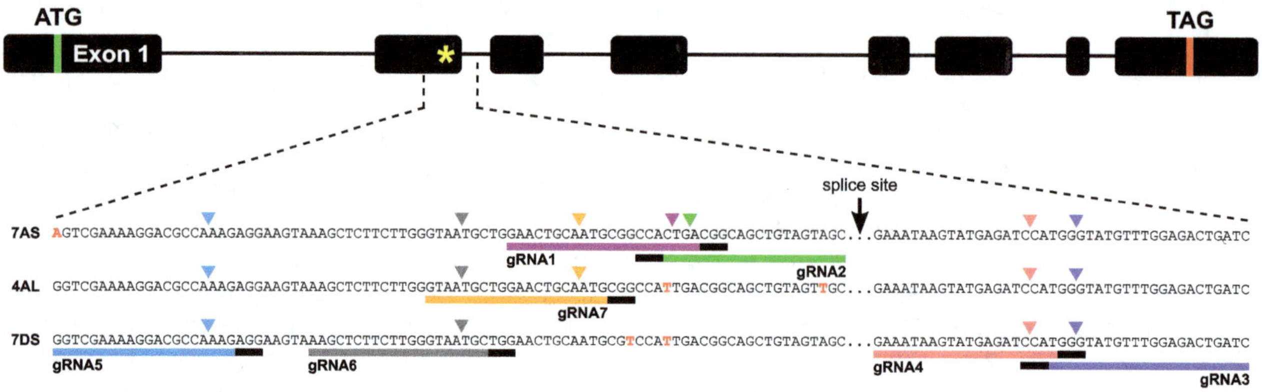

Fig. 1 Target sites for seven gRNAs targeting *EPSPS*. The gene structure of *EPSPS* is shown, annotated with the universal mutation hotspot for glyphosate resistance (yellow asterisk in exon 2). The detail underneath shows partial, abbreviated sequences for the three homoeoalleles of *EPSPS* on chromosomes 7AS, 4AL and 7DS. Target sites are indicated by coloured bars. PAM sites (5′-NGG-3′) are indicated by black bars at the ends of the coloured bars. Downward-pointing arrow heads indicate the position of the canonical cut site and predicted specificity based on the number and distribution of homoeologous SNPs at the corresponding target site/PAM

found that the key to achieving high transient transformation efficiencies was to dilute the protoplasts to a concentration of 3.0×10^5 cells/mL (instead of 2.5×10^6 cells/mL as described in another protocol [47]) prior to transformation, and to avoid extended incubation of DNA with protoplasts prior to adding PEG (Additional file 2).

Assessment of gRNA activity and specificity via TIDE analysis of Sanger sequence traces

We obtained high-quality forward and reverse Sanger sequence reads (Additional files 3-10) of homoeoallele-specific amplicons (Additional file 11) derived from protoplasts treated with each of the seven *EPSPS*-specific gRNAs and one non-targeting (random guide sequence) negative control gRNA. As expected, some of the sequence traces for samples treated with *EPSPS*-specific gRNAs contained mixed peaks downstream of the predicted cut site, and these mixed peaks were detected by TIDE as an increase in the percent of aberrant sequence relative to the negative control (Fig. 2a). There was a strong correlation between the indel spectra/frequencies calculated by TIDE for forward and reverse sequence traces, with each decomposition result having a high goodness of fit ($R^2 \geq 0.93$) (Additional file 12). Individual indels (significant at $p < 0.001$) were detected at frequencies down to approximately 1% (e.g. gRNA2, Rep 3, 7DS in Additional file 12). The mean frequency of significant indels ranged from 0.0–23.3% depending on the gRNA and homoeoallele (Fig. 2b). gRNA5 was the most highly active gRNA on all three homoeoalleles, and the presence of a single mismatch at the PAM-distal end of the guide sequence (position 20) on 7AS did not reduce the activity of gRNA5 (Fig. 2b). gRNA2 was moderately active on 7AS, and off-target indels were also detected at low frequency on 7DS in the presence of a single mismatch at the PAM-proximal end of the guide sequence (position 1) (Fig. 2b). gRNA4 was also moderately active on 7AS, but the frequency of indels appeared to be lower on 4AL and 7DS, even though no mismatches were present (Fig. 2b). All other gRNAs exhibited low or no activity (Fig. 2b).

Assessment of gRNA activity and specificity via CRISPResso analysis of amplicon reads

As an alternative method for detecting indels produced through NHEJ, and to crosscheck the TIDE results, we subjected all of the samples to amplicon deep sequencing. Due to high sequence similarity between amplicon reads derived from the three homoeoalleles of *EPSPS*, CRISPResso was unable to accurately map the reads to their respective reference amplicon sequences. Therefore, we used pre-mapped amplicon reads [NCBI BioProject PRJNA420019] as input for the CRISPResso analyses. In CRISPResso, the total number of aligned (analysed) amplicon reads in each subanalysis (i.e. a replicate for a gRNA on a homoeoallele) ranged from 7067 to 35,668 (mean = 18,110). In general, the CRISPResso results (Fig. 2c and d, Additional files 13 and 14) were in agreement with the TIDE results, but there was less variation between replicates in the CRISPResso results, as indicated by smaller standard errors (Fig. 2d). Notably, in the CRISPResso results, the activity of gRNA2 on 7DS (off-target) was only 7% of that on 7AS (on-target), and the difference was statistically significant ($p < 0.05$) (Fig. 2d). Also, the activity of gRNA4 was more consistent across homoeoalleles (Fig. 2d). The frequency of indels in the negative control was ≤0.1% (mean = 0.005%).

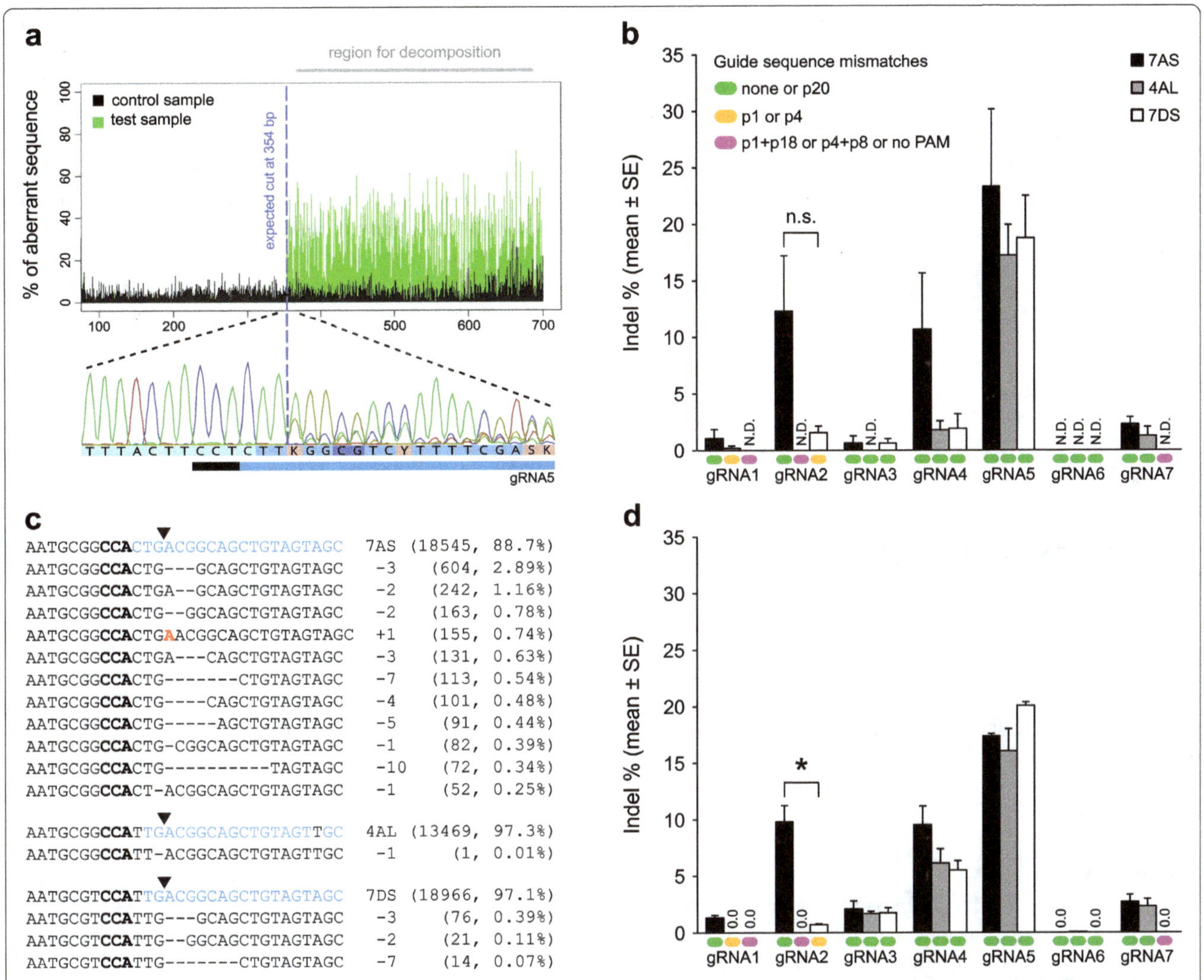

Fig. 2 Mutation detection and summary of editing efficiencies for seven gRNAs targeting *EPSPS* on chromosomes 7AS, 4AL and 7DS. **a** TIDE detection of mixed peaks in the reverse Sanger sequence read for gRNA5 on chromosome 7AS (replicate 1). **b** Summary of TIDE results. N.D., not detected. n.s., not statistically significant. Error bars represent the standard error of the mean ($n = 3$). **c** Alignment of representative mutant amplicon reads for gRNA2 on chromosomes 7AS, 4AL and 7DS (replicate 1). Bold black text, PAM; blue text, complementary to gRNA2 guide sequence; red text, inserted nucleotide. Downward-pointing arrow heads indicate the position of the canonical cut site. The number of reads and percent of total reads is shown in brackets. **d** Summary of CRISPResso results. Error bars represent the standard error of the mean ($n = 3$). * statistically significant ($p < 0.05$) based on a two-sample t-test assuming unequal variances. The keys in **b** also apply to **d**. In the key for guide sequence mismatches, p20 means position 20 in the guide sequence, etc.

Based on both TIDE and CRISPResso derived data, we determined that gRNA2 is likely to be the most effective gRNA for generating stable homoeoallele-specific (chromosome 7AS) edits in *EPSPS*.

In silico prediction of on-target gRNA activity

The seven gRNAs differed substantially in their in silico predicted on-target activity (Table 1). sgRNA Designer scores [34] ranged from 0.47–0.85 (potential range = 0–1). WU-CRISPR scores [32] ranged from < 50–85 (potential range = 0–100; scores < 50 are not output). There was some disagreement between the sgRNA Designer and WU-CRISPR scores. In particular, the top-scoring gRNA in sgRNA Designer (gRNA7) had a WU-CRISPR score of < 50. There did not appear to be any obvious correlation between experimentally determined and in silico predicted on-target gRNA activity (Table 1).

Analysis of large insertions

To detect large insertions (≥20 bp), we used unmapped amplicon reads [NCBI BioProject PRJNA420019] as input for a separate CRISPResso analysis. Large insertions were detected in the majority of samples. The third replicate of gRNA2 had the highest frequency of large insertions (8.5% of edited reads), all of which showed 100% sequence identity to components of the DNA vectors

Table 1 Experimentally determined versus in silico predicted on-target gRNA activity

	Target homoeoalleles	Indel frequency (%)[a]	sgRNA Designer score[b]	WU-CRISPR score[c]
gRNA1	7AS	1.3	0.59	58
gRNA2	7AS	9.8	0.58	60
gRNA3	7AS, 4AL, 7DS	1.8	0.64	82
gRNA4	7AS, 4AL, 7DS	7.0	0.52	< 50
gRNA5	7AS, 4AL, 7DS	17.8	0.68	85
gRNA6	7AS, 4AL, 7DS	0.0	0.47	< 50
gRNA7	7AS, 4AL	2.5	0.85	< 50

[a]Experimentally determined on-target gRNA activity, expressed as the mean proportion of edited amplicon reads derived from target homoeoalleles for three replicates. [b]Potential range = 0–1 (1 is highest predicted gRNA activity); **c** potential range = 0–100 (100 is highest predicted gRNA activity, scores < 50 are not output)

used for protoplast transformation (Fig. 3, Additional file 15). Similar frequencies of large insertions were observed for the third replicate of gRNA5 (5.8% of edited reads), and no large insertions were detected in the negative controls (Additional file 15). CRISPResso failed to correctly predict the size of the insertion when the insertion was accompanied by a deletion. For example, the + 42/− 31 and + 54/− 1 mutations (Fig. 3) were listed as + 21 and + 53 insertions respectively, in the CRISPResso allele frequency table (Additional file 15).

Discussion

We assessed the activity and specificity of seven gRNAs targeting *EPSPS* in wheat mesophyll protoplasts. Certain amino acid substitutions in EPSPS are known to confer resistance to the herbicide glyphosate [38], and therefore *EPSPS* is an ideal candidate for editing via HDR given that glyphosate resistance is a robust selectable marker in wheat tissue culture and during plant growth [48, 49]. Furthermore, in wheat it may be possible to avoid the severe yield penalty associated with homozygous amino acid substitutions and loss-of-function mutations in *EPSPS* [39–41], by performing homoeoallele-specific editing. However, given the inherent low frequency of HDR, and the potential for simultaneous editing of all three homoeoalleles, this application requires the use of a highly active and (ideally) homoeoallele-specific gRNA. With this in mind, our aim was to identify such a gRNA, and in doing so build on previous work [47] to develop an improved method for validating gRNAs in wheat and other polyploid species.

The wide range of on-target activities observed for the tested gRNAs in this study is consistent with previous reports of CRISPR-based genome editing using hexaploid wheat mesophyll protoplasts. A gRNA targeting *TaMLO-A1* caused indels at a frequency of 29% based on the PCR-restriction enzyme (PCR-RE) assay [9]. In a separate study, the same gRNA caused indels at a frequency of 36% based on a homoeoallele-specific T7E1 assay [50]. Similar editing efficiencies were attained with gRNAs targeting *TaGW2* and *TaGASR7* [51]. gRNAs targeting *TaDEP1*, *TaNAC2*, *TaPIN1* and *TaLOX2* were also evaluated by the PCR-RE assay [20], but the editing efficiencies were not calculated. Nevertheless, it could be seen that editing efficiencies varied substantially between gRNAs and were comparable with those presented here

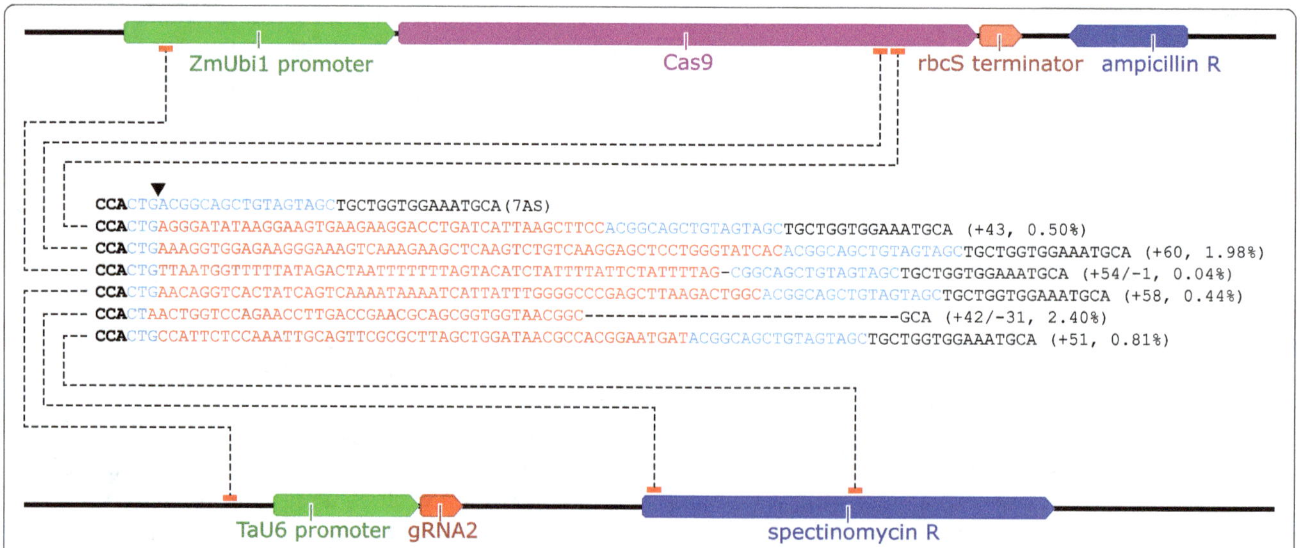

Fig. 3 Representative examples of large insertions (≥20 bp) for gRNA2 (replicate 3). Schematics of pUbi-Cas9-rbcS (top) and pCR8-U6-gRNA2 (bottom) are shown, annotated with links (dotted lines) indicating from where the inserted sequences originate. Bold black text, PAM; blue text, complimentary to gRNA2 guide sequence; red text, inserted sequence. The downward-pointing indicates the position of the canonical cut site. The type/size of the mutation is given in brackets, together with the allele frequency as a percent of edited amplicon reads

for *EPSPS*. gRNAs targeting *TaDREB2* and *TaERF3* caused indels at a frequency of 6.7 and 10.2% respectively, based on the T7E1 assay [52]. It has been reported that approximately three or four gRNAs out of ten induce indels at a frequency of > 20% in wheat protoplasts (whereas in rice protoplasts the number is approximately seven or eight gRNAs out of ten) [47]. Out of seven gRNAs, we found that one induced indels at a frequency of almost 20%, two induced indels at a frequency of 7–10%, and four induced indels at a frequency of < 3% (based on data presented in Table 1). Taken together, this limited dataset suggests that gRNAs with high activity in wheat are likely the exception rather than the rule. Moreover, we did not find any obvious correlation between experimentally determined and in silico predicted on-target gRNA activity, which suggests that further improvements to bioinformatics tools for gRNA design are needed. For these reasons, we consider it prudent to carry out gRNA validation prior to commencing experiments for the production of stably edited wheat plants.

We observed low levels of gRNA activity at off-target homoeoalleles of *EPSPS*. Our results are consistent with established models of gRNA specificity [27–30] in which: a) the absence of a canonical PAM site (5′-NGG-3′) greatly reduces or abolishes gRNA activity; b) mismatches within the PAM-proximal 8–12 nucleotide seed region of the guide sequence reduce gRNA activity to a greater degree than mismatches outside the seed region; and c) additional mismatches further reduce gRNA activity. Importantly, a single mismatch within the seed region (at position 4 in gRNA1 and position 1 in gRNA2) greatly reduced but did not abolish gRNA activity. However, when the mismatch within the seed region was accompanied by another mismatch (at position 8 in gRNA1 and position 18 in gRNA2), gRNA activity was further reduced to levels that were undetectable with Sanger sequencing and barely detectable with amplicon deep sequencing. These results are consistent with previous reports, which showed that off-target mutations can occur in plants when there is only a single mismatch in the seed region [20, 25]. In these studies, frequencies of off-target mutagenesis were approximately 50–80% lower than frequencies of on-target mutagenesis [20, 25]. By contrast, the gRNA targeting *TaMLO-A1* (mentioned above) did not appear to generate any off-target mutations in homoeoalleles (*TaMLO-B1* and *TaMLO-D1*) in wheat protoplasts or transgenic T_0 plants, due to the presence of a single mismatch at position 2 [50]. The apparent greater reduction in gRNA activity in the presence of a mismatch at position 1, 2 or 4 (compared with a mismatch at position 7, 8 or 9) may be due to the existence of five nucleotide 'core' within the seed region at the PAM-proximal end of the guide sequence [29, 30]. These results suggest that although off-target mutations are substantially reduced in the presence of a single mismatch in the seed region, they are often not eliminated. Therefore, ideally, potential off-target sites should lack a PAM, or else contain multiple mismatches, including at least one in the core of the seed region [24]. Where this is not possible, higher specificity may be achieved through the use of a truncated gRNA [53] and/or high fidelity variant of Cas9 [54]. Surprisingly, the activity of gRNA4 was apparently reduced on 4AL and 7DS, even though no mismatches were present. The reason for this is unknown. However, given that the reductions were less pronounced in the amplicon deep sequencing data (Fig. 2d), it would seem that this unexpected result is at least partly explained by the PCR and/or sequencing method used.

gRNA2 exhibited several properties that make it potentially suitable for the development of non-transgenic glyphosate resistant wheat. First, the canonical cut site for gRNA2 is adjacent to the universal mutation hotspot for resistance to glyphosate. This is important because the frequency of HDR tends to decrease as the distance between the DSB and the site of the desired mutation increases [55]. Second, gRNA2 was active at its target site on 7AS, although the activity was moderate. Third, gRNA2 was relatively specific for *EPSPS* on 7AS, which is the most highly transcribed copy of *EPSPS* in at least some wheat cultivars [45]. This high specificity would facilitate the creation of an *EPSPS* mutant that is edited on 7AS and wild type on 4AL/7DS. Such a mutant would have the desired trait (glyphosate resistance), and the yield penalty that could otherwise result from simultaneous modification or knockout of *EPSPS* on 4AL/7DS would be avoided.

One somewhat surprising finding in this study was the relatively high frequency of insertions (up to 8.5% of edited amplicon reads) that show 100% sequence identity to components of the DNA vectors used for transformation. These insertions are almost certainly vector-derived, and some (e.g. synthetic Cas9 sequences) are undoubtedly vector-derived. Recently, it was reported that DNA vector-derived insertions occur at very low frequencies (0.06–0.14% of edited amplicon reads) in Arabidopsis protoplasts transiently transformed with CRISPR-Cas9 vectors [56], although the authors state that the frequencies were likely underestimated because insertions of > 50 bp were excluded from the analysis. In addition to experimental differences, species-specific differences in NHEJ [57] may help to explain the much higher frequencies of DNA vector-derived insertions in wheat. If so, then sequence knockin via MMEJ may be a particularly effective genome editing strategy in wheat [19, 50]. On a related note, if DNA vector-derived sequences were to be found in an unregulated genome-edited crop (i.e. at an uncharacterised off-target site), it would set back both the regulatory authority and the industry [56]. Fortunately, recent advances in crop genome editing with in vitro transcribed RNA [20] and Cas9 ribonucleoproteins [51, 58, 59] suggest that the potential for unintended insertions of DNA vector-derived sequences can be eliminated in many cases. Where these methods cannot be used, unwanted DNA-vector derived insertions could be detected by whole genome sequencing and, if not linked to the target modification, removed by backcrossing.

Finally, this study demonstrates the utility of two methods for detecting and quantifying on-target and off-target indels: TIDE analysis of Sanger sequence traces, and CRISPResso analysis of amplicon reads. Although restriction enzyme-based assays such as PCR-RE and T7E1 are rapid, cheap and widely used for mutation detection, they suffer from several drawbacks and may not be suitable in some circumstances. For example, if the target site lacks a restriction site at the canonical cut site (as was the case with gRNA6 in this study), then the PCR-RE assay is impractical. On the other hand, the T7E1 enzyme recognises and cleaves mismatched DNA produced through denaturation and re-annealing of wild type and mutant PCR amplicons, so it does not require a restriction site. However, the T7E1 assay may produce results that are difficult to interpret if the wild type amplicons are polymorphic (a potential problem in polyploids) [60]. Moreover, restriction enzyme-based assays provide essentially no information about the indel spectrum or sequences of mutant alleles. Such information can be useful for the purpose of selecting gRNAs for particular applications [61, 62]. The sequencing-based mutation detection methods used in this study overcome the limitations described above, and come with the added benefit of greater sensitivity. Sanger sequencing combined with TIDE analysis takes only a few days and can be cost-effective for low/medium throughput screening, considering that sequencing of the opposite strand is unnecessary if the forward sequence trace is high quality. Amplicon deep sequencing combined with CRISPResso analysis takes longer and is only cost-effective for high throughput screening. In some cases, significant expertise in bioinformatics may be required for the analysis of amplicon reads derived from polyploid species. A number of other mutation detection methods have been established [63, 64]. Ultimately, the choice of mutation detection method should be made on a case-by-case basis, taking into consideration the goals of the experiment and the available resources [60].

Conclusions

In summary, this study demonstrates that gRNA validation is an essential step in the application of the CRISPR-Cas9 system in wheat. gRNA validation should be carried out prior to commencing transformation and tissue culture experiments for the production of stably edited wheat plants. We have established a rapid and reliable method for assessing gRNA activity and specificity in hexaploid wheat. The method is based on an improved wheat protoplast transformation protocol, as well as the use of sequencing-based mutation detection techniques that overcome many of the limitations of commonly used enzyme-based assays. The method was used to identify a gRNA that could potentially be used for the production of non-transgenic glyphosate resistant wheat lines. Our approach is applicable to any plant species amenable to protoplast transformation, and should facilitate the adoption of CRISPR-Cas9 technology for genome editing in wheat and other polyploid crops.

Methods

Cloning and sequencing of *EPSPS* in wheat cv. Fielder

A full-length *Triticum aestivum EPSPS* cDNA consensus sequence (1789 bp) was retrieved from GenBank [EU977181] and used as the query for a BLASTN search against the *T. aestivum* EST database. The returned ESTs were assembled de novo into contigs using the Geneious Assembler in Geneious v9. The cDNA consensus sequence and EST-derived contigs were then aligned with genomic sequences from the TGACv1 wheat genome assembly (scaffold_569503_7AS:9611–10,115, scaffold_290435_4AL:41770–42,544 and scaffold_623048_7DS:39649–41,774), using the MUSCLE Alignment tool in Geneious. Based on this multiple sequence alignment, primers were designed (Additional file 16) to amplify a > 2 kb region of the three homoeologous copies of *EPSPS* in cv. Fielder. Amplicons were TOPO cloned into pCR8 (Invitrogen), and two independent pCR8-TaEPSPS-FL clones derived from each homoeoallele (based on diagnostic restriction enzyme digest) were validated by Sanger sequencing (Australian Genome Research Facility). The Sanger sequence reads were aligned to produce a consensus sequence for each homoeoallele. The consensus sequences were then incorporated into the multiple sequence alignment and used as the basis for gRNA design.

gRNA design

Seven gRNAs were manually designed to target *EPSPS*. Target sites were 20–22 nucleotides in length, and were located immediately 5′ of a PAM sequence (5′-NGG-3′). An extra G nucleotide was appended to the 5′ end of gRNA6 in order to ensure efficient transcription of the gRNA expressed under the U6 promoter [65].

In silico prediction of on-target gRNA activity

On-target gRNA activity was predicted using the sgRNA Designer [34, 66] and WU-CRISPR [32, 67] tools, according to the developers' guidelines.

Vector design and construction

All vectors were designed using Geneious software. To construct the gRNA vector, the gRNA expression cassette [9] consisting of the TaU6 promoter and a non-targeting gRNA was synthesised (GenScript) and TOPO cloned into pCR8 (Invitrogen). The *BbsI* site in the pCR8 backbone was then removed by digestion with *NheI* and self-ligation, resulting in pCR8-U6-NCgRNA (negative control for editing). To insert guide sequences into pCR8-U6-NCgRNA, the guide sequence oligos (Additional file 16) were first annealed by combining 1 µL of each oligo (100 µM) with 1X T4 DNA ligase buffer (Invitrogen) in a total reaction volume of 10 µL. The reaction was heated to 95 °C for 5 min and then left at room temperature for 30 min. Annealed oligos were inserted into pCR8-U6-NCgRNA by simultaneous digestion/ligation using 1 µL annealed oligos, 50 ng pCR8-U6-NCgRNA, 1X NEBuffer 2.1, 2 units *BbsI* (New England Biolabs), 1X T4 DNA ligase buffer, and 0.5 units T4 DNA ligase (Invitrogen) in a total reaction volume of 10 µL. Cycling conditions were as follows: 37 °C for 1 h, 15 °C for 1 min, 20 °C for 10 min (2 cycles), and finally 37 °C for 1 h. Positive clones of pCR8-U6-gRNA1/2/3/4/5/6/7 were identified by diagnostic double digest with *BbsI* and *EcoRI-HF* (New England Biolabs), and validated by Sanger sequencing (Australian Genome Research Facility).

To construct the Cas9 vector, the rice codon-optimised SpCas9 gene with N- and C-terminal nuclear localisation signals [9] was synthesised (GenScript) and inserted into the generic vector pUbi-rbcS as an *NcoI–AscI* fragment between the maize Ubiquitin 1 promoter [68, 69] and the wheat rbcS Class II terminator [70], resulting in pUbi-Cas9-rbcS.

To construct the YFP vector, the EYFP gene was inserted into pUbi-rbcS in the same manner as above, resulting in pUbi-YFP-rbcS.

Protoplast isolation and transformation

Protoplast isolation and transformation was carried out as described [47], with several modifications. Seedlings of *T. aestivum* cv. Fielder were grown in potted soil within a growth chamber at 24 °C with a photoperiod of 12 h light (~ 100 µmol m^{-2} s^{-1}) and 12 h dark, for 7–8 days. Only vigorous seedlings (five to eight in total) were used for protoplast isolation. A razor blade was used to make a shallow cut across the adaxial surface of the primary leaf, from which the abaxial epidermis was peeled off. Leaf peels were placed abaxial side down in a petri dish containing 0.6 M mannitol for 15 min. Leaf peels were then placed abaxial side down in a petri dish containing 10 mL of cell wall-dissolving enzyme solution [20 mM MES-KOH (pH 5.7), 1.5% (wt/vol) cellulase Onozuka RS, 0.75% (wt/vol) macerozyme R10, 0.6 M mannitol, 10 mM KCl, 10 mM $CaCl_2$, 0.1% (wt/vol) BSA] for 3–4 h with very gentle agitation. After addition of one volume of W5 solution [2 mM MES-KOH (pH 5.7), 154 mM NaCl, 125 mM $CaCl_2$, 5 mM KCl] [71], protoplasts were filtered through a 100 µm nylon mesh into a petri dish and then carefully transferred to a 30 mL round-bottom tube (Sarstedt 55.517). Protoplasts were centrifuged for 3 min at 80 x *g*, resuspended in 15 mL of W5 solution, and incubated on ice for 30 min. The W5 solution was removed, and the protoplasts were resuspended in 500 µL MMG solution [4 mM MES-KOH (pH 5.7), 0.4 M mannitol, 15 mM $MgCl_2$] [71]. The protoplast concentration was determined by cell counting on a hemocytometer, and subsequently adjusted to 3.0×10^5 cells/mL using MMG solution.

In an empty 2 mL tube, Ubi-Cas9-rbcS (20 µg, 3.5 pmol) was mixed with either pCR8-U6-gRNA1/2/3/4/5/6/7 (gRNAs targeting *EPSPS*) (20 µg, 10.5 pmol), pCR8-U6-NCgRNA (negative control for editing) (20 µg, 10.5 pmol), or pUbi-YFP-rbcS (positive control for transformation) (20 µg, 5.7 pmol). Transformation was carried out by adding (in quick succession) 200 µL of protoplasts and then 200 µL of PEG solution [40% (wt/vol) PEG-4000, 0.2 M mannitol, 100 mM $CaCl_2$] to the tube containing pre-mixed DNA. The DNA/protoplast/PEG mixture was homogenised by gently flicking the tube, and then incubated for 15 min at room temperature. The transformation reaction was stopped by adding 840 µL of W5 solution and gently inverting the tube three times. The protoplasts were centrifuged for 2 min at 100 x *g*. The supernatant was removed and the protoplasts were resuspended in 500 µL W5 solution. The protoplasts were then transferred to 12-well plates (Sarstedt 83.3921.500) coated with 5% vol/vol fetal bovine serum (Sigma-Aldrich F4135), and incubated at 23 °C in the dark for 48 h.

The experiment was repeated twice more, from the seed planting step. Thus, there were three biological replicates for each treatment and control.

Microscopy

After 16–24 h of incubation, protoplasts co-transformed with pUbi-Cas9-rbcS and pUbi-YFP-rbcS (positive control for transformation) were imaged using a Nikon Ni-E microscope equipped with a 490–500 nm excitation filter and a 520–560 nm emission filter (Adelaide Microscopy Waite Facility). Transformation efficiencies were calculated as the proportion of spherical protoplasts ($n = 100$, bright field image) that emitted yellow fluorescence (dark field image).

Flow cytometry

After 20 h of incubation, protoplasts co-transformed with pUbi-Cas9-rbcS and pUbi-YFP-rbcS were subjected to flow cytometry using a BD Accuri C6. For the negative control, water was used instead of DNA.

gDNA extraction

At the end of the 48 h incubation period, protoplasts were transferred to 2 mL tubes and centrifuged for 2 min at 100 x *g*. The supernatant was removed and gDNA was extracted from the protoplast pellet using the DNeasy Plant Mini Kit (QIAGEN) according to manufacturer's instructions. DNA was eluted from the spin column with 28 μL of elution buffer.

Sanger sequencing and TIDE analysis

To obtain amplicons for Sanger sequencing, a genomic region (1781 bp on 7AS, 1572 bp on 4AL, and 1701 bp on 7DS) containing all seven target sites was amplified by PCR using homoeoallele-specific primers (Additional file 16). PCR was performed using 30–40 ng gDNA template, 0.8 μM primers, 200 μM dNTPs, 1X Phusion HF buffer, and 0.6 units Phusion Hot Start Flex DNA Polymerase (New England Biolabs) in a total reaction volume of 50 μL. gDNA obtained from nulli-tetrasomic lines of *T. aestivum* cv. Chinese Spring was used as template in control PCR reactions to confirm that amplification was homoeoallele-specific. Cycling conditions for touchdown PCR were as follows: initial denaturation at 98 °C for 1 min, denaturation at 98 °C for 5 s, annealing at 68–63 °C (7AS and 7DS) or 66–61 °C (4AL) for 15 s, extension at 72 °C for 55 s, and final extension at 72 °C for 5 min. The starting annealing temperature was decreased by 0.5 °C each cycle for 10 cycles, followed by 30 cycles at the final annealing temperature. The PCR product was run on 1% agarose gel, from which amplicons were extracted using the NucleoSpin Gel and PCR Clean-up kit (Macherey-Nagel) according to the manufacturer's instructions. DNA was eluted from the spin column with 15 μL of diluted (1 in 40) elution buffer and quantified using a NanoDrop 1000 spectrophotometer.

To detect targeted indels produced via NHEJ, homoeoallele-specific amplicons from each PCR reaction were subjected to Sanger sequencing (Australian Genome Research Facility) in the forward and reverse directions with nested homoeoallele-specific primers (Additional file 16). The 3730xl DNA Analyzer (Applied Biosystems) was used for sequencing, and bases were called with KB Basecaller v1.4.1.8. Output AB1 files for treated and untreated (negative control) samples were uploaded to the online TIDE analysis tool [42]. In TIDE, minor adjustments to the decomposition window were made based on information provided on the online TIDE analysis tool Troubleshooting webpage. All other TIDE settings were the default. The indel frequency for each gRNA/homoeoallele/replicate was calculated as the mean percent of sequences containing significant indels ($p < 0.001$) for the forward and reverse reads.

Amplicon deep sequencing and CRISPResso analysis

To obtain amplicons for deep sequencing, two rounds of PCR were carried out. In the first round of PCR, a genomic region (269 bp on 7AS, and 270 bp on 4AL/7DS) containing all seven target sites was amplified using conserved primers containing 5′ universal tail sequences (Additional file 16) to which Illumina index primers anneal in the second round of PCR. PCR was performed using 20–40 ng gDNA template, 0.25 μM primers, 200 μM dNTPs, 1X Phusion HF buffer, and 0.2 units Phusion Hot Start Flex DNA Polymerase in a total reaction volume of 20 μL. Cycling conditions for touchdown PCR were as follows: initial denaturation at 98 °C for 1 min, denaturation at 98 °C for 5 s, annealing at 62–57 °C for 15 s, extension at 72 °C for 10 s, and final extension at 72 °C for 2 min. The starting annealing temperature was decreased by 0.5 °C each cycle for 10 cycles, followed by 25 cycles at the final annealing temperature. The PCR product was purified using Agencourt AMPure XP beads (Beckman Coulter) according to the manufacturer's instructions. The second round of PCR was performed using 10 ng DNA template (purified amplicons from the first round of PCR), 0.3 μM primers (Illumina Nextera XT), 200 μM dNTPs, 1X Phusion HF buffer, and 0.2 units Phusion Hot Start Flex DNA Polymerase in a total reaction volume of 10 μL. Cycling conditions were as follows: initial denaturation at 98 °C for 1 min, denaturation at 98 °C for 5 s, annealing at 60 °C for 15 s, extension at 72 °C for 6 s, and final extension at 72 °C for 2 min (7 cycles in total). The indexed PCR products were purified using Agencourt AMPure XP beads.

Indexed PCR products were quantified by qPCR, diluted to 4 nM, pooled in equal volumes, spiked with 10% PhiX Control v3, and then sequenced on the Illumina MiSeq platform using the MiSeq Reagent Kit v3 600 cycle (Australian Genome Research Facility). The 300 bp unpaired raw reads from each sample were mapped to the three homoeologous amplicon reference sequences in two phases using Bowtie 2 [72]. The aim of the first phase was to map unedited reads using the following parameters: --end-to-end --very-sensitive --np 0 --mp 6, 1 --rdg 999,999 --rfg 999,999 --score-min L,-6,0. Unmapped reads from the first phase were used as input for the second phase, where reads with indels (deletions of up to 51 bp or insertions of up to 4 bp) and some low quality mismatches were mapped using the following parameters: --end-to-end --very-sensitive --np 0 --mp 76,1 --rdg 24,1 --rfg 9,14 --score-min L,-75,0. Next, the resulting two BAM files were sliced for the reads that mapped to the respective three amplicons and merged together using SAMtools [73]. An in-house bash script was used to extract the mapped unedited/edited reads from the merged BAM files, and these sequence files in FASTQ format were used as input for CRISPResso [43] analysis using the following parameters: --w 20 --hide_mutations_outside_window_NHEJ --save_also_png --trim_sequences -q 30 --exclude_bp_from_left 5 --exclude_bp_from_right 5 --ignore_substitutions. Allele frequencies shown in Fig. 2c were calculated by summing the values in the %Reads column of the CRISPResso allele frequency table, after applying an Excel text filter (Custom AutoFilter) to only show rows where the aligned sequence contains the allele sequence.. For each of the three biological replicates, indel frequencies were calculated as the %NHEJ for the gRNA minus the %NHEJ for the negative control, based on data from the CRISPResso pie charts.

Analysis of large insertions

To detect large insertions (≥20 bp), a separate CRISPResso analysis was carried out using unmapped amplicon deep sequencing reads as input, with the same CRISPResso settings as above. Data in CRISPResso allele frequency tables were sorted based on insertion size (largest to smallest), and then filtered to exclude aligned sequences containing insertions of < 20 bp. Reads containing insertions of ≥20 bp were aligned to the cv. Fielder consensus sequences in Geneious using the MUSCLE alignment tool. The sequences of the insertions were then searched for in the sequences of pUbi-Cas9-rbcS and pCR8-U6-gRNA2. Allele frequencies shown in Fig. 3 were calculated by summing the values in the %Reads column of the CRISPResso allele frequency table, after applying an Excel text filter (Custom AutoFilter) to only show rows where the aligned sequence contains the allele sequence.

Supplementary information

Additional file 1. Bright field and dark field microscopy of protoplasts co-transformed with pUbi-Cas9-rbcS and pUbi-YFP-rbcS (A, C, E, G, I, K), and untransformed protoplasts (B, D, F, H, J, L). Three replicates (A-D, E-H, and I-L) for each treatment are shown, along with the transformation efficiency for each replicate (% of protoplasts transformed). Scale bar = 100 μm.

Additional file 2. Flow cytometry of protoplasts co-transformed with pUbi-Cas9-rbcS and pUbi-YFP-rbcS. Protoplasts were diluted to different concentrations (50,000–500,000 cells per 200 μL) for transformation, and either incubated with the DNA for 10 min prior to the addition of PEG or not incubated with the DNA (PEG added immediately to DNA/protoplast mixture). The percent of protoplasts expressing YFP is indicated. MFI, mean fluorescence intensity of YFP-expressing protoplasts.

Additional file 3. Forward and reverse Sanger sequence reads of homoeoallele-specific amplicons derived from protoplasts treated with gRNA1.

Additional file 4. Forward and reverse Sanger sequence reads of homoeoallele-specific amplicons derived from protoplasts treated with gRNA2.

Additional file 5. Forward and reverse Sanger sequence reads of homoeoallele-specific amplicons derived from protoplasts treated with gRNA3.

Additional file 6. Forward and reverse Sanger sequence reads of homoeoallele-specific amplicons derived from protoplasts treated with gRNA4.

Additional file 7. Forward and reverse Sanger sequence reads of homoeoallele-specific amplicons derived from protoplasts treated with gRNA5.

Additional file 8. Forward and reverse Sanger sequence reads of homoeoallele-specific amplicons derived from protoplasts treated with gRNA6.

Additional file 9. Forward and reverse Sanger sequence reads of homoeoallele-specific amplicons derived from protoplasts treated with gRNA7.

Additional file 10. Forward and reverse Sanger sequence reads of homoeoallele-specific amplicons derived from protoplasts treated with a non-targeting (random guide sequence) negative control gRNA.

Additional file 11 Homoeoallele-specific amplification of *EPSPS* on chromosomes 7AS, 4AL and 7DS following transient co-expression of Cas9 and gRNA in wheat protoplasts. Nulli-tetra (–) genomic DNA template does not contain the target chromosome. Nulli-tetra (+) genomic DNA template does contain the target chromosome. NT, no template; *gRNA, non-targeting (random guide sequence) gRNA. Three replicates were performed. All bands to the right of the ladder were gel purified and Sanger sequenced in the forward and reverse directions.

Additional file 12 TIDE indel spectra/frequencies for gRNAs 1–7 targeting *EPSPS* in wheat protoplasts. Shown are results for forward and reverse Sanger sequence reads of homoeoallele-specific amplicons derived from chromosomes 7AS, 4AL, and 7DS. Three replicates were performed.

Additional file 13. CRISPResso allele frequency tables used for analysis of indels induced by gRNA2 in replicate 1.

Additional file 14. CRISPResso NHEJ pie charts.

Additional file 15. CRISPResso allele frequency tables used for analysis of large insertions (≥20 bp) induced by gRNA2 and gRNA5 in replicate 3.

Additional file 16. Primers and oligonucleotides used in this study.

Abbreviations

Cas9: CRISPR-associated protein 9; CRISPR: clustered regularly-interspaced short palindromic repeats; DSB: Double-strand break; EPSPS: 5-enolpyruvylshikimate-3-phosphate synthase; gRNA: guide RNA; HDR: Homology directed repair; NHEJ: Non-homologous end joining; PAM: Protospacer adjacent motif; PCR-RE: PCR-restriction enzyme; SNP: Single nucleotide polymorphism; TIDE: Tracking of Indels by DEcomposition

Acknowledgements

We are grateful to Matthew Tucker (The University of Adelaide) for critical discussions and reading of the manuscript, Adam Croxford (The University of Adelaide) for providing helpful advice and technical assistance with amplicon deep sequencing, Anzu Okada (The University of Adelaide) for providing technical assistance with qPCR and amplicon deep sequencing, Gwen Mayo (Adelaide Microscopy Waite Facility) for providing microscopy assistance, and Nikolai Borisjuk (Huaiyin Normal University) for providing support in the beginning stages of the project. We are grateful for the support provided by The University of Adelaide, Australian Research Council, Grains Research and Development Corporation, and the Government of South Australia.

Authors' contributions

TA conceived the project. TA, PL, UB and RW designed the experimental work. TA carried out the experimental work and analysed the data. TA, NS, UB and NW-H designed the bioinformatics work. TA, NS and NW-H carried out the bioinformatics work. TA and NS wrote the paper. PL, UB and RW critically revised the paper. All authors read and approved the final manuscript.

Funding
This work was funded by the Australian Centre for Plant Functional Genomics (ACPFG) Pty Ltd. The ACPFG had no role in the study other than providing financial support. TA acknowledges financial support received through an Australian Government Research Training Program Scholarship.

Ethics approval and consent to participate
Not applicable.

Consent for publication
Not applicable.

Competing interests
The authors declare that they have no competing interests.

Author details
[1]Present address: CSIRO, Agriculture and Food, Canberra, ACT, Australia. [2]Present address: New South Wales Department of Primary Industries, Research Excellence, Orange, NSW, Australia. [3]School of Agriculture, Food & Wine, The University of Adelaide, Waite Campus, Urrbrae, SA 5064, Australia. [4]Present address: Bioinformatics Hub, School of Biological Sciences, The University of Adelaide, Adelaide, SA 5005, Australia.

References

1. Puchta H, Fauser F. Gene targeting in plants: 25 years later. Int J Dev Biol. 2013;57:629–37.
2. Jinek M, Chylinski K, Fonfara I, Hauer M, Doudna JA, Charpentier E. A programmable dual-RNA-guided DNA endonuclease in adaptive bacterial immunity. Science. 2012;337:816–21.
3. Gasiunas G, Barrangou R, Horvath P, Siksnys V. Cas9-crRNA ribonucleoprotein complex mediates specific DNA cleavage for adaptive immunity in bacteria. Proc Natl Acad Sci U S A. 2012;109:E2579–86.
4. Cong L, Ran FA, Cox D, Lin S, Barretto R, Habib N, et al. Multiplex genome engineering using CRISPR/Cas systems. Science. 2013;339:819–23.
5. Mali P, Yang L, Esvelt KM, Aach J, Guell M, DiCarlo JE, et al. RNA-guided human genome engineering via Cas9. Science. 2013;339:823–6.
6. Cho SW, Kim S, Kim JM, Kim J-S. Targeted genome engineering in human cells with the Cas9 RNA-guided endonuclease. Nat Biotechnol. 2013;31:230–2.
7. Jinek M, East A, Cheng A, Lin S, Ma E, Doudna J. RNA-programmed genome editing in human cells. Elife. 2013;2:e00471.
8. Puchta H. The repair of double-strand breaks in plants: mechanisms and consequences for genome evolution. J Exp Bot. 2005;56:1–14.
9. Shan Q, Wang Y, Li J, Zhang Y, Chen K, Liang Z, et al. Targeted genome modification of crop plants using a CRISPR-Cas system. Nat Biotechnol. 2013;31:686–8.
10. Feng Z, Zhang B, Ding W, Liu X, Yang D-L, Wei P, et al. Efficient genome editing in plants using a CRISPR/Cas system. Cell Res. 2013;23:1229–32.
11. Mao Y, Zhang H, Xu N, Zhang B, Gou F, Zhu J-K. Application of the CRISPR-Cas system for efficient genome engineering in plants. Mol Plant. 2013;6: 2008–11.
12. Feng Z, Mao Y, Xu N, Zhang B, Wei P, Yang D-L, et al. Multigeneration analysis reveals the inheritance, specificity, and patterns of CRISPR/Cas-induced gene modifications in *Arabidopsis*. Proc Natl Acad Sci U S A. 2014; 111:4632–7.
13. Li J, Norville JE, Aach J, McCormack M, Zhang D, Bush J, et al. Multiplex and homologous recombination–mediated genome editing in *Arabidopsis* and *Nicotiana benthamiana* using guide RNA and Cas9. Nat Biotechnol. 2013;31: 688–91.
14. Baltes NJ, Gil-Humanes J, Cermak T, Atkins PA, Voytas DF. DNA replicons for plant genome engineering. Plant Cell. 2014;26:151–63.
15. Svitashev S, Young JK, Schwartz C, Gao H, Falco SC, Cigan AM. Targeted mutagenesis, precise gene editing, and site-specific gene insertion in maize using Cas9 and guide RNA. Plant Physiol. 2015;169:931–45.
16. Li Z, Liu Z-B, Xing A, Moon BP, Koellhoffer JP, Huang L, et al. Cas9-guide RNA directed genome editing in soybean. Plant Physiol. 2015;169:960 70.
17. Sun Y, Zhang X, Wu C, He Y, Ma Y, Hou H, et al. Engineering herbicide-resistant rice plants through CRISPR/Cas9-mediated homologous recombination of acetolactate synthase. Mol Plant. 2016;9:628–31.
18. Endo M, Mikami M, Toki S. Biallelic gene targeting in rice. Plant Physiol. 2016;170:667–77.
19. Nakade S, Tsubota T, Sakane Y, Kume S, Sakamoto N, Obara M, et al. Microhomology-mediated end-joining-dependent integration of donor DNA in cells and animals using TALENs and CRISPR/Cas9. Nat Commun. 2014;5:5560.
20. Zhang Y, Liang Z, Zong Y, Wang Y, Liu J, Chen K, et al. Efficient and transgene-free genome editing in wheat through transient expression of CRISPR/Cas9 DNA or RNA. Nat Commun. 2016;7:12617.
21. Andersson M, Turesson H, Nicolia A, Fält AS, Samuelsson M, Hofvander P. Efficient targeted multiallelic mutagenesis in tetraploid potato (*Solanum tuberosum*) by transient CRISPR-Cas9 expression in protoplasts. Plant Cell Rep. 2017;36:117–28.
22. Ryder P, McHale M, Fort A, Spillane C. Generation of stable nulliplex autopolyploid lines of *Arabidopsis thaliana* using CRISPR/Cas9 genome editing. Plant Cell Rep. 2017;36:1005–8.
23. Baltes NJ, Gil-Humanes J, Voytas DF. Genome engineering and agriculture: opportunities and challenges. Prog Mol Biol Transl Sci. 2017;149:1–26.
24. Cho SW, Kim S, Kim Y, Kweon J, Kim HS, Bae S, et al. Analysis of off-target effects of CRISPR/Cas-derived RNA-guided endonucleases and nickases. Genome Res. 2014;24:132–41.
25. Lawrenson T, Shorinola O, Stacey N, Li C, Østergaard L, Patron N, et al. Induction of targeted, heritable mutations in barley and *Brassica oleracea* using RNA-guided Cas9 nuclease. Genome Res. 2015;16:258.
26. Wolt JD, Wang K, Sashital D, Lawrence-Dill CJ. Achieving plant CRISPR targeting that limits off-target effects. Plant Genome. 2016;9. https://doi.org/10.3835/plantgenome2016.05.0047.
27. Hsu PD, Scott DA, Weinstein JA, Ran FA, Konermann S, Agarwala V, et al. DNA targeting specificity of RNA-guided Cas9 nucleases. Nat Biotechnol. 2013;31:827–32.
28. Sternberg SH, Redding S, Jinek M, Greene EC, Doudna JA. DNA interrogation by the CRISPR RNA-guided endonuclease Cas9. Nature. 2014;507:62–7.
29. Wu X, Scott DA, Kriz AJ, Chiu AC, Hsu PD, Dadon DB, et al. Genome-wide binding of the CRISPR endonuclease Cas9 in mammalian cells. Nat Biotechnol. 2014;32:670–6.
30. O'Geen H, Henry IM, Bhakta MS, Meckler JF, Segal DJ. A genome-wide analysis of Cas9 binding specificity using ChIP-seq and targeted sequence capture. Nucleic Acids Res. 2015;43:3389–404.
31. Doench JG, Hartenian E, Graham DB, Tothova Z, Hegde M, Smith I, et al. Rational design of highly active sgRNAs for CRISPR-Cas9–mediated gene inactivation. Nat Biotechnol. 2014;32:1262–7.
32. Wong N, Liu W, Wang X. WU-CRISPR: characteristics of functional guide RNAs for the CRISPR/Cas9 system. Genome Biol. 2015;16:218.
33. Graham DB, Root DE. Resources for the design of CRISPR gene editing experiments. Genome Biol. 2015;16:260.
34. Doench JG, Fusi N, Sullender M, Hegde M, Vaimberg EW, Donovan KF, et al. Optimized sgRNA design to maximize activity and minimize off-target effects of CRISPR-Cas9. Nat Biotechnol. 2016;34:184–91.
35. Oliveros JC, Franch M, Tabas-Madrid D, San-León D, Montoliu L, Cubas P, et al. Breaking-Cas—interactive design of guide RNAs for CRISPR-Cas experiments for ENSEMBL genomes. Nucleic Acids Res. 2016;44:W267–71.
36. Liu H, Ding Y, Zhou Y, Jin W, Xie K, Chen L-L. CRISPR-P 2.0: an improved CRISPR-Cas9 tool for genome editing in plants. Mol Plant. 2017;10:530–2.
37. Haeussler M, Schönig K, Eckert H, Eschstruth A, Mianné J, Renaud J-B, et al. Evaluation of off-target and on-target scoring algorithms and integration into the guide RNA selection tool CRISPOR. Genome Biol. 2016;17:148.
38. Sammons RD, Gaines TA. Glyphosate resistance: state of knowledge. Pest Manag Sci. 2014;70:1367–77.
39. Yu Q, Jalaludin A, Han H, Chen M, Sammons RD, Powles SB. Evolution of a double amino acid substitution in the 5-enolpyruvylshikimate-3-phosphate synthase in *Eleusine indica* conferring high-level glyphosate resistance. Plant Physiol. 2015;167:1440–7.
40. Li J, Meng X, Zong Y, Chen K, Zhang H, Liu J, et al. Gene replacements and insertions in rice by intron targeting using CRISPR-Cas9. Nat Plants. 2016;2: 16139.
41. Zhang HH, Zhang J, Wei P, Zhang B, Gou F, Feng Z, et al. The CRISPR/Cas9 system produces specific and homozygous targeted gene editing in rice in one generation. Plant Biotechnol J. 2014;12:797–807.

43. Pinello L, Canver MC, Hoban MD, Orkin SH, Kohn DB, Bauer DE, et al. Analyzing CRISPR genome-editing experiments with CRISPResso. Nat Biotechnol. 2016;34:695–7.
44. Lai K, Lorenc MT, Lee HC, Berkman PJ, Bayer PE, Visendi P, et al. Identification and characterization of more than 4 million intervarietal SNPs across the group 7 chromosomes of bread wheat. Plant Biotechnol J. 2015; 13:97–104.
45. Aramrak A, Kidwell KK, Steber CM, Burke IC. Molecular and phylogenetic characterization of the homoeologous EPSP synthase genes of allohexaploid wheat, *Triticum aestivum* (L.). BMC Genomics. 2015;16:844.
46. Zhou M, Xu H, Wei X, Ye Z, Wei L, Gong W, et al. Identification of a glyphosate-resistant mutant of rice 5-enolpyruvylshikimate 3-phosphate synthase using a directed evolution strategy. Plant Physiol. 2006;140:184–95.
47. Shan Q, Wang Y, Li J, Gao C. Genome editing in rice and wheat using the CRISPR/Cas system. Nat Protoc. 2014;9:2395–410.
48. Zhou H, Arrowsmith JW, Fromm ME, Hironaka CM, Taylor ML, Rodriguez D, et al. Glyphosate-tolerant CP4 and GOX genes as a selectable marker in wheat transformation. Plant Cell Rep. 1995;15:159–63.
49. Hu T, Metz S, Chay C, Zhou HP, Biest N, Chen G, et al. Agrobacterium-mediated large-scale transformation of wheat (*Triticum aestivum* L.) using glyphosate selection. Plant Cell Rep. 2003;21:1010–9.
50. Wang Y, Cheng X, Shan Q, Zhang Y, Liu J, Gao C, et al. Simultaneous editing of three homoeoalleles in hexaploid bread wheat confers heritable resistance to powdery mildew. Nat Biotechnol. 2014;32:947–51.
51. Liang Z, Chen K, Li T, Zhang Y, Wang Y, Zhao Q, et al. Efficient DNA-free genome editing of bread wheat using CRISPR/Cas9 ribonucleoprotein complexes. Nat Commun. 2017;8:14261.
52. Kim D, Alptekin B, Budak H. CRISPR/Cas9 genome editing in wheat. Funct Integr Genomics. 2018;18:31–41.
53. Fu Y, Sander JD, Reyon D, Cascio VM, Joung JK. Improving CRISPR-Cas nuclease specificity using truncated guide RNAs. Nat Biotechnol. 2014;32:279–84.
54. Kleinstiver BP, Pattanayak V, Prew MS, Tsai SQ, Nguyen NT, Zheng Z, et al. High-fidelity CRISPR–Cas9 nucleases with no detectable genome-wide off-target effects. Nature. 2016;529:490–5.
55. Paquet D, Kwart D, Chen A, Sproul A, Jacob S, Teo S, et al. Efficient introduction of specific homozygous and heterozygous mutations using CRISPR/Cas9. Nature. 2016;533:125–9.
56. Kim J, Kim J-S. Bypassing GMO regulations with CRISPR gene editing. Nat Biotechnol. 2016;34:1014–5.
57. Vu GTH, Cao HX, Reiss B, Schubert I. Deletion-bias in DNA double-strand break repair differentially contributes to plant genome shrinkage. New Phytol. 2017;214:1712–21.
58. Woo JW, Kim J, Kwon SI, Corvalán C, Cho SW, Kim H, et al. DNA-free genome editing in plants with preassembled CRISPR-Cas9 ribonucleoproteins. Nat Biotechnol. 2015;33:1162–4.
59. Svitashev S, Schwartz C, Lenderts B, Young JK, Cigan AM. Genome editing in maize directed by CRISPR-Cas9 ribonucleoprotein complexes. Nat Commun. 2016;7:13274.
60. Zischewski J, Fischer R, Bortesi L. Detection of on-target and off-target mutations generated by CRISPR/Cas9 and other sequence-specific nucleases. Biotechnol Adv. 2017;35:95–104.
61. Bae S, Kweon J, Kim HS, Kim J-S. Microhomology-based choice of Cas9 nuclease target sites. Nat Methods. 2014;11:705–6.
62. Vu GTH, Cao HX, Fauser F, Reiss B, Puchta H, Schubert I. Endogenous sequence patterns predispose the repair modes of CRISPR/Cas9-induced DNA double strand breaks in *Arabidopsis thaliana*. Plant J. 2017;92:57–67.
63. Yang Z, Steentoft C, Hauge C, Hansen L, Thomsen AL, Niola F, et al. Fast and sensitive detection of indels induced by precise gene targeting. Nucleic Acids Res. 2015;43:e59.
64. Mock U, Hauber I, Fehse B. Digital PCR to assess gene-editing frequencies (GEF-dPCR) mediated by designer nucleases. Nat Protoc. 2016;11:598–615.
65. Sander JD, Joung JK. CRISPR-Cas systems for editing, regulating and targeting genomes. Nat Biotechnol. 2014;32:347–55.
66. sgRNA Designer. https://portals.broadinstitute.org/gpp/public/analysis-tools/sgrna-design. Accessed 16 Feb 2017.
67. WU-CRISPR. http://crispr.wustl.edu/. Accessed 16 Feb 2017.
68. Christensen AH, Sharrock RA, Quail PH. Maize polyubiquitin genes: structure, thermal perturbation of expression and transcript splicing, and promoter activity following transfer to protoplasts by electroporation. Plant Mol Biol. 1992;18:675–89.
69. Christensen AH, Quail PH. Ubiquitin promoter-based vectors for high-level expression of selectable and/or screenable marker genes in monocotyledonous plants. Transgenic Res. 1996;5:213–8.
70. Sasanuma T. Characterization of the rbcS multigene family in wheat: subfamily classification, determination of chromosomal location and evolutionary analysis. Mol Gen Genomics. 2001;265:161–71.
71. Yoo S-D, Cho Y-H, Sheen J. *Arabidopsis* mesophyll protoplasts: a versatile cell system for transient gene expression analysis. Nat Protoc. 2007;2:1565–72.
72. Langmead B, Salzberg S. Fast gapped-read alignment with bowtie 2. Nat Methods. 2012;9:357–9.
73. Li H, Handsaker B, Wysoker A, Fennell T, Ruan J, Homer N, et al. The sequence alignment/map format and SAMtools. Bioinformatics. 2009;25: 2078–9.

10

Metabolic flux analysis for metabolome data validation of naturally xylose fermenting yeasts

Henrique C. T. Veras[1,2], Christiane G. Campos[2,3], Igor F. Nascimento[5], Patrícia V. Abdelnur[2,3], João R. M. Almeida[2,4] and Nádia S. Parachin[1,4*]

Abstract

Background: Efficient xylose fermentation still demands knowledge regarding xylose catabolism. In this study, metabolic flux analysis (MFA) and metabolomics were used to improve our understanding of xylose metabolism. Thus, a stoichiometric model was constructed to simulate the intracellular carbon flux and used to validate the metabolome data collected within xylose catabolic pathways of non-*Saccharomyces* xylose utilizing yeasts.

Results: A metabolic flux model was constructed using xylose fermentation data from yeasts *Scheffersomyces stipitis*, *Spathaspora arborariae*, and *Spathaspora passalidarum*. In total, 39 intracellular metabolic reactions rates were utilized validating the measurements of 11 intracellular metabolites, acquired by mass spectrometry. Among them, 80% of total metabolites were confirmed with a correlation above 90% when compared to the stoichiometric model. Among the intracellular metabolites, fructose-6-phosphate, glucose-6-phosphate, ribulose-5-phosphate, and malate are validated in the three studied yeasts. However, the metabolites phosphoenolpyruvate and pyruvate could not be confirmed in any yeast. Finally, the three yeasts had the metabolic fluxes from xylose to ethanol compared. Xylose catabolism occurs at twice-higher flux rates in *S. stipitis* than *S. passalidarum* and *S. arborariae*. Besides, *S. passalidarum* present 1.5 times high flux rate in the xylose reductase reaction NADH-dependent than other two yeasts.

Conclusions: This study demonstrated a novel strategy for metabolome data validation and brought insights about naturally xylose-fermenting yeasts. *S. stipitis* and *S. passalidarum* showed respectively three and twice higher flux rates of XR with NADH cofactor, reducing the xylitol production when compared to *S. arborariae*. Besides then, the higher flux rates directed to pentose phosphate pathway (PPP) and glycolysis pathways resulted in better ethanol production in *S. stipitis* and *S. passalidarum* when compared to *S. arborariae*.

Keywords: Xylose metabolism, MFA, Metabolome, Ethanol, Cofactor balance, Metabolomics

Background

Several non-*Saccharomyces* yeasts capable of naturally utilize xylose as carbon source have been identified [1–7]. Among them, *Scheffersomyces stipitis* is one of the most studied species and the *Spathaspora ssp.* has attracted attention in recent years [8–10]. The interest in the economic conversion of this pentose sugar, present in lignocellulosic biomass, to fuels and chemicals, motivated the study of xylose consumption in novel yeasts [4, 11–13]. However, fully understand the xylose metabolism is still a challenge to improve the use of this sugar, the second more abundant in nature, as a carbon source [14]. Therefore, the systems biology approach will be useful for the identification of principles and patterns that characterize the metabolism of xylose.

Metabolic flux analysis (MFA) is used to estimate the intracellular fluxes under a defined metabolic network [15, 16]. It gives insights on how metabolism is balanced,

* Correspondence: nadiasp@unb.br; nadiasp@gmail.com
[1]Grupo Engenharia de Biocatalisadores, Universidade de Brasília - UnB , Campus Darcy Ribeiro, Instituto de Ciências Biológicas, Bloco K, 1° andar, Asa Norte, Brasilia 70.790-900, Brazil
[4]Programa de Pós-Graduação em Biologia Microbiana, Instituto de Biologia, Universidade de Brasília - UnB, Brasilia, Brazil

that is, how organisms convert substrates into biomass and chemicals products [1, 17]. Thus, MFA is useful for the prediction of possible metabolic limitations. It can contribute to strain engineering towards high yields of lignocellulosic ethanol production [18–20]. The metabolic networks constructed for MFA, commonly, use the information available from genome annotation. A set of enzymatic reactions are identified and converted into a mathematical model [21]. Several bioinformatics tools are available to perform MFA. Among them, OptFlux is an open-source platform that allows in silico simulations of intracellular carbon fluxes distribution into a defined metabolic network [22]. The constraint-based flux analysis, included in the OptFlux platform, establishes a set of measured extracellular fluxes such as substrate uptake and products formation rates to determine the carbon flux distribution [15]. The number of measured fluxes determines the size of the network. Therefore, a higher number of measured fluxes results in a more accurate metabolic network.

The understating of a real state of a cell depends on a set of analyzes that provides dataset about the genome, transcriptome, proteome, and metabolome [21]. Among those, the metabolome dataset is advantageous since quantification of intracellular metabolites can be directly linked to the metabolic network reflecting the phenotype of an organism at that moment [23, 24]. The systems biology approach, considering the combination of two methods such as MFA and metabolomics, is a valuable strategy to predict intracellular metabolic fluxes distribution and to understand the behavior of a given metabolic network.

Nevertheless, the utilization of data from intracellular metabolite quantification is not routinely incorporated into MFA due to various technical challenges [18, 25]. Among them, there are the steps of data acquisition, such as sample preparation and metabolites extraction [26, 27], which are critical because of the high turnover rates of intracellular reactions [28]. Then, it needs the establishment of a sensitive and selective analytical method for detection and quantification the metabolites taking into consideration the low concentration of metabolites in a complex biological matrix [26, 28]. Finally, the amount of data generated demands statistical analysis so a trustable value can be applied to MFA [29, 30].

Thereby, the purpose of this work was to validate a dataset of 11 intracellular metabolites of naturally xylose-fermenting yeasts utilizing MFA. Thus, for the first time, a comparative evaluation of metabolic flux analysis with addition metabolome data was performed for *Scheffersomyces stipitis, Spathaspora passalidarum,* and *Spathaspora arborariae.*

Among them, 80% of total metabolites were confirmed with a correlation above 90% when compared to the stoichiometric model. Nevertheless, the metabolites phosphoenolpyruvate and pyruvate could not be validated in any studied yeasts. Finally, the three yeasts had the metabolic fluxes from xylose to ethanol compared. Xylose catabolism occurs at twice-higher flux rates in *S. stipitis* than *S. passalidarum* and *S. arborariae*. In yeasts *S. stipitis* and *S. passalidarum* is observed that after the xylose assimilation reactions, approximately 50% of the carbon flux rates are directed to pentose phosphate pathway (PPP) and 50% to glycolysis. Different from *S. arborariae*, where first, carbon flux is directed to reaction into oxidative-PPP. Besides, *S. passalidarum* present 1.5 times high flux rate in the xylose reductase reaction NADH-dependent than other two yeasts.

Results

MFA for xylose-fermenting yeasts *S. stipitis, S. arborariae,* and *S. passalidarum*

Initially, one stoichiometric model was constructed for *S. stipitis, S. arborariae,* and *S. passalidarum* containing the xylose catabolism, pentose phosphate pathway, glycolysis, and tricarboxylic acid cycle. The model has 39 reactions and 35 metabolites, including the cofactors NAD(P) H, $NAD(P)^+$, and ATP (Additional files 1, 2 and 3). The difference between the number of reactions and metabolites resulted in four degrees of freedom. Table 1 shows the measured extracellular rates included in the model. Different time points were used to calculate the rates because of the different growth rates between the yeasts. Therefore all samples were taken at the exponential phase at 28 h, 32 h, and 40 h for *S. stipitis*, S. *arborariae*, and *S. passalidarum*, respectively (Table 1).

The rates of extracellular metabolite were used as constraints to simulate the intracellular carbon flux distributions in the MFA model (Fig. 1). The xylose consumption rates of respectives yeasts are represents by a negative signals. The extracellular xylose consumption rate of *S. stipitis* is at least twice faster than observed in *S. arborariae*, while the *S. passalidarum* showed 1.5 times faster xylose consumption rate than *S. arborariae.*

Among product formation, ethanol was the major metabolite secreted by all the three yeasts evaluated. Therefore, the respective ethanol measurements rates were used to validate the intracellular flux simulations. The correlation between experimentally and calculated rates for ethanol production were above 90% (Fig. 2). Accordingly, we defined the respective intracellular carbon flux distributions as a metabolic flux calculated (MFA-calculated). In other words, the MFA-calculated is the one that the intracellular carbon flux distribution was simulated utilizing the extracellular rates.

In our analysis of MFA, we can identify that some metabolites influence the flux distribution. For example, the reactions for conversion the metabolites erythrose-4-phosphate and D-Ribulose-5-phosphate are 2.8 and 1.8 times

Table 1 Measured extracellular flux rates

ID	Extracellular flux measurements	Formula	Molar Weight $g.mol^{-1}$	*S. stipitis* (28h)	*S. arborariae* (32h)	*S. passalidarum* (40h)
	Biomass	$CH_{1.8}O_{0.5}N_{0.2}$	24.63	1.45	1.27	1.02
XYL[cons]	D-Xylose[cons]	$C_5H_{10}O_5$	150.13	-2.15 ± 0.48	-0.90 ± 0.07	-1.37 ± 0.40
ETOH[e]	Ethanol[e]	C_2H_6O	46.07	3.05 ± 0.25	0.89 ± 0.19	1.97 ± 0.74
ACE[e]	Acetate[e]	$C_2H_4O_2$	59.04	0.00 ± 0.00	0.00 ± 0.00	0.00 ± 0.00
CO2[e]	Carbon_dioxide[e]	CO_2	44.01	3.05 ± 0.25	0.89 ± 0.19	1.97 ± 0.74
XOL[e]	Xylitol[e]	$C_5H_{12}O_5$	152.15	0.00 ± 0.00	0.03 ± 0.01	0.00 ± 0.00
GROL[e]	Glycerol[e]	$C_3H_8O_3$	92.09	0.04 ± 0.01	0.02 ± 0.01	0.00 ± 0.00
PYR[e]	Pyruvate[e]	$C_3H_4O_3$	88.06	0.00 ± 0.00	0.01 ± 0.01	0.00 ± 0.00
SUC[e]	Succinic[e]	$C_4H_6O_4$	118.09	0.00 ± 0.00	0.00 ± 0.00	0.00 ± 0.00
	Growth Rates (μ)			0.11 ± 0.05	0.05 ± 0.01	0.11 ± 0.01
	Carbon Balance (%)			98	93	100
	Redox Balance (%)			100	94	100

Xylose consumption rates [cons] are represent by a negative signal, extracellular product formation [e] (mmol/gCDW.h^{-1}), Carbon balance (%), and redox balance (%). The experiments were performed with sample withdraw at the exponential growth phase in biological triplicates

lower in *S. arborariae* (R09 = 0.23) than observed in *S. stipitis* (R09 = 0.66) and *S. passalidarum* (R09 = 0.42), respectively. Also, it was found that ethanol production rates are inversely proportional to glycerol and xylitol metabolites formation.

Metabolomics of *S. stipitis, S. arborariae*, and *S. passalidarum* during growth on xylose as a carbon source

The quantification of the 11 metabolites concentration (mM) in all studied yeasts is shown in Fig. 3. The analysis of variance (ANOVA) could corroborate that the metabolome data are within a range of reliability (Additional file 4). The metabolite ACCOA could not be detected only in *S. stipitis* but was detected in *S. arborariae* and *S. passalidarum*, indicating that in *S. stipitis* less carbon was directed into respiratory metabolism. The RU5P (0.04 mM) is twice concentrated than R5P (0.02 mM) in both *S. stipitis* and *S. passalidarum*. While in *S. arborariae*, the R5P concentration (0.01 mM) was three times lower than RU5P (0.04 mM). The concentration of S7P is seven times higher in *S. passalidarum* than observed in *S. stipitis* and *S. arborariae*. The metabolite E4P was not detected only *S arborariae* but was detected in *S. stipitis* and *S. passalidarum*, indicating that E4P may rapidly be converted to G3P and F6P. The DHAP was about four times lower in *S. arborariae* (0.005 mM) than observed in *S. stipitis* (0.020 mM) and *S. passalidarum* (0.015 mM). The metabolites G6P (0.05 mM), F6P (0.06 mM) and PEP (0.06 mM) were at least twice concentrated in *S. passalidarum* than observed in *S. stipitis* (0.02 mM, 0.03 mM and 0.02 mM) and *S. arborariae* (0.02 mM, 0.03 mM and 0.02 mM), respectively. The metabolite PYR in *S. stipitis* (0.10 mM) is five times higher than observed in *S. arborariae* (0.02 mM) and *S. passalidarum* (0.02 mM). Finally, the concentration of MAL in *S. arborariae* (0.10 mM) was twice and three times lower than observed in *S. stipitis* (0.20 mM) and *S. passalidarum* (0.30 mM), respectively (Fig. 3).

Utilization of the NAD(P) H and $NAD(P)^+$ cofactors along with the metabolic network

The metabolic model constructed contains specific reaction rates cofactors NAD(P) H – $NAD(P)^+$ dependent. The NADPH/$NADP^+$ ratio is twice higher in *S. stipitis*, while in *S. arborariae* and *S. passalidarum* we did not observe differences. The high NADPH released can influence the higher biomass formation observed in *S. stipitis* in contrast with *S. arborariae* and *S. passalidarum*. Already the NAD^+/NADH ratio is 2.5 times high in *S. arborariae*, showing high NAD^+ released, which characterize an unbalance cofactor in this yeast. Among the simulations of intracellular carbon flux distribution, some reaction rates caught our attention (Fig. 4). For example, the reaction rate (R02) that converts XYL → XOL with NADH-dependent was three times higher than the reaction rate using NADPH. The reaction rate G6P + 2NADP → RU5P + 2NADPH (R10) in *S. arborariae* was four times higher than observed in *S. stipitis* and *S. passalidarum*. Showing that in the yeast *S. arborariae*, there is a higher necessity of regeneration the NADPH cofactor. Corroborating the high reaction rate (R01) that use NADPH in reaction conversion. Besides, the yeast *S. passalidarum* showed the reaction rate DHAP + NADH → GROL + NAD^+ (R15) negative, contrasting with *S. stipitis* and *S. arborariae*. This characteristic in *S.*

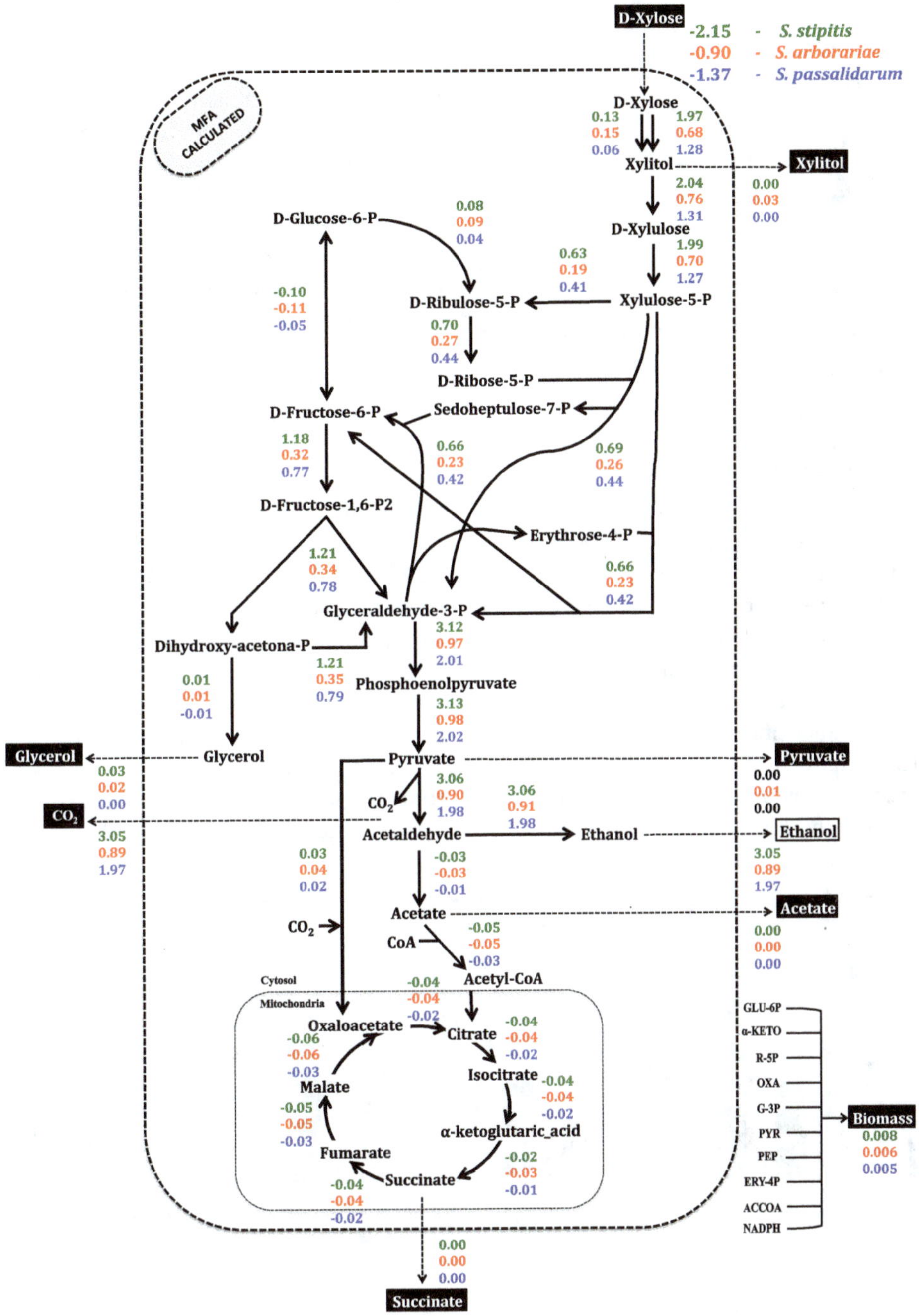

Fig. 1 Intracellular carbon fluxes distributions during xylose catabolism to ethanol production. *S. stipitis* (green), *S. arborariae* (red), *and S. passalidarum* (blue). Xylose consumption rates are represented by a negative signal. The first intracellular reaction (xylose to xylitol) shows two arrows; left represents reaction using NADPH, right represents reaction using NADH cofactor. The extracellular metabolites highlighted in black boxes had its flux rates used as constraints to the MFA-calculated

passalidarum agrees with the absence of glycerol formation and better cofactors balance. The reactions XYL + NADH → XOL + NAD$^+$ (R02), XOL + NAD$^+$ → XYLO + NADH (R03), GAP + NAD$^+$ → PEP + NADH (R16), and ACDH + NADH → ETOH + NAD$^+$ (R19) presented the highest rates. Indicating the importance of cofactor balance. The reaction rates with negative values indicate reversible reactions.

Validation of metabolome dataset utilizing MFA

The metabolite quantification was validated using MFA analysis. For that, the entire metabolome dataset was added to the MFA-calculated (Fig. 5). The metabolic flux

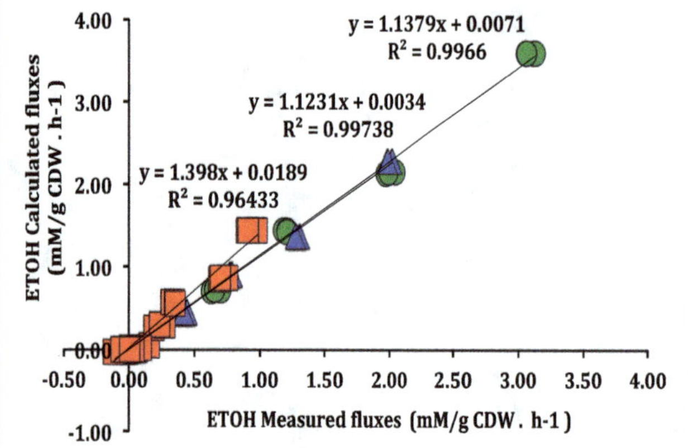

Fig. 2 Correlation between measured and calculated fluxes for ethanol production. *S. stipitis* (green cycle), *S. arborariae* (red square) and *S. passalidarum* (blue triangle)

model with metabolome data is defined as an MFA-measured. First, one metabolite was added per simulation and compared to the fluxes obtained experimentally in the MFA-Calculated. In *S. stipitis* and *S. arborariae*, ten intracellular metabolites were quantified and, consequently, ten simulations performed. While in *S. passalidarum*, 11 metabolites quantified, which resulted in 11 simulations of carbon flux distribution.

In the yeast *S. stipitis*, it was observed that from ten metabolites, eight (i.e., 80%) showed a correlation higher than 0.90 between calculated and measured fluxes (Additional file 5a). Only in *S. stipitis*, there was no simulation for metabolic flux distribution measured with ACCOA since it could not be detected experimentally. In its turn, in *S. arborariae*, it was observed that from ten metabolites, seven (i.e., 70%) showed a correlation higher than 0.90 between calculated and measured fluxes (Additional file 5b). Only on *S. arborariae*, no simulation occurred with metabolite E4P since it could not be detected experimentally. While in *S. passalidarum*, it was observed that from 11 metabolites, nine (i.e., 82%) showed a correlation higher than 0.90 between calculated and measured fluxes (Additional file 5c).

The consistency of intracellular metabolites measurements were verified using the Pearson correlation test (R^2) (Fig. 6). The correlation encountered between experimental data and calculated flux obtained for *S. stipitis*, *S. arborariae*, and *S. passalidarum* were above 90%. Overall, from 20 metabolic flux rates involved in xylose conversion to ethanol, only four of them were not predicted accurately in the metabolic model proposed. Therefore the metabolic flux distribution measured compared with metabolic flux calculated ones confirmed the accuracy of the metabolome data. Although most of the metabolites had a correlation above 90%, it is noted that the measured and calculated metabolic fluxes for the metabolites PEP (29%) and PYR (69%) has a

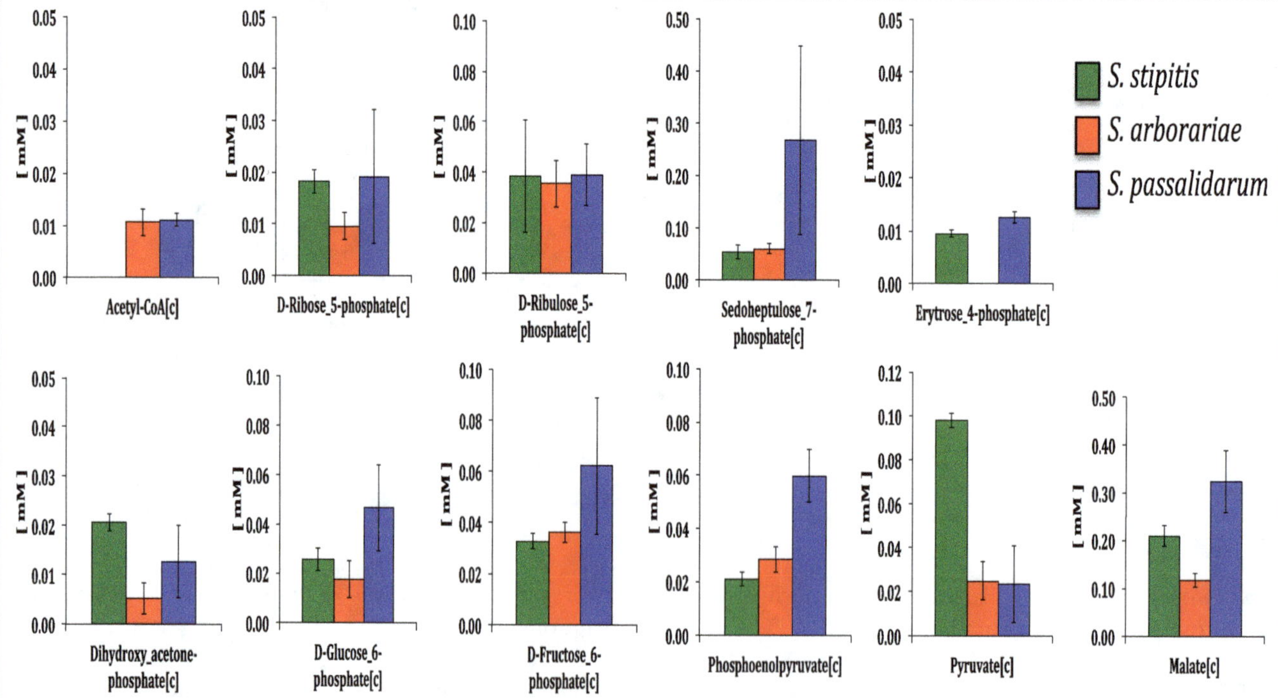

Fig. 3 Intracellular metabolites concentrations (mM). Yeast *S. stipitis* (green column), *S. arborariae* (red column), and *S. passalidarum* (blue column). All experiments performed in biological triplicates. [c] represent the cytosol metabolites. The presented values are the average of ANOVA analysis for three biological replicates and nine technical replicates

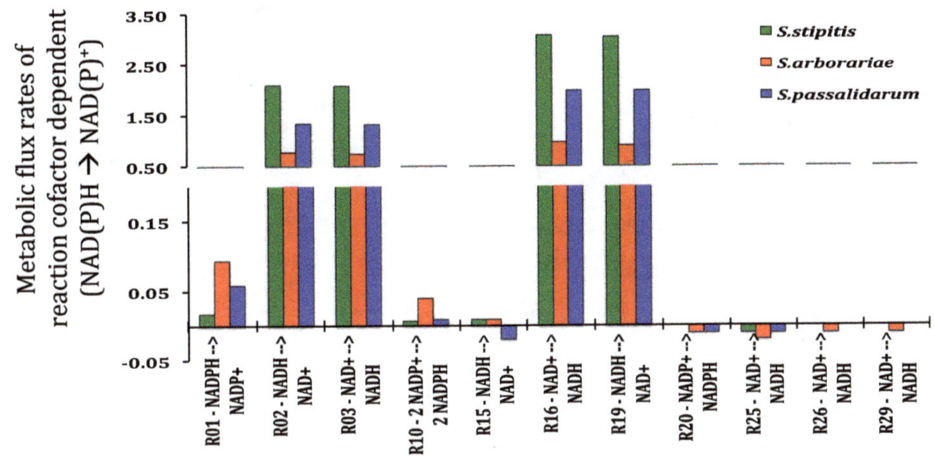

Fig. 4 Metabolic reaction rates using NAD(P) H / NAD(P)$^+$ cofactors. *S. stipitis* (green), *S. arborariae* (red), and *S. passalidarum* (blue). R01 – XYL to XOL; R02 - XYL to XOL; R03 – XOL to XYLU; R10 – G6P to RU5P; R15 – DHAP to GOL; R16 – GAP to PEP; R19 – ACCOA to ETOH; R20 – ACDH to ACE; R25 – ISO to AKG; R26 – AKG to SUC; and R29 – MAL to OXA

weak correlation in three tested xylose-fermenting yeasts.

The flux rates of *S. stipitis* and *S. arborariae* in the xylose assimilation pathway showed that the reaction XYL + NADPH → XOL + NADP (R01) have an error of 85 and 60%, respectively (Fig. 7). Figure 7 highlighted only the reaction rates that errors were above 10%. The reaction rate using cofactor NADPH influenced the xylitol accumulation and interfered the flux analysis.

The reaction rate G6P → RU5P (R10), responsible for regenerating NADPH in oxidative pentose phosphate pathway and the reversible reaction G6P ←→ F6P, also showed error higher than 60%. In contrast, in *S. passalidarum,* the reaction rate G6P → F6P (R11) showed at least 2.5 times less error when compared to *S. stipitis* and *S. arborariae.* In general, our analysis was able to predict 80% of intracellular carbon flux rates with an accuracy above 90% in relationship to calculated and measured flux rates from xylose until ethanol formation.

Discussion

Metabolome dataset for the intracellular carbon flux distribution

Metabolome data increase the precision of the actual state of cell metabolism [21]. Its measurements can be directly linked to the metabolic network since it enables the identification and quantification a large number of metabolites simultaneously under a specific condition [24, 29]. The metabolomic analyses result in the generation of a complex dataset. Some of the technical challenges are processing a large amount of data and performing statistical analyses, and then it can be linked to the studied biological system [30]. Therefore, it is crucial to develop an approach that is capable of validating metabolome data after statistical treatment.

Previous metabolomic studies did not succeed in quantifying the sugar-phosphate as such as G6P and F6P [31, 32]. Nevertheless, here, both isomers could be detected and quantified in the three yeasts *S. stipitis, S. arborariae,* and *S. passalidarum.* Also, the method based on UHPLC-MS/MS employed in this study enabled the separation and quantification of RU5P and R5P [26], overcoming the limitation observed in a previous metabolomics analysis for xylose fermentation performed by *S. passalidarum* [33].

Some metabolites could not be detected in particular yeasts; for example, E4P could not be detected only in *S. arborariae.* Since the flux rate formation for E4P in this yeast is at least twice slowly (reaction S7P + GAP → E4P + F6P = 0.23) when compared to *S. stipitis* and *S. passalidarum* (reaction S7P + GAP → E4P + F6P = 0.65 and 0.42). Therefore this metabolite was below the detection limit. Also, the absence of E4P is associated with low carbon flux rate into the pentose phosphate pathway in *S. arborariae,* when compared to *S. stipitis* and *S. passalidarum.* The ACCOA could not be detected in *S. stipitis* since it is observed that the carbon flux is preferably directed to ethanol formation.

Consequently, less carbon flux is directed to respiratory metabolism in the tricarboxylic acid cycle. Also, the difficulty of accurately quantifying metabolic flux influenced the detection of the ACCOA metabolite. Possibly, this limitation can be solved with the addition of more metabolic reactions in the stoichiometric model. Moreover, the ACCOA is involved in 34 compartmentalized metabolic reactions and used for acetylation of macromolecules [34]. To cell balance the use of this precursor metabolite, cells have evolved several levels of tight regulation, especially to control the biosynthesis of amino

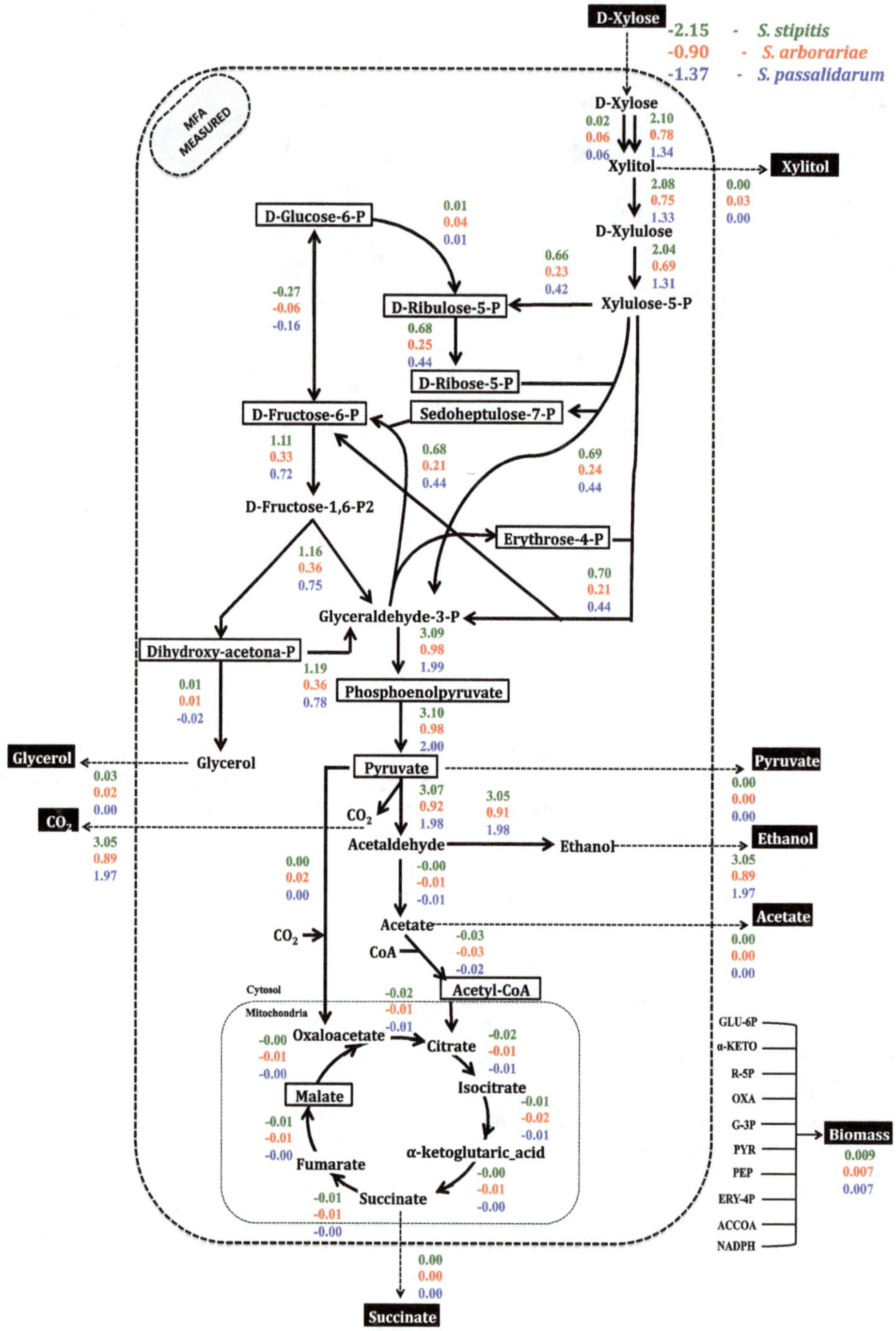

Fig. 5 Intracellular carbon flux distribution using measured data. *S. stipitis* (green), *S. arborariae* (red), *and S. passalidarum* (blue). Xylose consumption rates are represented by a negative signal. The first reaction (xylose to xylitol) shows two arrows; left represents reaction using NADPH, right represents reaction using NADH cofactor. Extracellular metabolites highlighted in black boxes had the flux rates used as constraints. Intracellular metabolites highlighted in white boxes limited the reaction rates with its concentration in MFA-measured

acids, lipids, nucleotides, and carbohydrates needed for cell growth, homeostasis, and maintenance [34].

The metabolites PEP and PYR could not be validated using MFA in any studied yeast. However, a previous study that assessed the metabolome of *S. passalidarum* in xylose fermentation encountered similar concentrations that were quantified here [33]. Possibly, as PYR is a branch-point metabolite involved in the respiration (mitochondria) and alcoholic fermentation (cytosol), this may have influenced the metabolomics analysis since the

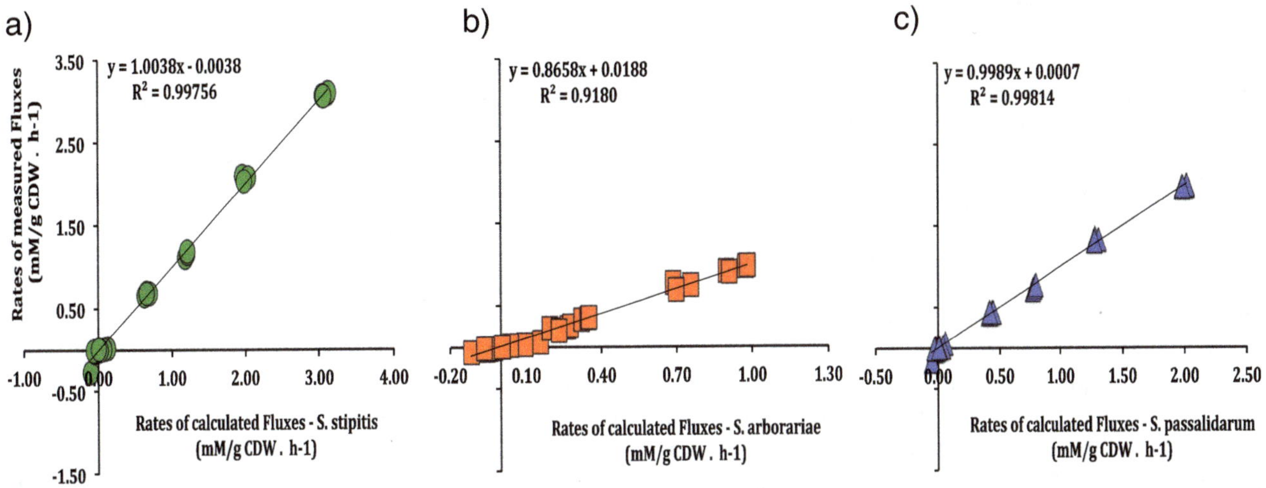

Fig. 6 Correlation test (R^2) between calculated and measured flux rates (mM/gCDW.h^{-1}). The relationship assessed with all metabolites measured that present a correlation higher than 0.90 for *S. stipitis* (**a**), *S. arborariae* (**b**), and *S. passalidarum.* (**c**)

metabolomics is not able to separate the compounds from different compartments cellular [33]. Moreover, the precise quantification of such metabolites is limited due to its presence in more than one cellular compartment [35].

The changes in metabolite concentration do not readily allow conclusions on metabolic fluxes or the direction of the flux changes. An increase in metabolite concentration can both be indicative of the increased activity of metabolite producing enzymes, but also decreased the activity of metabolite consuming enzymes [36]. Nevertheless, the addition of the metabolite concentration in the MFA model can be useful to validated metabolome data. Here a correlation between carbon flux distribution measured and calculated could be done for 80% of studied metabolites. Thus, the stoichiometric network model and intracellular carbon fluxes distribution could be estimated more precisely. They were integrating MFA with metabolome data of xylose-fermenting yeasts. Therefore, demonstrating that metabolite concentration determines the intracellular metabolic flux distribution in the central metabolism of yeast [37].

Intracellular carbon flux distribution of xylose metabolism in yeasts

The first step of xylose reduction to xylitol realized by xylose reductase (XR) enzyme using both NADPH and NADH as cofactors [1, 7, 8, 38]. However, the XR enzymes present different cofactor preference [7, 39–41]. Therefore, two reactions represent this first step in xylose metabolism in the model. First, the reaction is XYL + NADPH → XOL + NADP. Second reaction is XYL + NADH → XOL + NAD^+. Xylose reduction reaction NADH-dependent showed approximately twice higher flux rate in *S. stipitis* (1.97) and *S. passalidarum* (1.28) that is observed in *S. arborariae* (0.68). In *S. arborariae,* the flux rate of conversion xylose to xylitol

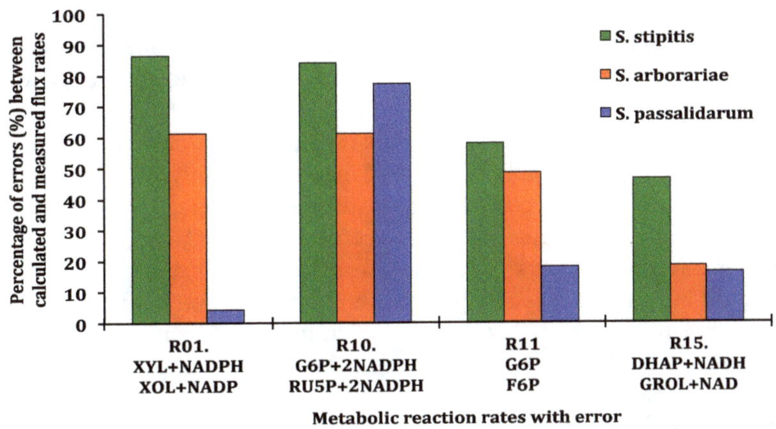

Fig. 7 Percentage of errors between calculated and measured flux rates. Columns are showing that the errors are less than 10% for most of the metabolic reaction rates from xylose to ethanol: *S. stipitis* (green), *S. arborariae* (red), and *S. passalidarum* (blue)

has employed the cofactor NADPH preferentially [7]. This feature observed in *S. arborariae* is associated with xylitol accumulation [7, 8]. Our results, corroborated with it and also demonstrated that only *S. arborariae* present xylitol production (Table 1). This characteristic may be associated with low xylose transportation capability [42]. The *S. arborariae* have a flux rate for oxidative-PPP (reaction G6P → RU5P) in 3.0 times higher than observed in *S. passalidarum*. This observation indicates the need to regenerate the NADPH cofactor in *S. arborariae*. Also, it noted that *S. arborariae* has a slower flux rate to consume xylose and a smaller growth rate. Due to the decreased cell growth, the requirement of NADPH has been reduced and caused the down-regulation of fluxes through the pentose phosphate pathway [13, 25].

On the contrary, for *S. passalidarum* already showed that it has two XR (genes XIL1.1 and XIL1.2), and one of them uses NADH preferentially as a cofactor [8]. Its enzymatic activity presents higher NADH-dependent XR [7]. Besides then, the conversion of G6P → RU5P in *S. passalidarum* was 3.0 times slower than *S. stipitis* and *S. arborariae.* Demonstrating less need for carbon flux to oxidative-PPP and carbon flux directed more to PPP and glycolysis pathways.

The enzymatic activities for xylose reductase (XR) of *S. stipitis* and *S. arborariae* cell extract are twice higher using NADPH as cofactor when compared to NADH-dependent activity [7, 8, 23]. Nevertheless, here, the calculated flux rates of the XR reaction using NADH was at least four times higher. Into MFA-measured was observed that the carbon flux distribution, preferably the use of NADH as the cofactor in XR reaction. This difference between enzymatic activities and calculated flux rates can be explained by optimal condition and concentration determined in enzymatic activities, not necessarily these occur in vivo. As observed previously, an MFA study showed the same result with flux distribution preferably using XR reaction with NAD-dependent in a recombinant xylose-utilizing *Saccharomyces cerevisiae* [23]. Therefore, the MFA showed that higher the reaction rate of XR NADH preference, more ethanol formation is observed (*S. stipitis* and *S. passalidarum*).

On the other hand, the XR NADPH preference showed xylitol accumulation (*S. arborariae*), confirming the literature data and our previous study [7, 8, 23]. The cofactor imbalance in XR and XDH reactions result in xylitol production, as observed in *S. arborariae* and prior studies [23]. The results of intracellular flux rates showed in MFA models are in good agreement with previous work that showed that higher NADH dependent XR activity resulted in less xylitol production [23].

Our results demonstrate a positive correlation between glycolytic flux rate and ethanol production. The low glycolytic flux seems to limit xylose utilization. These results are in agreement with a previous study that also applied metabolic flux analysis in genetic engineered *S. cerevisiae* [43, 44]. It has been previously suggested that the low glycolytic flux towards glyceraldehyde-3-phosphate and consequently, pyruvate may limit the consumption of xylose [43]. Therefore, increasing the metabolic reaction activities that direct carbon to glycolysis may be a valuable strategy to improve the metabolism of xylose. Take together, the metabolic flux along with the metabolome data, which increased the prediction accuracy, showed that in *S. stipitis* the glycolytic reaction (G6P ← → F6P) is 4.5 times faster than *S. arborariae*, and 2.7 times faster than observed in *S. passalidarum*. The faster metabolic fluxes in the glycolytic pathway observed in *S. stipitis* resulted in better ethanol production, and it was the main characteristic that differentiated it from the other evaluated yeasts.

Conclusions

The present study elucidated for the first time a stoichiometric model from xylose until ethanol to estimate the carbon flux distribution in *Spathaspora* yeasts for the first time. The metabolic flux model validated the quantification of 11 metabolites, where up to 80% of intracellular carbon flux rates could be correlated with an accuracy above 90%. The flux analysis corroborated that *S. stipitis* and *S. passalidarum* are the two yeasts with better metabolic characteristics towards xylose fermentation. These characteristics include higher xylose consumption rates, a higher reaction rate of XR with NADH preference, higher flux rates directed to PPP and glycolysis pathways, and less need to directed carbon flux to oxidative-PPP for the regeneration of NADPH. Characteristics that would enable better NADH/NAD^+ balance, thus allowing improves ethanol production from xylose.

Methods

Yeast strains and cultivation conditions

The yeasts used in this study were *Scheffersomyces stipitis* (NRRL Y-7124), *Spathaspora arborariae* (NRRL Y-48658), and *Spathaspora passalidarum* (NRRL Y-27907). These were kindly provided by the ARS (NRRL) culture collection (Peoria, USA). All are preserved in 30% glycerol at − 80 °C. As described previously, were performed all cultivations in the bioreactors [7]. Briefly, the fermentations were carried out in bioreactors (Multifors 2, Infors HT) with 500 mL of the defined mineral medium [45], supplemented with 40 g L^{-1} xylose as a carbon source. The fermentation started with an optical density of 600 nm (OD_{600}) equals 0.5. The temperature set up at 28 °C and stirred was kept at 400 rpm, pH was maintained at 5.5 by addition 3 M KOH. Oxygen was supplied at limited conditions. At those conditions, dissolved oxygen (DO) was kept below 10% with airflow of only 0.05 L/min throughout the cultivation in the bioreactors. All fermentations were carried out in biological triplicates. Samples were withdrawn to determine xylose consumption and product formation at regular intervals of time (approximately every 8 h) of cultivation. Extra- and intracellular quantification of metabolites was done using samples in the middle exponential growth phase where a pseudo-steady state is assumed, and therefore, all rates at that time-points were considered as constant [15]. The time point of during exponential growth was 28 h, 32 h, and 40 h for *S. stipitis*, *S. arborariae*, and *S. passalidarum*, respectively.

Determination of extracellular fluxes

The extracellular metabolites such as xylose, ethanol, xylitol, glycerol, acetate, pyruvate, and succinate concentrations were determined by High-Performance Liquid Chromatography (HPLC) as previously described [7]. Briefly, culture samples were collected in the middle exponential growth phase, centrifuged and the supernatant analyzed by an HPLC system (Acquity UPLC H Class, Waters) equipped with a refractive index detector. The metabolites were separated on an HPX-87 H column (Bio-Rad Laboratories) with a 5 mM sulfuric acid mobile phase at a flow rate of 0.6 mL/min and a temperature of 45 °C. After that, the extracellular concentration values are divided by biomass and time at that fermentation point. The data show the average ± standard deviation in mM/gCDW. h^{-1} of three biological triplicates (Table 1). The carbon balance and degree of reduction were calculated by taking the ratio of products in C-mole and consumed substrate in C-mole [46]. Biomass was measured through OD_{600} using a spectrophotometer (SpectraMax M3, Molecular Devices). For each collected point, cell dry weight (CDW) was performed through 5 mL of pre-inoculum and stationary growth phase of all yeasts in fermentative processes. Samples were withdrawn and centrifuged (12,000×g, 5 min). Before weighing, the cells were placed in a glass tube and incubated to dry at 60 °C at least 48 h. Therefore, it established a correlation between OD_{600} values and CDW. Approximately, for each OD_{600} = 1.0 were obtained 0.5 g L^{-1} of cells dry weight.

Standard metabolites and solvents

Acetate (ACE), acetyl coenzyme A (ACCOA), alpha-ketoglutaric acid (AKG), dihydroxyacetone phosphate (DHAP), erythrose-4-phosphate (E4P), ethanol (EtOH), fructose-6-phosphate (F6P), glucose (GLU), glucose-6-phosphate (G6P), glyceraldehyde-3-phosphate (GAP), glycerol (GOL), malate (MAL), phosphoenolpyruvate (PEP), pyruvate (PYR), ribose-5-phosphate (R5P), ribulose-5-phosphate (RU5P), sedoheptulose-7-phosphate (S7P), succinate (SUC), xylitol (XOL), xylose (XYL), xylulose (XYLU) and all solvents as such as sulphuric acid, tributylamine, acetonitrile and methanol used in HPLC and UHPLC-MS/MS analysis were purchased from Sigma-Aldrich (St. Louis, MO, USA) in their highest purity. Ultrapure water (18.2 MΩ) was obtained from a Direct 16 Milli-Q purification system (Millipore, Bedford, USA).

Metabolomics analysis

The experimental setup for determination and quantification of metabolome data is in Additional file 6. As mentioned before, all data was originated from the three replicates samples collected in the middle of the exponential growth phase under oxygen-limited conditions. This data point was the same used to calculate the extracellular flux rates, the carbon recovered, and redox balance. The sample preparation protocol and analytical data acquisition were previously described and optimized [26, 47, 48]. Briefly, preparation of the samples involved the steps of quenching and metabolites extraction using cold methanol (– 40 °C) followed by boiling ethanol (96 °C). The analytical method was based on UHPLC-MS/MS for metabolite separation and quantification [48]. Details of this analysis are previously showed [26, 48]. The MS methodology was carried out on an AcQuity^{tm} UPLC system (Waters, Milford, MA, USA) coupled to a triple quadrupole mass spectrometer (Xevo TQD, Waters) equipped with an electrospray ionization source. UPLC it performed on a Hydrophilic Interaction Liquid Chromatography (HILIC) with a BEH amide column (2.1 × 150 mm × 1.7 μm) (Waters Corporation, Milford, MA, USA) and Ion-Pairing Chromatography (IPC) with a reverse phase column, HSS-T3 (2.1 × 150 mm × 1.8 μm) (Waters Corporation, Milford, MA, USA). The metabolites detected in both environment extra and intracellular, for example, xylose, xylitol, and glycerol were not included into the metabolome dataset since the quantification of these compounds presents high variance, and it was not possible to define how much was intra- and extracellular.

Statistical analysis of intracellular metabolite concentration

It performed in biological triplicates all cultivations in bioreactors. For each replicate, it extracted the intracellular metabolites in three-time points within the exponential growth phase of yeasts. The intracellular extraction of each time point was analyzed by UHPLC-MS/MS method in three technical replicates, giving a total of 27 samples measurements. Therefore, to compile all resulting data into a single value, the metabolome dataset was statistically processed through a measured repeated ANOVA design. RStudio software (http://www.rstudio.com, https://www.rstudio.com/products/rstudio/

download/) was used to construct the ANOVA model. The following mathematical equation represents how the ANOVA model test was performed.

$$y\_ijk = \mu + \alpha\_i + \beta\_(j(i)) + \tau\_k + (\alpha\tau)\_ik + e\_ijk$$

$y_{ijk} = \mu + \alpha_i + \beta_{j(i)} + \tau_k + (\alpha\tau)_{ik} + e_{ijk}$Using this linear model, it was assumed that the data for class (*i*) for yeast (*j*) at time (*k*) is equal to an overall mean (μ) plus the treatment effect (αi), the effect of the yeast within that class ($\beta_{j(i)}$) the effect of time (τk), the effect of the interaction between time and class ($(\alpha\tau)_{ik}$) and the error (eijk).

Such that:

- μ = overall mean
- α_i = effect of class *i*
- $\beta_{j(i)}$ = random effect of yeast *j* receiving class *i*
- τ_k = effect of time *k*
- $(\alpha\tau)_{ik}$ = class by time interaction
- ε_{ijk} = experimental error

The Additional file 4 shows the concentrations obtained after statistical analysis from the metabolome data.

Stoichiometric model construction

An overview of the metabolic model is shown in Additional file 1. The stoichiometric model was constructed based on previous studies [13, 23, 49]. The model composed of 39 reactions within the xylose assimilation pathway, pentose phosphate pathway, glycolysis, and TCA. It included the TCA cycle, but the compartmentalization into mitochondria and cytosol was not considered due to an equipment limitation. The probe of oxygen used covers only the measurements of dissolved oxygen in the medium. Thereby, it was not possible to measure the oxygen released, data necessary for the metabolic flux calculation. However, similar models have proved efficient to support understanding sugar metabolism in yeasts [23, 49]. Biomass equation was determined as previously described [23]. It consists of the macromolecules components of the cell (i.e., proteins, nucleic acid, and polysaccharides) [50]. The stoichiometric model was constructed based on the information available at The Kyoto Encyclopedia of Genes and Genomes (KEGG). It used as the reference genomic and biochemical information of *S. stipitis* (Entry T01023). The genes encoding for the enzymes on the carbohydrate metabolism present in the respective genome could be determined using KEGG pathway.

Metabolic flux analysis using OptFlux

The model uploaded into OptFlux from an Excel file (Additional file 2). The degree of freedom of the metabolic network was calculated using the properties of the stoichiometric model. The accurate number of degrees of freedom obtained by the difference between the number of metabolites of the system and the number of linearly independent equations [17]. For differentiate internal and external reactions, external metabolites were identified with "[e]" and intracellular metabolites occurring in cytosolic subsystems with "[c]." It used a biomass reaction as an objective function [22, 23, 51]. Thus, added the extracellular measured flux rates obtained from of middle exponential growth phase to the model. The extracellular measurements used are xylose consumption rate, xylitol, acetate, glycerol, pyruvate, and succinate production rates. The simulations were run using the algebraic method with least square fitting as properties.

Carbon flux distribution using extracellular flux rates measurements

The extracellular flux rates measurements were utilized to simulate the metabolic fluxes and to calculate the carbon fluxes distribution. The model classified as an overdetermined system containing 39 reactions, 35 metabolites, 27 genes, and 04 degrees of freedom. Nine extracellular flux rates were measured (xylose consumption, biomass, ethanol, carbon dioxide, xylitol, glycerol, acetate, pyruvate, and succinate productions) and used to limit the initial metabolic model resolution. Considering the principle of mass conservation and molarity, this ensures that the total amount of compounds produced must be equal to the total amount being consumed [17, 36, 52].

Between them, for each yeast, xylose consumption, and biomass production rates were maintained fixed during all simulations. Xylose consumption rates were fixed with following values − 2.15, − 0.90, and − 1.37 (mM/gCDW.h^{-1}), whereas biomass production rates were fixed with 1.45, 1.27, and 1.01 (g. L^{-1}) for *S. stipitis, S. arborariae,* and *S. passalidarum,* respectively. The extracellular ethanol flux rates were used to validate the carbon flux distribution in the calculated model. First, the simulations were performed without the ethanol flux rate, and then, the flux rate obtained through this simulation was compared with the ethanol rate obtained experimentally.

Validation of metabolome data using metabolic flux analysis

The statistical analysis resulted in a single value of intracellular metabolites concentration. Those were added to the stoichiometric model. After that, a simulation was performed to determine the flux distribution within the metabolic network. Initially, the simulations performed with the addition of one metabolite by time. Thus, 11 measured fluxes distribution were obtained, one for each measured metabolite. Then, the fluxes derived from stoichiometric calculations and the ones with the addition of metabolome data were compared using the Pearson correlation coefficient (R^2). The correlation coefficient is useful to find the relationship between the calculated and measured fluxes distributions. The metabolites, whose correlation was above 90%, were used in the stoichiometric model for a new round of carbon flux simulations. Thus it was possible to estimate the percentage of error, and consequently, identify hits between calculated and measured fluxes for each measured metabolite.

Additional files

Additional file 1: Overview of metabolic network from xylose to ethanol. The metabolic model showns the directions of intracellular metabolic reactions (continuos arrows), xylose consumption and products formation (dashed arrows), and cofactors ($NADPH/NADP^+$; $NADH/NAD^+$; ATP) utilized in some reactions. (PDF 169 kb)

Additional file 2: The stoichiometric model. Metabolic reactions added to the OptFlux. (PDF 41 kb)

Additional file 3: List of metabolites. The intracellular and extracellular metabolites added to the OptFlux. (PDF 39 kb)

Additional file 4: Intracellular metabolites concentrations (μg/mL). Average and standart desviation of metabolites concentrations obtained after statistical analysis (ANOVA) from the metabolome data. (PDF 52 kb)

Additional file 5: **a** Correlation (R2) between calculated and measured fluxes - *S. stipitis*. Acetyl-CoA (ACCOA), dihydroxy-acetone-phosphate (DHAP), erythrose-4-phosphate (E4P), fructose-6-phosphate (F6P), glucose-6-phosphate (G6P), malate (MAL), phosphoenolpyruvate (PEP), pyruvate (PEP), ribose-5-phosphate (R5P), ribulose-5-phosphate (RU5P), and sedoheptulose-7-phosphate (S7P) were the metabolites measured. (X-axis) show the calculated fluxes using the constrained values of products formation. (Y-axis) show measured fluxes with respectively metabolites concentrations. Graphics in square presents a correlation higher than 0.9. Flux rates are in $mmol/gCDW.h^{-1}$. **b** Correlation (R2) between calculated and measured fluxes - *S. arborariae*. Acetyl-CoA (ACCOA), dihydroxy-acetone-phosphate (DHAP), erythrose-4-phosphate (E4P), fructose-6-phosphate (F6P), glucose-6-phosphate (G6P), malate (MAL), phosphoenolpyruvate (PEP), pyruvate (PYR), ribose-5-phosphate (R5P), ribulose-5-phosphate (RU5P), and sedoheptulose-7-phosphate (S7P) were the metabolites measured. (X-axis) show the calculated fluxes using the constrained values of products formation. (Y-axis) show measured fluxes with respectively metabolites concentrations. Graphics in square presents a correlation higher than 0.9. Flux rates are in $mmol/gCDW.h^{-1}$. **c** Correlation (R2) between calculated and measured fluxes - *S. passalidarum*. Acetyl-CoA (ACCOA), dihydroxy-acetone-phosphate (DHAP), erythrose-4-phosphate (E4P), fructose-6-phosphate (F6P), glucose-6-phosphate (G6P), malate (MAL), phosphoenolpyruvate (PEP), pyruvate (PYR), ribose-5-phosphate (R5P), ribulose-5-phosphate (RU5P), and sedoheptulose-7-phosphate (S7P) were the metabolites measured. (X-axis) show the calculated fluxes using the constrained values of products formation. (Y-axis) show measured fluxes with respectively metabolites concentrations. Graphics in square presents a correlation higher than 0.9. Flux rates are in $mmol/gCDW.h^{-1}$. (ZIP 183 kb)

Additional file 6: Experimental design for metabolomics data. Three species of xylose-fermenting yeasts *S. stipitis*, *S. arborariae*, and *S. passalidarum*. The times of replicates (T1, T2, T3) and the technical replicates (R1, R2, R3) are repeated for each biological replicate into an oxygen-limited condition (A, B, C). (PDF 135 kb)

Abbreviations

ACCOA: Acetyl coenzyme A; ACE: Acetate; AKG: Alpha Ketoglutaric Acid; ANOVA: Analysis of variance; CDW: Cell dry weight; DHAP: Dihydroxy Acetone Phosphate; DR: Degree of reduction; E4P: Erythrose-4-phosphate; ETOH: Ethanol; F6P: D-Fructose-6-Phosphate; G6P: D-Glucose-6-Phosphate; GAP: Glyceraldehyde-3-phosphate; GLU: D-Glucose; GOL: Glycerol; HILIC: Hydrophilic Liquid Chromatography; HPLC: High-Performance Liquid Chromatography; IPC: Ion-Paring Chromatography; KEGG: Kyoto Encyclopedia of Genes and Genomes; KOH: Potassium Hydroxide; LC: Liquid Chromatography; MAL: Malate; MFA: Metabolic flux analysis; NAD^+: Nicotinamide Adenine Dinucleotide; NADH: Nicotinamide Adenine Dinucleotide Reduced; $NADP^+$: Nicotinamide Adenine Dinucleotide Phosphate; NADPH: Nicotinamide Adenine Dinucleotide Phosphate Reduced; OD_{600}: Optical Density of 600 nm; PEP: Phosphoenolpyruvate; PPP: Pentose phosphate pathway; PYR: Pyruvate; R^2: Person Correlation; R5P: D-Ribose-5-Phosphate; RU5P: D-Ribulose-5-Phosphate; S7P: Sedoheptulose-7-Phosphate; SUC: Succinate; TCA: Tricarboxylic acid; UHPLC-MS/MS: Ultra-High Performance Liquid Chromatography coupled to Tandem Mass Spectrometry; XDH: Xylitol Dehydrogenase; XOL: D-Xylitol; XR: Xylose reductase; XYL: D-Xylose; XYLU: D-Xylulose

Acknowledgments

The authors gratefully acknowledge José Antônio de Aquino Ribeiro, Patrícia Pinto Kalil Gonçalves Costa, Katiúscia Pereira Araújo, and Clenilson Martins Rodrigues of EMBRAPA Agroenergia for their support in the development of the UHPLC-MS/MS method for the quantification of intracellular metabolites.

Authors' contributions

HCTV participated in the design of the study, performed all experiments, analyzed the data, and wrote the manuscript. CGC performed the metabolome experiments, analysis and commented on the manuscript. IFN performed statistical analysis. PVA participated in the design of the metabolomics study and commented on the manuscript. JRMA and NSP participated in the design of the study and commented on the manuscript. All authors read and approved the manuscript.

Funding

We greatly appreciate the financial support from the EMBRAPA (grant number 02.12.01.006.00.00), the Coordination for the Improvement of Higher Education Personnel (CAPES), the National Council for Scientific and Technological Development (CNPq), the University of Brasilia (UnB), and the Federal University of Goiás (UFG). The funders role was providing funding for the study. All other aspects of project as design of the study and collection, analysis, and interpretation of data and writing of the manuscript was executed by the authors.

Ethics approval and consent to participate

Not applicable.

Consent for publication

Not applicable.

Competing interests

The authors declare that they have no competing interests.

Author details

[1]Grupo Engenharia de Biocatalisadores, Universidade de Brasília - UnB , Campus Darcy Ribeiro, Instituto de Ciências Biológicas, Bloco K, 1° andar, Asa Norte, Brasilia 70.790-900, Brazil. [2]Empresa Brasileira de Pesquisa Agropecuária, EMBRAPA Agroenergia, Brasília-DF, Brazil. [3]Instituto de Química, Universidade Federal de Goiás - UFG, Goiânia, Brazil. [4]Programa de Pós-Graduação em Biologia Microbiana, Instituto de Biologia, Universidade de Brasília - UnB, Brasilia, Brazil. [5]Programa de Pós-Graduação em Administração, Universidade de Brasília - UnB, Brasília, Brazil.

References

1. Skoog K, Hahn-Hagerdal B. Effect of oxygenation on xylose fermentation by *Pichia stipitis*. Appl Environ Microbiol. 1990;56:3389–94.

2. Su Y, Willis LB, Jeffries TW. Effects of aeration on growth, ethanol and polyol accumulation by *Spathaspora passalidarum* NRRL Y-27907 and *Scheffersomyces stipitis* NRRL Y-7124. Biotechnol Bioeng. 2014;112:457–69.
3. Cadete RM, Santos RO, Melo MA, Mouro A, Goncalves DL, Stambuk BU, et al. *Spathaspora arborariae* sp. nov., a d-xylose-fermenting yeast species isolated from rotting wood in Brazil. FEMS Yeast Res. 2009;9(8):1338–42.
4. Trausinger G, Gruber C, Krahulec S, Magnes C, Nidetzky B, Klimacek M. Identification of novel metabolic interactions controlling carbon flux from xylose to ethanol in natural and recombinant yeasts. Biotechnol Biofuels. 2015;8:1–13.
5. du Preez JC, van Driessel B, Prior BA. D-xylose fermentation by *Candida shehatae* and *Pichia stipitis* at low dissolved oxygen levels in fed-batch cultures. Biotechnol Lett. 1989;11:131–6.
6. Wilkins MR, Mueller M, Eichling S, Banat IM. Fermentation of xylose by the thermotolerant yeast strains *Kluyveromyces marxianus* IMB2, IMB4, and IMB5 under anaerobic conditions. Process Biochem. 2008;43(4):346–50.
7. Veras HCT, Parachin NS, Almeida JRM. Comparative assessment of fermentative capacity of different xylose-consuming yeasts. Microb Cell Factories. 2017;16(1):153.
8. Cadete RM, de Las Heras AM, Sandstrom AG, Ferreira C, Girio F, Gorwa-Grauslund MF, et al. Exploring xylose metabolism in *Spathaspora* species: XYL1.2 from *Spathaspora passalidarum* as the key for efficient anaerobic xylose fermentation in metabolic engineered *Saccharomyces cerevisiae*. Biotechnol Biofuels. 2016;9:167:1–14.
9. Harner NK, Wen X, Bajwa PK, Austin GD, Ho CY, Habash MB, et al. Genetic improvement of native xylose-fermenting yeasts for ethanol production. J Ind Microbiol Biotechnol. 2015;42(1):1–20.
10. Hou X. Anaerobic xylose fermentation by *Spathaspora passalidarum*. Appl Microbiol Biotechnol. 2012;94(1):205–14.
11. Sonderegger M, Jeppsson M, Hahn-Hägerdal B, Sauer U. Molecular basis for anaerobic growth of *Saccharomyces cerevisiae* on xylose, investigated by global gene expression and metabolic flux analysis. Appl Environ Microbiol. 2004;70(4):2307–17.
12. Balagurunathan B, Jonnalagadda S, Tan L, Srinivasan R. Reconstruction and analysis of a genome-scale metabolic model for *Scheffersomyces stipitis*. Microb Cell Factories. 2012;11:27.
13. Liang M, Damiani A, He QP, Wang J. Elucidating xylose metabolism of *Scheffersomyces stipitis* for lignocellulosic ethanol production. ACS Sustain Chem Eng. 2013;2(1):38–48.
14. Osiro KO, Brink DP, Borgstrom C, Wasserstrom L, Carlquist M, Gorwa-Grauslund MF. Assessing the effect of d-xylose on the sugar signaling pathways of *Saccharomyces cerevisiae* in strains engineered for xylose transport and assimilation. FEMS Yeast Res. 2018;18(1). https://doi.org/10.1093/femsyr/fox096.
15. Stephanopoulos GN, Aristidou AA, Nielsen J. Metabolic engineering: principles and methodologies. San Diego: Academic; 1998. p. 725.
16. Bideaux C, Montheard J, Cameleyre X, Molina-Jouve C, Alfenore S. Metabolic flux analysis model for optimizing xylose conversion into ethanol by the natural C5-fermenting yeast *Candida shehatae*. Appl Microbiol Biotechnol. 2016;100(3):1489–99.
17. Carreira R, Evangelista P, Maia P, Vilaça P, Pont M, Tomb J, et al. CBFA: phenotype prediction integrating metabolic models with constraints derived from experimental data. BMC Syst Biol. 2014;8:123.
18. Sánchez BJ, Nielsen J. Genome scale models of yeast: towards standardized evaluation and consistent omic integration. Royal Soc Chem. 2015;7:846–58.
19. Jones JA, Toparlak OD, Koffas MA. Metabolic pathway balancing and its role in the production of biofuels and chemicals. Curr Opin Biotechnol. 2015;33:52–9.
20. Unrean P, Nguyen NHA. Metabolic pathway analysis of *Scheffersomyces stipitis (Pichia) stipitis*: effect of oxygen availability on ethanol synthesis and flux distributions. Appl Microbiol Biotechnol. 2012;94:1387–98.
21. O'Brien EJ, Monk JM, Palsson BO. Using genome-scale models to predict biological capabilities. Cell. 2015;161(5):971–87.
22. Rocha I, Maia P, Evangelista P, Vilaça P, Soares S, Pinto JP, et al. OptFlux: an open-source software platform for in silico metabolic engineering. BMC Syst Biol. 2010;4:45.
23. Wahlbom CF, Eliasson A, Hahn-Hägerdal B. Intracellular fluxes in a recombinant xylose-utilizing *Saccharomyces cerevisiae* cultivated anaerobically at different dilution rates and feed concentrations. Biotechnol Bioeng. 2001;72(3):289–96.
24. Bogaerts P, Mhallem Gziri K, Richelle A. From MFA to FBA: defining linear constraints accounting for overflow metabolism in a macroscopic FBA-based dynamical model of cell cultures in bioreactor. J Process Control. 2017;60:34–47.
25. Quiros M, Martinez-Moreno R, Albiol J, Morales P, Vazquez-Lima F, Barreiro-Vazquez A, et al. Metabolic flux analysis during the exponential growth phase of *Saccharomyces cerevisiae* in wine fermentations. PLoS One. 2013;8(8):e71909.
26. Campos CG, Veras HCT, de Aquino Ribeiro JA, Costa P, Araujo KP, Rodrigues CM, et al. New protocol based on UHPLC-MS/MS for quantitation of metabolites in xylose-fermenting yeasts. J Am Soc Mass Spectrom. 2017;28(12):2646–57.
27. Abdelnur PV, Caldana C, Martins MCM. Metabolomics applied in bioenergy. Chem Biol Technol Agric. 2014;1:22.
28. Granucci N, Pinu FR, Han TL, Villas-Boas S. Can we predict the intracellular metabolic state of a cell based on extracellular metabolite data? Mol BioSyst. 2015;11:3297.
29. Riekeberg E, Powers R. New frontiers in metabolomics: from measurement to insight. F1000Research. 2017;6:1148.
30. Johnson CH, Ivanisevic J, Siuzdak G. Metabolomics: beyond biomarkers and towards mechanisms. Nat Rev Mol Cell Biol. 2016;17(7):451–9.
31. Kato H, Izumi Y, Hasunuma T, Matsuda F, Kondo A. Widely targeted metabolic profiling analysis of yeast central metabolites. J Biosci Bioeng. 2012;113(5):665–73.
32. Matsuda F, Toya Y, Shimizu H. Learning from quantitative data to understand central carbon metabolism. Biotechnol Adv. 2017;35(8):971–80.
33. Long TM, Su YK, Headman J, Higbee A, Willis LB, Jeffries TW. Cofermentation of glucose, xylose, and cellobiose by the beetle-associated yeast *Spathaspora passalidarum*. Appl Environ Microbiol. 2012;78(16):5492–500.
34. Nielsen J, Keasling JD. Engineering cellular metabolism. Cell. 2016;164(6):1185–97.
35. Österlund T, Nookaew I, Bordel S, Nielsen J. Mapping condition-dependent regulation of metabolism in yeast through genome-scale modeling. BMC Syst Biol. 2013;7:36.
36. Buescher JM, Antoniewicz MR, Boros LG, Burgess SC, Brunengraber H, Clish CB, et al. A roadmap for interpreting (13) C metabolite labeling patterns from cells. Curr Opin Biotechnol. 2015;34:189–201.
37. Hackett SR, Zanotelli VR, Xu W, Goya J, Park JO, Perlman DH, et al. Systems-level analysis of mechanisms regulating yeast metabolic flux. Science. 2016; 354:aaf2786. https://doi.org/10.1126/science.aaf2786.
38. Petschacher B, Nidetzky B. Altering the coenzyme preference of xylose reductase to favor utilization of NADH enhances ethanol yield from xylose in a metabolically engineered strain of *Saccharomyces cerevisiae*. Microb Cell Factories. 2008;7:9.
39. Bruinenberg PM, de Bot PHM, van Dijken JP, Scheffers WA. NADH-linked aldose reductase - the key to anaerobic alcoholic fermentation of xylose by yeasts. Appl Microbiol Biotechnol. 1984;19:256–60.
40. Bengtsson O, Hahn-Hagerdal B, Gorwa-Grauslund MF. Xylose reductase from *Pichia stipitis* with altered coenzyme preference improves ethanolic xylose fermentation by recombinant *Saccharomyces cerevisiae*. Biotechnol Biofuels. 2009;2:9:1–10.
41. Jeffries TW, Shi NQ. Genetic engineering for improved xylose fermentation by yeasts. Adv Biochem Eng Biotechnol. 1999;65:118–61.
42. Stambuk BU, Eleutherio ECA, Florez-Pardo LM, Souto-Maior AM, Bon EPS. Brazilian potential for biomass ethanol: challenge of using hexose and pentose co-fermenting yeast strains.pdf. Sci Ind Res. 2008;67:918–26.
43. Wasylenko TM, Stephanopoulos G. Metabolomic and 13C-metabolic flux analysis of a xylose-consuming *Saccharomyces cerevisiae* strain expressing xylose isomerase. Biotechnol Bioeng. 2015;112:470–83.
44. Feng X, Zhao H. Investigating xylose metabolism in recombinant *Saccharomyces cerevisiae* via 13C metabolic flux analysis. Microb Cell Factories. 2013;12:114.
45. Verduyn C, Postma E, Scheffers WA, Van Dijken JP. Effect of benzoic acid on metabolic fluxes in yeasts: a continuous-culture study on the regulation of respiration and alcoholic fermentation. Yeast. 1992;8:501–17.
46. Hagman A, S T, Compagno C, Piskur J. Yeast "make-accumulate-consume" life strategy envolved as a multi-step process that predates the whole genome duplication. PLoS One. 2013;8(7):e68734.
47. Bergdahl B, Heer D, Sauer U, Hahn-Hagerdal B, van Niel EW. Dynamic metabolomics differentiates between carbon and energy starvation in recombinant *Saccharomyces cerevisiae* fermenting xylose. Biotechnol Biofuels. 2012;5(1):34.
48. Campos CG, Ribeiro JAA, Almeida JRM, Quirino BF, Abdelnur PV. Targeted metabolomics of xylose-fermenting yeasts based on mass spectrometry. Microbial Metabolomics. Methods Mol Biol. 2018;1859:155–69.
49. Almeida JR, Bertilsson M, Hahn-Hagerdal B, Liden G, Gorwa-Grauslund MF. Carbon fluxes of xylose-consuming *Saccharomyces cerevisiae* strains are

affected differently by NADH and NADPH usage in HMF reduction. Appl Microbiol Biotechnol. 2009;84(4):751–61.

50. Senger RS. Biofuel production improvement with genome-scale models: the role of cell composition. Biotechnol J. 2010;5(7):671–85.
51. Wahlbom CF, Eliasson A, Hahn-Hagerdal B. Intracellular fluxes in a recombinant xylose-utilizing *Saccharomyces cerevisiae* cultivated anaerobically at different dilution rates and feed concentration. Biotechnol Bioeng. 2001;72(3):289–96.
52. Tummler K, Klipp E. The discrepancy between data for and expectations on metabolic models: how to match experiments and computational efforts to arrive at quantitative predictions? Curr Opin Syst Biol. 2018;8:1–6.

11

Genome-wide response on phytosterol in 9-hydroxyandrostenedione-producing strain of *Mycobacterium* sp. VKM Ac-1817D

Eugeny Y. Bragin[1,2*], Victoria Y. Shtratnikova[4], Mikhail I. Schelkunov[3,5], Dmitry V. Dovbnya[1,2] and Marina V. Donova[1,2]

Abstract

Background: Aerobic side chain degradation of phytosterols by actinobacteria is the basis for the industrial production of androstane steroids which are the starting materials for the synthesis of steroid hormones. A native strain of *Mycobacterium* sp. VKM Ac-1817D effectively produces 9α-hydroxyandrost-4-ene-3,17-dione (9-OH-AD) from phytosterol, but also is capable of slow steroid core degradation. However, the set of the genes with products that are involved in phytosterol oxidation, their organisation and regulation remain poorly understood.

Results: High-throughput sequencing of the global transcriptomes of the *Mycobacterium* sp. VKM Ac-1817D cultures grown with or without phytosterol was carried out. In the presence of phytosterol, the expression of 260 genes including those related to steroid catabolism pathways significantly increased. Two of the five genes encoding the oxygenase unit of 3-ketosteroid-9α-hydroxylase (*kshA*) were highly up-regulated in response to phytosterol (55- and 25-fold, respectively) as well as one of the two genes encoding its reductase subunit (*kshB*) (40-fold). Only one of the five putative genes encoding 3-ketosteroid-Δ^1-dehydrogenase (KstD_1) was up-regulated in the presence of phytosterol (61-fold), but several substitutions in the conservative positions of its product were revealed.
Among the genes over-expressed in the presence of phytosterol, several dozen genes did not possess binding sites for the known regulatory factors of steroid catabolism. In the promoter regions of these genes, a regularly occurring palindromic motif was revealed. The orthologue of TetR-family transcription regulator gene *Rv0767c* of *M. tuberculosis* was identified in *Mycobacterium* sp. VKM Ac-1817D as *G155_05115*.

Conclusions: High expression levels of the genes related to the sterol side chain degradation and steroid 9α-hydroxylation in combination with possible defects in KstD_1 may contribute to effective 9α-hydroxyandrost-4-ene-3,17-dione accumulation from phytosterol provided by this biotechnologically relevant strain. The TetR-family transcription regulator gene *G155_05115* presumably associated with the regulation of steroid catabolism. The results are of significance for the improvement of biocatalytic features of the microbial strains for the steroid industry.

Keywords: Steroid catabolism, Bioconversion, Phytosterol, 9α-hydroxyandrostenedione, *Mycobacterium*, Transcriptome

Background

Steroids are terpenoid lipids that contain a gonane core of four cycloalkane rings (A-D) in their structure. This class of organic molecules includes the compounds which fulfil essential vital functions in higher organisms. Sterols such as cholesterol, β-sitosterol, stigmasterol, campesterol etc. are steroid 3β-alcohols with different aliphatic side chains at C17.

The molecular mechanisms of sterol degradation have been intensively studied, mainly due to their essential role in the pathogenesis of *Mycobacterium tuberculosis* [1, 2] and the wide application of non-pathogenic mycobacteria capable of partial catabolism of sterols to produce intermediates for the pharmaceutical industry, such as androst-4-ene-3,17-dione (AD), androsta-1,4-diene-3,17-dione (ADD) and 9α-

* Correspondence: bragory@yandex.ru
[1]Institute of Biochemistry and Physiology of Microorganisms, Federal Research Center "Pushchino Center for Biological Research of the Russian Academy of Sciences", Nauki, 5, Pushchino, Russian Federation142290
[2]Pharmins Ltd., Institutskaya, 4, Pushchino, Russian Federation142290

hydroxyandrost-4-ene-3,17-dione (9-OH-AD). 9-OH-AD is extensively used as a starting compound for the synthesis of various steroid drugs bearing a halogen in position 9, such as dexamethasone, fluoxymesterone, triamcinolone, betha-metasone and others [3].

Steroid-degrading bacteria are widespread in the environment, and play an important role in the global carbon cycle [4]. Bacterial sterol degradation represents cascades of reactions which can be conventionally divided according to the parts of the steroid molecule: aliphatic side chain, rings A/B and rings C/D degradation (Fig. 1). Dozens of enzymes are involved in this process. The groups of putative genes engaged in sterols degradation have been characterised in the genomes of several strains such as *Mycobacterium neoaurum* VKM Ac-1815D, *Gordonia neofelifaecis* NRRL B-59395, *M. tuberculosis* H37Rv, *Rhodococcus jostii* RHA1, *M. neoaurum* NRRL 3805B, *Nocardioides simplex* VKM Ac-2033D and *M. smegmatis* mc^2 155 [2, 5–9]. It had been proposed that steroid catabolism pathways were conserved in certain different *Actinobacteria* taxa [10].

The first reaction of the pathway is modification of the 3β-hydroxy-5-ene to 3-keto-4-ene- moiety (Fig. 1). Cholesterol oxidases (ChOs) and 3β-hydroxysteroid dehydrogenases (3-HsDs) are able to carry out this reaction (for review, see [11]). In some strains leading role is played by ChOs [12], while in several mycobacteria species the reaction was proposed to be mainly catalysed by 3-HsDs [9, 13]. Side chain degradation is initiated with hydroxylation at C-26(27) with cytochrome P450 monooxygenase encoded by *cyp125* [14, 15]. Similar activity has been confirmed for the proteins encoded by *cyp124* and

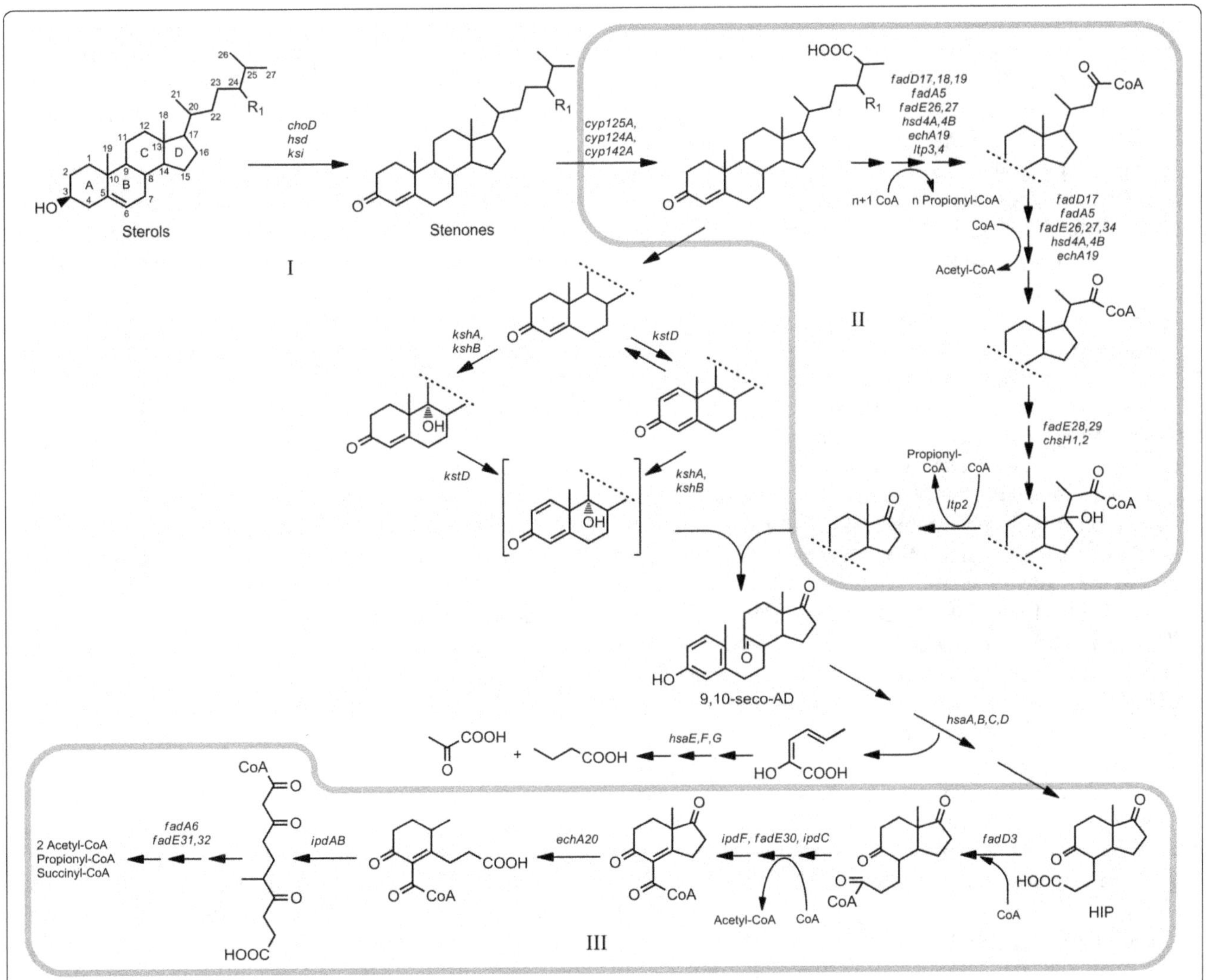

Fig. 1 General scheme of sterol catabolism organisation in Actinobacteria: stages and the identified genes involved. I – initial steroid core oxidation (rings A-B), II – side chain degradation, III – deep steroid core oxidation (rings C-D). HIP – 3aα-H-4α(3'-propanoate)-7aβ-methylhexahydro-1,5-indanedione; 9,10-seco-AD – 3-hydroxy-9,10-secoandrost-1,3,5(10)-triene-9,17-dione; COCHEA-CoA – 2-(2-carboxyethyl)-3-methyl-6-oxocyclohex-1-ene-1-carboxyl-CoA; R1 – H, CH3 or C2H5 for various sterols

cyp142 [16]. Degradation of the C24-branched chain sterols was shown to occur via the aldolytic cleavage with aldol lyases encoded by *ltp3* and *ltp4* [17].

The pathway of the following side chain degradation of sterols is generally similar to the fatty acids β-oxidation. CoA is attached to the carboxyl group of cholestenoate with acyl-CoA synthetase encoded by *fadD19* [18]. The CoA-thioester formed is further subjected to three successive cycles of β-oxidation to complete the degradation of the side chain, resulting in the production of C17-keto androstanes [14, 19, 20].

Steroid core degradation proceeds through the well-known "9(10)-seco pathway" with the key steps catalysed by 9α-hydroxylase (Ksh) and 3-ketosteroid-Δ^1-dehydrogenase (KstD) [21, 22]. Their cooperative action results in the ring B opening to form 3-hydroxy-9,10-seconandrost-1,3,5(10)-triene-9,17-dione (3-HSA) (Fig. 1). Metabolic blocks preventing activity of KstD and/or Ksh may provide effective accumulation of the useful intermediates such as AD, ADD, or 9-OH-AD [1, 23, 24].

Further rings A/B degradation is catalysed by the enzymes encoded by *hsaA, hsaB, hsaC, hsaD, hsaE, hsaF, hsaG* [25–27]. Most of the genes coding for the enzymes of the side chain and A/B rings degradation are regulated with KstR transcription factor [28].

As has been proven for several actinobacteria strains, the side chain and rings A/B degradation may occur simultaneously [21, 22, 29]. Moreover, some cholesterol ring-degrading enzymes were shown to exhibit much higher activities towards the side-chain degradation intermediates than towards the corresponding C-17 ketosteroids with fully degraded side chain, e.g. steroid CoA-thioesters represent physiological substrates for Ksh of *M. tuberculosis* [21]. These results were further confirmed for *M. neoaurum* ATCC 25795: greater activities of Ksh (KshA1, KshA2) have been demonstrated in vitro towards 23,24-bisnorchol-1,4-diene-22-oic acid (1,4-BNC) than towards AD [24, 30]. It coincides with the structural features of KshA (for review, see [31]) [21, 32] and the resulting transcriptional changes of *kshA1* and *kshA2*: unlike cholesterol, AD was shown to be a poor inducer for both enzymes in *M. neoaurum* ATCC 25795 [30]. At the same time, KshAs demonstrate lower activity towards cholestenone than towards C19-steroids such as AD, ADD, or testosterone [30, 33].

The KstD enzyme may also be involved in different stages of steroid catabolism, in particular, KstD activity has been identified for both AD and partially oxidised side chain steroids, such as 22-hydroxy-23,24-bisnorchola-4-en-3-one [34].

The C/D rings of the steroid core are degraded by the products of the genes regulated with KstR2. KstR2 is a transcription repressor which similar to KstR belongs to TetR family. As previously shown, the effector of KstR2 is HIP-CoA which is a thioester of cholesterol metabolite HIP (3aα-*H*-4α(3-propanoate)-7aβ-methylhexahydro-1,5-indanedione) [35]. Regulon KstR2 in *M. smegmatis* mc^2 155 was reported to consist of 15 genes (*MSMEG_5999—MSMEG_6004, MSMEG_6008, MSMEG_6009, MSMEG_6011—MSMEG_6017*) including *kstR2* itself (*MSMEG_6009*) [36]. For many genes of the KstR2 regulon, the participation in HIP catabolism has been already proven [37–39], while the functions of some genes of this regulon have not been cleared yet.

Whole transcriptome analysis is being applied increasingly more often in sterol biodegradation research in order to reveal the complete set of the genes whose expression is induced in the presence of sterols, such as cholesterol or phytosterols [2, 9, 40, 41]. This approach allows the identification of those genes whose products are directly or indirectly involved in the sterol degradation pathways including the genes that have not been known before as related to sterol catabolism. Moreover, this method makes it possible to determine how the specific genes are grouped within the genome, as well as to estimate the features of their regulation.

One of the first investigations which exploited whole genome approach showed that 572 genes in *R. jostii* RHA1 increased the level of their expression more than 2-fold in the presence of cholesterol. These genes were grouped into 6 clusters, but only two of them included the known genes related to sterol catabolism [2]. In *M. smegmatis* mc^2 155, the expression of 89 genes increased more than 3-fold in response to cholesterol [9]. Most of these genes were grouped into 3 clusters including the genes involved in sterol catabolism. Study of the global transcriptome of *M. smegmatis* mc^2 155 unraveled 454 genes increased their expression in the presence of cholesterol as compared to the control. Eleven and sixteen gene clusters were induced by cholesterol when compared with glycerol, or androstenedione, respectively [40]. However, in spite of the growing body of research on bacterial steroid degradation, many aspects of the catabolic pathways regulation remain unknown, especially those related to phytosterol oxidation by saprotrophic fast-growing mycobacteria which were recently suggested to re-classify as *Mycolicibacterium* in accordance with modern taxonomy [42].

Unlike other cholesterol-degrading or phytosterol-transforming mycobacteria such as *M. tuberculosis, M. smegmatis* mc^2 155, *M. neoaurum* VKM Ac-1815D, VKM Ac-1816D, NRRL 3805B or NRRL 3683B, the saprotrophic fast-growing wild-type strain of *Mycobacterium* sp. VKM Ac-1817D effectively produces 9-OH-AD from phytosterol [5, 23]. The identification of side-chain degradation intermediates showed the presence of 9α-hydroxyl function in their structures, thus suggesting the action of Ksh at the early stages of phytosterol degradation [23]. However, effective 9-OH-AD accumulation by this strain was shown to be accompanied with slow steroid degradation due to residual 3-ketosteroid-Δ^1-dehydrogenase activity [43].

Earlier, we had carried out a comparative study of the sequences of steroid catabolism genes in the whole genome scales in *Mycobacterium* sp. VKM Ac-1817D (*Myc* 1817) and two *M. neoaurum* strains, namely, VKM Ac-1815D and 1816D converting phytosterols to AD and ADD, respectively [5]. The complete genome of *Myc* 1817 was later assembled [44]. In comparison with *M. neoaurum*, *Myc* 1817 possesses a larger genome (6.35 Mbp) containing more homologues of the genes with putative role in sterol catabolism, including several copies of *kshA*, *kshB* and *kstD*. Based on 16S rRNA phylogeny, *Myc* 1817 is in the close relationship with *M. gilvum* and *M. smegmatis* [5]. These data, in combination with specific catalytic features of *Myc* 1817D strain, and mainly the ability for 9α-hydroxylation of all the intermediates of sterol side chain degradation with accumulation of 9-OH-AD as a major product, allowed us to predict the differences in the induction pattern.

In this study, we estimated genome-wide response on phytosterol in *Mycobacterium* sp. VKM Ac-1817D based on RNA-sequencing and whole-transcriptome analysis in order to understand peculiar properties of this organism.

Results

Growth and phytosterol bioconversion

Phytosterol as a sole carbon and energy source provided very slow initial growth and bioconversion activity of *Mycobacterium* sp. VKM Ac-1817D (data not shown). To facilitate growth, both the control and induction media were supplemented with glycerol as a co-substrate. The growth curves are given in Fig. 2. The glycerol-grown cells were harvested in the late active growth stage after 18 h when reaching 1.76 ± 0.05 g/l (d.c.w.) In the induction medium, the culture converted phytosterol mainly to 9-OH-AD as a major product (87% relative abundance (r.a.) after 48 h of incubation, mol/mol) and 9-hydroxylated 3-keto-4-ene steroids with partly, or completely degraded side chain: 9-hydroxy-22-carboxy-23,24-bisnorchol-4-ene-3-one (9-HCBC, r.a 6.5%), 9, 22-dihydroxy-23,24-bisnorchol-4-en-3-one (9,22-DHBC, r.a. 4.4%), 9,24-dihydroxychol-4-en-3-one (9,24-DHC, r.a. 2.2%) and 9-hydroxy-testosterone (9-HT, r.a. 1.9%). AD was observed at a noticeably smaller level – 0.12% r.a. (Fig. 3).

The harvesting point for the phytosterol-induced cells was decided on the basis of high specific activity of phytosterol oxidation suggesting maximum level of expression of the specific transcripts. During initial 9 h the conversion rate considerably increased and stabilised at the level of 90–100 μmol/l h g (d.c.w.) until depletion of phytosterol after approx. 30 h. The induced culture was harvested for RNA isolation at the age of 18–21 h. The

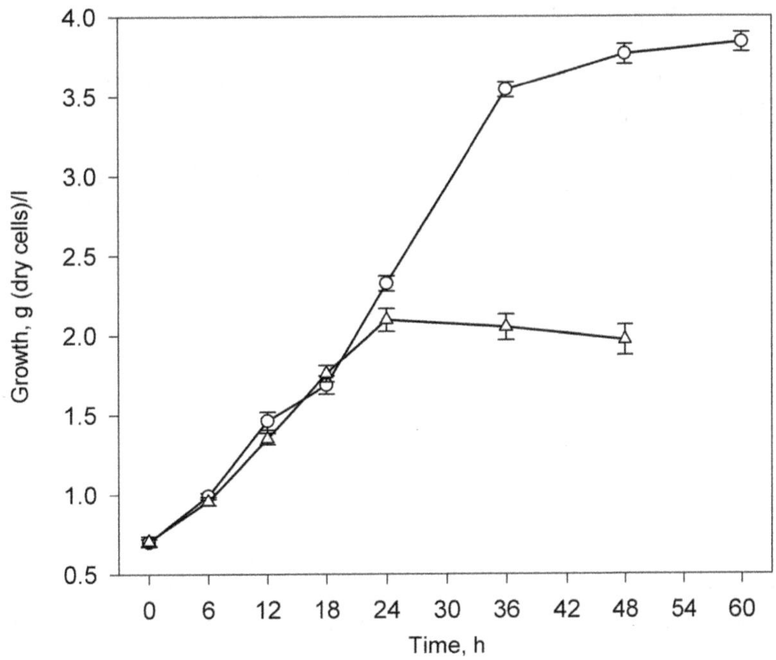

Fig. 2 Growth of *Mycobacterium* sp. VKM Ac-1817D on the control medium with 5 g/l glycerol as the carbon source (triangles) and induction medium supplemented with 12 mmol/l phytosterol (circles). The values represent the average of at least three independent experiments; error bars represent the standard deviations

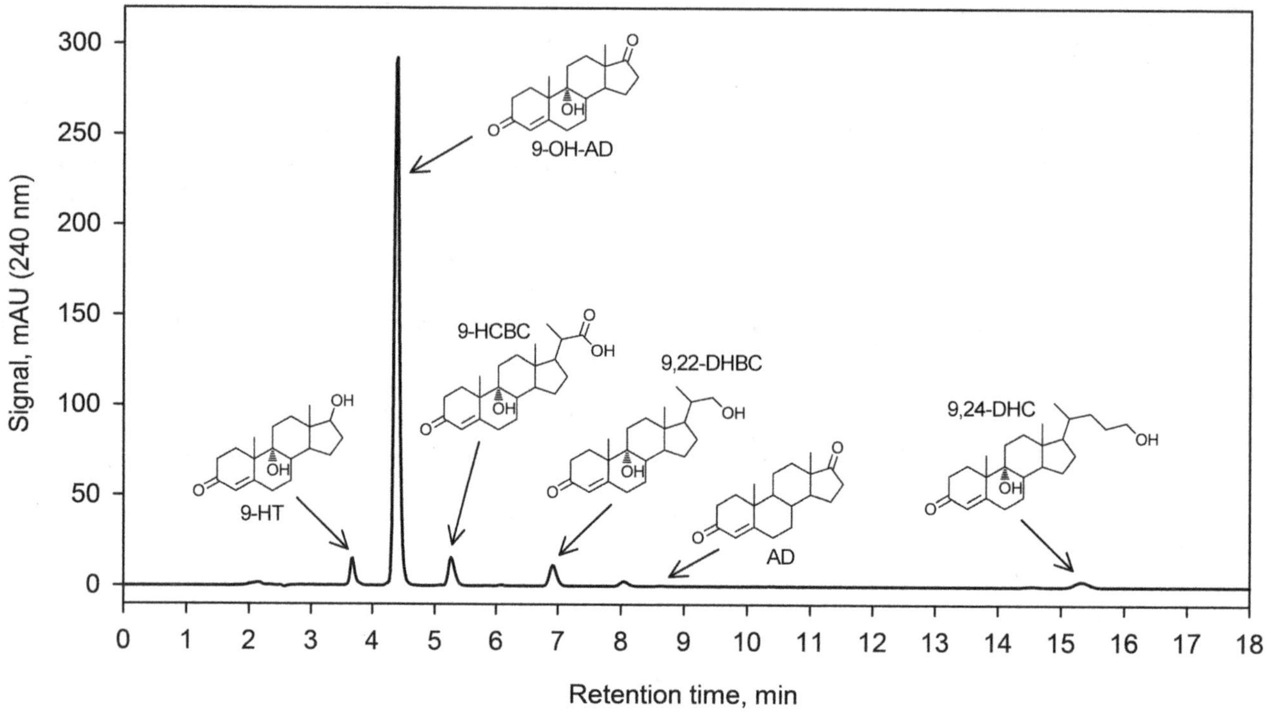

Fig. 3 HPLC profile and structures of major steroid products accumulated by *Mycobacterium* sp. VKM Ac-1817D in the induction medium as a result of phytosterol bioconversion

growth at harvesting point corresponded to 1.7–2.3 g/l (d.c.w.), or approx. 2.5×10^8 CFU/ml.

Transcriptome sequencing

We studied the differential expression of *Myc* 1817 genes during its growth in medium with phytosterol in comparison with the control medium without any steroids. Statistical parameters of the sequencing results are given in Additional file 1.

For the analysis, we selected the genes whose expression increased by no less than 3-fold at a q-value (false discovery rate) ≤ 0.01. In summary, 260 genes were revealed. The full list of these genes is presented in Additional file 2.

Sterol-induced genes were distributed irregularly within the *Myc* 1817 genome: they formed some large clusters, as well as a number of relatively small groups. The distribution of the genes which increased their expression in the presence of phytosterol is shown in Fig. 4.

Real-time PCR

For the expression analysis with real-time PCR, the following single-copy genes with known functions were chosen: (i) *ltp3* and *fadE29/chsE2* related to cholesterol side-chain cleavage [17, 20]; (ii) *kstD* and *kshB* encoding the enzymes accounted for the key steps of steroid core degradation [22, 32]; (iii) *fadD3* and *kstR2* that belongs to KstR2-regulon and related to the C/D ring degradation pathway [35]. The three housekeeping genes *rpoB*, *rpoD* and *ftsQ* known for insignificant expression deviations at the transcriptomic studies [45] were used as the reference genes. The results of qRT-PCR qualitatively coincided with the transcriptomic data. The expression of all six genes significantly increased as outlined below from both the transcriptomic and qRT-PCR results (Table 1). Both methods showed similar quantitative results for 4 genes, while the difference was 2-fold for *fadE29/chsE2* and the qRT-PCR indicated 5.9-fold higher up-regulation for *kstR2*. The latter may be attributed to the rather unstable expression of the KstR2-regulon genes under the induction conditions, since its putative effector is HIP, but not the early sterol intermediates.

Identification of transcriptional factors binding sites

Putative KstR binding sites were detected for 57 operons including 43 operons with the genes whose expression increased in the presence of phytosterol. Most of the genes related to steroid catabolism were the parts of these operons.

Putative KstR2 binding sites were detected for 8 operons. The genes in all of them were up-regulated on phytosterol. Along with 15 orthologues of the genes putatively involved in the rings C/D degradation [36], we found also additional inducible operon which included the gene of amino hydrolase family *G155_10290* with a

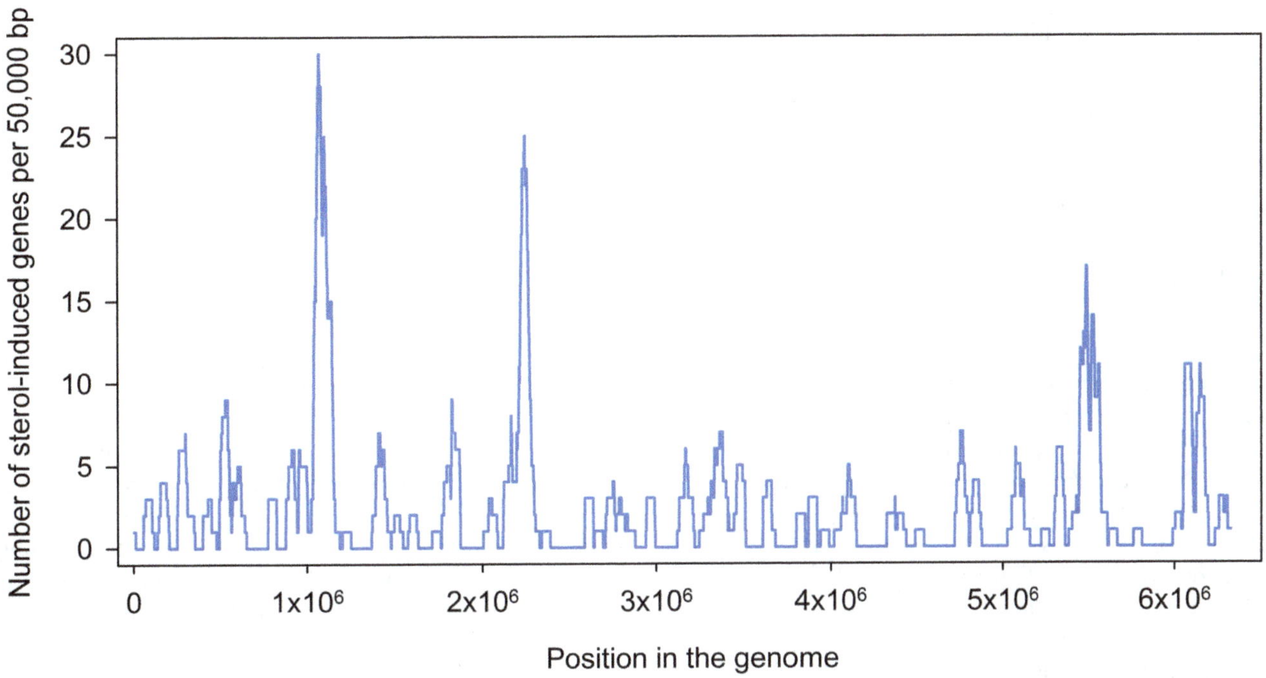

Fig. 4 Distribution of phytosterol-induced genes over genome of *Mycobacterium* sp. VKM Ac-1817D. The numbers of sterol-induced genes on a 50.000 bp interval are shown

putative KstR2-binding site in the promoter (Additional file 3). However, this site was found in only one of the two used programs (UGENE), while the score of this site only slightly exceeds the threshold value.

Among the genes over-expressed in the presence of phytosterol, we found several dozen genes without binding sites of KstR and KstR2. Among them, some orthologues of the known steroid catabolism genes were identified. In the promoters of these genes, a regularly occurring palindromic motif (motif X) was observed (Fig. 5). This motif was very similar to the binding site of *M. tuberculosis* transcription factor Rv0767c [46] with a Q-value of a similarity of 2×10^{-10} (Fig. 6). This means that the similarity was most likely not accidental. The orthologue of *Rv0767c* in *Myc* 1817 is the TetR-family transcription regulator gene *G155_05115*. There was also a sequence that corresponded to the motif X near the start codon of *G155_05115*. This may be a sign of autoregulation, further suggesting that the motif X indeed belongs to this transcription factor.

We identified 14 operons which adjacent with the motif X and presumably regulated with *G155_05115* (Additional file 3). The expression of the genes in 12 of these operons in response to phytosterol increased more than 3-fold, and one operon (*G155_19665—G155_19675*) raised expression 2.7-fold, on average. The operons with an adjacent site that is corresponding to motif X included the genes of putative steroid Δ-isomerase (*G155_05080*), some cytochromes, alcohol and aldehyde dehydrogenases, short chain dehydrogenases and TetR-family transcriptional regulators: *G155_05115* itself, *G155_10530*, and *G155_29025*. Among the genes putatively encoded cytochromes, four different genes (namely, *G155_05095*, *G155_05100*, *G155_05110* and *G155_26175*) were presumably coded for cytochrome P450-monooxygenases. Interestingly, that in two cases the sites that corresponded to the motif X were found together with KstR binding sites. For example, the binding sites for both KstR and G155_05115 were found near the promoter of the gene *G155_26175*. Probably, the enzymes which are putatively under regulation of the protein encoded by *G155_05115* with motif X may play role in steroid catabolism.

Table 1 Phytosterol inducible expression changes of 6 sterol catabolism genes with qRT-PCR data

Gene name	Expression increasing, transcriptomic data	Expression increasing, qRT-PCR data
fadE29/chsE2 (G155_26165)	76	152
ltp3 (G155_0479)	97	82
kshB (G155_26490)	40	40
kstD (G155_04625)	61	65
fadD3 (G155_26340)	10	13
kstR2 (G155_26275)	16	95

Steroid catabolism genes up-regulated in the presence of phytosterol

Among the phytosterol-inducible genes of *Myc* 1817, genes were found whose orthologues were proved, or

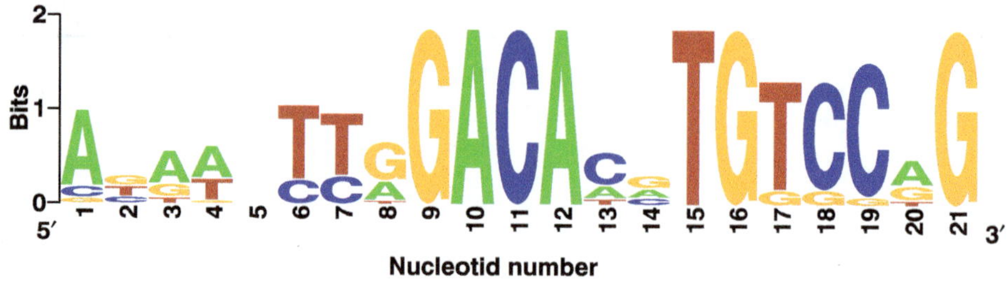

Fig. 5 Sequence of motif X, putative binding site of candidate transcription factor *G155_05115*

assumed, to have a function in sterol catabolism (Table 2). These are the genes putatively encoding the side chain oxidation enzymes including starting oxygenases, as well as the genes involved in β-oxidation (encoded acyl-CoA-synthases, enoyl-CoA-hydratases, acyl-CoA-dehydrogenases, aldol lyases, acetyl-CoA-acetyltransferases and homologs of *igr*-operon genes), the genes related to steroid core destruction, uptake of steroids (*mce4*-operon) and the known regulators of steroid catabolism (Table 2) [2, 9, 40, 41]. The genes related to the side chain oxidation significantly (50–136-fold) increased their expression (Table 2).

As described above, Ksh and KstD are the key enzymes of the rings A/B degradation. In our previous study, five genes homologous to oxygenase subunit of 3-ketosteroid-9α-hydroxylase (*kshA*) were revealed in *Myc* 1817 [5]. In this work, we found that two of them are phytosterol-induced: the expression of *kshA_1* (*G155_04755*) and *kshA_2* (*G155_24375*) was increased by 55-fold and 25-fold, respectively (Table 2).

KshA_1 is similar to KshA of *M. smegmatis* (*MSMEG_5925*) by 84%, and to KshA1N of *M. neoaurum* ATCC 25795 — by 85%. KshA1N was shown to express activity towards different 3-oxosteroids including 23,24-bisnorchol-1,4-diene-22-oic acid (1,4-BNC), 4-cholestene-3-one (cholesterone), AD, hydrocortisone, 4-pregnene-3, 20-dione (progesterone) and others [30]. KshA_2 showed similarity to MSMEG_5925 by 67%, and KshA1N from *M. neoaurum* ATCC 25795 — by 65%. At the same time, only one of two homologous *kshB* genes of Ksh reductase subunit, namely *kshB_1* (*G155_26490*) was found to be phytosterol-induced: its expression increased 40-fold in response to phytosterol (Table 2).

Among the five putative *kstD* genes encoding 3-ketosteroid Δ^1-dehydrogenases, only *kstD_1* (*G155_04625*) was over-expressed (in 61-fold) in the presence of phytosterol (Table 2). Expression of the gene *G155_08770* increased in 10 times in the presence of phytosterol. This gene is of 99% similarity to the putative 3-oxo-5α-steroid Δ^4-dehydrogenase (Δ4(5α)-KSTD) XA26_17580 annotated earlier in *Mycobacterium fortuitum* CT6 genome (CP011269.1) [57]. It is worth noting that *G155_08770*

```
G155_05115     1 MSSDAVSAVTEPA---SEPGKES-PRNRRQEETFRKVLTAGTEMLRESSY  46
                   ||||.:  ||.||   :||..|: .|||||:.||||||.|....|||.||
Rv0767c        1 MSSDVL--VTTPAQRQTEPHAEAVSRNRRQQATFRKVLAAAMATLREKSY  48

G155_05115    47 ADLTVRAVAARAGVAPATAYTYFSSKNHLIAEMYLDLVSRVPYFTDVNDT  96
                   ||||||.|||||.|||||||||||||||||||:|||||.:||..||||..
Rv0767c       49 ADLTVRLVAARAKVAPATAYTYFSSKNHLIAEVYLDLVRQVPCVTDVNVP  98

G155_05115    97 MTTRVDAALRALTLVVADEPEVAAACTTALLGGGSDPAVRKVRDRIGAEI 146
                   |..||.::||.|.||||||||:.||||.|||.||:|||||.|||||||||
Rv0767c       99 MPIRVTSSLRHLALVVADEPEIGAACTAALLDGGADPAVRAVRDRIGAEI 148

G155_05115   147 HKRIRSAVGPDADSRVVSALEMTFFGALVNAGSGAFTYHQIADRLTYVVG 196
                   |:||.||:||.||...|.||||.||||||.||||.||||:|||||.||||
Rv0767c      149 HRRITSAIGPGADPGTVFALEMAFFGALVQAGSGTFTYHEIADRLGYVVG 198

G155_05115   197 LVL-GEDK------- 203
                   |:| |.::
Rv0767c      199 LILAGANEPSTGGSE 213
```

Fig. 6 Sequence alignment of the transcription factor G155_05115 from *Mycobacterium* sp. VKM Ac-1817D and its orthologue Rv0767c from *Mycobacterium tuberculosis* H37Rv. Vertical lines represent matching amino acids (green), colons represent similar amino acids (yellow), and dots represent dissimilar (red); gaps/insertions are blue. Different amino acids are called similar if their score in the BLOSUM62 matrix is positive

Table 2 Phytosterol inducible expression changes of known sterol catabolism genes

Gene name	Locus	Expression increasing	Function of ortholog	References
ltp3	*G155_04795*	96.89	Aldol lyase	[17]
ltp4	*G155_04800*	96.38	Aldol lyase	[17]
echA19	*G155_04835*	68.59	Enoyl-CoA hydratase	[2]
fadD19	*G155_04840*	97.53	Long-chain fatty-acid-CoA ligase	[47]
fadD17	*G155_04880*	81.64	Long-chain fatty-acid-CoA ligase	[18]
chsE5(fadE27)	*G155_04885*	74.14	Acyl-CoA dehydrogenase	[48]
chsE4(fadE26)	*G155_04890*	89.21	Acyl-CoA dehydrogenase	[48]
ltp2	*G155_26150*	69.08	Putative thiolase	[49]
chsH1	*G155_26155*	66.32	Steroid enoyl-CoA hydratase	[50]
chsH2	*G155_26160*	62.40	Steroid enoyl-CoA hydratase	[50]
chsE2(fadE29)	*G155_26165*	76.45	Acyl-CoA dehydrogenase	[20]
chsE1(fadE28)	*G155_26170*	58.19	Acyl-CoA dehydrogenase	[20]
cyp125	*G155_26175*	107.58	C26-steroid monooxygenase	[14]
fadA5	*G155_26180*	136.07	Acetyl-CoA acetyltransferase	[19]
chsE3(fadE34)	*G155_26510*	50.22	Acyl-CoA dehydrogenase	[51]
kstD	*G155_04625*	61.43	3-oxosteroid 1-dehydrogenase	[52]
4kstD	G155_08770	10.09	3-oxosteroid 1-dehydrogenase	[53]
hsaE	*G155_04630*	11.57	2-keto-4-pentenoate hydratase	[25]
hsaG	*G155_04635*	9.95	Acetaldehyde dehydrogenase	[25]
hsaF	*G155_04640*	7.13	4-hydroxy-2-oxovalerate aldolase	[25]
kshA_1	*G155_04755*	55.51	3-ketosteroid-9-alpha-hydroxylase oxygenase subunit	[21]
kshA_2	*G155_24375*	24.58	3-ketosteroid-9-alpha-hydroxylase oxygenase subunit	[21]
hsaB	*G155_26470*	26.88	Monooxygenase component B	[26]
hsaC	*G155_26475*	26.18	2,3-dihydroxybiphenyl 1,2-dioxygenase	
hsaD	*G155_26480*	32.34	2-hydroxy-6-oxo-6-phenylhexa-2,4-dienoate hydrolase	[27]
hsaA	*G155_26485*	34.01	Monooxygenase component A	[26]
kshB	*G155_26490*	39.90	3-ketosteroid-9-alpha-hydroxylase reductase subunit	[21]
hsd	G155_22700	83.36	3-hydroxysteroid dehydrogenase	[54]
Hsd4A	G155_04900	11.12	17β- hydroxysteroid dehydrogenase	[55]
echA20	*G155_26220*	9.82	Putative enoyl-CoA hydratase	[38]
ipdA	*G155_26225*	8.61	Putative CoA-transferase subunit alpha	[38]
ipdB	*G155_26230*	8.35	Putative CoA-transferase subunit beta	[38]
fadA6	*G155_26270*	16.55	3-ketoacyl-CoA thiolase, acetyl-CoA acetyltransferase	[38]
fadE30	*G155_26335*	10.49	Acyl-CoA dehydrogenase	[38]
fadD3	*G155_26340*	9.84	Fatty-acid-CoA ligase	[38]
fadE31	*G155_26345*	8.60	Acyl-CoA dehydrogenase	[38]
fadE32	*G155_26350*	7.51	Acyl-CoA dehydrogenase	[38]
fadE33	*G155_26355*	8.45	Acyl-CoA dehydrogenase	[38]
kstR	*G155_26515*	13.17	Transcriptional regulator, TetR family	[28]
kstR2	*G155_26275*	15.81	Transcriptional regulator, TetR family	[36]
yrbEA	*G155_04905*	3.58	Conserved hypothetical integral membrane protein	[56]
yrbEB	*G155_04910*	3.71	Conserved hypothetical integral membrane protein	[56]
mceA	*G155_04915*	3.21	MCE-family protein	[56]
mceB	*G155_04920*	3.63	MCE-family protein	[56]

Table 2 Phytosterol inducible expression changes of known sterol catabolism genes *(Continued)*

Gene name	Locus	Expression increasing	Function of ortholog	References
mceC	*G155_04925*	3.67	MCE-family protein	[56]
mceD	*G155_04930*	3.40	MCE-family protein	[56]
mceE	*G155_04935*	3.87	MCE-family protein	[56]
mceF	*G155_04940*	3.17	MCE-family protein	[56]
mas4B	*G155_04950*	3.07	MCE-associated protein	[56]

does not show any similarity with the known genes of Δ4(5α)-KSTDs whose protein function has been experimentally confirmed, such as *ro05698* of *Rhodococcus jostii* RHA1 and *Rv1817* of *M. tuberculosis* [53].

The expression of orthologues of the genes whose products are involved in further A/B ring degradation, i.e. *hsaABCD, hsaEFG* and others, increased in 7–34 times (Table 2). Cholesterol oxidases have not been identified among the phytosterol-induced genes, but the *hsd* gene (*G155_22700*) encoding 3β-hydroxysteroid dehydrogenase was over-expressed in the presence of phytosterol by 83-times. The genes of the only one of at least nine *mce*-operons were slightly (in 3–4 times) up-regulated on phytosterol. This operon is similar to the *mce4* operon of *M. smegmatis* mc^{2}155 that takes part in sterol transport [56]. Interestingly, the expression of genes of the ATPases associated with the *mce*-operons (G155_06965 and G155_02045) was not increased in response to phytosterol in *Myc* 1817.

Among phytosterol-inducible genes with putative KstR binding sites in *Myc* 1817, the genes of short chain dehydrogenases, 3α(20β)-hydroxysteroid dehydrogenase (*G155_16070)*, aldehyde- and alcohol dehydrogenases, 2 hydride-transferase fragments and many orthologous genes related to fatty acid β-oxidation were revealed (Additional file 3). In *M. smegmatis*, the expression of many orthologues was shown to be increased, e.g., alcohol dehydrogenase *G155_00805*, acyltransferase *G155_01310*, aldehyde dehydrogenase *G155_01350*, 2-nitropropane dioxygenase *G155_16090*, acyl-CoA ligase *fadD8 G155_25385*, and thiolase *G155_17455* [9]. Their products possibly take part in side chain cleavage, or another parts of the phytosterol degradation pathway.

Genes related to the C/D ring degradation pathway that controlled by KstR2 were also up-regulated in phytosterol. In particular, the genes involved in the C/D ring degradation (*echA20, ipdA, ipdB, fadA6, fadD3, fadE30, fadE31, fadE32, fadE33)* showed increased expression by 7–16 times, with an average of 10 times (Table 2).

Mutations in KshAs, KshB and KstD_1

In order to evaluate whether the capability of *Myc* 1817 to accumulate 9-OH-AD as a major product from phytosterol is associated with malfunctioning of any proteins, we estimated degrees of conservation of the key enzymes of the A/B ring degradation. Amino acid substitutions in evolutionarily conserved positions of KstD_1, KshA, and KshB were evaluated. The known phytosterol degrading strain *M. smegmatis* mc^2 155 was used for comparison (NCBI Reference Sequence: NC_008596.1). The results are presented in Table 3.

The amount of substitutions in the most conserved positions (rank 1) in KshA_1 and KshB in *Myc* 1817 was *on a par* similar with that in *M. smegmatis* mc^2 155. Of particular note, KshA_2 in *M. smegmatis* mc^2 155 has fewer changes in the conservative positions than its orthologue in *Myc* 1817. The greater number of substitutions in the most conserved positions was observed in KstD_1 of *Myc* 1817 as compared with the orthologous KstD in *M. smegmatis* mc^2 155 (5 and 2, respectively) (Table 3).

Discussion

The saprotrophic, fast growing native strain *Myc* 1817 is of industrial interest due to its ability to fully transform phytosterol at the high loads with forming 9-OH-AD as a major product [58]. Steroid metabolite profiling of *Myc* 1817D grown in the presence of phytosterol showed the accumulation of 9α-hydroxylated intermediates with a partially oxidised side chain, thus confirming that 9α-hydroxylation precedes full side chain degradation by the strain (Fig. 7). The simultaneous presence of 9α-hydroxylated products with C_3 and C_5 side chain in the form of carboxylic acid (9-HCBC) and alcohols (9,22-DHBC, 9,24-DHC) may indicate that its metabolic pathway differs from those known for *M. smegmatis* mc^2 155 [9, 59] and *M. neoaurum* ATCC25795 [55], but seems to

Table 3 Number of amino acid substitution in conservative and non-conservative positions of KshA, KshB and KstD proteins of strain *Myc* 1817 and *M. smegmatis* mc^2 155

Protein	Degree of conservation	1	2	3	4	5	6	7	8	9
KshA	KshA_1 (G155_04755)	2	6	9	7	9	14	6	7	48
	KshA_2 (G155_24375)	1	8	5	11	11	11	10	7	65
	KshA_1 (MSMEG_5925)	2	5	3	7	14	7	11	8	44
	KshA_2 (MSMEG_2870)	0	2	3	3	11	3	8	11	40
KshB	KshB (G155_26490)	1	6	8	12	10	11	6	4	36
	(MSMEG_6039)	1	6	11	9	11	8	10	3	32
KstD	KstD (G155_04625)	5	7	7	16	15	12	6	7	55
	KstD (MSMEG_5941)	2	2	15	11	12	10	13	6	52

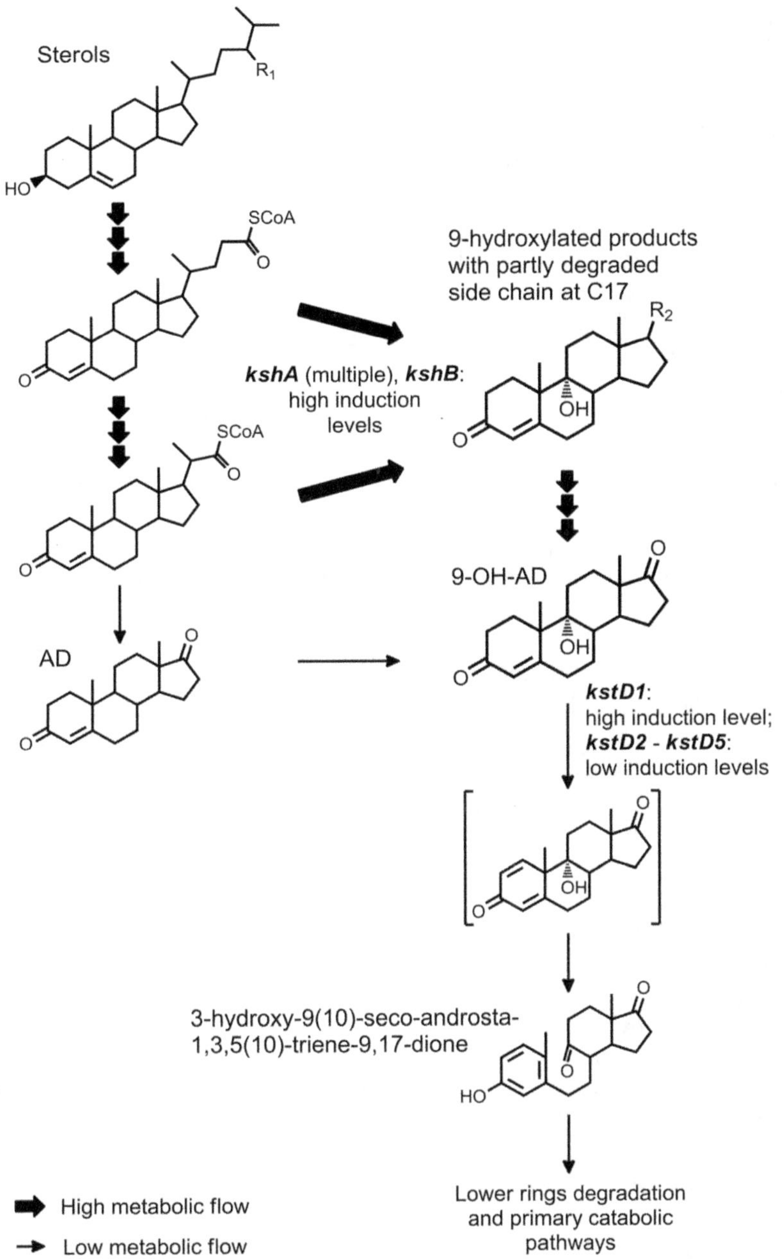

Fig. 7 Sterol catabolism and gene expression peculiarities in *Mycobacterium* sp. VKM Ac-1817D. R_1 – H, CH_3 or C_2H_5 (various sterols); R_2 – partly degraded side chains. The thickness of arrows reflects suggested relative level of the catabolic activity

generally correspond to that suggested for *M. tuberculosis* [21], and some other *Mycobacterium* strains [23, 24, 33].

It had been postulated that accumulation of C_{22}– steroids, such as 4-BNC, or 1,4-BNC with some mycobacteria occurs due to the absence, or inefficient 9α-hydroxylation [59]. Accumulation of the corresponding 9α-hydroxyderivatives at the phytosterol conversion with *Myc* 1817D allows to suggest that 9α-hydroxylation is rather not so critical for the elimination of the last isopropyl group of sterol side chain.

Differential transcriptome analysis showed 260 genes that were up-regulated in response to phytosterol in *Myc* 1817, including numerous genes of steroid catabolism. A suite of phytosterol-induced genes related to side-chain, A/B ring and C/D ring degradation was generally similar to that in other phytosterol/cholesterol digesting mycobacteria [2, 9, 40]. Interestingly, the genes which putatively encoded ChOs did not show increased expression in response to phytosterol, while *hsd* (*G155_22700*) was significantly (83 times) up-regulated. This corresponds to the literature data - a similar pattern is observed in *M. smegmatis* [9]. Insignificant upregulation of *hsd4A* homolog whose product may play a role as 17β-hydroxysteroid dehydrogenase, or β-hydroxyacyl-CoA-dehydrogenase [55] is generally coincided with small amounts of 9-HCBC and 9,22-DHBC observed among the metabolites at phytosterol conversion with *Myc* 1817, and allows to propose that Hsd4A may be active towards 9α-hydroxylated steroids.

Along with the genes involved in the side chain degradation and rings A/B oxidation, the genes related to the lower degradation pathway (rings C/D degradation) were also up-regulated on phytosterol. It is clear that the strain has the ability to fully degrade the steroid core. However, relatively high increase in expression of the rings C/D degradation genes was a bit surprising. These genes belong to KstR2-regulon, which is induced with HIP-CoA, – the compound formed after degradation of the rings A and B [37]. Since *Myc* 1817 mainly accumulates 9-OH-AD, the formation of HIP-CoA was expected to occur much slower than in the sterol-degrading strains, thus predicting attenuation of the KstR2-regulon genes induction.

Effective 9-OH-AD accumulation is generally stipulated with high 3-ketosteroid-9α-hydroxylase (Ksh) and low (or fully blocked) 3-ketosteroid-Δ^1-dehydrogenase (KstD) activities. In *Myc* 1817, two genes of the oxygenase subunit (*kshA_1* and *kshA_2)* and one gene of the reductase subunit (*kshB_1*) of Ksh significantly (in dozens of times) increased their expression in the presence of phytosterol, while among five putative *kstDs,* an only *kstD_1* was phytosterol-inducible. The mutational analysis showed, however, that the product of this gene contained a significant number of nucleotide substitutions in the conserved positions, thus allowing us to propose either the absence, or low levels of activity. Four remained *kstDs* which did not up-regulated on phytosterol may, nevertheless, be expressed constitutively, thus contributing to the observed steroid core oxidation by the strain. Expression increasing of the KstR2-regulon genes probably also occurs through the activity of these 3-ketosteroid-Δ^1-dehydrogenases.

A new putative transcriptional regulator G155_05115 involved in the control of steroid catabolism was identified in *Myc* 1817, while most of the genes putatively controlled by this factor did not increased their expression in *M. smegmatis* mc^2 155 and *R. jostii* RHA1 [2, 9]. In *Myc* 1817, many genes presumably related to the oxidation of the side chain of sterols were among the putative regulon of this factor.

It is known that in addition to the genes involved in the catabolism of cholesterol and sitosterol, actinobacteria usually have clusters of the genes involved in the catabolism of other steroid compounds and, respectively, regulated by other transcription factors. In particular, there is a cluster of catabolism genes of the cholate [60], and the so-called C-19 cluster, whose genes are involved in the catabolism of yet unknown steroid compounds [61].

Because phytosterol contains β-sitosterol and other plant sterols which differ from cholesterol by branched side chains, it is reasonable to assume that the genes putatively regulated by G155_05115 may play a role in the oxidation of the side chains of the plant sterols of the phytosterol. However, there is a significant number of genes whose relationship with the side chain oxidation, or steroid catabolism remains uncertain in the putative regulon of G155_05115. This indicates that the function of G155_05115-regulon apparently is not limited to the oxidation of the sterols side chain.

Generation of the engineered mycobacterial strains producing valuable steroids is often complicated by homolog multiplicity of steroid catabolism genes [13, 47, 52, 62]. Moreover, various genes capable of performing the same function may have different roles in the catabolism of sterols, as demonstrated for *kstDs*, *hsd4A*, *hsd*, *choD* and others [13, 19, 34, 55, 63]. On the other hand, the known steroid core degradation genes present in the genomes of different mycobacterial strains including those having different metabolic blocks and accumulating specific intermediates of phytosterol degradation process. Knowledge of gene expression regulation in response to phytosterol allows a role for the specific homologous genes in sterol catabolism to be predicted.

Methods

Materials

Phytosterol was obtained from S&D Chemicals, UK. Androst-4-ene-3,17-dione (AD) was purchased from Steraloids, USA. Statistically methylated β-cyclodextrin (MCD) was purchased from Wacker Chemie, Germany. Androst-4-ene-3,17-dione (AD) was purchased from Steraloids, USA.9-OH-AD and other 9-hydroxylated steroids (9-HCBC, 9,22-DHBC, 9,24-DHC and 9-HT) with a purity of no less than 98% were obtained at the Institute of Biochemistry and Physiology of Microorganisms, Russian Academy of Sciences (IBPM RAS). The suppliers of other reagents are indicated below.

Microorganism and cultivation

The strain *Mycobacterium* sp. VKM Ac-1817D was obtained from All-Russian Collection of Microorganisms (VKM IBPM RAS) and pre-cultured as described earlier [64]. The strain was cultured in glycerol-mineral (control) medium [64] and the same medium supplemented with 12 mmol phytosterol (induction medium). Both media contained 24.1 mM MCD. Phytosterol powder and MCD were added to the medium before autoclaving; the sterilized media were sonicated during 2 min on an ultrasonic bath (100 W, 35 kHz) and incubated on an orbital shaker (200 rpm, 30 °C) overnight before inoculation. Experiments were carried out in the two independent biological replicates in 750-ml shake flasks containing 100 ml media.

Growth estimations

The growth was followed gravimetrically [65] due to high cell-to-cell and cell-to-phytosterol aggregation. Briefly, the samples of the cultivation broth were sedimented by centrifugation at 6500×g for 15 min, then the cakes were washed twice with 40 ml of 10% (w/v) aqueous MCD for phytosterol removal and then twice with 40 ml of distilled water. The washed cells were dried at 70 °C. For viable cell counts, broth samples were serially diluted with 1 g/l aqueous Tween 80 under vigorous agitation and plated on the solid minimal medium. The growth experiments were carried out in three replicates.

Analytical methods

Phytosterol and steroid metabolites were analysed as described earlier [65] by isocratic reversed-phase HPLC using Waters Symmetry (USA) 250 mm × 4.6 mm (5 μm) column at 50 °C and flow rate 1 ml/min. For phytosterol analysis, the samples were diluted with a mixture of 2-propanol and acetonitrile (45:50, v/v); the analysis was performed using 2-propanol:acetonitrile:deionised water (45:50:5, v/v) as a mobile phase with detection at 200 nm. Total areas of peaks relevant to plant sterols (6–12 min) were used for phytosterol quantification.

For steroid metabolites analysis, the samples were diluted 1:50 with 50% (v/v) aqueous acetonitrile; the analysis was performed using acetonitrile:water:acetic acid (52:48: 0.01, v/v) as a mobile phase with detection at 240 nm.

Isolation of mRNA

The cells were harvested by centrifuge at 8000×g for 10 min and immediately ground in a porcelain mortar under liquid nitrogen. Total RNA was isolated with Qiagen RNeasy mini kit (Qiagen, Netherlands), DNAse I and Ribo-Zero rRNA Removal Kit (Epicentre, USA) according to the protocols of the suppliers.

High-throughput sequencing

We used TruSeq RNA Sample prep kit v.2 (Illumina, USA) for sample preparation of mRNA for high-throughput sequencing. Sequencing was performed with HiSeq 2000 (50-nucleotide single-read run) according to the protocols of the manufacturer (Illumina, USA).

Accession numbers

The reads have been deposited in NCBI Sequence Read Archive (SRA) under the accessions numbers: SAMN05941438, and SAMN05941440.

Transcriptome analysis

For reads mapping and analysis of differential expression the Rockhopper 2.03 have been used [66]. A gene was considered differentially expressed between the conditions with and without phytosterol, if its expression increased, or decreased more than threefold with q-value less, or equal to 0.01.

Genome annotation

The genome was annotated with RAST (http://rast.theseed.org/FIG/rast.cgi) and PGAP [67]. To find putative genes encoding the enzymes of steroid catabolism and analyse orthologous relations between the genes of *Myc* 1817 and other actinobacteria, we used orthogroups between the genes of *Myc* 1817 and other species of actinobacteria including their plasmids that were constructed by OrthoMCL 2.09 with the OrthoMCL inflation parameter of 1.5, all other parameters were set to the default values.

The operons with the found genes were identified with the internet-service FgenesB (http://linux1.softberry.com/berry.phtml?topic=fgenesb&group=programs&-subgroup=gfindb), and with Rockhopper 2.03 on the base of transcriptomic data [52].

Analysis of transcription factors and binding sites

In order to find binding sites (BSs) of transcription factors which may regulate steroid metabolism, we analysed regions of 500 bp upstream plus 50 bp downstream with respect to start codons of the genes that changed their expression more than threefold with q-value less, or equal to 0.01. The analysis was performed for the up-regulated and down-regulated genes separately. We searched for the over-represented motifs in these regions using MEME 4.10 [68]. Motifs were allowed to be from 8 to 50 bp long. The 20 top-scoring motifs in a form of position-weight matrices (PWMs) were compared with the known motifs of mycobacterial steroid metabolism regulators KstR [28] and KstR2 [36]. The motifs which corresponded to KstR and KstR2 were determined in the lists obtained by MEME. Then, possible motifs of other putative transcription factors were estimated among the remaining motifs in the lists.

Criteria for a motif to be considered as a putative motif of a transcription factor were as follows: 1) the motif must be statistically significant with E-value $\leq 10^{-5}$; 2) the motif must contain a palindrome (palindromic motifs are typical for bacterial transcription factors; and 3) the motif should not be a simple repeat, such as CGCGCGCG..., which is unusual for bacterial transcription factor binding motifs [69]. A single motif that conformed to these criteria was found in a set of the genes that increased their expression and further indicated as "motif X".

To compare this motif with the known motifs of the transcription factors from TetR family, we replicated a methodology used in [46] to determine motifs of various TetR-family proteins in mycobacteria. The only change in the methodology was that besides 10 species used in that work, we added *Myc* 1817 and other species of our interest, *M. neoaurum* Ac-1815D, to the analysis. After determining the putative motifs of the transcription factors, motif X was compared with them by TOMTOM [70], which is a special tool for motif comparisons from the MEME suite. Then, to seek for all sites of KstR, KstR2 and the transcription factor of motif X in *Myc* 1817 genome we scanned the regions 500 bp upstream plus 50 bp downstream with respect to the start codons of all genes using KstR, KstR2 motifs and motif X found in the previous step. The scan was performed by a tool FIMO from MEME suite [71]. The sites determined with false discovery rate less, or equal to 0.01 (q-value, estimated by FIMO using Benjamini-Hochberg technique) were considered as putative binding sites. For comparison, we also scanned the same regions with the same position weight matrices by UGENE 1.13.1 [72], considering a site as a putative binding site of a transcription factor if its score is no less than 85% of a maximum possible score for the motif of that factor (85% is the default value in UGENE). The binding sites that were presented in the results obtained with both tools (or in the results obtained with one tool, but located before the up-regulated operons) were used in further analyses. The employment of two tools allowed the number of false-positive predictions to be reduced. Detection of a binding site before a first gene of a putative operon predicted by Rockhopper was considered in the analyses as a sign that this binding site regulates all genes in the operon.

The amino acid sequences of the transcription factor G155_05115 from *Myc* 1817 and its orthologue Rv0767c from *M. tuberculosis* H37Rv (GenBank accession number CCP43514) were aligned by EMBOSS Needle 6.6.0 with the default parameters [73].

Real time PCR

cDNA synthesis was performed using the MMLV reverse transcription kit (Evrogen, Russian Federation) with 0.5 μg of total RNA in accordance with the manufacturer's instructions. A real-time PCR was carried out using AriaMx Real-time PCR system (Agilent, USA) with Eva Green I M-439 kit (Syntol, Russian Federation). The nucleotide sequences of the primers used in this study for the target and reference genes are listed in Additional file 4. Each sample was run twice and the experiment was performed in duplicate. The amplification was performed as follows: 95 °C for 5 min (1 cycle), 95 °C for 10 s, and 60 °C for 30 s (40 cycles). Gene expression levels were calculated using the ddCq method [74].

Search of mutations in *kshA*, *kshB* and *kstD* genes

To evaluate amino acid substitutions in proteins KshA_1, KshA_2, KshB and KstD of *Myc* 1817 and their orthologous proteins in *M. smegmatis* mc^2 155, we used ConSurf server [75]. For each protein under analysis, it looks for its homologs in 150 close species, performs multiple alignments of their amino acid sequences, and then estimates amino acid conservation for each position. The positions are ranked by ConSurf from the most conserved (rank 1) to the least conserved (rank 9). The most phylogenetically conserved positions in a protein are likely to be crucial for its functioning.

Conclusions

High expression levels of the genes related to the sterol side chain degradation and steroid 9α-hydroxylation in combination with possible defects in KstD_1 may contribute to effective 9α-hydroxyandrost-4-ene-3,17-dione accumulation from phytosterol provided by this biotechnologically relevant strain. The TetR-family transcription regulator gene *G155_05115* was presumably associated with the regulation of steroid catabolism.

Additional files

Additional file 1: Table S1. Primary results of high-throughput sequencing: number of reads and percentage of rRNA-reads. (XLSX 7 kb)

Additional file 2: Table S2. Phytosterol-induced and phytosterol-repressed genes of *Mycobacterium* sp. VKM Ac-1817D. (XLSX 32 kb)

Additional file 3: Table S3. Operons of *Mycobacterium* sp. VKM Ac-1817D with putative binding sites of sterol catabolism transcription regulators. (XLSX 23 kb)

Additional file 4: Table S4. Primers for qRT-PCR. (XLSX 7 kb)

Abbreviations

1,4-BNC: 23,24-bisnorchol-1,4-diene-22-oic acid; 3-HSA: 3-hydroxy-9,10-seconandrost-1,3,5(10)-triene-9,17-dione; 3-HsDs: 3β-hydroxysteroid dehydrogenases; 9,22-DHBC: 9,22-dihydroxy-23,24-bisnorchol-4-en-3-one; 9,24-DHC: 9,24-dihydroxychol-4-en-3-one; 9-HCBC: 9-hydroxy-22-carboxy-23,24-bisnorchol-4-ene-3-one; 9-HT: 9-hydroxy-testosterone; 9-OH-AD: 9α-hydroxyandrost-4-ene-3,17-dione; AD: Androst-4-ene-3,17-dione; ADD: Androsta-1,4-diene-3,17-dione; CFU: Colony forming unit; ChOs: Cholesterol oxidases; d.c.w: Dry cells weight; HIP: 3aα-*H*-4α(3-propanoate)-7aβ-methylhexahydro-1,5-indanedione; KstD: 3-ketosteroid-Δ^1-dehydrogenase; MCD: Statistically methylated β-cyclodextrin; r.a: Relative abundance; Δ4(5α)-KSTD: 3-oxo-5α-steroid Δ^4-dehydrogenase

Acknowledgements

We are grateful to Proof-Reading-Service.com for professional checking the English language of the article.

Authors' contributions

EB designed the study, performed the experiments and wrote the manuscript. VS participated in the high-throughput sequencing, PCR performance and data processing. MS participated in primary bioinformatic processing of sequencing results and analysis of transcription factor binding sites. DD participated in induction experiments, growth estimation, steroid analyses and writing of the final manuscript. MD participated in the design of research, writing and revision of the manuscript. All authors read and approved the final manuscript.

Funding
This work was supported by the Ministry of Science and Higher Education of the Russian Federation, project ID RFMEFI58818X0008. This funding organisation was not involved in the design of the study and collection, analysis, and interpretation of data, or in writing the manuscript.

Ethics approval and consent to participate
Not applicable.

Consent for publication
Not applicable.

Competing interests
The authors declare that they have no competing interests.

Author details
[1]Institute of Biochemistry and Physiology of Microorganisms, Federal Research Center "Pushchino Center for Biological Research of the Russian Academy of Sciences", Nauki, 5, Pushchino, Russian Federation142290. [2]Pharmins Ltd., Institutskaya, 4, Pushchino, Russian Federation142290. [3]Skolkovo Institute of Science and Technology, Nobelya, 3, Moscow, Russian Federation121205. [4]A.N. Belozersky Research Institute of Physico-Chemical Biology, M.V. Lomonosov Moscow State University, Leninskye gory, 1, building 40, Moscow, Russian Federation119992. [5]Institute for Information Transmission Problems, Russian Academy of Sciences, Bolshoy Karetny, 19, build. 1, Moscow, Russian Federation127051.

References

1. Brzostek A, Sliwiński T, Rumijowska-Galewicz A, Korycka-Machała M, Dziadek J. Identification and targeted disruption of the gene encoding the main 3-ketosteroid dehydrogenase in *Mycobacterium smegmatis*. Microbiol Read Engl. 2005;151:2393–402.
2. Van der Geize R, Yam K, Heuser T, Wilbrink MH, Hara H, Anderton MC, Sim E, Dijkhuizen L, Davies JE, Mohn WW, Eltis LD. A gene cluster encoding cholesterol catabolism in a soil actinomycete provides insight into *Mycobacterium tuberculosis* survival in macrophages. Proc Natl Acad Sci U S A. 2007;104(6):1947–52.
3. Donova MV, Egorova OV. Microbial steroid transformations: current state and prospects. Appl Microbiol Biotechnol. 2012;94:1423–47.
4. Holert J, Cardenas E, Bergstrand LH, Zaikova E, Hahn AS, Hallam SJ, Mohn WW. Metagenomes reveal global distribution of bacterial steroid catabolism in natural, engineered, and host environments. MBio. 2018. https://doi.org/10.1128/mBio.02345-17.
5. Bragin EY, Shtratnikova VY, Dovbnya DV, Schelkunov MI, Pekov YA, Malakho SG, Egorova OV, Ivashina TV, Sokolov SL, Ashapkin VV, Donova MV. Comparative analysis of genes encoding key steroid core oxidation enzymes in fast-growing *Mycobacterium* spp. strains. J Steroid Biochem Mol Biol. 2013;138:41–53.
6. Ge F, Li W, Chen G, Liu Y, Zhang G, Yong B, Wang Q, Wang N, Huang Z, Li W, Wang J, Wu C, Xie Q, Liu G. Draft genome sequence of *Gordonia neofelifaecis* NRRL B-59395, a cholesterol-degrading actinomycete. J Bacteriol. 2011;193:5045–6.
7. Rodríguez-García A, Fernández-Alegre E, Morales A, Sola-Landa A, Lorraine J, Macdonald S, Dovbnya D, Smith MCM, Donova M, Barreiro C. Complete genome sequence of "*Mycobacterium neoaurum*" NRRL B-3805, an androstenedione (AD) producer for industrial biotransformation of sterols. J Biotechnol. 2016;224:64–5.
8. Shtratnikova VY, Schelkunov MI, Fokina VV, Pekov YA, Ivashina T, Donova MV. Genome-wide bioinformatics analysis of steroid metabolism-associated genes in *Nocardioides simplex* VKM ac-2033D. Curr Genet. 2016;62(3):643–56.
9. Uhía I, Galán B, Kendall SL, Stoker NG, García JL. Cholesterol metabolism in *Mycobacterium smegmatis*: cholesterol pathway. Environ Microbiol Rep. 2012;4:168–82.
10. Bergstrand LH, Cardenas E, Holert J, Van Hamme JD, Mohn WW. Delineation of steroid-degrading microorganisms through comparative genomic analysis. mBio. 2016;7:e00166. https://doi.org/10.1128/mBio.00166-16.
11. Kreit J. Microbial catabolism of sterols: focus on the enzymes that transform the sterol 3β-hydroxy-5-en into 3-keto-4-en. FEMS Microbiol Lett. 2017; 364(3). https://doi.org/10.1093/femsle/fnx007.
12. Yao K, Wang FQ, Zhang HC, Wei DZ. Identification and engineering of cholesterol oxidases involved in the initial step of sterols catabolism in *Mycobacterium neoaurum*. Metab Eng. 2013;15:75–87.
13. Ivashina TV, Nikolayeva VM, Dovbnya DV, Donova MV. Cholesterol oxidase ChoD is not a critical enzyme accounting for oxidation of sterols to 3-keto-4-ene steroids in fast-growing *Mycobacterium* sp. VKM ac-1815D. J Steroid Biochem Mol Biol. 2012;129:47–53.
14. Capyk JK, Kalscheuer R, Stewart GR, Liu J, Kwon H, Zhao R, Okamoto S, Jacobs WR Jr, Eltis LD, Mohn WW. Mycobacterial cytochrome P450 125 (Cyp125) catalyzes the terminal hydroxylation of C27 steroids. J Biol Chem. 2009;284:35534–42.
15. Rosłoniec KZ, Wilbrink MH, Capyk JK, Mohn WW, Ostendorf M, van der Geize R, Dijkhuizen L, Eltis LD. Cytochrome P450 125 (CYP125) catalyses C26-hydroxylation to initiate sterol side-chain degradation in *Rhodococcus jostii* RHA1. Mol Microbiol. 2009;74:1031–43.
16. Johnston JB, Ouellet H, de Montellano PRO. Functional redundancy of steroid C26-monooxygenase activity in *Mycobacterium tuberculosis* revealed by biochemical and genetic analyses. J Biol Chem. 2010;285:36352–60.
17. Wilbrink MH, van der Geize R, Dijkhuizen L. Molecular characterization of *ltp3* and *ltp4*, essential for C24-branched chain sterol-side-chain degradation in *Rhodococcus rhodochrous* DSM 43269. Microbiology. 2012;158:3054–62.
18. Casabon I, Swain K, Crowe AM, Eltis LD, Mohn WW. Actinobacterial acyl coenzyme a synthetases involved in steroid side-chain catabolism. J Bacteriol. 2014;196:579–87.
19. Nesbitt NM, Yang X, Fontán P, Kolesnikova I, Smith I, Sampson NS, Dubnau E. A thiolase of *Mycobacterium tuberculosis* is required for virulence and production of androstenedione and androstadienedione from cholesterol. Infect Immun. 2010;78:275–82.
20. Thomas ST, Sampson NS. *Mycobacterium tuberculosis* utilizes a unique heterotetrameric structure for dehydrogenation of the cholesterol side chain. Biochem Mosc. 2013;52:2895–904.
21. Capyk JK, Casabon I, Gruninger R, Strynadka NC, Eltis LD. Activity of 3-ketosteroid 9α-hydroxylase (KshAB) indicates cholesterol side chain and ring degradation occur simultaneously in *Mycobacterium tuberculosis*. J Biol Chem. 2011;286:40717–24.
22. Petrusma M, Hessels G, Dijkhuizen L, van der Geize R. Multiplicity of 3-ketosteroid-9α-hydroxylase enzymes in *Rhodococcus rhodochrous* DSM43269 for specific degradation of different classes of steroids. J Bacteriol. 2011;193: 3931–40.
23. Donova MV, Gulevskaya SA, Dovbnya DV, Puntus IF. *Mycobacterium* sp. mutant strain producing 9alpha-hydroxyandrostenedione from sitosterol. Appl Microbiol Biotechnol. 2005;67:671–8.
24. Yao K, Xu LQ, Wang FQ, Wei DZ. Characterization and engineering of 3-ketosteroid-Δ^1-dehydrogenase and 3-ketosteroid-9α-hydroxylase in *Mycobacterium neoaurum* ATCC 25795 to produce 9α-hydroxy-4-androstene-3,17-dione through the catabolism of sterols. Metab Eng. 2014; 24:181–91.
25. Carere J, McKenna SE, Kimber MS, Seah SYK. Characterization of an aldolase-dehydrogenase complex from the cholesterol degradation pathway of *Mycobacterium tuberculosis*. Biochem Mosc. 2013;52:3502–11.
26. Dresen C, Lin LY-C, D'Angelo I, Tocheva EI, Strynadka N, Eltis LD. A Flavin-dependent monooxygenase from *Mycobacterium tuberculosis* involved in cholesterol catabolism. J Biol Chem. 2010;285:22264–75.
27. Lack NA, Yam KC, Lowe ED, Horsman GP, Owen RL, Sim E, Eltis LD. Characterization of a carbon-carbon hydrolase from *Mycobacterium tuberculosis* involved in cholesterol metabolism. J Biol Chem. 2010;285:434–43.
28. Kendall SL, Withers M, Soffair CN, Moreland NJ, Gurcha S, Sidders B, Frita R, Ten Bokum A, Besra GS, Lott JS, Stoker NG. A highly conserved transcriptional repressor controls a large regulon involved in lipid degradation in *Mycobacterium smegmatis* and *Mycobacterium tuberculosis*. Mol Microbiol. 2007;65:684–99.

29. Dovbnya DV, Egorova OV, Donova MV. Microbial side-chain degradation of ergosterol and its 3-substituted derivatives: a new route for obtaining of deltanoids. Steroids. 2010;75:653–8.
30. Liu H-H, Xu L-Q, Yao K, Xiong L-B, Tao X-Y, Liu M, Wang F-Q, Wei D-Z. Engineered 3-ketosteroid 9α-hydroxylases in *Mycobacterium neoaurum*: an efficient platform for production of steroid drugs. Appl Environ Microbiol. 2018;84:e02777–17. https://doi.org/10.1128/AEM.02777-17.
31. Szaleniec M, Wojtkiewicz AM, Bernhardt R, Borowski T, Donova M. Bacterial steroid hydroxylases: enzyme classes, their functions and comparison of their catalytic mechanisms. Appl Microbiol Biotechnol. 2018;102(19):8153–71.
32. Capyk JK, D'Angelo I, Strynadka NC, Eltis LD. Characterization of 3-ketosteroid 9α-hydroxylase, a Rieske oxygenase in the cholesterol degradation pathway of *Mycobacterium tuberculosis*. J Biol Chem. 2009;284:9937–46.
33. Li H, Wang X, Zhou L, Ma Y, Yuan W, Zhang X, Shi J, Xu Z. Enhancing expression of ketosteroid-9α-hydroxylase oxygenase, an enzyme with broad substrate range and high hydroxylation ability, in *Mycobacterium* sp. LY-1. Appl Biochem Biotechnol. 2018. https://doi.org/10.1007/s12010-018-2876-2.
34. Zhang R, Liu X, Wang Y, Han Y, Sun J, Shi J, Zhang B. Identification, function, and application of 3-ketosteroid Δ^1-dehydrogenase isozymes in *Mycobacterium neoaurum* DSM 1381 for the production of steroidic synthons. Microb Cell Factories. 2018;17(1):77.
35. Casabon I, Zhu S-H, Otani H, Liu J, Mohn WW, Eltis LD. Regulation of the KstR2 regulon of *Mycobacterium tuberculosis* by a cholesterol catabolite. Mol Microbiol. 2013;89:1201–12.
36. Kendall SL, Burgess P, Balhana R, Withers M, ten Bokum A, Lott JS, Gao C, Uhia-Castro I, Stoker NG. Cholesterol utilization in mycobacteria is controlled by two TetR-type transcriptional regulators: *kstR* and *kstR2*. Microbiology. 2010;156:1362–71.
37. Casabon I, Crowe AM, Liu J, Eltis LD. FadD3 is an acyl-CoA synthetase that initiates catabolism of cholesterol rings C and D in actinobacteria: role of FadD3 in cholesterol catabolism. Mol Microbiol. 2013;87:269–83.
38. Crowe AM, Casabon I, Brown KL, Liu J, Lian J, Rogalski JC, Hurst TE, Snieckus V, Foster LJ, Eltisa LD. Catabolism of the last two steroid rings in *Mycobacterium tuberculosis* and other bacteria. MBio. 2017. https://doi.org/10.1128/mBio.00321-17.
39. van der Geize R, Grommen AWF, Hessels GI, Jacobs AC, Dijkhuizen L. The steroid catabolic pathway of the intracellular pathogen *Rhodococcus equi* is important for pathogenesis and a target for vaccine development. PLoS Pathog. 2011;7:e1002181. https://doi.org/10.1371/journal.ppat.1002181.
40. Li Q, Ge F, Tan Y, Zhang G, Li W. Genome-wide transcriptome profiling of *Mycobacterium smegmatis* mc^2 155 cultivated in minimal media supplemented with cholesterol, androstenedione or glycerol. Int J Mol Sci. 2016;17:689.
41. Liu M, Xiong LB, Tao X, Liu QH, Wang FQ, Wei DZ. Integrated transcriptome and proteome studies reveal the underlying mechanisms for sterol catabolism and steroid production in *Mycobacterium neoaurum*. J Agric Food Chem. 2018;66(34):9147–57.
42. Gupta RS, Lo B, Son J. Phylogenomics and comparative genomic studies robustly support division of the genus *Mycobacterium* into an emended genus *Mycobacterium* and four novel genera. Front Microbiol. 2018;9:67. https://doi.org/10.3389/fmicb.2018.00067.
43. Sukhodolskaya GV, Nikolayeva VM, Khomutov SM, Donova MV. Steroid-1-dehydrogenase of *Mycobacterium* sp. VKM ac-1817D strain producing 9alpha-hydroxy-androst-4-ene-3,17-dione from sitosterol. Appl Microbiol Biotechnol. 2007;74:867–73.
44. Shtratnikova VY, Schelkunov MI, Dovbnya DV, Pekov YA, Bragin EY, Ashapkin VV, Donova MV. Complete genome sequence of *Mycobacterium* sp. strain VKM ac-1817D, capable of producing 9α-hydroxy-androst-4-ene-3,17-dione from phytosterol. Genome Announc. 2015;3:e01447–14. https://doi.org/10.1128/genomeA.01447-14.
45. Rocha DJP, Santos CS, Pacheco LGC. Bacterial reference genes for gene expression studies by RT-qPCR: survey and analysis. Antonie Van Leeuwenhoek. 2015;108:685–93.
46. Balhana RJC, Singla A, Sikder MH, Withers M, Kendall SL. Global analyses of TetR family transcriptional regulators in mycobacteria indicates conservation across species and diversity in regulated functions. BMC Genomics. 2015;16:479. https://doi.org/10.1186/s12864-015-1696-9.
47. Wilbrink MH, Petrusma M, Dijkhuizen L, van der Geize R. FadD19 of *Rhodococcus rhodochrous* DSM43269, a steroid-coenzyme a ligase essential for degradation of C-24 branched sterol side chains. Appl Environ Microbiol. 2011;77:4455–64.
48. Wipperman MF, Yang M, Thomas ST, Sampson NS. Shrinking the FadE proteome of *Mycobacterium tuberculosis*: insights into cholesterol metabolism through identification of an $\alpha_2\beta_2$ heterotetrameric acyl coenzyme a dehydrogenase family. J Bacteriol. 2013;195(19):4331–41. https://doi.org/10.1128/JB.00502-13.
49. Gilbert S, Hood L, Seah SYK. Characterization of an aldolase involved in cholesterol side chain degradation in *Mycobacterium tuberculosis*. J Bacteriol. 2017;200. https://doi.org/10.1128/JB.00512-17.
50. Yang M, Guja KE, Thomas ST, Garcia-Diaz M, Sampson NS. A distinct MaoC-like enoyl-CoA hydratase architecture mediates cholesterol catabolism in *Mycobacterium tuberculosis*. ACS Chem Biol. 2014;9(11):2632–45.
51. Yang M, Lu R, Guja KE, Wipperman MF, St Clair JR, AC B, Garcia-Diaz M, Sampson NS. Unraveling cholesterol catabolism in *Mycobacterium tuberculosis*: ChsE4-ChsE5 $\alpha_2\beta_2$ acyl-CoA dehydrogenase initiates β-oxidation of 3-oxo-cholest-4-en-26-oyl CoA. ACS Infect Dis. 2015;1(2):110–25.
52. van der Geize R, Hessels GI, Dijkhuizen L. Molecular and functional characterization of the kstD2 gene of *Rhodococcus erythropolis* SQ1 encoding a second 3-ketosteroid delta(1)-dehydrogenase isoenzyme. Microbiol Read Engl. 2002;148:3285–92.
53. van Oosterwijk N, Knol J, Dijkhuizen L, van der Geize R, Dijkstra BW. Structure and catalytic mechanism of 3-ketosteroid-Delta4-(5α)-dehydrogenase from *Rhodococcus jostii* RHA1 genome. J Biol Chem. 2012;287(37):30975–83.
54. Drzyzga O, Fernandez de las Heras L, Morales V, Navarro Llorens JM, Perera J. Cholesterol degradation by *Gordonia cholesterolivorans*. Appl Environ Microbiol. 2011;77:4802–10.
55. Xu LQ, Liu YJ, Yao K, Liu HH, Tao XY, Wang FQ, Wei DZ. Unraveling and engineering the production of 23,24-bisnorcholenic steroids in sterol metabolism. Sci Rep. 2016;6:21928.
56. García-Fernández J, Papavinasasundaram K, Galán B, Sassetti CM, García JL. Molecular and functional analysis of the mce4 operon in *Mycobacterium smegmatis*. Environ Microbiol. 2017;19(9):3689–99.
57. Costa KC, Bergkessel M, Saunders S, Korlach J, Newman DK. Enzymatic degradation of phenazines can generate energy and protect sensitive organisms from toxicity. MBio. 2015;6(6):e01520–15. https://doi.org/10.1128/mBio.01520-15.
58. Donova MV, Dovbnya DV, Kalinichenko AN, Vagabova LM, Arinbasarova AYu, Morozova ZV, Koshcheyenko KA. Method of 9-alpha-hydroxyandrost-4-ene-3,17-dione preparing 1997; Patent RU2077590.
59. Galán B, Uhía I, García-Fernández E, Martínez I, Bahíllo E, de la Fuente JL, Barredo JL, Fernández-Cabezón L, García JL. *Mycobacterium smegmatis* is a suitable cell factory for the production of steroidic synthons. Microb Biotechnol. 2017;10:138–50.
60. Mohn WW, Wilbrink MH, Casabon I, Stewart GR, Liu J, van der Geize R, Eltis LD. Gene cluster encoding cholate catabolism in *Rhodococcus* spp. J Bacteriol. 2012;194(24):6712–9.
61. Fernandez-Cabezon L, Galan B, Garcia JL. Unravelling a new catabolic pathway of C-19 steroids in *Mycobacterium smegmatis*. Environ Microbiol. 2018;20(5):1815–27.
62. Knol J, Bodewits K, Hessels GI, Dijkhuizen L, van der Geize R. 3-keto-5alpha-steroid Delta(1)-dehydrogenase from *Rhodococcus erythropolis* SQ1 and its orthologue in *Mycobacterium tuberculosis* H37Rv are highly specific enzymes that function in cholesterol catabolism. Biochem J. 2008;410:339–46.
63. García JL, Uhía I, Galán B. Catabolism and biotechnological applications of cholesterol degrading bacteria. Microb Biotechnol. 2012;5:679–99.
64. Donova MV, Nikolayeva VM, Dovbnya DV, Gulevskaya SA, Suzina NE. Methyl-beta-cyclodextrin alters growth, activity and cell envelope features of sterol-transforming mycobacteria. Microbiology. 2007;153:1981–92.
65. Shtratnikova VY, Schelkunov MI, Dovbnya DV, Bragin EY, Donova MV. Effect of methyl-β-cyclodextrin on gene expression in microbial conversion of phytosterol. Appl Microbiol Biotechnol. 2017;101(11):4659–67. https://doi.org/10.1007/s00253-017-8288-3.

66. Tjaden B. *De novo* assembly of bacterial transcriptomes from RNA-seq data. Genome Biol. 2016;16:1. https://doi.org/10.1186/s13059-014-0572-2.
67. Zhao Y, Wu J, Yang J, Sun S, Xiao J, Yu J. PGAP: pan-genomes analysis pipeline. Bioinformatics. 2012;28(3):416–8.
68. Bailey TL, Elkan C. Fitting a mixture model by expectation maximization to discover motifs in biopolymers. Proc Int Conf Intell Syst Mol Biol. 1994;2:28–36.
69. Rodionov DA. Comparative genomic reconstruction of transcriptional regulatory networks in bacteria. Chem Rev. 2007;107:3467–97.
70. Gupta S, Stamatoyannopoulos JA, Bailey TL, Noble WS. Quantifying similarity between motifs. Genome Biol. 2007;8:R24. https://doi.org/10.1186/gb-2007-8-2-r24.
71. Grant CE, Bailey TL, Noble WS. FIMO: scanning for occurrences of a given motif. Bioinforma Oxf Engl. 2011;27:1017–8.
72. Okonechnikov K, Golosova O, Fursov M. Unipro UGENE: a unified bioinformatics toolkit. Bioinformatics. 2012;28:1166–7.
73. Rice P, Longden I, Bleasby A. EMBOSS: the European molecular biology open software suite. Trends Genet. 2000;16:276–7.
74. Livak KJ, Schmittgen TD. Analysis of relative gene expression data using real-time quantitative PCR and the $2^{-\Delta\Delta C_T}$ method. Methods. 2001;25:402–8.
75. Ashkenazy H, Abadi S, Martz E, Chay O, Mayrose I, Pupko T, Ben-Tal N. ConSurf 2016: an improved methodology to estimate and visualize evolutionary conservation in macromolecules. Nucleic Acids Res. 2016;44:344–50.

12

Use of the *kojA* promoter, involved in kojic acid biosynthesis, for polyketide production in *Aspergillus oryzae*: implications for long-term production

Koichi Tamano[1,2*], Mahoko Kuninaga[3], Naoshi Kojima[4], Myco Umemura[2,3], Masayuki Machida[1] and Hideaki Koike[3]

Abstract

Background: *Aspergillus oryzae*, a useful industrial filamentous fungus, produces limited varieties of secondary metabolites, such as kojic acid. Thus, for the production of valuable secondary metabolites by genetic engineering, the species is considered a clean host, enabling easy purification from cultured cells. *A. oryzae* has been evaluated for secondary metabolite production utilizing strong constitutive promoters of genes responsible for primary metabolism. However, secondary metabolites are typically produced by residual nutrition after microbial cells grow to the stationary phase and primary metabolism slows. We focused on a promoter of the secondary metabolism gene *kojA*, a component of the kojic acid biosynthetic gene cluster, for the production of other secondary metabolites by *A. oryzae*.

Results: A *kojA* disruptant that does not produce kojic acid was utilized as a host strain for production. Using this host strain, a mutant that expressed a polyketide synthase gene involved in polyketide secondary metabolite production under the *kojA* gene promoter was constructed. Then, polyketide production and polyketide synthase gene expression were observed every 24 h in liquid culture. From days 0 to 10 of culture, the polyketide was continuously produced, and the synthase gene expression was maintained. Therefore, the *kojA* promoter was activated, and it enabled the continuous production of polyketide for 10 days.

Conclusions: The combined use of the *kojA* gene promoter and a *kojA* disruptant proved useful for the continuous production of a polyketide secondary metabolite in *A. oryzae*. These findings suggest that this combination can be applied to other secondary metabolites for long-term production.

Keywords: *Aspergillus oryzae*, Kojic acid, Secondary metabolism, Transcriptional regulation, Polyketide

Background

Aspergillus oryzae is a useful industrial filamentous fungus; it has been utilized for the production of Japanese alcohol (*sake*), soy sauce (*shoyu*), soybean paste (*miso*), and other products since ancient periods in Japan. It has a strong ability to produce hydrolases, such as amylase, protease, and lipase. It is also beneficial with respect to safety owing to its lack of mycotoxin production and formally approved as a safe microorganism with the GRAS (Generally Recognized as Safe) status by FDA (Food and Drug Administration of USA). The whole genome sequence of the wild-type *A. oryzae* strain RIB40 was reported in 2005 [1], providing a basis for targeted gene disruption or overexpression [2, 3]. In addition to enzymes, industrially valuable metabolite production has been attempted using *A. oryzae* by metabolic engineering via genetic modification technology [4–6].

A. oryzae produces a limited array of secondary metabolites, such as kojic acid [7, 8], and does not produce

* Correspondence: tamano-k@aist.go.jp
[1]Bioproduction Research Institute, National Institute of Advanced Industrial Science and Technology (AIST), 2-17-2-1 Tsukisamu-Higashi, Toyohira-ku, Sapporo, Hokkaido 062-8517, Japan
[2]AIST-Waseda University Computational Bio Big-Data Open Innovation Laboratory (CBBD-OIL), National Institute of Advanced Industrial Science and Technology (AIST), 5-20, Building 63, Nishi-waseda campus, Waseda University, 3-4-1 Okubo, Shinjuku-ku, Tokyo 169-8555, Japan

mycotoxins, such as aflatoxin and cyclopiazonic acid [9, 10]. In fact, heterologous production of secondary metabolites of polyketides, terpenes, and non-ribosomal peptides has ever been attempted in *A. oryzae*, and from the data presented here, it is indicated that none or only one peak of compounds with similar physical properties exist in the wild-type strain [11–14]. Thus, for the heterologous production of valuable secondary metabolites, we proposed *A. oryzae* to be adequate with respect to the ease of purification from cultured cells. In other words, it should be expected to be a clean host with limited endogenous secondary metabolites. Furthermore, knockout mutants of kojic acid biosynthesis genes are considered more effective hosts for secondary metabolite production than the wild-type strain because they do not produce any kojic acid.

Fungal secondary metabolites are required in the cells or spores for protection against ultraviolet (UV) and other microorganisms like bacteria, or for development [15]. Some of them possess medicinal properties and are actually used as pharmaceuticals such as penicillin, lovastatin, and cyclosporin [15, 16]. Most secondary metabolites are categorized into three groups: polyketides, small peptides (e.g. non-ribosomal peptides), and terpenes [15, 17]. Polyketides and terpenes are synthesized from acetyl-CoA and malonyl-CoA, and thus are characteristically composed of only carbon, oxygen, and hydrogen, but not nitrogen. Non-ribosomal peptides are synthesized from amino acids and thus contain nitrogen as well. Kojic acid shares the components of polyketides and terpenes; however, it is considered not to be categorized as either of them and considered to be synthesized through a different metabolic pathway [18].

In fact, for the first time, we previously identified genes responsible for kojic acid biosynthesis in *A. oryzae* [19]. They comprise *kojA* (AO090113000136), *kojR* (AO090113000137), and *kojT* (AO090113000138) and form a gene cluster on the chromosome (Fig. 1a) [19]. To date, it is revealed that the *kojR*-translated product, KojR, acts as a transcription factor and can activate promoters of other cluster genes, namely *kojA* and *kojT* [20]. The *kojA*-translated product, KojA, is predicted to be an oxidoreductase and directly contributes to kojic acid biosynthesis using some sort of metabolite as a substrate. However, it does not have a motif sequence of polyketide synthase (PKS) and P450 that usually exist in polyketide and terpene biosynthesis gene clusters, respectively. The *kojT*-translated product, KojT, is predicted to function as a transporter located on the cell membrane and transport kojic acid to the extracellular

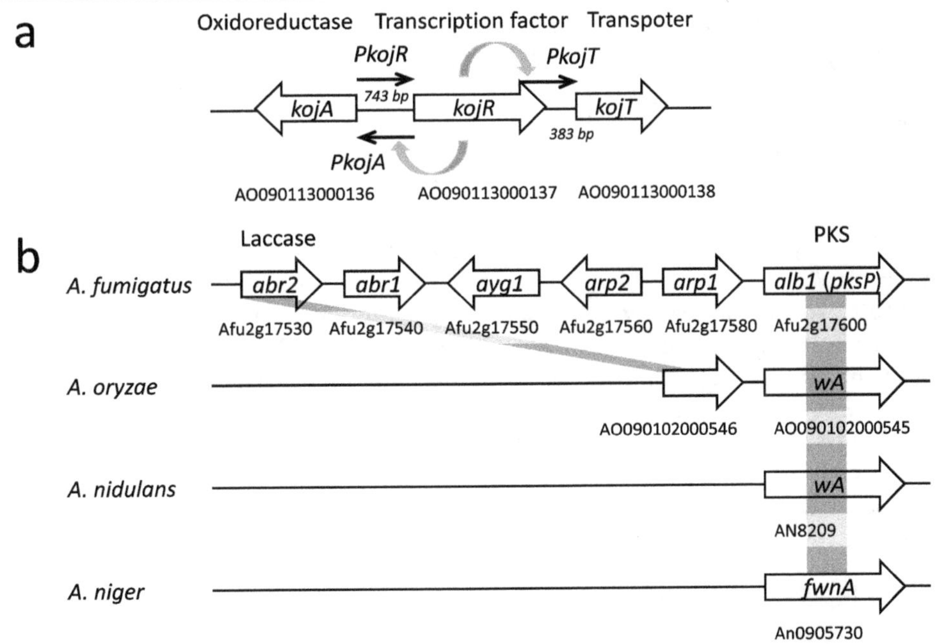

Fig. 1 Maps of the kojic acid biosynthesis cluster and the polyketide synthase (PKS) gene with the surrounding genes associated with spore pigment biosynthesis in *Aspergillus* species. Each gene is shown by box arrow. The estimated function, systematic gene name, and gene ID are denoted above, in, and below the box arrow, respectively. (**a**) The kojic acid biosynthesis cluster composed of three genes. Each promoter is shown by arrow. (**b**) Four species that are *A. fumigatus, A. oryzae, A. nidulans*, and *A. niger* are used for the mapping. The PKS genes are conserved among all four species, whereas genes of the other orthologs composing the spore pigment biosynthesis cluster in *A. fumigatus* are not almost located around the PKS gene in the other species. Only the laccase gene in *A. oryzae* is exceptionally located adjacent to the PKS gene

environment. These findings also indicate that kojic acid biosynthesis will be independent of polyketide and terpene biosyntheses.

We also analyzed the culture conditions under which this biosynthesis is activated. Kojic acid was continuously synthesized from culture day 3 to day 10 in kojic acid production liquid medium using residual glucose as a substrate after the nitrogen source was completely catabolized [19]. This implies that kojic acid biosynthesis is suppressed in the presence of a nitrogen source. In fact, the *kojA* gene expression was found to be repressed until the nitrogen source was used up [19]. Kojic acid is not essential for cell growth and thus is considered a secondary metabolite. Therefore, the promoters in the kojic acid biosynthesis gene cluster should exhibit lower activity levels than those of primary metabolism genes. However, because kojic acid is continuously produced, promoters of the genes involved in its biosynthesis are expected to be activated for a long period during production. In the stationary growth phase of *A. oryzae*, the primary metabolism diminishes, and kojic acid accumulates in the cells owing to its continuous production using the residual carbon source, glucose [19]. In such a situation, selective production of a metabolite seems to come into play, which will contribute to the easy and efficient recovery from cells. Therefore, in this study, we focused on the long-term production characteristics of kojic acid, and a promoter of *kojA* was evaluated for the long-term production of other secondary metabolites in *A. oryzae*. The polyketide synthase gene *wA* of *A. oryzae* was chosen for the evaluation, because polyketide has the same composition as kojic acid and could be easily quantified by a specific ultraviolet–visible (UV-Vis) absorbance.

Results

Production of a polyketide YWA1 synthesized by *A. oryzae wA* under the *kojA* promoter in the *kojA* disruptant

To begin with, we validated our assumption that the expression of primary metabolism genes attenuates toward the end of culture. The *enoA* gene encoding the enolase enzyme in glycolysis is considered constitutive gene and has ever been used for heterologous gene expression [21]. It was selected for the validation. The expression was certainly decreased as the culture of RIB40 proceeded in the kojic acid production medium (Additional file 1: Figure S1). Thus, our assumption seemed adequate, and as mentioned above, we then chose a polyketide secondary metabolite for evaluating the applicability of the *kojA* promoter to the production. It is because the *kojA* promoter activation is dependent on a nitrogen-limited culture environment. Polyketide does not contain nitrogen in the molecule, and thus it was considered appropriate for evaluating the expression system by the *kojA* promoter. In the *A. oryzae* genome, orthologues of genes involved in the biosynthesis of the spore pigment were predicted (Fig. 1b). A PKS encoded by *wA* (AO090102000545), an ortholog of *alb1* (Afu2g17600) in *Aspergillus fumigatus* Af293, *fwnA* (An09g05730) in *Aspergillus niger* CBS 513.88, and *wA* (AN8209) in *Aspergillus nidulans* A4, is likely involved in the initial step of biosynthesis of spore pigments using malonyl-CoAs as substrates. According to a previous study based on *A. nidulans* (Fig. 2), the product of *A. nidulans wA* is a spore pigment precursor; however, it has a yellow color and is named YWA1 [22]. The product of *A. oryzae wA* was expected to be similar to that of *A. nidulans wA* and expected to be measured by absorbance easily in the experimental technique. Thus, by replacing the promoter of *wA* with that of *kojA*, production of the polyketide derived from *A. oryzae wA* under the regulation of the *kojA* promoter was attempted in *A. oryzae*.

A *kojA* disruptant constructed in our previous study and named *ΔkojA_niaD-* [19] was used as a parental host strain for constructing the mutant strain expressing *wA* under the control of the *kojA* promoter. The *ΔkojA_niaD-* strain does not produce kojic acid and should produce only small amounts of other secondary metabolites. The *ΔkojA_niaD-* strain is also an auxotroph of nitrate due to a mutation in a nitrate reductase gene (*niaD*, AO090012001035); thus, *niaD* was used as a selectable marker. A DNA cassette for the expression of *A. oryzae wA* under the *kojA* promoter (Additional file 1: Figure S2) was prepared by fusion PCR as described in the Methods section. It was introduced to the *ΔkojA_niaD-* strain by transformation, and the transformant colonies were subjected to single spore isolation and verified by both PCR and Southern hybridization. Homologous recombination was confirmed at only the targeted *wA* locus by Southern hybridization (Fig. 3a-b). The mutant with promoter replacement was named *ΔkojA_PkojA::wA*. Similarly, the entire intact *niaD* gene was prepared by PCR using RIB40 chromosomal DNA as a template (Additional file 1: Figure S3), and was introduced to the *ΔkojA_niaD-* strain to construct a control mutant strain. The *niaD*-complemented *kojA* disruptant was confirmed by Southern hybridization (Fig. 3c-d) and was named *ΔkojA*. Both mutant strains were cultured on kojic acid production agar. The *ΔkojA* strain showed yellow-green spores, similar to the spores of the wild-type strain, whereas the *ΔkojA_PkojA::wA* strain showed white spores (Fig. 4a, upper row). In contrast, the agar became yellowish only for the *ΔkojA_PkojA::wA* strain (Fig. 4A, lower row). When cultured in kojic acid production liquid medium, the culture supernatant of the *ΔkojA_PkojA::wA* strain became more yellowish than the supernatant of the *ΔkojA* strain (Fig. 4b). Considering that this

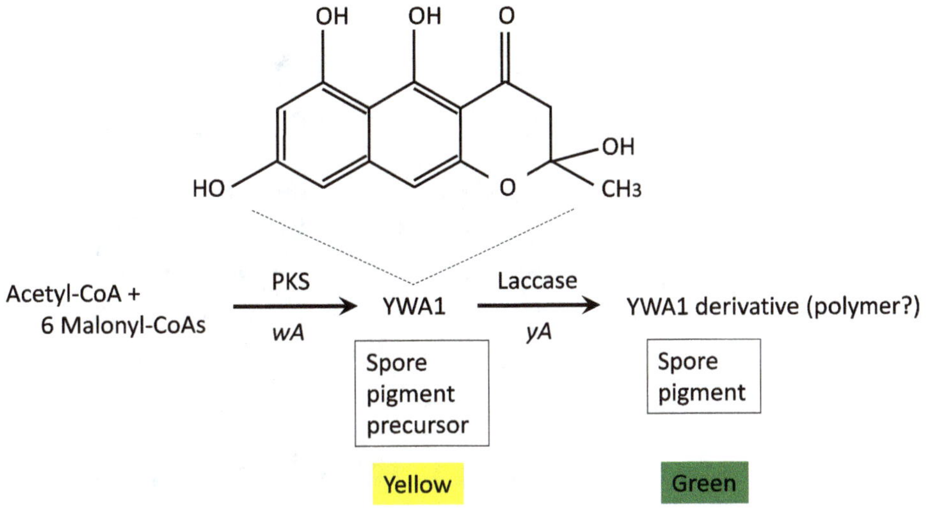

Fig. 2 Illustration of biosynthetic pathway of spore pigment in *A. nidulans*. The first step is performed by PKS encoded by *A. nidulans wA*. The polyketide named YWA1 is synthesized and the structure is drawn above. The second step is performed by a laccase encoded by *yA*. The similar pathway seems to be there in *A. oryzae*

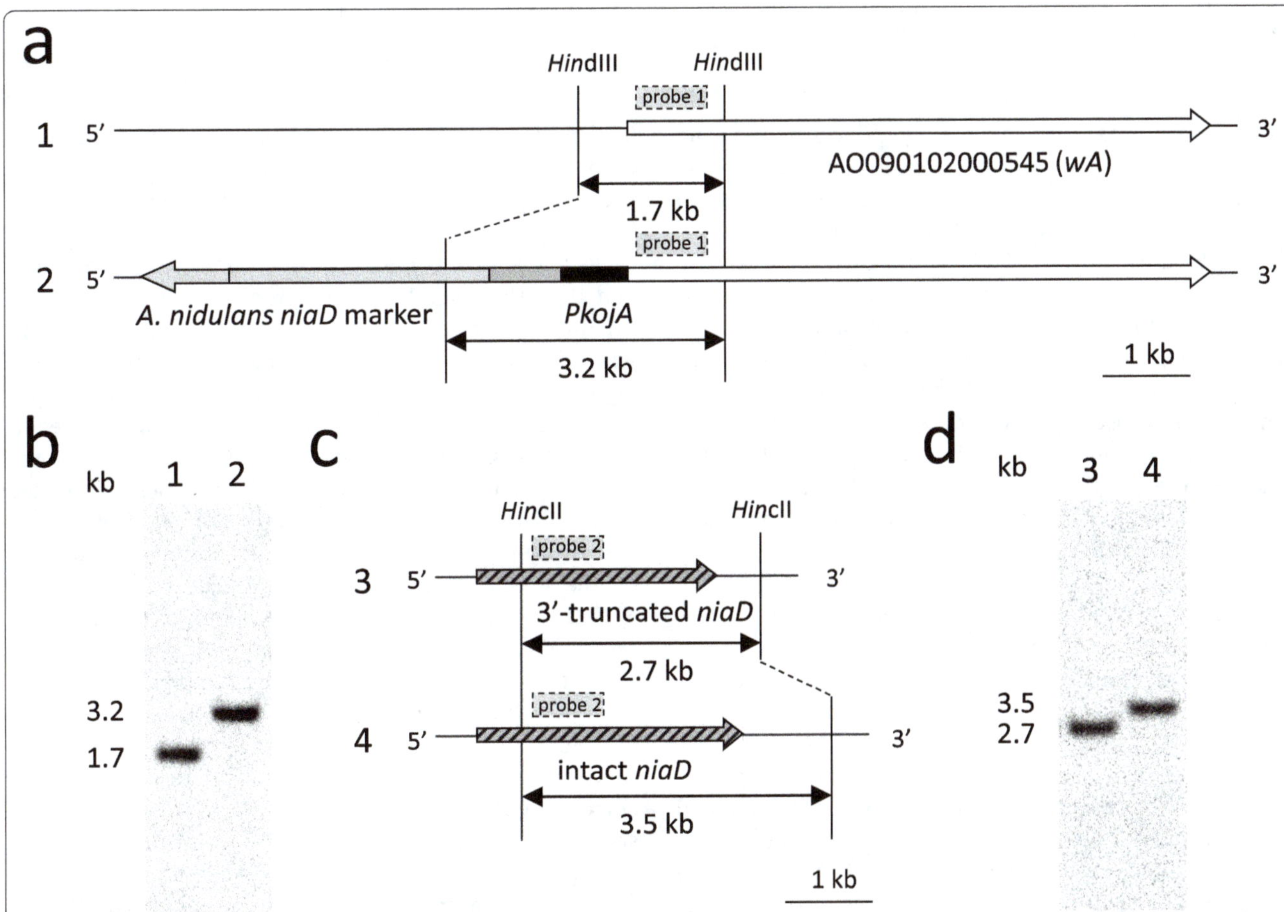

Fig. 3 Construction and confirmation of the Δ*kojA_PkojA::wA* and Δ*kojA* strains. (**a**) Change in the *wA* promoter region during construction of the Δ*kojA_PkojA* strain. (**b**) Construction of the Δ*kojA_PkojA* strain was confirmed by Southern hybridization. The site where the probe 1 hybridizes is shown. 1: Δ*kojA*, 2: Δ*kojA_PkojA::wA*. (**c**) Complementation of the *niaD* during construction of the Δ*kojA* strain. (**d**) Construction of the Δ*kojA* strain was confirmed by Southern hybridization. The site where the probe 2 hybridizes is shown. 3: Δ*kojA_niaD-*, 4: Δ*kojA*

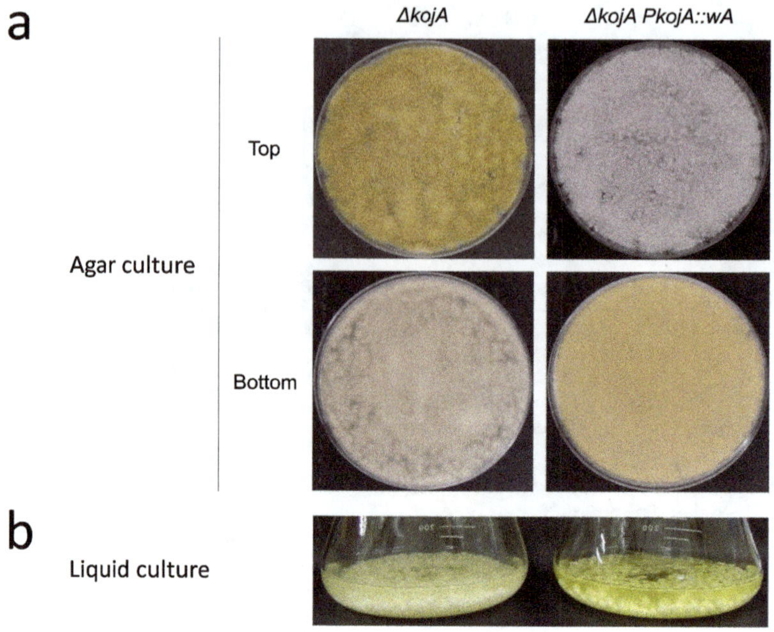

Fig. 4 Images of the Δ*kojA* (left) and the Δ*kojA_PkojA::wA* (right) strains. (**a**) Image obtained after incubation at 30 °C for 7 days of the CD agar plates where 5×10^5 spores were spread-inoculated. Images of plates taken from the top (upper row) or the bottom (lower row) are shown. (**b**) Image of the liquid culture after cultivation at 30 °C at 200 rpm for 5 days

yellow color was derived from PKS, extraction and confirmation of the yellow-colored compound were attempted using ethyl acetate. The UV-Vis spectrum corresponded to that of the YWA1 standard (Additional file 1: Figures S4). A UPLC-MS analysis was also performed for the sample and the standard. The retention time of the major peak at 4.0 min observed at 280 nm by UPLC superimposed with that of YWA1 standard (Fig. 5a). In addition, the MS chromatogram of sample compounds whose m/z was 277 was identical to that of the YWA1 standard (Fig. 5b). Hence, taken together, the yellow-colored compound produced by the *ΔkojA_PkojA::wA* strain was attributed to the polyketide synthase encoded by *wA* and concluded to be YWA1.

Time-course shifts in *wA* expression under the *kojA* promoter and the polyketide yield

The activation of the *kojA* promoter was evaluated in the *ΔkojA_PkojA::wA* strain. The expression level of *wA* and the YWA1 yield were measured every 24 h in a stepwise manner from the beginning of the culture period until day 10. As determined by qRT-PCR, *wA* remained active from day 2 to day 10 of the culture period (Fig. 6). The expression level exhibited a peak on day 5 and decreased after day 9, but the differences were minimal. Therefore, the *kojA* promoter was active up to day 10 of the culture period. In the control *ΔkojA* strain, the expression level of *wA* showed a similar trend. As a result, *wA* expression of the *ΔkojA* strain was approximately 0.05 fold or less of that of *ΔkojA_PkojA::wA* strain in each culture period. This suggests that *wA* was overexpressed by the *kojA* promoter in the *ΔkojA_PkojA::wA* strain.

Next, YWA1 was measured by estimates of absorbance, and then the YWA1 amount produced by unit gram of dried cells was evaluated. An absorbance peak at 406 nm was detected by UV-Vis spectrometry (Additional file 1: Figure S4); accordingly, absorbance at 406 nm in the culture supernatant was measured every 24 h from the beginning of the culture period until day 10. Absorbance of the culture medium before inoculation was then subtracted from that on each culture period to remove background, and the subtracted absorbance was subsequently divided by the weight of dried cells to evaluate the amount of YWA1 produced by the same number of cells. It was observed that YWA1 was continuously produced from day 1 to day 10; production was maintained during long-term liquid culture (Fig. 7). The YWA1 amount per gram of dried cells in the *ΔkojA_PkojA::wA* strain was 10.6-fold more than that in the control strain. The results corresponded to the enhanced *wA* expression level as shown in Fig. 6, indicating that the *kojA* promoter is useful for high and long-term production. As in the case of *enoA* (Additional file 1: Figure S1), expression of constitutive genes is generally considered to attenuate toward the stationary growth phase. Though we do not directly compare the case of *wA* expression under between *kojA* promoter and *enoA* promoter, because high and long-term

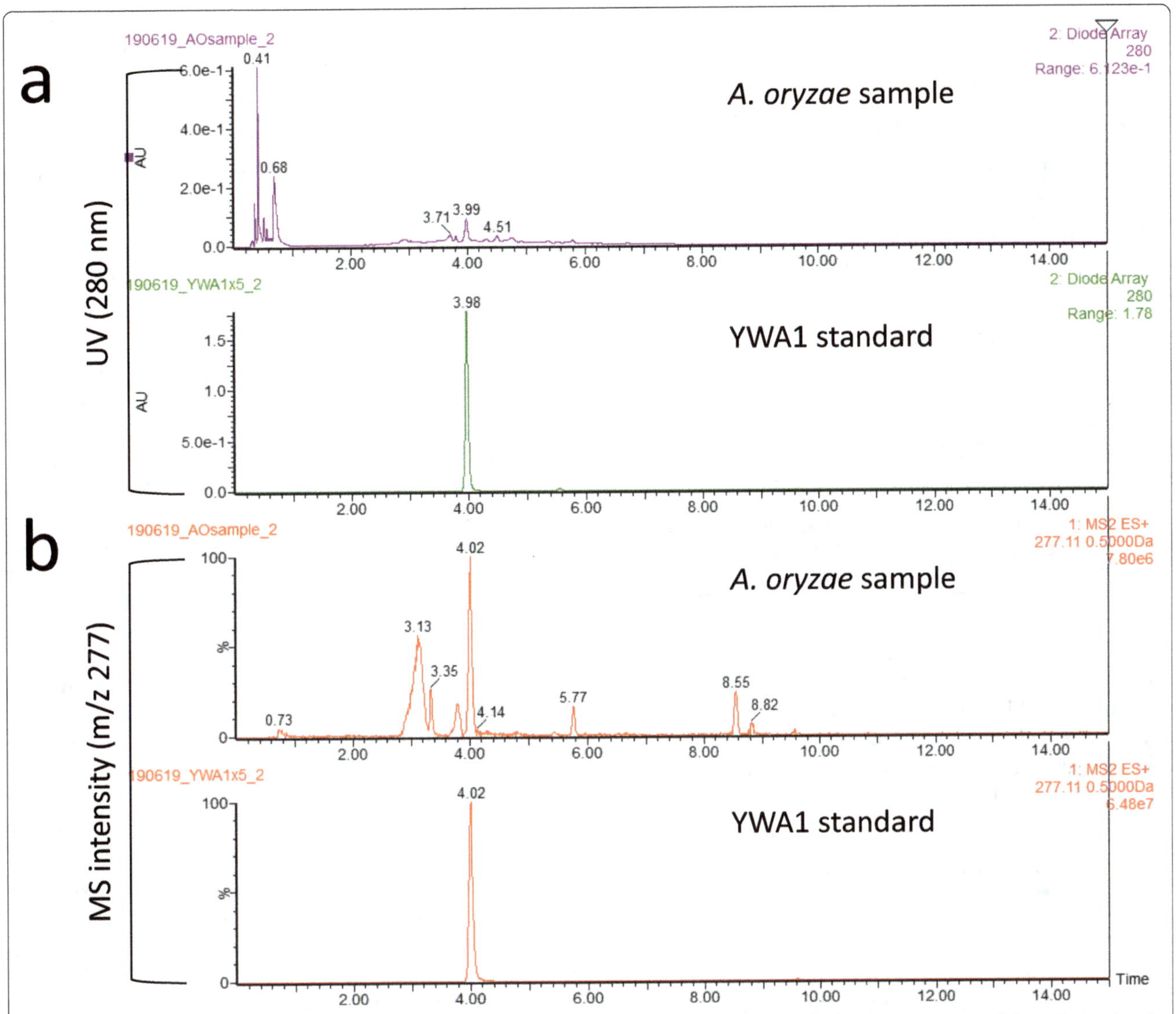

Fig. 5 Analysis of the yellow-colored compound produced by the Δ*kojA_PkojA::wA* strain by UPLC-MS. The chromatograms of the yellow-colored compound denoted as *A. oryzae* sample and the YWA1 standard by UPLC (**a**) and MS (**b**) are shown. The UPLC chromatograms are taken at 280 nm, while the MS chromatograms are taken at m/z: 277. X-axes; retention time (min): Y-axes; absorbance at 280 nm (**a**) and signal intensity (**b**)

production was observed in case of YWA1 in this study, the *kojA* promoter seems to be useful for such a purpose.

Discussion

In this study, we showed that the *kojA* promoter enabled the continuous production of polyketide YWA1 encoded by *wA* for 10 days in *A. oryzae*. As shown in Fig. 1b, in spite of conservation of the *wA* orthologs between four *Aspergillus* species, the other orthologs included in the melanin synthesis cluster of *A. fumigatus* do not exist or are located far apart from the *wA* ortholog on the genome in the other three *Aspergillus* species. Only the laccase gene ortholog (AO090102000546) exists adjacent to *wA* on the *A. oryzae* genome. However, another gene involved in the spore pigment biosynthesis named *yA* (AN6635) is present on the *A. nidulans* genome, which is located far away from *A. nidulans wA* on the genome. The *yA* encodes laccase as well and is reported to generate green-colored spore pigments, probably via polymerization of YWA1 through the laccase function (Fig. 2). However, the *yA* homolog of *A. oryzae* is AO090011000755 and is different from the ortholog. Thus, there seem to be two laccase candidate genes involved in the spore pigment biosynthesis in *A. oryzae*. A fact which predominantly functions in it should be researched hereafter.

Because of the white spores of the *ΔkojA_PkojA::wA* strain, the *kojA* promoter may be active only in hyphae or even in spores at lower level than the RIB40. Because the production of valuable enzymes and compounds by *A. oryzae* is performed in liquid culture, the phenomenon of

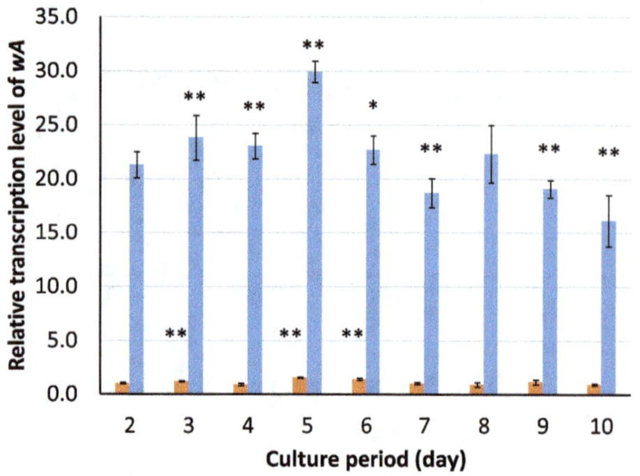

Fig. 6 Time-course shift in *A. oryzae wA* expression. Expression levels of *wA* in the wild-type RIB40 (orange) and the Δ*kojA_PkojA::wA* (blue) strains at each time point relative to that on day 2 of RIB40 are shown. Error bars represent mean ± standard deviation ($n = 3$). The result of a significance test (Student's t-test) between day 2 and the other date is shown above each bar. *and ** represent $p < 0.05$ and $p < 0.01$, respectively

whitish spores will not be a drawback when using the *kojA* promoter for production.

Because YWA1 does not contain nitrogen, the production system using the *kojA* promoter in the *kojA* disruptant could be applied to the production of other compounds lacking nitrogen, like polyketides. On the other hand, because this production system is undertaken in nitrogen source-deficient condition, it was considered difficult to be applied to the production of compounds containing nitrogen, such as non-ribosomal peptides. However, unexpectedly, YWA1 production by the *ΔkojA_PkojA::wA* strain was observed from day 1 of the culture period (Fig. 7). We expected production to begin on day 3, after the nitrogen source was catabolized. In fact, *wA* expression level under the *kojA* promoter at day 2 of the culture period was already high in the *ΔkojA_PkojA::wA* strain, while *kojA* expression level under the *kojA* promoter at the day 2 was quite low in the RIB40 strain (Additional file 1: Figure S5). The data would imply that the expression in the *ΔkojA_PkojA::wA* strain was actually independent of nitrogen starvation. The reason for this unexpected result is unclear and should be evaluated in future studies. There might be additional regulatory mechanisms apart from that utilizing the nitrogen source. For example, the *kojA* promoter was attached to the *wA* locus, where epigenetic modifications might be different from those at the original *kojA* locus. However, with respect to metabolite production, expression from the early stage of culture is advantageous and is not an issue. In addition, since the *kojA* promoter seemed active even at culture period day 1 when nitrogen source remained, although it was not exemplified in this study, the expression system with the *kojA* promoter could be applied to non-ribosomal peptide production where plenty of nitrogen source will be required in culture.

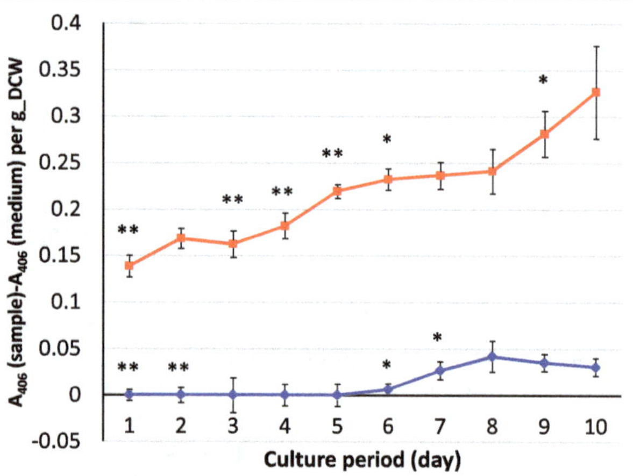

Fig. 7 Time-course shift in the *A. oryzae wA* product YWA1 secreted by the Δ*kojA_PkojA::wA* strain. The product yield was measured by specific absorbance at 406 nm of the culture supernatant. Absorbance of the culture supernatant at each culture period was subtracted by that of culture medium without inoculation, followed by division by dry cell weight of hyphae. The calculated value was monitored for both the Δ*kojA_PkojA::wA* strain (red) and the control Δ*kojA* strain (blue). Error bars represent mean ± standard deviation (n = 3). The result of a significance test (Student's t-test) between the day 0 and the other date is shown above each bar. * and ** represent $p < 0.05$ and $p < 0.01$, respectively

Polyketide final products are generally synthesized not only by PKS but also by several modifying enzymes. Thus, for increase in production, several genes of enzymes including PKS involved in the synthesis will need to be overexpressed. In such an attempt, multiple uses of the *kojA* promoter for expressing these genes will be one possible solution. If the *kojA* promoter was used redundantly, its transcriptional factor, KojR, would get shortage in the cell. In such a case, it will be effective to use the *kojR* gene multiply as well. Fortunately, as shown in Fig. 1a, the *kojA* promoter shares the locus with the *kojR* promoter but in the opposite direction. In other words, it is a bidirectional promoter. Thus, only extending the DNA region used for replacement with the *kojA* promoter in this study to the end of *kojR* coding sequence will enable not only the replacement but also additional *kojR* arrangement on the genome.

Conclusions

The combination of the *kojA* gene promoter and the *kojA* disruptant was useful for continuous production of a polyketide secondary metabolite until day 10 of the culture period. The results suggested that this combination is applicable to other secondary metabolites for long-term production in *A. oryzae*.

Methods

Fungal strains and culture

A. oryzae wild-type strain RIB40 and its mutants were used. The RIB40 strain was distributed by the National Research Institute of Brewing (NRIB, Hiroshima, Japan) and approved of use for research purposes [1]. For genetic engineering, an *A. oryzae ΔkojA_niaD-* strain with the genotype *ΔligD::ptrA ΔpyrG::AnsC ΔkojA::AnpyrG* (*niaD-*), derived from NS4 [23, 24], the *niaD-* and *sC-* auxotrophic mutant of RIB40, and constructed in our previous study [19], was used. The *ΔkojA_niaD-* strain was originally removed capability of non-homologous end joining (NHEJ) due to knockout of a ligase gene *ligD* responsible for NHEJ [23]. Thus, only homologous recombination can be performed in the strain and the efficiency is quite high.

Each *A. oryzae* strain constructed by genetic engineering was maintained at 30 °C on Czapek–Dox (CD) agar medium [25], and the spores were stored in 20% (w/v) glycerol at − 80 °C until use. To test YWA1 production under the *kojA* promoter, the spores were inoculated on kojic acid production agar medium (0.25% yeast extract, 0.1% K_2HPO_4, 0.05% $MgSO_4$-$7H_2O$, 10% glucose, and 1.5% agar; pH adjusted to 6.0 before autoclaving) [19] and harvested. Then, 2.5×10^7 spores were inoculated into 50 mL of the kojic acid production liquid medium and cultured at 30 °C and 200 rpm for up to 10 days.

Construction of the *A. oryzae* mutant strains

All primers used in this study are listed in Additional file 1: Table S1. The promoter-replaced strain, *ΔkojA_PkojA::wA*, was constructed as follows. The *ΔkojA_niaD-* strain, a *kojA* disruptant lacking *niaD* constructed in our previous study [19], was transformed with the DNA fragment for *wA* expression under the *kojA* promoter. The DNA fragment was constructed by PCR and the subsequent fusion PCR using the KOD-PLUS DNA polymerase (Toyobo, Osaka, Japan) [26]. Two 1-kb DNA fragments of the promoter and the coding sequence from the start codon of *wA* were amplified with LU/LL and RU/RL primer pairs from the RIB40 genomic DNA template. Similarly, a 1-kb DNA fragment of the *kojA* promoter was amplified with PU_PkojA/PL_PkojA primer pair from the RIB40 genomic DNA. A 4.7-kb DNA fragment of *AnniaD* including the promoter and the terminator was amplified as a selectable marker with PU_AnniaD/PL_AnniaD primer pair from *A. nidulans* A851 genomic DNA. These four DNA fragments were mixed and were subjected to fusion PCR with LU/RL primer pair (Additional file 1: Figure S2). The resultant fragment harboring the *wA* gene portion in which the promoter was replaced with the *kojA* promoter was introduced to the *ΔkojA_niaD-* strain by transformation. The *ΔkojA_PkojA::wA* stain was selected by picking transformants grown on CD agar medium and confirmed by both PCR with LU/RL primer pair and Southern hybridization.

The control strain, *ΔkojA*, was made by complementing the *niaD* gene obtained from *A. oryzae* with the *ΔkojA_niaD-* strain. A 5.4-kb DNA fragment of *niaD* including 0.9–1 kb regions of the promoter and the terminator was amplified with PU_niaD/PL_niaD primer pair from *A. oryzae* RIB40 genomic DNA and used for the transformation of *ΔkojA_niaD-* (Additional file 1: Figure S3). The *ΔkojA* strain was selected by picking transformants grown on CD agar medium and confirmed by PCR with PU_niaD/PL_niaD primer pair.

Southern hybridization

Aliquots (3 μg) of genomic DNA were digested with the restriction enzyme (*Hin*dIII or *Hinc*II), fractionated on a 0.7% agarose gel. Digoxigenin (DIG)-labeled probe (approximately 800 bp) was prepared using a PCR DIG Probe Synthesis Kit (Roche Applied Science, Mannheim, Germany) with the primers SBU and SBL. The probe preparation, hybridization, and signal detection were performed as described previously [3].

Extraction of the yellow-colored compound, confirmation by UV-Vis spectrometry and UPLC-MS, and the quantification

To analyze the yellow-colored compound secreted from the *ΔkojA_PkojA::wA* strain, 30 mL of culture supernatant after liquid culture was extracted with 20 mL of ethyl acetate. The yellow-colored compound was extracted in the ethyl acetate fraction. The separated ethyl acetate fraction was concentrated in vacuo using a rotary evaporator, and the residue was dissolved in ethanol (1.5 mL). A small portion was added to 1 mL of Milli-Q water, and then UV-Vis spectrometry was performed by monitoring absorbance in the wavelength range from 200 nm to 500 nm using a UV-Vis spectrophotometer (Shimadzu UV-1800; Tokyo, Japan). Then, another small portion was redissolved with ethyl acetate and 1 μL of the solution was evaluated by ultraperformance liquid chromatography (UPLC)-mass spectrometry (MS) using the Waters Xevo TQD, Acquity UPLC H class equipped with an Waters Acquity UPLC BEH C18 (1.7 μm, 2.1 × 50 mm) column (Waters, Milford, MA, USA). 15% acetonitrile in Milli-Q water over 0–0.5 min and a subsequent linear gradient of 15 to 100% acetonitrile in Milli-Q water containing 0.1% formic acid over 0.5–15 min at a flow rate of 0.4 mL/min were used as the mobile phase. The column temperature was set to 40 °C. The yellow-colored compound was detected by monitoring absorbance at 280 nm. Because other compounds with absorbance at 280 nm, such as nucleic acids and proteins, were removed at the extraction step with ethyl acetate, the yellow-colored compound was detected with high sensitivity.

For the quantification of the yellow-colored compound, 500 μL of the culture supernatant was sampled and filtered using Ultrafree-MC-HV Centrifugal Filters Durapore-PVDF 0.45 μm (Millipore, Burlington, MA, USA). Then, absorbance at 406 nm was measured according to the second largest peak of UV spectrum at this wavelength. Since the culture supernatant was directly measured, this wavelength was used for precise quantification.

As a standard, the naphthopyrone compound YWA1, a product of polyketide synthase encoded by *wA* of *A. nidulans,* was similarly evaluated by both UV-Vis spectrometry and UPLC-MS [22].

Quantitative reverse transcription polymerase chain reaction (qRT-PCR)

Primers named "real_fwd" and "real_rev" were used for the quantification of gene expression levels. Expression level of the histone gene (AO090012000496) was used for normalizing the expression level of each gene [27]. RNA extraction and the subsequent qRT-PCR were performed as described previously [3].

Supplementary information

Additional file 1: Figure S1. Time-course shift in *enoA* expression in the wild-type RIB40 strain. Expression level at each time point relative to that on day 2 of RIB40 is shown. Error bars represent mean ± standard deviation ($n = 3$). The result of a significance test (Student's t-test) between day 2 and the other date is shown above each bar. ** represent $p < 0.01$. **Figure S2.** Construction of the DNA fragment for replacement of *wA* promoter with *kojA* promoter in *A. oryzae*. The 7794 bp long DNA fragment was constructed for the replacement. Primers used for the construction and clone check are shown as arrows. **Figure S3.** Construction of the DNA fragment for complementing the *niaD* gene to the *A. oryzae* Δ*kojA_niaD-* where 769 bp long region was deleted at the original *niaD* locus. The 5398 bp long DNA fragment was constructed. Primers used for the construction and clone check are shown as arrows. **Figure S4.** UV-Vis spectra of the YWA1 standard (A) and yellow-colored compound produced by the Δ*kojA_PkojA::wA* strain (B). **Fig. S5.** Time-course shift of the *kojA* promoter activities. The *kojA* promoter activity evaluated as *kojA* expression level at the *kojA* locus in the wild-type RIB40 strain (orange) and the one evaluated as *wA* expression level at the *wA* locus in the Δ*kojA_PkojA::wA* strain (blue) are compared. Both expression levels are normalized by the expression level of the histone gene as an internal standard and then compensated expression level at each time point relative to that on day 2 of RIB40 is shown. Error bars represent standard deviation ($n = 3$). Result of a significance test (Student's t-test) between day 2 of RIB40 and the other date is shown above each bar. ** represent $p < 0.01$. **Table S1.** DNA primers used in this study. Tails of primers are shown in lower case letters.

Abbreviations

A. nidulans: *Aspergillus nidulans*; *A. niger*: *Aspergillus niger*; *A. oryzae*: *Aspergillus oryzae*; CD: Czapek-Dox; cDNA: complementary DNA; DIG: Digoxigenin; HPLC: High performance liquid chromatography; PCR: Polymerase chain reaction; PKS: Polyketide synthase; qRT-PCR: quantitative reverse transcription polymerase chain reaction; UPLC-MS: Ultraperformance liquid chromatography-mass spectrometry; UV-Vis: Ultraviolet–visible

Acknowledgements

The authors would like to thank Dr. Isao Fujii for kindly providing the YWA1 standard. The authors are also grateful to Dr. Tomomi Toda for experimental assistance.

Authors' contributions

KT, MU, MM, and HK conceived the experiments. KT, MK, NK, and MU performed experiments and analyzed the data. KT, NK, MU, and HK wrote the manuscript. All authors read and approved of the final manuscript.

Funding

This work was supported by a Grant-in-Aid for Scientific Research on Innovative Areas (JSPS KAKENHI Grant Number 22108007) and a Grant-in-Aid for Scientific Research (JSPS KAKENHI Grant Number 16 K01930), from the Ministry of Education, Culture, Sports, Science and Technology (MEXT) of Japan. The funding body had no role in the design of the study, the collection, analysis, and interpretations of data and in writing the manuscript.

Ethics approval and consent to participate

Not applicable.

Consent for publication

Not applicable.

Competing interests

The authors declare that they have no competing interests.

Author details

[1]Bioproduction Research Institute, National Institute of Advanced Industrial Science and Technology (AIST), 2-17-2-1 Tsukisamu-Higashi, Toyohira-ku, Sapporo, Hokkaido 062-8517, Japan. [2]AIST-Waseda University Computational Bio Big-Data Open Innovation Laboratory (CBBD-OIL), National Institute of Advanced Industrial Science and Technology (AIST), 5-20, Building 63, Nishi-waseda campus, Waseda University, 3-4-1 Okubo, Shinjuku-ku, Tokyo 169-8555, Japan. [3]Bioproduction Research Institute, National Institute of Advanced Industrial Science and Technology (AIST), Central 6, 1-1-1 Higashi, Tsukuba, Ibaraki 305-8566, Japan. [4]Biomedical Research Institute, National Institute of Advanced Industrial Science and Technology (AIST), Central 6, 1-1-1 Higashi, Tsukuba, Ibaraki 305-8566, Japan.

References

1. Machida M, Asai K, Sano M, Tanaka T, Kumagai T, Terai G, Kusumoto K, Arima T, et al. Genome sequencing and analysis of *Aspergillus oryzae.* Nature. 2005;438:1157–61.
2. Maruyama J, Kitamoto K. Targeted gene disruption in Koji mold *Aspergillus oryzae*. Methods Mol Biol. 2011;765:447–56.
3. Tamano K, Miura A. Further increased production of free fatty acids by overexpressing a predicted transketolase gene of the pentose phosphate pathway in *Aspergillus oryzae faaA* disruptant. Biosci Biotechnol Biochem. 2016;80:1829–35.
4. Brown SH, Bashkirova L, Berka R, Chandler T, Doty T, McCall K, McCulloch M, McFarland S, et al. Metabolic engineering of *Aspergillus oryzae* NRRL 3488 for increased production of L-malic acid. Appl Microbiol Biotechnol. 2013;97:8903–12.
5. Ichinose S, Tanaka M, Shintani T, Gomi K. Improved α-amylase production by *Aspergillus oryzae* after a double deletion of genes involved in carbon catabolite repression. Appl Microbiol Biotechnol. 2014;98:335–43.
6. Tamano K, Cox RS 3rd, Tsuge K, Miura A, Itoh A, Ishii J, Tamura T, Kondo A, et al. Heterologous production of free dihomo-γ-linolenic acid by *Aspergillus oryzae* and its extracellular release via surfactant supplementation. J Biosci Bioeng. in press.
7. Frisvad JC, Møller LLH, Larsen TO, Kumar R, Arnau J. Safety of the fungal workhorses of industrial biotechnology: update on the mycotoxin and

secondary metabolite potential of *Aspergillus niger*, *Aspergillus oryzae*, and *Trichoderma reesei*. Appl Microbiol Biotechnol. 2018;102:9481–515.
8. Sano M. *Aspergillus oryzae nrtA* affects kojic acid production. Biosci Biotechnol Biochem. 2016;80:1776–80.
9. Tominaga M, Lee YH, Hayashi R, Suzuki Y, Yamada O, Sakamoto K, Gotoh K, Akita O. Molecular analysis of an inactive aflatoxin biosynthesis gene cluster in *Aspergillus oryzae* RIB strains. Appl Environ Microbiol. 2006;72:484–90.
10. Tokuoka M, Seshime Y, Fujii I, Kitamoto K, Takahashi T, Koyama Y. Identification of a novel polyketide synthase-nonribosomal peptide synthetase (PKS-NRPS) gene required for the biosynthesis of cyclopiazonic acid in *Aspergillus oryzae*. Fungal Genet Biol. 2008;45:1608–15.
11. Kan E, Katsuyama Y, Maruyama JI, Tamano K, Koyama Y, Ohnishi Y. Production of the plant polyketide curcumin in *Aspergillus oryzae*: strengthening malonyl-CoA supply for yield improvement. Biosci Biotechnol Biochem. 2019;83:1372–81.
12. Okada M, Matsuda Y, Mitsuhashi T, Hoshino S, Mori T, Nakagawa K, Quan Z, Qin B, et al. Genome-based discovery of an unprecedented cyclization mode in fungal Sesterterpenoid biosynthesis. J Am Chem Soc. 2016;138:10011–8.
13. Tagami K, Liu C, Minami A, Noike M, Isaka T, Fueki S, Shichijo Y, Toshima H, et al. Reconstitution of biosynthetic machinery for indole-diterpene paxilline in *Aspergillus oryzae*. J Am Chem Soc. 2013;135:1260–3.
14. Yoshimi A, Yamaguchi S, Fujioka T, Kawai K, Gomi K, Machida M, Abe K. Heterologous production of a novel cyclic peptide compound, KK-1, in *Aspergillus oryzae*. Front Microbiol. 2018;9:690.
15. Keller NP. Fungal secondary metabolism: regulation, function and drug discovery. Nat Rev Microbiol. 2019;17:167–80.
16. Tamano K. Enhancing microbial metabolite and enzyme production: current strategies and challenges. Front Microbiol. 2014;5:718.
17. Shen B. Polyketide biosynthesis beyond the type I, II and III polyketide synthase paradigms. Curr Opin Chem Biol. 2003;7:285–95.
18. Bentley R. From miso, saké and shoyu to cosmetics: a century of science for kojic acid. Nat Prod Rep. 2006;23:1046–62.
19. Terabayashi Y, Sano M, Yamane N, Marui J, Tamano K, Sagara J, Dohmoto M, Oda K, et al. Identification and characterization of genes responsible for biosynthesis of kojic acid, an industrially important compound from *Aspergillus oryzae*. Fungal Genet Biol. 2010;47:953–61.
20. Marui J, Yamane N, Ohashi-Kunihiro S, Ando T, Terabayashi Y, Sano M, Ohashi S, Ohshima E, et al. Kojic acid biosynthesis in *Aspergillus oryzae* is regulated by a Zn(II)(2)Cys(6) transcriptional activator and induced by kojic acid at the transcriptional level. J Biosci Bioeng. 2011;112:40–3.
21. Takaya T, Koda R, Adachi D, Nakashima K, Wada J, Bogaki T, Ogino C, Kondo A. Highly efficient biodiesel production by a whole-cell biocatalyst employing a system with high lipase expression in *Aspergillus oryzae*. Appl Microbiol Biotechnol. 2011;90:1171–7.
22. Watanabe A, Fujii I, Sankawa U, Mayorga ME, Timberlake WE, Ebizuka Y. Re-identification of *Aspergillus nidulans wA* gene to code for a polyketide synthase of naphthopyrone. Tetrahedron Lett. 1999:91–4.
23. Mizutani O, Kudo Y, Saito A, Matsuura T, Inoue H, Abe K, Gomi K. A defect of LigD (human Lig4 homolog) for nonhomologous end joining significantly improves efficiency of gene-targeting in *Aspergillus oryzae*. Fungal Genet Biol. 2008;45:878–89.
24. Yamada O, Lee BR, Gomi K. Transformation system for *Aspergillus oryzae* with double auxotrophic mutations, *niaD* and *sC*. Biosci Biotechnol Biochem. 1997;61:1367–9.
25. Tamano K, Satoh Y, Ishii T, Terabayashi Y, Ohtaki S, Sano M, Takahashi T, Koyama Y, et al. The beta-1,3-exoglucanase gene *exgA* (*exg1*) of *Aspergillus oryzae* is required to catabolize extracellular glucan, and is induced in growth on a solid surface. Biosci Biotechnol Biochem. 2007;71:926–34.
26. Amberg DC, Botstein D, Beasley EM. Precise gene disruption in *Saccharomyces cerevisiae* by double fusion polymerase chain reaction. Yeast. 1995;11:1275–80.
27. Tamano K, Sano M, Yamane N, Terabayashi Y, Toda T, Sunagawa M, Koike H, Hatamoto O, et al. Transcriptional regulation of genes on the non-syntenic blocks of *Aspergillus oryzae* and its functional relationship to solid-state cultivation. Fungal Genet Biol. 2008;45:139–51.

13

Lack of specificity associated with using molecular beacons in loop mediated amplification assays

Patrick Hardinge* and James A. H. Murray

Abstract

Background: Loop mediated isothermal amplification of nucleic acid templates is a rapid, sensitive and specific method suitable for molecular diagnostics. However the complexity of primer design and the number of primers involved can lead to false positives from non-specific primer interactions. Standard methods of LAMP detection utilise the increasing concentrations of DNA or inorganic pyrophosphate and therefore lack specificity for identifying the desired LAMP amplification. Molecular beacons used in PCR reactions are target specific and may enhance specificity with LAMP.

Results: We present a potential molecular beacon approach to LAMP detection targeting the single stranded region between loops, and test this for LAMP molecular beacons targeting the 35S promoter and NOS terminator sequences commonly used in GM crops. From these studies we show that molecular beacons used in LAMP, despite providing a change in fluorescent intensity with amplification, appear not to anneal to specific target sequences and therefore target specificity is not a benefit of this method. However, molecular beacons demonstrate a change in fluorescence which is indicative of LAMP amplification products. We identify the LAMP loop structure as likely to be responsible for this change in signal.

Conclusions: Molecular beacons can be used to detect LAMP amplification but do not provide sequence specificity. The method can be used to determine effectively LAMP amplification from other primer-driven events, but does not discriminate between different LAMP amplicons. It is therefore unsuitable for multiplex LAMP reactions due to non-specific detection of LAMP amplification.

Keywords: Loop-mediated amplification (LAMP), Molecular beacons, DNA, Genetically modified (GM) crops, Fluorescence

Background

Isothermal nucleic acid amplification methods represent an attractive alternative to PCR technologies which are reliant on thermocycling equipment. There are numerous isothermal methods [1, 2], but the most prevalent in the scientific literature is Loop-mediated amplification (LAMP) first described over a decade ago [3].

The LAMP reaction utilises 4 or more primers complementary to 6 or more target sequences. At the initiation of the amplification hairpin forming forward and backward inner primers (FIP and BIP) strand invade the double stranded DNA target. These primers anneal and are extended by a polymerase with strong displacement activity. The nascent strands are displaced by further primers (F3 and B3) enabling the subsequent formation of double hairpin structures from which amplification can cycle. Further primers can be added targeting the single stranded DNA hairpin loop structures [4] or the stem regions between loops [5] to accelerate the DNA amplification. LAMP amplification has been used for many pathogen detection strategies due to rapid reaction times, high sensitivity, robustness to contaminants and specificity.

The specificity is inherent from the use of multiple primers but this can also mean that non-specific primer interaction is an issue. The false positives that these

*Correspondence: hardingep@cardiff.ac.uk
Cardiff School of Biosciences, Cardiff, Museum Avenue, CF10 3AX Cardiff, UK

interactions can produce are primarily a result of LAMP primer design and the reaction conditions, but also the non-specific nature of some of the detection methods employed. Methods for detection of LAMP amplification have mainly focused on determining the increasing concentrations of DNA or inorganic pyrophosphate for both real time and end point assays and will inherently detect LAMP and potentially also non-specific DNA synthesis. Such methods include SYBR Green [6], BART bioluminescent reporter [7], turbidimetry [8] and specific dyes [9].

Recent work investigating the use of molecular beacons with LAMP [10] suggested that the problem of non-specific amplification could be solved for real-time assays through their use. An endpoint low melt temperature molecular beacon with LAMP assay has also been shown [11]. Molecular beacons for use in PCR were first described in 1996 [12] in which the beacon has complementary arms 5 nucleotide long and a 15 nucleotide long probe loop. A fluorophore attached to the 5' end is quenched due to fluorescence resonance energy transfer (FRET) by the proximity of a quencher at the 3' end. In the presence of specific target DNA the probe anneals and undergoes conformational change and fluorescence. No change is observed in the absence of specific target DNA and the fluorophore would remain quenched.

The thermodynamics and kinetics of molecular beacons for PCR have been studied [13, 14] and recommendations for arm length, probe length, melt temperature and GC content for different applications [15] have been provided. In thermocycling, the PCR primers and molecular beacon bind at the annealing temperature to the target, and before the end of this stage the fluorescence is generated and acquired. Increasing the temperature to the extension temperature liberates the molecular beacon allowing PCR cycling to continue. The denaturing stage renders the DNA single stranded to continue the process, and the increase in fluorescence with each cycle is observed in the presence of increasing amounts of specific template DNA.

In LAMP amplification this process is different due to two factors; namely the single temperature throughout the assay and the displacement polymerase. The recent study [10] with MB-LAMP showed that fluorescence increased in real time together with LAMP template amplification. This was achieved with molecular beacon lengths of 25-45 base pairs, concentrations of 0.6-1.0 picomoles/microlitre and reaction temperatures of 60-65 degrees C.

The mechanism by which increased specificity is conferred from the annealing of the LAMP molecular beacon to the specific target sequence is however unclear. Since assays involve a constant temperature and use a displacement polymerase that would likely displace the molecular beacon. We show here that molecular beacons indeed detect LAMP amplification but this does not require specific annealing to the specific target sequence. However the increase in fluorescence does appear to be linked to increasing concentration of LAMP amplicons and therefore the specificity is greater than methods that report on DNA amplification only since a positive signal is indicative of successful LAMP amplification. The problem of posssible non-specific amplification remains due to the failure of the LAMP molecular beacons to anneal specifically to the target sequence. Nevertheless, the low fluorescence output from the LAMP molecular beacon method enables the quantification of a wider dynamic range of target concentrations to be assayed in comparison to detection with an intercalating fluorescent dye.

The detection and quantification of genetically modified materials in food is essential to comply with legislation and two of the most common markers of GMO contamination are the cauliflower mosaic virus 35S promoter and nopaline synthase terminator from *Agrobacterium tumefaciens*, used to express the transgenes. Here we target these transgenic elements in maize event Bt11 in a mass fraction with wild type maize. We have designed molecular beacons for the 35S promoter and NOS terminator sequences to complement LAMP primers previously established for amplification [16–18].

Our work initially focused on probing the forward loop structure of LAMP amplicons, but this precluded the use of accelerating loop primers [4]. We therefore concentrated on the stem region between loops to enable the forward loop primer to be included for faster amplification. This work revealed that whilst molecular beacons detect LAMP amplification products, they do so irrespective of amplified sequence. They can therefore be used to detect LAMP amplicon, but not the nature of the amplified sequence.

Methods

DNA templates

The plasmid pART7 [19] containing a cauliflower mosaic virus 35S promoter (35Sp) sequence variant was linearised and initially quantified by NanoDrop spectrophotometry and Agilent Bioanalyzer for a defined target concentration. A further plasmid containing a retroviral LTR sequence from Moloney Murine Leukaemia Virus (accession number: J02258.1) was targetted by LTR LAMP primers. The sequences of the plasmids for the LAMP primers are displayed in Fig. 1. Transgenic GM maize event Bt11 certified reference material (CRM) from the Institute for Reference Materials (Geel, Belgium) was purchased from Fluka GMBH (Buchs, Switzerland) with a defined mass fraction of 5.0 percent w/w with wild type maize. Bt11 contains the 35S promoter sequence (35Sp) from the cauliflower mosaic virus and the nopaline synthase terminator (NOSt) from *Agrobacterium*

Fig. 1 Position of LAMP primers for 35Sp **a** and LTR **b** with plasmid template DNA sequences. Displacement primers denoted F3 and B3 are labelled in green with the loop primers LoopF and LoopB labelled yellow. The two components of the LAMP primers FIP and BIP are labelled red. The grey highlights for the pART7 sequence show the mismatches of this variant of the 35S promoter sequence

tumefaciens. The genomic DNA was extracted using the Promega Wizard genomic DNA purification kit (Madison, United States) and quantified using a Qubit fluorimeter (Life Technologies, UK).

Two further DNA templates were synthesised by IDT (Coralville, Iowa, USA) firstly to create a simplified target for the 35S promoter LAMP primers using a sequence length and structure appropriate to the FIP, BIP, LoopF and LoopB primers only. The second oligonucleotide was designed to match the length of the first primer but without loop forming sequences. The sequences are shown in Table 1.

Table 1 Sequence information for artificial 35Sp LAMP templates

Target, Notation, Version	Template Sequence (5′ to 3′)
35Sp BIP Dumbbell, LOOPS, PH	CACGTCTTCAAAGCAAGTGGGGATAGTGGGAT
	TGTGCGTCCCCTTACGTCAGTGGACCACTTGC
	TTTGAAGACGTGGTCTAGATTCCACGATGCTC
	CTCGGGGGTCCATCTTTGGGCCACTGTCGGCA
	GAGGCGAGGAGCATCGTGGAA
35Sp partial sequence, LINEAR, PH	TTGCGATAAAGGAAAGGCCATCGTTGAAGATG
	CCTCTGCCGACAGTGGTCCCAAAGATGGACCC
	CCACCCACGAGGAGCATCGTGGAAAAAGAAGA
	CGTTCCAACCACGTCTTCAAAGCAAGTGGATT
	GATGTGATATCTCCACTGACGT

35Sp templates synthesised and PAGE purified by IDT (Coralville, Iowa, USA). Both oligonucleotides are 150 bases in length, diluted to 10 micromolar and stored at minus 20 degrees C

LAMP amplification

Loop-mediated amplification uses primers with different functions to enable isothermal DNA amplification. The initiation of the process requires the target strand invasion by the forward and backward inner primers (FIP and BIP) with synthesis of the nascent DNA from these primers utilising the displacement activity of Bst polymerase. The initiation event also requires the displacement primers F3 and B3 to displace the products of FIP and BIP extension which can in turn start the cycling between dumbbell LAMP structures (a dynamic structure with single stranded DNA loops at each end). Additional forward and backward loop primers, LoopF and LoopB, anneal and extend from the target loop structures to increase the synthesis of DNA and inorganic pyrophosphate for accelerated detection strategies. This role can also be met by the incorporation of STEM primers that bind between the F1 and B1 primer sequences. The position of the different sets of LAMP primers is shown for the two plasmid targets in Fig. 1.

Primer design and synthesis

Oligonucleotide primers for LAMP nucleic acid amplification were synthesized and HPLC purified by Sigma Aldrich (Poole, UK). Primers were hydrated with molecular grade water to 100 micromolar and stored at minus 20 degrees C. LAMP primers for Moloney Murine Leukaemia Virus LTR were designed with the assistance of PrimerExplorer v4 from Eiken. The primer sequences are shown in Table 2.

LAMP molecular beacons

The design of the 35Sp 5nt LoopF MB for LAMP followed in part the guidelines for molecular beacon design for PCR amplification; arm length 5 to 7 nucleotides, probe sequence 15 to 30 nucleotides, Tm above the annealing temperature of the assay, 5' fluorophore not attached to a guanine nucleotide with a 3' non-fluorescent quencher. The length of the probe sequence was restricted by the single stranded elements of the LAMP loop or STEM. The 35Sp LoopF 5nt MB is similar in structure to the first described molecular beacon by Tyagi and Kramer in 1996 [12], with 5nt arms and 15nt probe sequence. The primer sequences are shown in Table 3 with associated melt temperatures (Tm), GC content (percentage) and nucleotide length. The LAMP molecular beacons were synthesized and HPLC purified by Sigma Aldrich (Poole, UK).

Table 2 LAMP primers for 35S promoter, NOS terminator and LTR sequences

Target, Type, Notation, Version	Primer Sequence (5' to 3')
35Sp, Displacement, F3, GK	CTTATATAAGAGGAAGGGTCT
35Sp, Displacement, B3, GK	GATAAAGGAAAGGCTATCATT
35Sp, LAMP, FIP, GK	CCACGTCGGCAAAGCAAGTGG-TTTT-GGATAGTGGGATTGTGCGTC
35Sp, LAMP, BIP, DL	TTCCACGATGCTCCTCG-TTTT-CCTCTGCCGACAGTGG
35Sp, Loop, LoopF, DL	TCCACTGACGTAAGGG
35Sp, Loop, LoopB, DL	GGGGTCCATCTTTGGG
NOSt, Displacement, DL	F3,CGCGATAATTTATCCTAGTTTG
NOSt, Displacement, DL	B3,CGTTCAAACATTTGGCAAT
NOSt, LAMP, FIP, PH	GCATGACGTTATTTATGAGA-TTTT-TCGCGCTATATTTTGTTTTCTA
NOSt, LAMP, BIP, PH	CATGCTTAACGTAATTCAACA-TTTT-TGAATCCTGTTGCCGGTC
NOSt, Loop, LoopF, DL	GATTAGAGTCCCGCAATTATAC
NOSt, Loop, LoopB, DL	AAATTATATGATAATCATCGCAA
LTR, Displacement, F3, PH	ACAGATGGTCCCCAGATGC
LTR, Displacement, B3, PH	TCAATCGGAGGACTGGCG
LTR, LAMP, FIP, PH	ATTTCAGGTCCTTGGGGCACCCCAGCCCTCAGCAGTTTC
LTR, LAMP, BIP, PH	ACCAATCAGTTCGCTTCTCGCTGAGGGGTTGTGGGCTCTT
LTC, Loop, LoopF, PH	CTGATGGTTCTCTAGA
LTC, Loop, LoopB, PH	TTCGCGCGCTTCTGCT

Primers targeting the 35S promoter and NOS terminator have been previously described and are attributable to David Lee (DL) [16] and Guy Kiddle (GK) [17]. The Moloney Murine Leukaemia Virus LTR LAMP primers are described for the first time

Fluorescent LAMP amplification

DNA samples were amplified and detected in real time on a Qiagen (Hilden, Germany) Rotor-Gene thermal cycler acquiring to the green channel. Unless otherwise stated, all reagents are supplied by Sigma Aldrich (Poole, UK). The reaction chemistry for LAMP and amplification detection was 1X isothermal buffer (NEB, Massachusetts, United States), 300 micromolar each dNTP, 0.8 micromolar Betaine, 0.5 micromolar SYTO9 Green, 0.32 units per microlitre Bst polymerase v2.0 warm start (NEB), 0.8 micromolar each LAMP primer, 0.4 micromolar each Loop primer, 0.2 micromolar each displacement primer and molecular grade water for a reaction volume of 20 microlitre. The parameters were set for 60 cycles of 60 seconds at 60 degrees C unless otherwise stated. Melt temperature analysis provided data between 60 and 95 degrees C. LAMP amplification detection with SYTO9 was substituted with LAMP molecular beacons and supplemented with molecular grade water if required.

PCR amplification

DNA samples were amplified using PCR on a Qiagen Rotor-Gene thermal cycler with detection of SYTO9 on the green channel and detection of the 35Sp STEM 5nt

Table 3 LAMP molecular beacons designed for 35S promoter and NOS terminator sequences

Target, Type, Vers., Tm, GC, length	Primer Sequence (5′ to 3′)
ompW, LBP, WL, 69.8, 69.2, 25	[6FAM]AGCGGCTGCAGCCCTACTAGCCGCT[Dabcyl]
35Sp LoopF, 5nt FAM, 76.8, 66.6, 24	[6FAM]CCGAGCCACTGACGTAAGGCTCGG[OQA]
35Sp stem, 4nt FAM, 63.1, 61.1, 18	[6FAM]CCACTGGAACGTCTGTGG[BHQ1]
35Sp stem, 5nt FAM, 72.7, 65.0, 20	[6FAM]CGCACTGGAACGTCTGTGCG[BHQ1]
35Sp stem, 5nt JOE, 72.7, 65.0, 20	[JOE]CGCACTGGAACGTCTGTGCG[BHQ1]
35Sp stem, 6nt FAM, 77.9, 68.1, 22	[6FAM]CGGCACTGGAACGTCTGTGCCG[BHQ1]
35Sp stem, 7nt FAM, 82.0, 70.8, 24	[6FAM]CCGGCACTGGAACGTCTGTGCCGG[BHQ1]
NOSt stem, 5nt FAM, 66.0, 43.4, 23	[6FAM]CGCACTTACATGTTAATTGTGCG[BHQ1]
NOSt stem, 5nt FAM, 66.0, 43.4, 23	[JOE]CGCACTTACATGTTAATTGTGCG[BHQ1]

The fluorophores used were 6-carboxyfluoroscein (6FAM) and 4′, 5′-dichloro-2′, 7′-dimethoxy-5(6)-carboxyfluorescein (JOE). The quenchers used were 4-(dimethylaminoazo)benzene-4-carboxylic acid (Dabcyl), Onyx Quencher A (OQA) and Black Hole Quencher 1 (BHQ1). Molecular beacon version WL was designed by Liu et al. 2017 [10] to detect part of the *ompW* gene from *Vibrio cholerae*

JOE MB on the yellow channel. The primers shown in Table 4 where used to flank the JOE labelled MB without overlapping target sequences. The reaction chemistry was 1X Qiagen Taq PCR mastermix, 0.5 micromolar SYTO9 Green, 1.0 micromolar 35Sp 5nt JOE MB and 2.0 micromolar each PCR primer with molecular grade water for a reaction volume of 10 microlitre. The thermocycling used the following conditions; initial denaturation for 3 min at 94 degrees C, followed by 40 cycles of 30 seconds at 94 degrees C, 30 seconds at 55 degrees C and 60 seconds at 72 degrees C. Melt curve analysis acquired stepwise between 30 and 95 degrees C. Fluorescence acquisition for the JOE labelled MB was on the yellow channel at the end of the annealing step. SYTO9 detection used the green channel with acquisition at the end of the extension step.

Data analysis

Sequences were aligned using CLC sequence viewer 7 and graphically displayed in pdf format. The Basic Local Alignment Search Tool (BLAST) from NCBI was used for sequence specificity and LAMP primers and LAMP molecular beacons were designed with the assistance of Eiken Primer Explorer v4 and Primer3 [22] respectively. Rotor-Gene thermocycling results were analysed using Rotor-Gene 6000 software v1.7, converted with TeeChart Office and graphically displayed using GraphPad Prism 7. Statistical analysis used Microsoft Excel to calculate the average Ct and standardard deviation results and GraphPad Prism 7 for the coefficient of determination (R^2) values.

Table 4 PCR primers for 35S promoter amplification. The M35F primer is derived from Wu et al. 2014 [20] and the M3R is from Fernandez et al. 2005 [21]. Both primers were synthesised by Sigma with HPLC purification

Target, Type, Notation	Primer Sequence (5′ to 3′)
35Sp, Forward, M25F	CATCATTGCGATAAAGGAAAGGC
35Sp, Reverse, M3R	TCTTGCGAAGGATAGTGGGATT

Results

Specificity of LAMP molecular beacon targeting the forward loop sequence

We designed a LAMP molecular beacon targeted to the forward loop sequence of a variant of the 35S promoter from Cauliflower mosaic virus (Fig. 1), with a probe sequence of 14 nucleotides and complementary arms of 5 nucleotides. Using this, we noted an increase in fluorescence with a 125 nanomolar concentration of the MB for the positive sample in relation to the negative sample (Fig. 2; 1A). The use of this primer in a 63 degrees C 35Sp LAMP assay at a concentration of 320 nanomolar with the omission of the LoopF primer from the reaction mix resulted in a clear difference between the positive and negative samples (Fig. 2: 2A). A further experiment with the inclusion of the LoopF primer, which targets the same sequence as the LAMP molecular beacon, produced a clearer separation between the positive and negative samples. Moreover the average Ct values improved by almost 4 min, presumably due to the increased amplification from the LoopF primer despite possible interference from the MB binding the same sequence. The dynamic environment of LAMP amplification which involves strand displacement by Bst polymerase, makes it unlikely that the molecular beacon remains annealed to the Loop sequence, given the positive effect of the LoopF primer on LAMP kinetics. Melt curve analysis revealed that the positive samples have a lower melt temperature than the negative samples; the difference is about 2 degrees C for the assay without the LoopF primer (Fig. 2; 2B) and as an average 3 degrees C for the assay with the LoopF primer (Fig. 2; 3B).

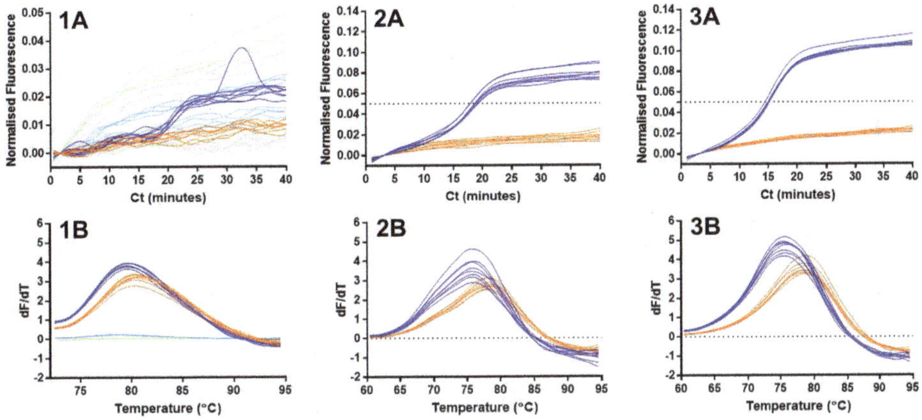

Fig. 2 LAMP amplification with 35Sp LoopF molecular beacon. LAMP primers designed to amplify part of the 35S promoter sequence detected by fluorescence from a LAMP molecular beacon targeting the LoopF sequence. 1A shows the fluorescence for dilutions of the 35Sp LAMP MB; 125 nanomolar dark blue, 12.5 nanomolar light blue, 1.25 nanomolar light green, 125 picomolar pink and NTC in orange. 2A denotes the fluorescent output against cycles of 1 min at 63 degrees C for 35Sp LAMP amplification excluding the LoopF oligonucleotide (positive samples with template in blue and NTC in orange). 3A charts the fluorescent output against cycles of 1 min at 63 degrees C for 35Sp LAMP amplification with the inclusion of the LoopF oligonucleotide. LAMP molecular beacon used at a concentration of 0.32 micromolar for these two assays. Melt curve analysis shown in 1B to 3B with peaks inverted

LAMP molecular beacons targeting the STEM region and optimal reaction temperature

To allow for rapid amplification from the unencumbered use of Loop primers, we therefore designed LAMP molecular beacons to target the region of the LAMP amplicon called the STEM region, lying between the hairpin loops. The probe length of these molecular beacons remained fixed at 10 nucleotides but the length of the self-complementary arms was increased from 4 to 7 nucleotides incrementally. Positive and negative samples were amplified using LAMP reactions containing each of these molecular beacons over a temperature range of 55 to 65 degrees C. All negative template controls showed negative results below the 0.05 normalised fluorescence threshold level set for the assays. The negative results followed a similar line with the exception of assays at 55 degrees C in which the fluorescence from the molecular beacons was reduced (Fig. 3).

The LAMP molecular beacon results from the positive samples showed stronger signals from the 35Sp 5nt STEM MB (Fig. 3; A) and 35Sp 6nt STEM MB (Additional file 1: Figure S1; 2A) than for the 35Sp 4nt STEM MB (Additional file 1: Figure S1; 1A) and 35Sp 7nt STEM MB (Figure S1;3A). The optimum temperature giving the fastest and most differentiated result compared to the negative samples was 64 degrees C. The dynamic equilibrium between quenched fluorescence and unquenched fluorescence due to the interaction between the arms of the molecular beacons is expected to cause increased baseline fluorescence at higher temperature. This appears to be less evident as the length of the arms increases therefore increasing the stability of the quenched form. The results for the positive samples indicate the molecular

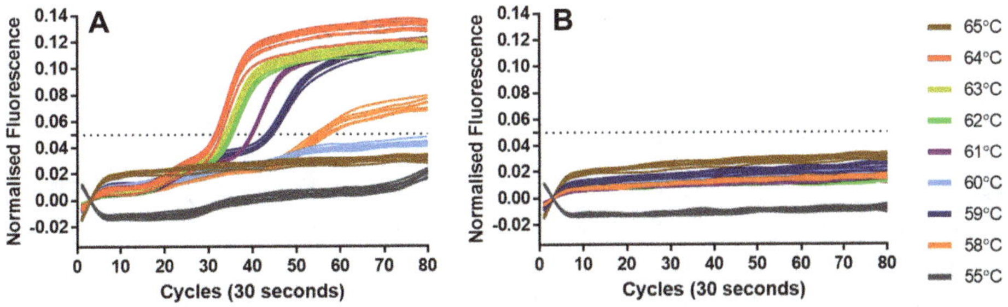

Fig. 3 Amplification curves of 35Sp 5nt STEM molecular beacon at different temperatures. LAMP assay temperature incrementally from 55 degrees C to 65 degrees C for positive template and NTC, (**a**) shows the increase in fluorescence of the FAM fluorophore attached to the 35Sp 5nt STEM molecular beacon at various temperatures and (**b**) shows the corresponding NTCs for each assay. The positive and negative samples at various temperatures for the 35Sp 4nt STEM MB, 35Sp 6nt STEM MB and for 35Sp 7nt STEM MB are shown in Additional file 1: Figure S1

beacon with 5 nucleotide arms gives the strongest fluorescent signal relative to background and was selected for further optimisation.

Limitations of molecular beacons with LAMP

It is assumed that the LAMP molecular beacons add increased specificity to the LAMP assay by exhibiting fluorescence when associated with a particular amplified target sequence. In a series of experiments displayed in Fig. 4, this assumption was investigated.

35S promoter LAMP primers were used to amplify a standard concentration of template under identical conditions with the exception of SYTO9 (Fig. 4; 1A), 35Sp 5nt STEM MB (Fig. 10; 1B) and NOSt 5nt STEM MB (Fig. 4; 1C) incorporated for amplification detection. As expected the LAMP amplification was detected by the SYTO9 dye and by the LAMP molecular beacon designed for the 35Sp LAMP STEM sequence. However, unexpectedly the 35Sp LAMP amplification was also detected by the LAMP molecular beacon designed to a NOS terminator LAMP STEM sequence not present in the target. The probe sequence of this primer is 5'-TTACATGTTAATT-3' compared to 5'-TGGAACGTCT-3' for the probe sequence of the 35Sp LAMP STEM molecular beacons. Further investigation into this lack of specificity used LAMP amplification of a retroviral LTR sequence to provide an alternative LAMP template for the various LAMP MBs. Detection of the LTR LAMP amplification was detected by SYTO9 (Fig. 10; 2A), by the 35Sp 5nt LoopF MB (Fig. 4; 2B), by the NOSt 5nt STEM MB (Fig. 4; 2C) and by 35Sp STEM molecular beacons 4nt, 5nt and 6nt (Fig. 4); 2D to 2F respectively). We conclude the detection of LAMP amplicons by this type of molecular beacons is not sequence dependent.

Specificity of LAMP molecular beacon targeting the backward loop sequence

Liu et al. 2017 [10] described LAMP molecular beacons targeting the forward loop sequence of the *ompW* gene from *Vibrio chlorerae*. Their optimised molecular beacon had a 13 nucleotide probe sequence with two complementary arms of 6 nucleotides terminated with a 5' FAM fluorophore and a 3' DABCYLl quencher. The optimal temperature was 63 degrees C at a concentration of 320 nanomolar. We used this molecular beacon with the 35Sp LAMP amplification of a 35Sp artificial template to investigate the specificity of this method (Fig. 5).

The high concentration of the 35Sp artificial template was rapidly amplified using 35Sp LAMP primers with detection of fluorescence from the LBP LAMP molecular beacon. The fluorescence from the non template control remained sub-threshold and there was clear separation

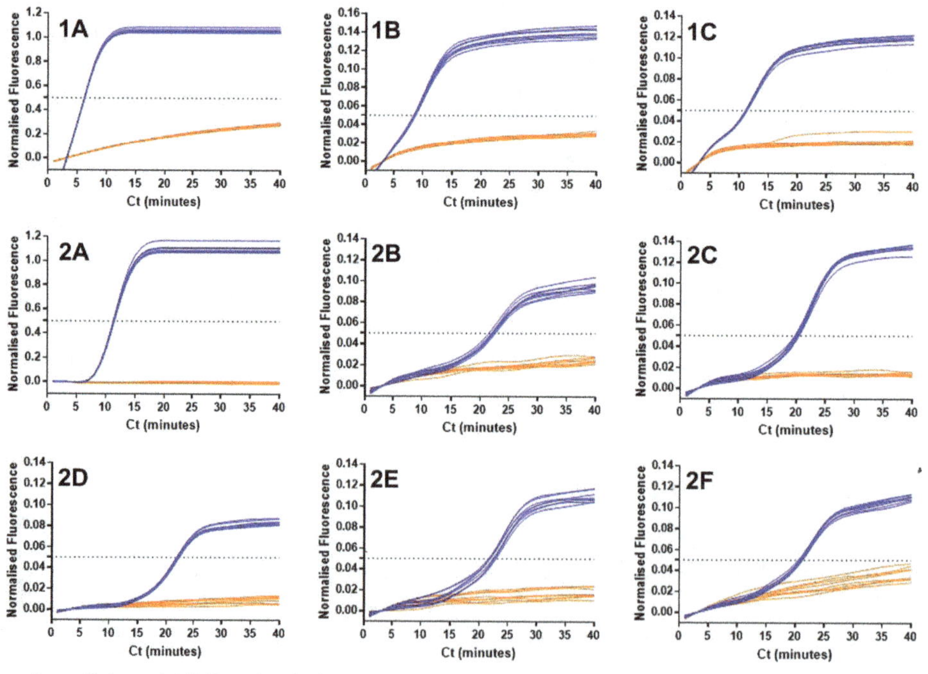

Fig. 4 Lack of enhanced specificity with LAMP molecular beacons. 1A denotes 35S promoter LAMP amplification of an optimised 35Sp template with amplification detection using SYTO9 dye (positive template control in blue and NTC in orange), 1B shows detection of 35Sp LAMP amplification with 35Sp 5nt MB and 1C shows detection of 35Sp LAMP amplification with NOSt 5t MB. 2A denotes LTR LAMP amplification of a plasmid using SYTO9 dye, 2B shows detection of the LTR LAMP amplification with 35Sp LoopF MB, 2C shows LTR LAMP with NOSt 5nt STEM molecular beacon detection, 2D shows LTR LAMP with 35Sp 4nt STEM MB, 2E with 35Sp 5nt STEM MB and 2F with 35Sp 6nt STEM MB

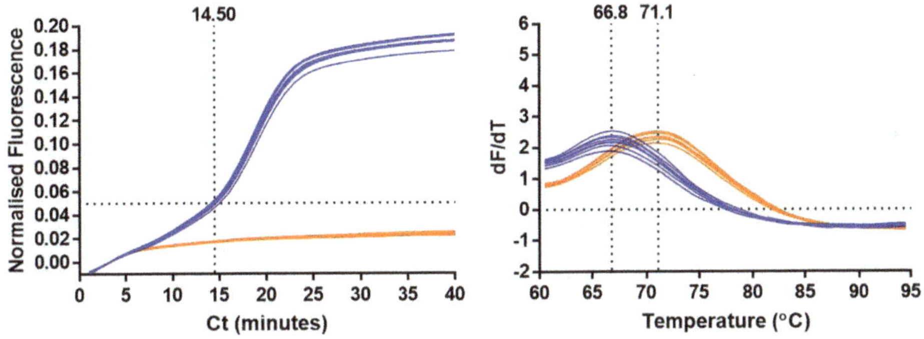

Fig. 5 Lack of enhanced specificity with LBP LAMP molecular beacon. LAMP amplification of 35Sp template in blue and non-template control in orange with detection of amplification with a FAM fluorophore labelled molecular beacon designed for the ompW gene of *Vibrio chlorerae* [10]. Fluorescent output against cycles of 60 seconds at 60 degrees C and melt curve analysis post amplification

between the positive and negative samples after less than 10 min. The melt curve fluorescence showed a lower temperature for the molecular beacon with LAMP amplification of 66.8 degrees C compared to the molecular beacon without LAMP amplification of 71.1 degrees C.

LAMP molecular beacons with looped or linear DNA templates

Two oligonucleotide templates and ten-fold dilutions were combined separately with 0.8M betaine, 1x isothermal buffer (NEB) and the LBP LAMP molecular beacon. The fluorescence from the FAM fluorophore was acquired after 60 seconds for a total of 5 min at 60 degrees C. The oligonucleotide designed to form loops, based on one of the 35Sp dumbbell structures, showed an increase in fluorescence when compared to the NTCs with undiluted and diluted template (Fig. 6 1A, 1B). The melt curve analysis for these assays showed a reduction in melt temperature of template compared to NTC. The increase in real-time fluorescence and decrease in melt temperature were reduced by template dilution.

The other oligonucleotide template designed as a partial sequence of the 35S promoter without loop forming sequences showed a decrease in fluorescence for both the undiluted and diluted template compared to the NTCs during real-time acquisition, and an increase in the melt temperature comparison (Fig. 6 2A, 2B). The increased fluorescence from the LAMP molecular beacon in the presence of the 'dumbbell' template and the reduced melt temperature match the results seen in Fig. 5 with increasing concentrations of loop DNA structures during LAMP amplification. This effect was not observed with the linear DNA template.

Optimal concentration for the 35Sp 5nt STEM molecular beacon at 64 degrees C

Standard LAMP reactions at 64 degrees C using controlled positive and negative samples were used to optimise the concentration of the 35Sp 5nt STEM MB at 0.5, 1.0, 2.0 or 4.0 micromolar. The results indicated faster results for the 0.5 and 1.0 micromolar concentrations (Fig. 7) with no signal from the negative template controls.

To further investigate the optimum concentration for this LAMP molecular beacon dilution range, a serial dilution of the 35Sp plasmid template was prepared from 10^{-1} to 10^{-5} from stock concentration of 10^{6} copies/microlitre. All four LAMP molecular beacon concentrations previously tested (0.5 micromolar and 1.0 micromolar are shown in Fig. 8 and 2.0 micromolar and 4.0 micromolar are shown in the Additional file 1: Figure S2) were used for 35Sp LAMP assays and compared to identical assays with the molecular beacon replaced with the intercalating dye SYTO9 to report amplification. The results showed a reduced amplification frequency for the most dilute sample using the lowest concentration of 35Sp 5nt STEM molecular beacon and this concentration also had poor separation between the results for the higher concentrations of template.

Low variation between replicates was observed for the 10^{-1}, 10^{-2} and 10^{-3} dilutions with LAMP amplification detected by SYTO9 fluorescence and all the dilutions of the LAMP molecular beacon. The increase in signal above baseline was observable for SYTO9 assays and MB assays 0.5 and 1.0 micromolar MB, but was delayed with higher MB concentrations of 2.0 and 4.0 micromolar, for all target DNA concentrations.

Quantification using 35Sp 5nt MB

The 35S promoter artificial LAMP template was serially diluted and the range 10^{-3} to 10^{-10} of the 10^{12} copies/microlitre stock, was investigated for the potential of the 35Sp LAMP molecular beacon method for quantification in comparison to the SYTO9 detection method. The results are shown in Fig. 9.

The LAMP molecular beacon results show clear NTCs throughout the course of the assay and positive results

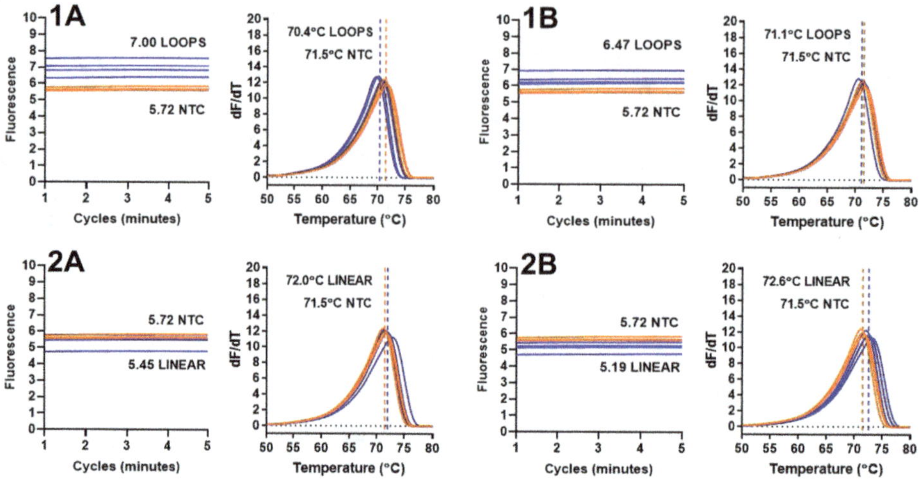

Fig. 6 Fluorescent signal from MB-LAMP with looped or linear DNA.35Sp oligonucleotide designed for 'dumbbell' looped structure undiluted (1A) and ten-fold dilution (1B) in blue, NTC in orange. Fluorescence from LBP molecular beacon in real-time at 60 degrees C for 5 min and melt curve analysis. Average fluorescence and melt temperature derived from replicates. 35Sp oligonucleotide of the same length without loops undiluted (2A) and ten-fold diluted (2B) in blue with NTC in orange. Fluorescence acquisition from LBP molecular beacon at 60 degrees C for 5 min and melt curve analysis

distinguishable from the fluorescence baseline (Fig. 9; 1A). For each dilution there appears to be low variation between replicates and dilutions are clearly separate by their timing of signal increase. This is supported in the graph of average Ct against template dilution graph (Fig. 9; 1C) with the standard deviation between replicates shown as error bars and correlation of the data points to the semi-logarithmic trend line approaching an R^2 value of 0.99. The melt curve analysis (Fig. 9; 1B) showed a range of average melt temperatures from 72.0 degrees C for the NTC and higher concentrations of template, down to 70.5 degrees C for the lowest concentration. LAMP amplification with detection using SYTO9 fluorescence of the same template dilution showed low variation between replicates and good separation between template dilutions. The wide ranging intensities of the fluorescence between the NTC and the most concentrated template dilution were problematic for the instrumentation with normalised fluorescence values ranging from approximately 0.5 for the 10^{-3} dilution to greater than 1.0 for the 10^{-6} to 10^{-10}

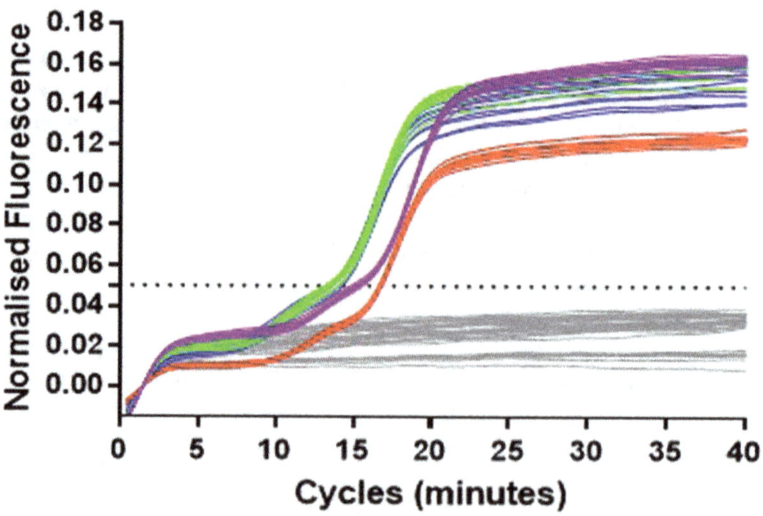

Fig. 7 Optimal concentration of the 35Sp 5nt STEM molecular beacon. Standard LAMP assay at 64 degrees C with linearised plasmid template and negative template samples for each molecular beacon concentration. Molecular beacon concentrations: 4.0 micromolar purple, 2.0 micromolar red, 1.0 micromolar green, 0.5 micromolar blue and NTC in grey

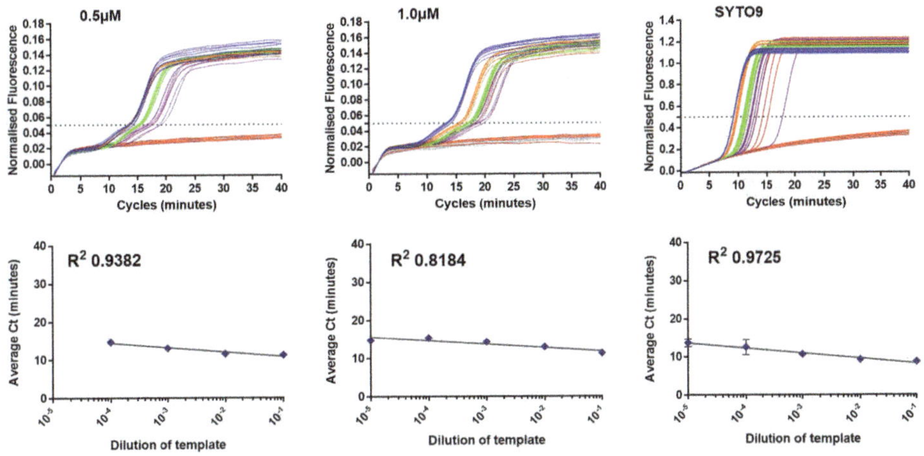

Fig. 8 Variation and separation between replicates in quantification. Fluorescent output from 35Sp 5nt STEM MB or SYTO9 with 35Sp LAMP amplification of a serial dilution of plasmid template. Template dilutions: 10^{-1} dark blue, 10^{-2} orange, 10^{-3} light green, 10^{-4} purple, 10^{-5} red and NTC in grey. For each concentration of the molecular beacon the Ct value is derived from a set threshold of 0.05 (for the SYTO9 assay this value is 0.5). See Additional file 1: Figure S2 for 2.0 and 4.0 micromolar MB assays

dilutions. With the threshold set at 0.4 the average Ct values were plotted against the template dilutions (Fig. 6; 2C) and the highest concentration was excluded from further analysis due to the extremely high fluorescence producing an anomalous result. Lower variation was observed between replicates and lower average Ct values across the range than the LAMP molecular beacon results but the correlation of the data points to the semi-logarithmic trend line is lower (calculation excludes the data from the 10^{-3} dilution). We conclude that MBs can provide wide dynamic range quantification of LAMP amplification.

Sensitivity

The quantified linear plasmid stock containing the 35S promoter was diluted to the equivalent of a single DNA copy per partition, and assayed by LAMP amplification using fluorescence detection with either 1 micromolar LAMP molecular beacon or SYTO9. The four LAMP molecular beacons designed to the STEM region were used to compare the sensitivity to the more established technique of intercalating fluorescent dye detection (Table 5).

The amplification frequency or percentage of positives from the number of replicates, was lowest for the 35Sp 7nt STEM MB at 18.3 percent, but the other LAMP MBs and the SYTO9 results were in a narrow range between 25.8 percent and 32.4 percent (Table 5). This is below the statistically predicted 63.2 percent amplification frequency for single copies per partition (0.3 to 0.4 copies per partition is indicated by these percentages). The results show that

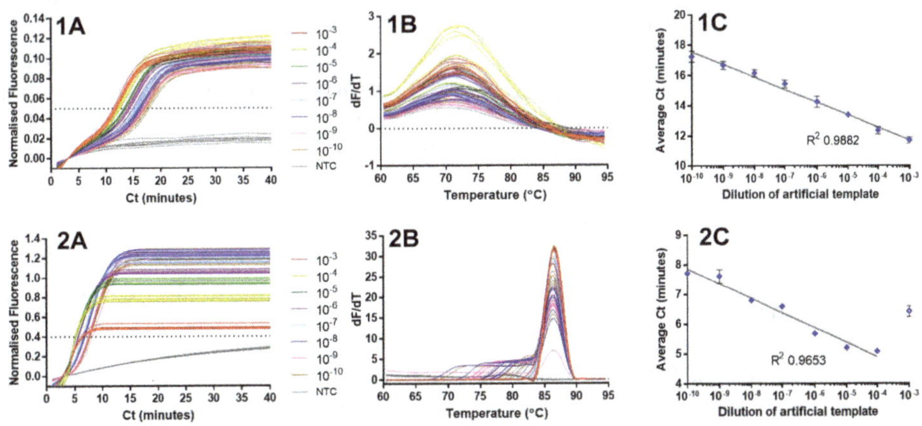

Fig. 9 LAMP quantification with molecular beacon and SYTO9. The 35S promoter STEM 5nt molecular beacon was used with a serial dilution of an optimised artificial template in (1) and the same template dilutions were used with SYTO9 detection in (2). Normalised fluorescence for the dilutions series and NTC (A), Melt curve analysis for the dilution series and NTC (B) and average Ct against template dilution for the dilution series (C). The threshold was set at 0.05 for the LAMP MB and 0.4 for the SYTO9 detection. The peaks in 1B for the LAMP molecular beacon melt curve were inverted

Table 5 35S promoter STEM molecular beacons and SYTO9 detection of low copy number assay

Detection Method	Amp. Freq.	Average Ct.	St. dev.	Fastest Ct.	Slowest Ct.
4nt STEM MB	25.80	31.06	12.74	21.91	83.33
5nt STEM MB	26.80	32.90	12.74	24.10	80.06
6nt STEM MB	32.40	35.26	10.65	25.02	65.16
7nt STEM MB	18.30	46.29	13.17	34.29	75.74
SYTO9	28.20	20.59	15.61	9.83	64.25

Diluted linear plasmid DNA containing the 35S promoter sequence assayed by 35Sp STEM molecular beacons of various lengths in comparison to the detection with SYTO9 fluorescence. LAMP assay at 64 degrees C with 1.0 micromolar LAMP molecular beacons and 72 replicates of 1 copy per partition. SYTO9 LAMP assay at 60 degrees C. The cycle parameters were set for 90 cycles of 60 seconds each cycle acquiring for FAM/SYTO9

the LAMP MB and SYTO9 methods have the equivalent sensitivity to low copy number LAMP amplification. The average Ct and the fastest Ct both show that the 35Sp 7nt STEM MB produces the slowest results and SYTO9 is the faster method of detection. The standard deviation from the replicates for each assay were high within the range of 10.65 to 15.61; the highest value associated with the SYTO9 results. The slowest Ct values for each assay were all above 60 min highlighting the range of times that are observed with LAMP amplification at low copy number.

LAMP molecular beacons end point assay

Molecular beacons have been used with LAMP to distinguish from false positives in an end-point assay (Liang Wan et al. 2017) [11]. We investigated whether the molecular beacons described here could be used for end-point LAMP amplification (Fig. 10).

The clearest separation between positive template and negative NTCs at 75 degrees C was for the 35Sp LAMP amplification with detection using the 35Sp LoopF MB (Fig. 10: 1B). Post amplification analysis of these samples showed that the fluorescence profile of the molecular beacon with template was different from that of the molecular beacon without template. Melt curve analysis showed that the MB without template had an average melt temperature of 76.7 degrees C close to the predicted Tm of 76.8 degrees C and yet the Tm for the MB with LAMP amplification of template was slightly reduced to 75.4 degrees C (Fig. 10: 1D). Lower average melt temperatures were also observed with the 35Sp STEM 5nt MB (predicted Tm of 72.7 degrees C) of 69.9 degrees C compared to 70.4 degrees C, and the NOSt STEM MB (predicted Tm of 66.0 degrees C) of 65.0 degrees C compared to 67.4 degrees C.

PCR with LAMP molecular beacon; real time and end point

The typical structure of a PCR molecular beacon includes 5-7 nucleotide GC rich arms with a 22-30 nucleotide probe sequence [23] but the LAMP molecular beacons already described [10] and herein require shorter probe sequences of 10-14 nucleotides. PCR reactions were set up to include SYTO9 to report DNA amplification and a LAMP molecular beacon. Further PCR assays with SYTO9 but without a LAMP MB were run to produce an amplicon for end point detection through LAMP MB addition. The assays were repeated with LAMP amplification replacing PCR amplification (Fig. 11).

The results showed that the real time increase in fluorescence with the LAMP MB was not evident with PCR amplification. With LAMP amplification of the same sample (Fig. 11: 1 and 2) the SYTO9 detection results with LAMP MB (Fig. 11: 1A) and without LAMP MB (Fig. 11: 1B) showed similar fluorescent profiles indicating that the inclusion of the LAMP MB in the reaction mix did not significantly interfere with the LAMP amplification or detection. In contrast the inclusion of the LAMP MB appeared to interfere with PCR amplification; Fig. 11: 4A shows PCR amplification in the absence of the LAMP MB with a clear separation between positive and negative results, however this separation is not seen in Fig. 11: 3A with the inclusion of the LAMP MB. The detection of the JOE labelled MB was only evident with LAMP amplification in Fig. 11: 1B and not with PCR (Fig. 11: 3B). Melt curve detection of the LAMP MB was observed with both LAMP and PCR amplification in Fig. 11: 1C and 3C respectively, and with the addition of the LAMP MB post amplification in Fig. 11: 2D and 4D. We had noted in all LAMP experiments involving a MB the melt temperatures were lower in the presence of amplicon than for the MB alone in NTCs. This was unexpected since the MB has 10 nucleotides of homology to the target and self complementary arms of only 5 nucleotides. Moreover we did not observe this with PCR amplified samples (Fig. 11; 4D).

Discussion

The use of a LAMP molecular beacon targeting the 35S promoter sequence and specifically the forward loop structure in the LAMP amplification was successful whether or not the LoopF primer was present, which would be competing for the sequence. Increased fluorescence and faster times were observed when the LoopF primer was included. The faster times can be explained by the accelerated LAMP reaction but the reporting of the faster time with increased fluorescence cannot be explained by the annealing of the probe and the primer to

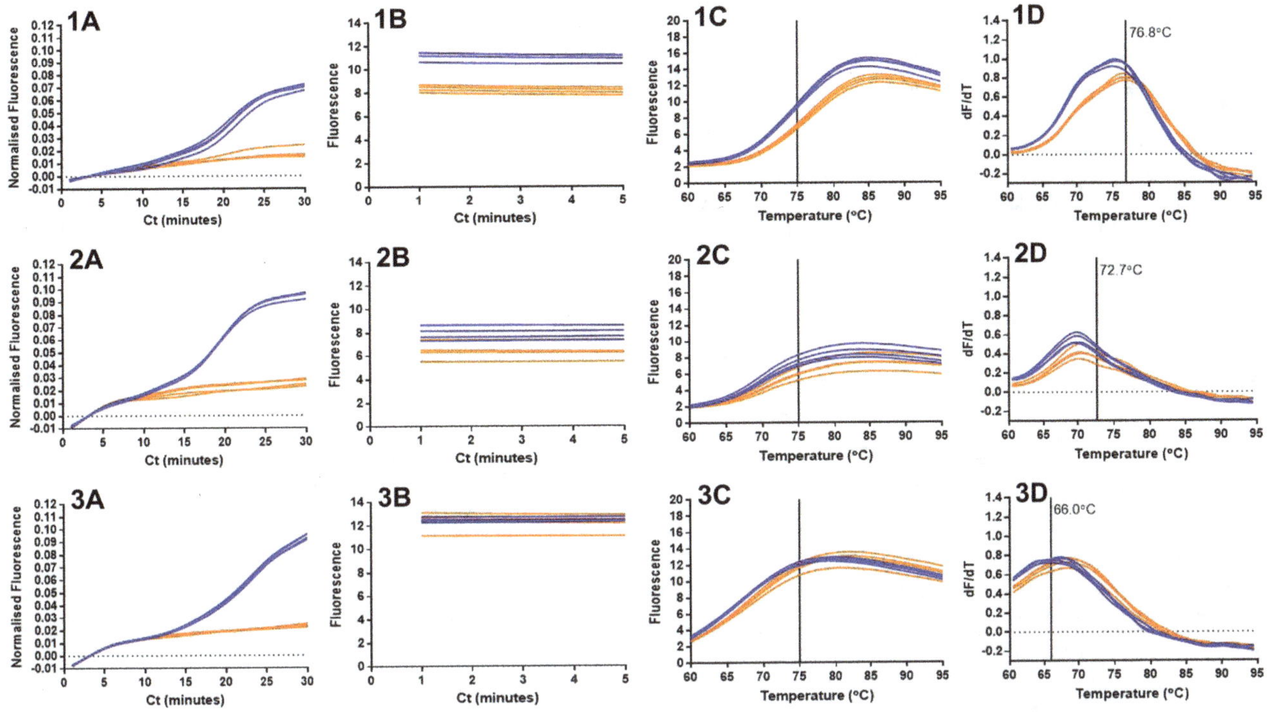

Fig. 10 End point LAMP molecular beacons at 75 degrees C. LAMP amplification at 64 degrees C for 30 cycles of 60 seconds, followed by 95 degrees C for 5 min and 75 degrees C for 5 min with fluorescence acquisition. Melt curve analysis between 60 degrees C and 95 degrees C. Template in blue, NTCs in orange. (1) 35Sp LAMP amplification with 35Sp LOOP MB, (2) 35Sp LAMP amplification with 35Sp STEM MB and (3) NOSt amplification with NOSt STEM MB. (A) real-time normalised fluorescence during amplification, (B) end point fluorescence at 75 degrees C, (C) melt profiles showing separation at 75 degrees C and (D) melt curve analysis with the melt temperature (Tm) of the molecular beacon

the same site due to the anticipated displacement activity of the polymerase. An alternative and potentially non-specific mechanism for the LAMP molecular beacon may be evident here.

We selected the single stranded DNA region between the loops in LAMP amplification to enable us to us to use the full set of primers and hence maximise the reaction kinetics and design MBs for 35S promoter and NOS terminator sequences commonly used in GM crops. Optimisation of these stem LAMP MBs resulted in a 64 degrees C reaction temperature, 1 micromolar probe concentration and 5 nucleotide arms (as predicted for PCR molecular beacons (Marras 2006 [15])). If it is the case that the LAMP molecular beacons do anneal to form a probe:target hybrid forcing the hybridised arms apart and increasing fluorescence, then the LAMP amplification would continue due to the displacement polymerase releasing the molecular beacon during extension. The dynamic process of LAMP amplification may explain why the fluorescence in the presence of template is low (but above that of the NTCs), because of the continuous annealing and displacement of the open molecular beacons which can then hybridise the arm sequences. A benefit of the low levels of fluorescence intensity for positive samples is the potentially high dynamic range of detection and quantification. Also the stem LAMP MB method achieved a similar high level of sensitivity in terms of amplification frequency for single copies per partition compared to SYTO9 detection.

The increased specificity of LAMP molecular beacons derives from the assumption that a fluorescent signal can only be generated with the target amplification product. More precisely that the molecular beacon can only fluoresce after conformational change brought on by probe:target hybridisation. Non-specific amplification from the interaction between primers should therefore remain undetected. However our investigations showed that the detection of LAMP amplification without a specific molecular beacon binding can occur. This was first shown with 35Sp LAMP amplification detected by a 5 nucleotide arm molecular beacon with a 13 nucleotide probe sequence specific for the stem region of NOS terminator LAMP amplification. This probe sequence is not present in the 35S promoter amplicon. Further analysis on LAMP molecular beacons was conducted on the LAMP amplification of a retroviral LTR sequence of a plasmid DNA template. All the LAMP molecular beacons should have non-specificity to the amplified products however all the 35Sp and NOSt designed MBs produced positive results in the presence of LTR LAMP amplification and

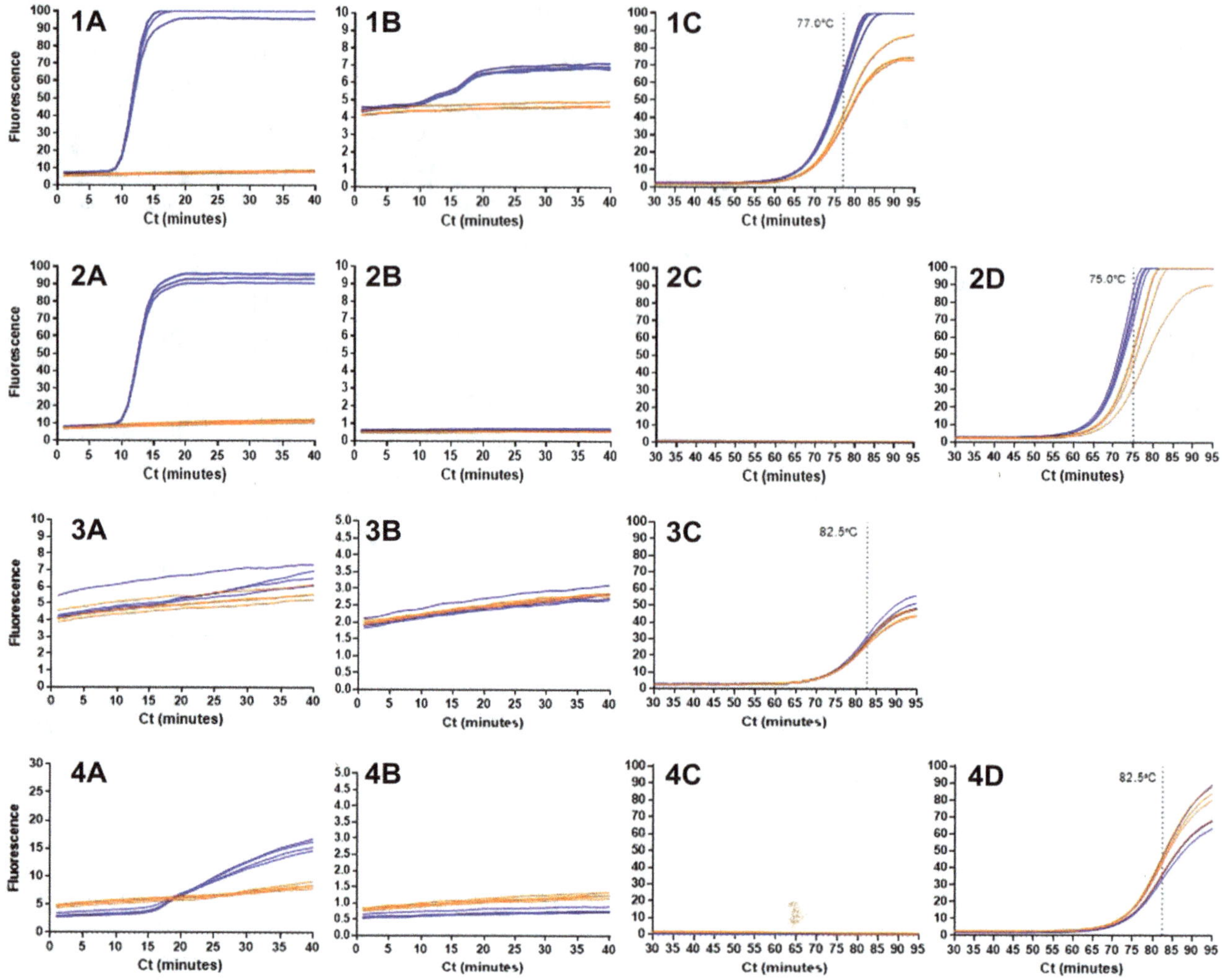

Fig. 11 LAMP MBs with PCR and LAMP amplification 1 and 2 represent LAMP assays whereas 3 and 4 are PCR results. 1 and 3 represent assays with SYTO9 and the 35Sp 5nt MB, 2 and 4 include SYTO9 but no LAMP MB (template results in blue and NTCs in orange). Fluorescent detection of SYTO9 are shown in A. Detection of the presence or absence of the 35Sp 5nt JOE MB are shown in B with corresponding melt curves in C. 2D and 4D show the melt curve for the amplified samples from 2 and 4 after the post amplification addition of the LAMP MB

negative results with the NTCs. From this data the LAMP molecular beacons are not binding specifically to the LAMP amplicons and therefore the mechanism for the increase in fluorescence is directed towards the increase in DNA (Fig. 12). It is unclear whether this is related to an increase in total DNA or to the mechanism of LAMP amplification, but it does appear that the loop structures of LAMP amplification are important and our experiments with linear and looped DNA templates support this.

End point LAMP amplification detection with LAMP MBs was previously described (Liang Wan et al. 2017) [11] with low melt temperatures molecular beacons to mitigate a perceived interference with LAMP amplification. We found that an increase in fluorescence at end point for molecular beacons with template over those with NTCs could be associated with a reduction in the melt temperature. Interaction between the products of LAMP amplification and the molecular beacons appear to give this marginal effect. However we would expect interaction with multiple copies of the target sequence and the specific molecular beacon to lead to a large increase in fluorescence due to unquenching, when compared to the quenched molecular beacon without template, this was not observed.

A LAMP molecular beacon in a PCR reaction failed to show increased fluorescence for the positive sample. The shorter probe sequence is likely to be a contributing factor to this. Conversely Liu et al. [10] showed that a typical PCR molecular beacon design failed to produce a fluorescent signal with LAMP amplification. This may be indicative of the annealing of the probe sequence

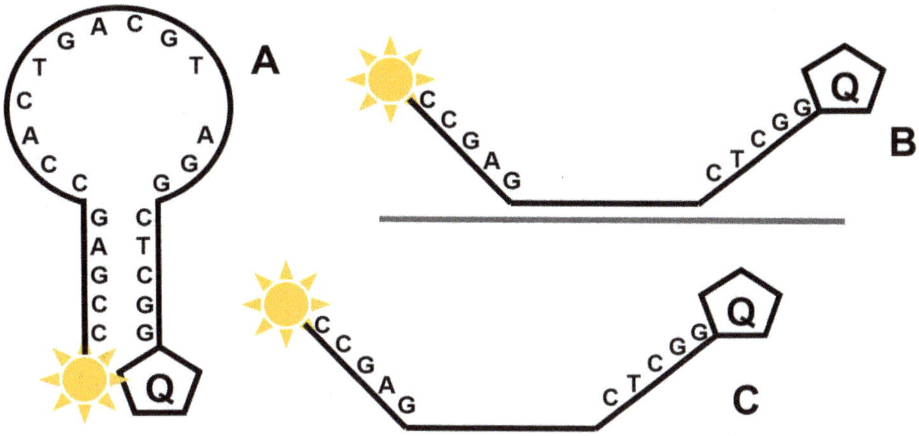

Fig. 12 Molecular beacon structure and potential conformations. Molecular beacons are single stranded DNA oligonucleotides with a 3′ fluorophore and 5′ quencher. Hybridised arms ensure fluorescence is quenched in the absence of specific template to the probe sequence in the loop (**a**). In (**b**) the presence of a specific single stranded DNA sequence forms a more thermodynamically stable probe – template hybrid pulling the quencher away from the fluorophore and enabling fluorescence to be emitted. In (**c**) the molecular beacon is in 'open' formation which may be due to the assay temperature in equilibrium with the 'closed' form, or by interaction with LAMP amplification products

to the amplicon interfering with LAMP amplification in real time. Furthermore the melt curve analysis from the PCR and LAMP amplifications with the LAMP molecular beacons showed an atypical profile for PCR molecular beacons in that at low temperature the presence of template did not increase proposed fluorescence from the molecular beacon.

It would be surprising that a molecular beacon could be used successfully in LAMP amplification. Perhaps therefore the increased fluorescence from the molecular beacons is the result of interactions with the LAMP concatemers. Whilst we cannot be sure that this increase in fluorescent signal with LAMP molecular beacons is a general effect applying in all combinations of LAMP amplicons and MBs, it seems clear that whilst they can, at appropriate concentration levels, provide an indication of amplification this does not result from sequence specificity but rather a more general feature of the LAMP amplicon.

Conclusion

Molecular beacons for LAMP amplification targeting the stem region between loops are effective but non-specific reporters of LAMP amplification with both 35S promoter and NOS terminator primers. We found the optimum temperature for our assays to be 64°C and achieved sensitivity comparable to the use of SYTO9. The mechanism by which the fluorescence from molecular beacons increases with LAMP amplification is unclear, although it appears that LAMP loop structures are important, but it is clear that there is no gain in specificity from the LAMP molecular beacons used in this study.

Abbreviations

35Sp: 35S promoter; BART: Bioluminescent assay in real time; BLAST: Basic local alignment search tool; CRM: Certified reference material; Ct: Cycle threshold; DNA: Deoxyribonucleic acid; GC content: Guanine-cytosine content; HPLC: High-performance liquid chromatography; F3 and B3; Forward and backward displacement primers; FIP and BIP: Forward and backward inner primers; FRET: Fluorescence resonance energy transfer; GMO: Genetically modified organism; LAMP: Loop-mediated amplification; LoopF and LoopB: Forward and backward loop primers; LTR: Long terminal repeat; MB: Molecular beacons; NOSt: Nopaline synthase terminator; nt: Nucleotide; NTC: No template control; PCR: Polymerase chain reaction; Tm: Melt temperature

Acknowledgments

Not applicable

Authors' contributions

PH and JAHM conceived the experiments, PH conducted the experiments, PH and JAHM analysed the results and wrote the manuscript. Both authors reviewed the manuscript. Both authors read and approved the final manuscript.

Funding

The study design, experimental work, analysis and interpretation of the findings was supported by BBSRC Industrial CASE Award number BB/G017336/1 and subsequently by BBSRC Stand Alone LINK grant BB/L022346/1. The writing of the manuscript was part funded by the European Regional Development Fund.

Ethics approval and consent to participate

Not applicable.

Consent for publication

Not applicable.

Competing interests
The authors declare that they have no competing interests.

References

1. Gill P, Ghaemi A. Nucleic acid isothermal amplification technologies: a review. Nucleosides Nucleotides Nucleic Acids. 2008;27(3):224–43.
2. Asiello P. J, Baeumner A. J. Miniaturized isothermal nucleic acid amplification, a review. Lab Chip. 2011;11(8):1420–30.
3. Notomi T, Okayama H, Masubuchi H, Yonekawa T, Watanabe K, Amino N, Hase T. Loop-mediated isothermal amplification of DNA. Nucleic Acids Res. 2000;28(12):E63.
4. Nagamine K, Hase T, Notomi T. Accelerated reaction by loop-mediated isothermal amplification using loop primers. Mol Cell Probes. 2002;16(3): 223–9.
5. Gandelman O, Jackson R, Kiddle G, Tisi L. Loop-mediated amplification accelerated by stem primers. Int J Mol Sci. 2011;12(12):9108–24.
6. Iwamoto T, Sonobe T, Hayashi K. Loop-mediated isothermal amplification for direct detection of Mycobacterium tuberculosis complex, M. avium, and M. intracellulare in sputum samples. J Clin Microbiol. 2003;41(6):2616–22.
7. Gandelman O. A, Church V. L, Moore C. A, Kiddle G, Carne C. A, Parmar S, Jalal H, Tisi L. C, Murray J. A. H. Novel bioluminescent quantitative detection of nucleic acid amplification in real-time. PLoS ONE. 2010;5(11):.
8. Mori Y, Kitao M, Tomita N, Notomi T. Real-time turbidimetry of LAMP reaction for quantifying template DNA. J Biochem Biophys Methods. 2004;59(2):145–57.
9. Fischbach J, Xander N. C, Frohme M, Glokler J. F. Shining a light on LAMP assays–a comparison of LAMP visualization methods including the novel use of berberine. Biotechniques. 2015;58(4):189–94.
10. Liu W, Huang S, Liu N, Dong D, Yang Z, Tang Y, Ma W, He X, Ao D, Xu Y, Zou D, Huang L. Establishment of an accurate and fast detection method using molecular beacons in loop-mediated isothermal amplification assay. Sci Rep. 2017;7.
11. Wan L, Chen T, Gao J, Dong C, Wong A. H. H, Jia Y, Mak P. I, Deng C. X, Martins R. P. A digital microfluidic system for loop-mediated isothermal amplification and sequence specific pathogen detection. Sci Rep. 2017;7(1).
12. Tyagi S, Kramer F. R. Molecular Beacons: Probes that Fluoresce Upon Hybridization. Nat Biotechnol. 1996;14(3):303–8.
13. Bonnet G, Tyagi S, Libchaber A, Kramer F. R. Thermodynamic basis of the enhanced specificity of structured DNA probes. Proc Natl Acad Sci. 2002;96(11):6171–6.
14. Tsourkas A, Behlke M. A, Rose S. D, Bao G. Hybridization kinetics and thermodynamics of molecular beacons. Nucleic Acids Res. 2003;31: 1319–30.
15. Marras S. A. E, Tyagi S, Kramer F. R. Real-time assays with molecular beacons and other fluorescent nucleic acid hybridization probes. Clin Chim Acta. 2006;363(1-2):48–60.
16. Lee D, La Mura M, Allnutt T. R, Powell W. Detection of genetically modified organisms (GMOs) using isothermal amplification of target DNA sequences. BMC Biotechnol. 2009;9:7.
17. Kiddle G, Hardinge P, Buttigieg N, Gandelman O, Pereira C, McElgunn C. J, Rizzoli M, Jackson R, Appleton N, Moore C, Tisi L. C, Murray J. A. H. GMO detection using a bioluminescent real time reporter (BART) of loop mediated isothermal amplification (LAMP) suitable for field use. BMC Biotechnol. 2012;12(1):15.
18. Hardinge P, Kiddle G, Tisi L, Murray J. A. H. Optimised LAMP allows single copy detection of 35Sp and NOSt in transgenic maize using Bioluminescent Assay in Real Time (BART). Sci Rep. 2018;8(1).
19. Gleave A. P. A versatile binary vector system with a T-DNA organisational structure conducive to efficient integration of cloned DNA into the plant genome. Plant Mol Biol. 1992;20(6):1203–7.
20. Wu Y, Wang Y, Li J, Li W, Zhang L, Li Y, Li X, Zhu L, Wu G. Development of a general method for detection and quantification of the P35S promoter based on assessment of existing methods. Sci Rep. 2017;4.
21. Fernandez S, Charles-Delobel C, Geldreich A, Berthier G, Boyer F, Collonnier C, Coue-Philippe G, Diolez A, Duplan M. N, Kebdani N, Romaniuk M, Feinberg M, Bertheau Y. Quantification of the 35S promoter in DNA extracts from genetically modified organisms using real-time polymerase chain reaction and specificity assessment on various genetically modified organisms, part I: operating procedure. J AOAC Int. 2005;88(2):547–57.
22. Untergasser A, Cutcutache I, Koressaar T, Ye J, Faircloth B. C, Remm M, Rozen S. G. Primer3-new capabilities and interfaces. Nucleic Acids Res. 2012;40(15):e115.
23. Vet J. A, Marras S. A. Design and optimization of molecular beacon real-time polymerase chain reaction assays. Methods Mol Biol. 2005;288: 273–90.

Improved the expression level of active transglutaminase by directional increasing copy of *mtg* gene in *Pichia pastoris*

Xiaoping Song[1,3*], Changsheng Shao[3], Yugang Guo[2,3], Yajie Wang[1] and Jingjing Cai[1]

Abstract

Background: The microbial transglutaminase (MTG) is inactive when only the mature sequence is expressed in *Pichia pastoris*. Although co-expression of MTG and its N-terminal pro-peptide can obtain the active MTG, the enzyme activity was still low. One of the basic steps for strain improvement is to ensure a sufficient level of transcription of the heterologous gene, based on promoter strength and gene copy number. To date, high-copy-number recombinants of *P. pastoris* are achievable only by cloning of gene concatemers, so methods for rapid and reliable multicopy strains are therefore desirable.

Results: The coexpression strains harboring different copies *mtg* were obtained successfully by stepwise increasing Zeocin concentration based on the rDNA sequence of *P. pastoris*. The genome of coexpression strains with the highest enzyme activity was analyzed by real-time fluorescence quantitative PCR, and three copies of *mtg* gene (*mtg*-3c) was calculated according to the standard curve of *gap* and *mtg* genes (*gap* is regarded as the single-copy reference gene). The maximum enzyme activity of *mtg*-3c was up to 1.41 U/mL after being inducted for 72 h in 1 L flask under optimal culture conditions, and two protein bands were observed at the expected molecular weights (40 kDa and 5 kDa) by Western blot. Furthermore, among the strains detected, compared with *mtg*-2c, *mtg*-6c or *mtg*-8c, *mtg*-3c is the highest expression level and enzyme activity, implying that *mtg*-3c is the most suitable for co-expression pro-peptide and MTG.

Conclusions: This study provides an effective strategy for improving the expression level of active MTG by directional increasing of *mtg* copies in *P. pastoris*.

Keywords: Transglutaminase, *Pichia pastoris*, rDNA, Pro-peptide, Co-expression

Background

Transglutaminase (TG, EC2.3.2.13) is an effective protein cross-linker that transfers acyl between glutamine and acyl acceptors to improve the properties of proteins [1–3]. The microbial transglutaminase (MTG) is widely used in the food industry to improve the texture and nutrition [4, 5]. Besides, TG has broad potential applications in biopharmaceuticals [6–9],tissue engineering [3, 7], site-directed protein cross-linking [1, 3, 10], and antibody-drug conjugates [11].

The MTG was produced by *Streptomyces mobaraense* [12], *Streptomyces hygroscopicus* [13], and *Bacillus circulans* [14]. The catalytic active center of MTG consisted of catalytic triad C64-D255-H274 is located at the bottom of its active fracture (Additional file 1: Figure S1). The MTG could be secret to extracellular as a zymogen and the N-terminal peptide was removed by endogenous protease to activate the enzyme [15, 16]. It is hard to express MTG in engineering strains due to the activation mode [17–21]. Some researchers suggested that the MTG is inactive if the full or mature sequence is expressed in *P. pastoris* [17, 21]. To solve the problem, the pro-peptide and mature sequence of MTG could be expressed as two separate elements in *P. pastoris*. The activity of MTG reached 0.25 U/mL and 0.338 U/mL respectively in flasks [22], indicating that the pro-peptide

* Correspondence: sxp20081012@sina.com
[1]Department of Pharmacy, Anhui Medical College, Hefei 230061, China
[3]Anhui Engineering Research Center of Recombinant Protein Pharmaceutical Biotechnology, Hefei 230022, China
Full list of author information is available at the end of the article

sequence could promote and activate the mature enzyme. However, the activity was too low to meet the industrial requirements. Therefore, the increasing demand on MTG is urgently needed to improve the enzyme activity of MTG.

As we know, the yield of protein is related to the copy number of the target gene [23, 24]. This is a useful method to improve the yield of protein by increasing the copy number of the target gene in yeast [25, 26]. The repeated sequence of rDNA of *P. pastoris* was separated by the non-transcribed intergenic spacer (NTS) [26, 27]. It could be used as the recombination site to increase the copy number of the target gene. In our previous work, the gene of snake venom antiplatelet thrombolytic was integrated into the rDNA repeat sequence. The combinant antiplatelet thrombolytic shared similar physicochemical properties and biological activities with the natural antiplatelet thrombolytic [27]. Therefore, the strategy based on the repeated sequence of rDNA of *P. pastoris* to construct multiple copy of the target gene can be employed in this paper.

In order to improve the expression level and the enzyme activity of the recombinant MTG, genes coding MTG and pro-peptide (*mtg* and *pro*) were optimized to make their mRNA secondary structure more stable according to GC contents and the codon bias of *P. pastoris*. The genes (*pro* and *mtg*) were inserted sequentially into the chromosome of *P. pastoris* at ribosomal DNA repeat sites (rDNA) and the mutant histidinol dehydrogenase site (His 4), respectively, and regulated by the constitutive promoter (*GAP*) and the methanol-induced promoter (*AOX*1), respectively. A high coexpression of pro-peptide and MTG was obtained by high-pressure screening of antibiotic and optimization of fermentation conditions. To our knowledge, this is a new report of improving the expression level of active MTG by directional increase of the copy of *mtg* gene in *P. pastoris.*

Results

Construction balanced co-expression strains for pro-peptide and MTG

The genes (*pro* and *mtg*) were inserted into vector pGAP9 and pPICZα to construct the expression vector pGAP9-*pro*and pPICZα-rDNA-*mtg*, respectively (Fig. 1). Both *pro* and *mtg* genes could be integrated into the his4 site and non-coding rDNA sequence site when vectors were transformed into the host strain GS115, respectively. The genes (*pro* and *mtg*) were fused with the *S. cerevisiae* α-mating factor signal sequence and placed under the regulation of GAP and AOX1 promoters. The recombinants of co-expression strain GS115 (*pro*/rDNA-*mtg*) were analyzed by PCR (Additional file 4: Figure S4), of which the results implied that the *pro* and *mtg* genes were integrated into the desired location of *P. pastoris* genome.

To generate the coexpression strain of pro-peptide and mature MTG, the expression plasmid (pGAP-*pro*) was transformed into *P. pastoris* GS115 and selected by histidine-deficient minimal dextrose (MD) plates, while the clone (# 34–4) with the highest expression of pro-peptide was selected (Fig. 2a and b) and used for the host strain of the co-expressed (*pro*/rDNA-*mtg*). The clones with higher co-expression were selected in MD plate (contain 400 μg/mL, 600 μg/mL, 800 μg/mL

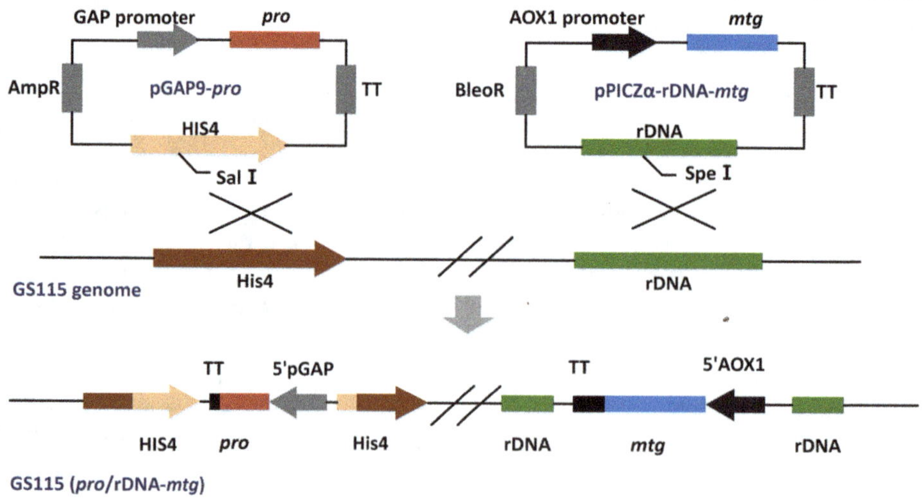

Fig. 1 Strategy for the construction of *mtg* and *pro* co-expression strains. This schematic map represents the constructed expression vectors for the *pro* and *mtg*, designated pGAP9/*pro* and pPICZα-rDNA/*mtg*, respectively. The his4 and rDNA non-coding sequences allow the vector to be inserted into the corresponding sites in the genome of strain GS115 through homologous recombination. The pPICZα-rDNA-*mtg* vector was constructed by introducing a non-coding rDNA (indicated as rDNAnc) sequence into the pPICZα vector. The expression of both *pro* and *mtg* were under the control of the GAP and AOX1 promoters, respectively

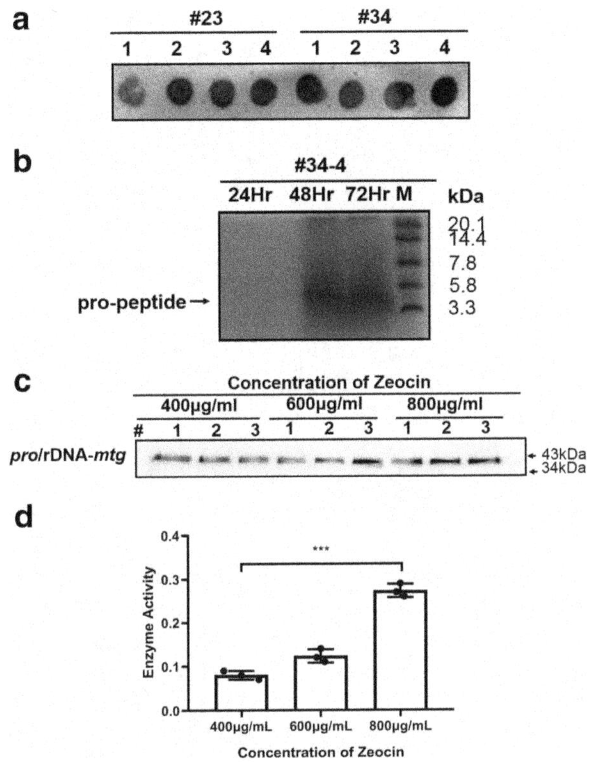

Fig. 2 Establishing balanced expression strains for pro-peptide and MTG. **a** The clones of GS115 (*pro*) were screened by Dot blot (anti-His-HRP).1–4 were four repeated of clones. **b** The protein (pro-peptide) of 20 μl culture supernatant was separated by Tricine-SDS-PAGE and stained with Coomassie blue under non-reducing conditions. **c** The co-expression of pro-peptide and MTG from screened colonies with various doses of Zeocin were experimented by Western blot under non-reducing conditions. **d** The enzyme activity of MTG from screened clones with various doses of Zeocin were experimented. Three individual clones were tested for each condition.***Represents a statistically significant difference of $P < 0.001$ compared with other two antibiotic dosage groups

Zeocin) and the expression of MTG was measured by Western blot with anti-MTG antibody employed. All selected clones rendered positive results by Western blot and showed an increasing trend with the increase of antibiotic dose (Fig. 2c). The enzyme activity of clone (contain 800 μg/mL Zeocin) reached up to 0.27 U/mL, higher than those of the other two antibiotic dosage groups (0.08 U/mL, 0.12 U/mL) (Fig. 2d).

Copy number of *mtg* gene in *P. pastoris* genome

The results of *mtg* copy number in *P. pastoris* genome are shown in Table 1. The amplification and melt curves of *gap* and *mtg* genes are shown in Fig. 3 while the standard curves are shown in Additional file 5: Figure S5. Both the melt curves of *gap* and *mtg* genes have only one single peak, indicating specificity of the PCR products (Fig. 3a, b). The amplification curves of *gap* and *mtg* genes were shown in Fig. 3c and d. The regression equations of both *gap* and *mtg* genes standard curve were $y = -3.6892x + 35.747$ ($R = 0.9985$), $y = -3.4511x + 33.968$ ($R = 0.9985$), respectively. By the slope of the two standard curves, the response efficiency of *gap* and *mtg* genes were calculated as 86.66 and 94.88%, respectively, indicating similar results of the two ones. The result of *mtg* gene copy number in *P. pastoris* (with the highest activity) is shown in Table 1, demonstrating that *mtg* gene existed in the genome in the form of 3.36 copies and that 3 copies *mtg* gene were directed to the rDNA site in *P. pastoris* and named *mtg*-3c. In addition, the copy numbers of *mtg* gene in the other three detected strains were 2.21, 5.72 and 7.62, respectively (named *mtg*-2c, *mtg*-6c, *mtg*-8c) (Table 1).

Effects of temperature and pH on growth and enzyme activities of co-expressing strains

The engineering strain *mtg*-3c (*pro*/rDNA-*mtg*) was cultured in 200 mL BMMY medium at different pH, with the result suggesting that the initial pH of the medium exerted minor influence on the cell growth but significant one on the enzyme activity of MTG as illustrated in Fig. 4a, b. The activity of MTG was higher at pH 6 and 7 than that at pH 5 or 8, indicating that the neutral environment was conducive to producing enzymes or the enzyme was more stable in neutral environment. Furthermore, the strain GS115 (*pro*/rDNA-*mtg*) was inducted at various temperatures as shown in Fig. 4c, d. The activity of MTG at 25 or 30 °C was higher than that at 20 °C within 48 h. However, the activity of MTG at 30 °C decreased after 48 h. The highest activity of MTG was 0.91, 1.12, and 0.81 U/mL at 20, 25, and 30 °C, respectively for 72 h, indicating that the optimum inducted temperature was 25 °C. The activity of MTG would increase significantly by controlling the fermentation conditions (the initial pH of BMMY medium and inducted temperatures).

Co-expression of pro-peptide and MTG in 1 L flask

The engineering strain *mtg*-3c was inoculated into BMGY medium for 24 h and then transferred into 200 mL BMMY medium. The pro-peptide and MTG in the supernatant have been evaluated by SDS-PAGE electrophoresis and Western blot, respectively (Fig. 5). The experimental results showed that one protein band was observed at the expected molecular weight (40 kDa) (Fig. 5d). In addition, another protein band was observed at the expected molecular weight (5 kDa) by Tricine-SDS-PAGE (Fig. 5b). Because the molecular weights of pro-peptide (5 kDa) and MTG (40 kDa) are significantly different from each other, it is difficult to display them on the same SDS-PAGE gel (Fig. 5a, c). The strategy of combining two homologous recombination sites (His4 and rDNA repeat site) could enable the co-expression of pro-peptide and MTG in

Table 1 Copy numbers of *gap* and *mtg* gene detected by real-time fluorescent quantitative PCR

Strain	Value Ct		Gene copy(10^n)		Number of copies of *mtg* gene in *P. Pichia* genome (*mtg* copy number/*gap* copy number)
	mtg gene	*gap* gene	*mtg* gene	*gap* gene	
1	14.14 ± 0.27	16.49 ± 0.17	5.58 ± 1.02	1.66 ± 0.19	3.36 ± 0.24
2	14.93 ± 0.28	16.67 ± 0.22	3.29 ± 0.50	1.48 ± 0.18	2.21 ± 0.11
3	14.97 ± 0.30	18.23 ± 0.23	3.20 ± 0.63	5.60 ± 0.71	5.72 ± 0.35
4	15.30 ± 0.41	19.05 ± 0.17	2.56 ± 0.55	3.36 ± 0.31	7.62 ± 1.11

Datas are presented as mean ± SD of triplicate observations. Strain 1: the clone with the highest enzyme activity. Strain 2,3,4:another three positive clones

P. pastoris. The enzyme activity of *mtg*-3c (culture supernatant) was 1.41 U/mL after being inducted for 72 h.

Discussion

The *P. pastoris* has many outstanding advantages as a host for the expression of foreign proteins [2, 18]. It enjoys the advantages of the strong inducing promoter AOX1, high protein expression, stable genetic traits, simple composition of the medium, etc. Various exogenous genes such as interferon, HBsAg, phytase, albumin, antibody and collagen were expressed in *P. pastoris* successfully [18, 21, 27]. Some researchers have attempted to express recombinant *Streptomyces* MTG in *P. pastoris*; however, the yield of MTG is very low [2, 18, 21]. We also encountered the problem of low yield of recombinant MTG in the previous research. In order to increase the yield of recombinant MTG in *P. pastoris*, we constructed 3 copies *mtg* of GS115 (*pro*/rDNA-*mtg*) with higher expression, and optimized such conditions of protein expression as pH and temperature to increase the yield of MTG.

The copy number of the target gene is an important factor for the yield of the target protein in *P. pastoris* [23]. The expression of the target protein could be effectively boosted by increasing the copy number to 4–9 of the gene [17]. When the copy number was more than nine, the yield of target gene might decrease because the

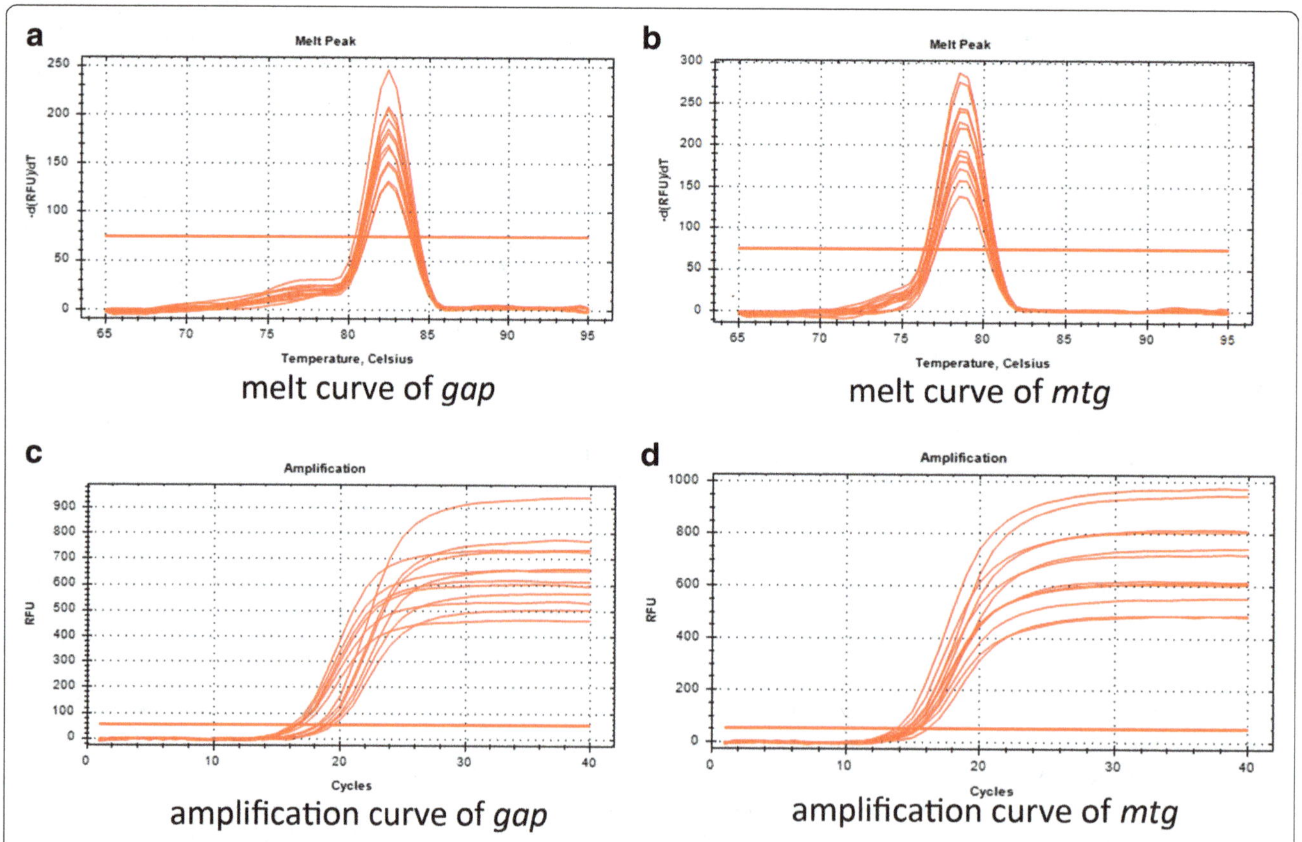

Fig. 3 Detection of *mtg* copy number in *P. pastoris* genome by absolute quantitative PCR. **a** and **b** The melt curves of *gap* and *mtg* genes. **c** and **d** The amplification curves of *gap* and *mtg* genes

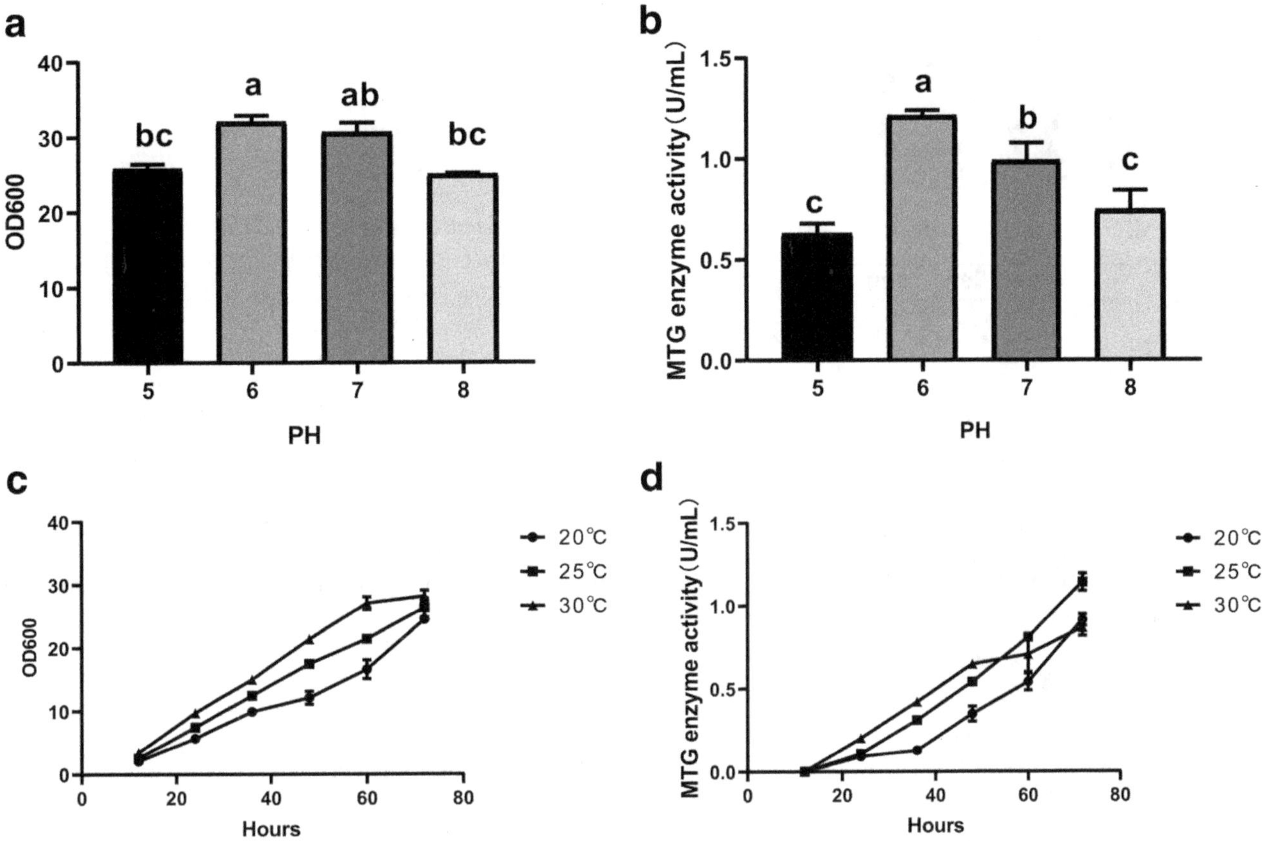

Fig. 4 Effects of inducted temperature and pH on growth of GS115 (*pro*/rDNA-*mtg*) and enzyme activities of MTG. **a** and **b** were the effect of different pH on the strain growth and enzyme activity of MTG. **c** and **d** were the effect of different induced temperature on the growth and enzyme activity of MTG. Datas of strain growth and enzyme activity are presented as mean ± SD of triplicate observations. Alphabets (**a**–**c**) in superscript donate significant difference ($P < 0.05$)

high concentration of target protein had a negative effect on the strain growth and metabolism [25, 28]. In this study, we constructed a high expression strain harboring different copies *mtg* by directionally increasing *mtg* gene copy in *P. pastoris,* and verified the influence of gene dosage on the expression level of MTG. The absolute quantitative PCR analysis was successfully applied to quantify *mtg* copies in the recombination strains. Among the clones being detected, *mtg* genes existed in the *P. pastoris* genome in the form of 2,3,6 or 8 copies, respectively (named *mtg*-2c,*mtg*-3c,*mtg*-6c, *mtg*-8c). Under the same culture conditions (200mLBMMY in shake flask), the effect of different copy *mtg* genes on protein expression and enzyme activity was shown in Additional file 6: Table S1 and Figure S6. The experimental datas showed that the expression levels of *mtg* gene varying with copy numbers were inconsistent, and fell into the range of was *mtg*-3c > *mtg*-2c > *mtg*-6c (Additional file 6: Figure S6). The results revealed that the recombinant strain produced the enzyme faster in case of no more than 3 copies of *mtg* gene, showing the positive effect of gene copy number. However, when gene copy number is more than 3, the enzyme production of the strain dwindled for the excessive protein load, resulting in more heteropolymeric protein and a sharp decrease in enzyme activity than one of the low-copy strain [28, 29]. This fact demonstrated the negative effect of gene copy number, which may be attributed to the burden of the excessive copy number on the cell, resulting in the metabolic imbalance of the strain itself. For this reason, the multiple copies may not result in a high expression level. These results imply that other factors limited the transcription, translation, and secretion of MTG, indicating that the copy number was not necessarily related to the expression level, and that the reasonable copy number was beneficial to improving the protein expression level. These results revealed that 3 copies of *mtg* gene were the most suitable for *pro* and *mtg* co-expression in this study.

The result of culture conditions optimization showed that the initial pH of medium has significant influence on the enzyme activity of MTG. The optimum pH of MTG expression is between 6.0 and 7.0 and the stability of MTG is poor under acid and alkaline conditions,

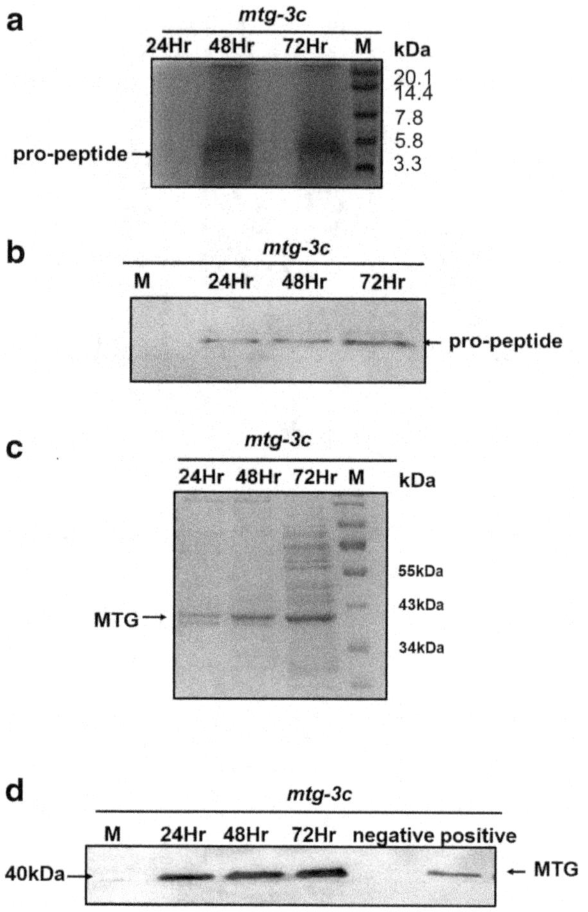

Fig. 5 Detection of coexpression pro-peptide and MTG in GS115 (*mtg*-3c). **a** and **b** The pro-peptide were detected by SDS-PAGE and Western blot under non-reducing conditions. The proteins (pro-peptide) of 20 μl culture supernatant were separated by Tricine-SDS-PAGE and subjected to Western analysis using anti-His-HRP. **c** and **d** The protein (MTG) of 20 μl culture supernatant were separated by SDS- PAGE and subjected to Western analysis (anti-MTG).positive control: MTG from *E.coli*

especially under acid conditions (pH < 6.0). The activity of protease produced by *P. pastoris* is suppressed at the pH 6.0–7.0, which could reduce the degradation of the target protein [28]. Furthermore, the temperature would affect the yield of MTG, which is lower at 30 °C than that at 20 or 25 °C, indicating that the rate of cell growth is lower at 20 and 25 °C, but the yield of MTG is higher in this case, attributable to the hypothermia reducing the residence time of the protein in the endoplasmic reticulum, which is conducive to the correct folding of the target protein [30, 31].

Finally, the engineering strain GS115 (*pro*/rDNA-*mtg*) was cultured in 200 mL BMMY medium for 72 h. The maximum enzyme activity of supernatant was 1.41 U/mL, 3.17 times higher than that of a coexpression strain as single-copy expression cassettes (0.338 U/mL) [22]. In addition, two protein bands were observed at the expected molecular weight (40 kDa and 5 kDa) by Western blot, indicating that the co-expression of pro-peptide and MTG can directly secrete active MTG. This is a new report of improving the expression level of active MTG by directional increasing copies of *mtg* gene in *P. pastoris* in the published articles .

The pro-peptide is essential for protein folding, playing the main roles of helping proteins form compact structures and thus reducing the steric hindrance of proteins, guiding proteins to correctly fold and forming active mature enzymes [21]. In this study, the expression level of MTG was mainly measured by enzyme activity. Therefore, the enhanced expression of pro-peptide will facilitate the correct folding of MTG, thus forming much more active MTG. However, the balance of co-expression of pro-peptide and MTG is a problem to be solved in this study. The balance expression of MTG and pro-peptide can be achieved by gradually increasing the concentration of Zeocin antibiotics.

Conclusions

Based on the non-transcriptional rDNA sequence of *P. pastoris*, the co-expression strain (*pro*/*rDNA-mtg*) with three copies *mtg* genes was constructed successfully. However, the expression in *P. pastoris* is at the milligram per liter level in this paper, which needs to be raised to grams per liter for industrial applications. Successful co-expression of MTG and pro-peptide as well as optimization of their ratios is a challenge, but worth investigating.

Methods

Strains and vectors

Strains and vectors were described in Table 2. The E. coli was cultured in Luria-Bertani (LB) medium at 37 °C, whereas the *P. pastoris* was cultured in buffered glycerol-complex medium (BMGY) or buffered methanol-complex medium (BMMY) at 28 °C or 25 °C, respectively.

Synthesis of *mtg* and *pro* genes

The *mtg* and *pro* genes were selected from the cDNA of *S.mobaraensis* (NO. DQ132977) [19, 21] and synthesized after the codon optimization based on the *P. pastoris* preference (Additional file 2 and Additional file 3: Figures S2 and S3). The primers were designed by Primer premier 5(http://www.premierbiosoft.com/primer-design/index.html) as shown in Table 3.

The restriction enzyme site (*Xho*I) and the Kex2 endopeptidase recognition site were added to the 5′-terminus of MTG and pro-peptide (primer P1 and P3). The restriction enzyme site of *Not*Iand the sequence of histidine tag were added to the 3′-terminus (primer P2 and P4).

Table 2 Strains and plasmids used in this study

Strains	Genotype	References
E. coli TOP10	F-*mcrA*Δ (*mrr-hsdRMS-mcrBC*) φ80 *lacZ*ΔM15 Δ*lacX74 recA1 ara*Δ139 Δ (*ara-leu*)7697 *galUgalKrpsL* (*StrR*) *endA1 nupG.* used for vector sub-cloning	Invitrogen
Pichia pastoris GS115	*his4*,host strain	Invitrogen
GS115 (*pro*)	The pro-peptide of MTG was integrated into his4 site of *Pichia pastoris* GS115	This study
GS115(*pro*/rDNA-*mtg*)	The mature sequence of MTGwas integrated into non-coding rDNA site ofGS115 (*pro*)	This study
Plasmids	Description	Reference
pPICZα-B	Containing AOX1 promoter for tightly regulated, methanol-induced expression of the gene, Zeocin^R	Invitrogen
pGAP9	pPIC9 derivative, GAP promoter instead of AOX1 promoter	Lab stock
pGAP9-*pro*	pGAP9 derivative carrying an internal 135 bp fragment of *pro* gene	This study
pPICZα-rDNA-*mtg*	pPICZα-*mtg* derivative carrying an internal 1153 bp fragment of rDNA gene, Zeocin^R	This study

The construction of rDNA-mediated *mtg* multi-copy expression vector

The structure of expression pPICZα-rDNA-*mtg* was shown in Fig. 1.The AOX1-terminator (primer P5and P6) and non-coding rDNA fragment (primer P7and P8) were amplified and fused to form the AOX1-rDNA by overlapping PCR as follows: a program consisting of 30 cycles at 95 °C for 40 s, 54 °C for 50 s and 72 °C for 120 s. Then the AOX1-rDNA and the vector pPICZa-*mtg* were digested by *Not*I and *Bam*HI and connected by T4 DNA ligase. The recombinant vector (pPICZα-rDNA-*mtg*) was transformed into *E. coli* TOP10 and sequenced to ensure base sequence was not mutated. The same method was used to construct the control group (pPICZα-*mtg).*

Table 3 PCR primers used in the present study

Primer name	Primer sequence (5′→3′)
P1	ACACTCGAGaaaagaGACTCCGACGACAGGGTCAC
P2	TATGCGGCCGCTCAATGGTGATGGTGATGATGCGGC CAGCCCTGC
P3	ACACTCGAGaaaagaGACAAT GGC GCG GGG GAAG
P4	TATGCGGCCGCTCAATGGTGATGGTGATGATGGGGA GCCCGGAACG
P5	TATGCGGCCGCCAGCTTTCTAGAACAAAAACTCATCTC
P6	CTTAAATATTAGGAAAAACGGTAACCTTAT CTCACTTAATCTTCTGTACTCTG
P7	CAGAGTACAGAAGATTAAGAGAGATAGT TAGGTTACCGTTTTTCCTAATATTTAAG
P8	CGCGGATCCCTTCCACCAACAGTCAACCACCAGTC
P9	GGTATTAACGGTTTCGGACGTATTG
P10	GATGTTGACAGGGTCTCTCTCTTGG
P11	TGAAGAAAGAATTGGCTAACGG
P12	AGCTGGTCTGAAAGCATCTGG

Remarks: The underlined sites are those for the digestion of restriction enzymes *Xho*I. The wavy line sites are those for the digestion of restriction enzymes *Not*I. The double corrugated underlined sites are the ones for the digestion of restriction enzymes *Bam*HI. 6xHis-tag label sequence is indicated by the dotted line. The Kex2-endopeptidase recognition site is marked with lowercase letters. The bold underlining represents the sequence of overlapping segments of the AOX1 terminator-rDNA fusion gene

Construction of the pro-peptide recombinant strain

The plasmid pGAP9 and pro-peptide gene were digested with *Xho*I and *Not*I, ligated by DNA T4 ligase. Then it was transferred into *E. coli* TOP10 competent cells to construct the vector pGAP9-*pro*. It was digested by *Sal*I and electro-transformed into host strain GS115. The recombinants were selected in MD plate without histidine and cultured in BMGY medium for 72 h. The pro-peptide in the supernatant was identified by Dot blot with the antibody anti-His-HRP. Furthermore, the pro-peptide was measured by Tricine-SDS-PAGE. The engineering strain was named GS115(*pro*) and used as the host strain of co-expressed pro-peptide and MTG.

Construction and selection of the co-expression strain

Recombinant vectors (pPICZα-rDNA-*mtg*) were linearized by *Spe*I, and transformed into the host strain GS115 (pro-peptide). The clones with higher expression were selected by stepwise increasing the Zeocin concentration by 400 μg/mL, 600 μg/mL and 800 μg/mL on MD plates, respectively, so as to strike a better balance of the co-expression of MTG and pro-peptide. The clone with the highest expression was selected in MD plate (contain 800 μg/mL Zeocin) and named GS115 (*pro*/rDNA-*mtg*), respectively. Then they were cultured in BMGY medium for 24 h until the OD600 reached 6.0. The cell was harvested and induced in 5 mL BMMY medium containing 1% methanol for 72 h. The MTG in the supernatant was identified by Western blot (the antibody: Anti-MTG) as described by "Molecular cloning: A laboratory manual". A clone with the highest enzyme activity was selected.

Analysis of *mtg copy* number by absolute quantitative PCR

The clones with the highest enzyme activity and other three positive ones were selected for analyzing copy number of *mtg* gene, which in *P. pastoris* genome were determined by real-time fluorescent quantitative PCR (qPCR) [32]. The housekeeping gene *gap* exists as a single copy in the genome of *P. pastoris* [29], so the copy number of *gap* gene can be used to represent the initial copy number of the genome in the template.

The standard curves of *gap* and *mtg* genes were prepared with the standard plasmids containing *gap* and *mtg* genes, respectively. Both *gap* and *mtg* genes were amplified by qPCR with the genome of *P.pastoris* transfected *mtg* gene used as template, respectively (*gap* primer: P9 and P10, *mtg* primer: P11 and P12). Each sample was detected three times, and three parallel reactions were set for each time. The *Ct* values of *gap* and *mtg* genes obtained substituted for the corresponding standard curves, and then the initial template copy number of *gap* and *mtg* genes in DNA samples was calculated. The ratio of initial template copy number of *mtg* and *gap* genes was the copy number of *mtg* gene in *P. pastoris* genome. The reaction system is as follows: a total of 10 μl reaction system, including SYRB Green Realtime PCR Master 5 μl, DNA template 1 μl, each primer 0.5 μl (10 $\mu mol \cdot L^{-1}$), H_2O 3 μl. The reaction procedures: 10 mins at 95 °C, followed by 40 cycles of 10 s at 95 °C, 30 s at 60 °C.

Optimization of culture conditions for engineering strain *pro*/rDNA-*mtg*

The strain GS115 (*pro*/rDNA-*mtg*) of higher enzyme activity was cultured in 100 mL BMGY medium for 24 h and controlled at 28 °C. Then the cell was harvested by centrifugation and inoculated into 200 mL BMMY medium at different initial pH (5.0, 6.0, 7.0 and 8.0) for 72 h and maintained at 25 °C. 1 M $Na_2HPO_4 \cdot 12H_2O$ (pH = 8.85) and 1 M KH_2PO_4 (pH = 4.01) were mixed in a certain proportion to form phosphate buffer with buffer range pH 4–9 and used to adjust the initial pH of BMMY medium. The cell density (OD600) and the activity of MTG were detected as described below. Besides, the cell density and the activity of MTG were measured once every 12 h at different inducted temperatures (i.e. 20, 25 and 30 °C).

The activity of MTG was measured as Yurimoto described [21]. The MTG enzyme activity was defined as follows: catalytic substrate (N-CBZ-Gln-Gly) to generate 1 μmol glutamine hydroxamate (hydroxamic acid) at 37 °C as one enzyme activity unit. Protein concentrations were determined by using a Bradford Protein Assay Kit (TaKaRa) and the standard protein was bovine serum protein (Thermo) [21].

Expression of engineering strain *pro*/rDNA-*mtg* in flask

The strain GS115 (*pro*/rDNA-*mtg*) of higher enzyme activity was cultured in 100 mL BMGY medium, as described above until the OD600 reached 6–7, and controlled at 28 °C. Then they were harvested and cultured in 200 mL of BMMY medium containing 1% methanol (initial OD600 reached 1.2–1.5) and maintained at 25 °C. 1% methanol (*V*/*V*) was added into BMMY medium once every 24 h for continuous induction for 72 h. The supernatant was collected once every 24 h, centrifuged at 12000 g for 30 mins, and analyzed by SDS-PAGE and Western blot under non-reducing condition. The enzyme activity was detected and verified as described above.

Additional files

Additional file 1: Figure S1. Three-dimensional structure of MTG (a) and pro-MTG (b) from *S. mobaraesis* as determined by PyMOL. The catalytic triad C64-D255-H274 was represented by light blue, red and dark blue, respectively. (TIF 275 kb)

Additional file 2: Sequences of *mtg* and *pro* genes. (DOCX 17 kb)

Additional file 3: Figures S2, S3. Secondary structure diagram of mRNA of *mtg* and *pro*. **Figure S2.** The secondary structure diagram of mRNA of *mtg*-WT and *mtg*-Opt. **Figure S3.** The secondary structure diagram of mRNA of *pro*-WT and *pro*-Opt. (ZIP 93 kb)

Additional file 4: Figure S4. The product of gene amplification of *pro* and *mtg* in GS115 (*pro*/ rDNA-*mtg*).1,3,5,7: *pro* was amplified by Fw-*pro*(P3) and Rv -*pro* (P4)primers; 2,4,6,8: *mtg* was amplified by Fw-*mtg*(P1) and 3'AOX primers; 9,10: negative controls; M: DNA Marker. (TIF 365 kb)

Additional file 5: Figure S5. Standard curves of *gap and mtg gene*. (DOCX 101 kb)

Additional file 6: Table S1, Figure S3. Effect on expression and enzyme activity of *mtg* gene copies. **Table S1.** The MTG activity in different copy strains. **Figure S6.** Detection of protein expression in strains with different *mtg* copy by Western blotting. The protein (MTG) in 20 μl of culture supernatant were separated by Western blot analysis (anti-MTG). Lane 1–3: *mtg*-2c, lane 4–6: *mtg*-3c, lane 7–9: *mtg*-6c. (ZIP 48 kb)

Abbreviations

Ct: Cycle threshold; rDNA: ribosomal DNA; rDNAnc: non-coding rDNA; RFU: relative fluorescence units

Acknowledgments

We are thankful for the technical support for WB analysis from Anhui Engineering Research Center of Recombinant Protein Pharmaceutical Biotechnology; the kind gifts of pGAP9 and GS115 from Prof. Weihua Xiao, School of Life Sciences, University of Science and Technology of China, Hefei Anhui, China.

Authors' contributions

XS, CS and YG conceived and designed the experiments. CS,YG, YW,and JC performed the experiments. XS,CS and YG analyzed the data. XS wrote the manuscript. All authors reviewed the manuscript.

Funding

This work was supported by the Anhui Provincial Natural Science Foundation (Grant No.1808085MC86), and the Education Science Foundation of Anhui Province (Grant Nos. KJ2016A378, KJ2017A695, KJ2015B02). These funding bodies had no role in the design of the study and collection, analysis, and interpretation of data and in writing the manuscript.

Ethics approval and consent to participate

Not applicable.

Consent for publication
Not applicable.

Competing interests
The authors declare that they have no competing interests.

Author details
[1]Department of Pharmacy, Anhui Medical College, Hefei 230061, China. [2]Institute of advanced technology, University of Science and Technology of China, Hefei 230031, China. [3]Anhui Engineering Research Center of Recombinant Protein Pharmaceutical Biotechnology, Hefei 230022, China.

References
1. Shleikin AG, Danilov NP. Evolutionary-biological peculiarities of transglutaminase: Structure, physiological functions, application. J Evol Biochem Phys. 2011;47(1):1–14.
2. Li H, Zhang L, Cui Y, Luo X, Xue C, Wang S. Expression of soluble recombinant transglutaminase from Zea mays in *Pichia pastoris*. World J Microbiol Biotechnol. 2013;29(5):939–47.
3. Martins IM, Matos M, Costa R, Silva F, Pascoal A, Estevinho LM, Choupina AB. Transglutaminases: recent achievements and new sources. Appl Microbiol Biot. 2014;98(16):6957–64.
4. Gharibzahedi SMT, Chronakis IS. Crosslinking of milk proteins by microbial transglutaminase: utilization in functional yogurt products. Food Chem. 2018;245:620–32.
5. Kieliszek M, Misiewicz A. Microbial transglutaminase and its application in the food industry: a review. Folia Microbiol. 2014;59(3):241–50.
6. Verma A, Mehta K. Tissue transglutaminase-mediated chemoresistance in cancer cells. Drug Resist Update. 2007;10(4):144–51.
7. Garcia Y, Wilkins B, Collighan RJ, Griffin M, Pandit A. Towards development of a dermal rudiment for enhanced wound healing response. Biomaterials. 2008;29(7):857–68.
8. Rosenthal AK, Derfus BA, Henry LA. Transglutaminase activity in aging articular chondrocytes and articular cartilage vesicles. Arthritis Rheum. 2014; 40(5):966–70.
9. Souza CFVD, Venzke JG, Flôres SH, Ayub MAZ. Enzymatic properties of transglutaminase produced by a new strain of *Bacillus circulans* BL32 and its action over food proteins. LWT Food Sci Technol. 2011;44(2):443–50.
10. Mu D, Lu J, Shu C, Li H, Li X, Cai J, Luo S, Yang P, Jiang S, Zheng Z. Improvement of the activity and thermostability of microbial transglutaminase by multiple-site mutagenesis. Biosci Biotechnol Biochem. 2018;82(1):106–9.
11. Dennler P, Chiotellis A, Fischer E, Brégeon D, Belmant C, Gauthier L, Lhospice F, Romagne F, Schibli R. Transglutaminase-based chemo-enzymatic conjugation approach yields homogeneous antibody-drug conjugates. Bioconjug Chem. 2014;25(3):569.
12. Sommer C, Hertel TC, Schmelzer CEH, Pietzsch M. Investigations on the activation of recombinant microbial pro-transglutaminase: in contrast to proteinase K, dispase removes the histidine-tag. Amino Acids. 2012;42(2–3): 997–1006.
13. Cui L, Du GC, Zhang DX, Chen J. Thermal stability and conformational changes of transglutaminase from a newly isolated *Streptomyces hygroscopicus*. Bioresour Technol. 2008;99(9):3794–800.
14. Suzuki S, Izawa Y, Kobayashi K, Eto Y, Yamanaka S, Kubota K, Yokozeki K. Purification and characterization of novel transglutaminase from *Bacillus subtilis* spores. Biosci Biotechnol Biochem. 2000;64(11):2344–51.
15. Cai D, Wei X, Qiu Y, Chen Y, Chen J, Wen Z, Chen S. High-level expression of nattokinase in*Bacillus licheniformis* by manipulating signal peptide and signal peptidase. J Appl Microbiol. 2016;121(3):704–12.
16. Cai D, Wang H, He P, Zhu C, Wang Q, Wei X, Nomura CT, Chen S. A novel strategy to improve protein secretion via overexpression of the SppA signal peptide peptidase in *Bacillus licheniformis*. Microb Cell Factories. 2017;16(1): 1–10.
17. Javitt G, Ben-Barak-Zelas Z, Jerabek-Willemsen M, Fishman A. Constitutive expression of active microbial transglutaminase in *Escherichia coli* and comparative characterization to a known variant. BMC Biotechnol. 2017; 17(23):1–10.
18. Liu S, Zhang D, Du G, Chen J. Progress in expression and molecular modification of microbial transglutaminase. Sheng Wu Gong Cheng Xue Bao. 2011;27(12):1681–9.
19. Salis B, Spinetti G, Scaramuzza S, Bossi M, Jotti GS, Tonon G, Crobu D, Schrepfer R. High-level expression of a recombinant active microbial transglutaminase in *Escherichia coli*. BMC Biotechnol. 2015;15(84):1–11.
20. Zhang L, Zhang L, Yi H, Du M, Ma C, Han X, Feng Z, Jiao Y, Zhang Y. Enzymatic characterization of transglutaminase from *Streptomyces mobaraensis* DSM 40587 in high salt and effect of enzymatic cross-linking of yak milk proteins on functional properties of stirred yogurt. J Dairy Sci. 2012; 95(7):3559–68.
21. Yurimoto H, Yamane M, Kikuchi Y, Matsui H, Kato N, Sakai Y. The pro-peptide of *Streptomyces mobaraensis* transglutaminase functions in cis and in trans to mediate efficient secretion of active enzyme from methylotropic yeasts. Biosci Biotechnol Biochem. 2004;68(10):2058–69.
22. Li P, Sun H, You L, Gong F, Chen Z, Zhang A, Zhu T. Direct secretory expression of active microbial transglutaminase in *Pichia pastoris*. Chin J Biotechnol. 2013;29(2):180–8.
23. Zhu T, Guo M, Tang Z, Zhang M, Zhuang Y, Chu J, Zhang S. Efficient generation of multi-copy strains for optimizing secretory expression of porcine insulin precursor in yeast *Pichia pastoris*. J Appl Microbiol. 2010; 107(3):954–63.
24. Khan FF, Carpenter D, Mitchell L, Mansouri O, Black HA, Tyson J, Armour JAL. Accurate measurement of gene copy number for human alpha-defensin DEFA1A3. BMC Genomics. 2013;14.
25. Marx H, Mecklenbrauker A, Gasser B, Sauer M, Mattanovich D. Directed gene copy number amplification in *Pichia pastoris* by vector integration into the ribosomal DNA locus. FEMS Yeast Res. 2009;9(8):1260–70.
26. Li H, Fan H, Li Y, Shi GY, Ding ZY, Gu ZH, Zhang L. Construction and application of multi-host integrative vector system for xylose-fermenting yeast. FEMS Yeast Res. 2017;17(6).
27. Guo YG, Wu J, Jia H, Chen W, Shao CS, Zhao L, Ma JJ, Li R, Zhong YJ, Fang F, et al. Balancing the expression and production of a heterodimeric protein: recombinant agkisacutacin as a novel antithrombotic drug candidate. Sci Rep-Uk. 2015;5(11730):1–10.
28. Juturu V, Wu JC. Heterologous protein expression in *Pichia pastoris*: latest research progress and applications. Chembiochem. 2018;19(1):7–21.
29. Waterham HR, Digan ME, Koutz PJ, Lair SV, Cregg JM. Isolation of the *Pichia pastoris* glyceraldehyde-3-phosphate dehydrogenase gene and regulation and use of its promoter. Gene. 1997;186(1):37–44.
30. Dragosits M, Stadlmann J, Albiol J, Baumann K, Maurer M, Gasser B, Sauer M, Altmann F, Ferrer P, Mattanovich D. The effect of temperature on the proteome of recombinant *Pichia pastoris*. J Proteome Res. 2009;8(3):1380–92.
31. Yu H, Huang H. Engineering proteins for thermostability through rigidifying flexible sites. Biotechnol Adv. 2014;32(2):308–15.
32. Johnson MR, Wang K, Smith JB, Heslin MJ, Diasio RB. Quantitation of dihydropyrimidine dehydrogenase expression by real-time reverse transcription polymerase chain reaction. Anal Biochem. 2000;278(2):175–84.

15

Standardized mounting method of (zebrafish) embryos using a 3D-printed stamp for high-content, semi-automated confocal imaging

David Simon Kleinhans and Virginie Lecaudey*

Abstract

Background: Developmental biology relies to a large extent on the observation and comparison of phenotypic traits through time using high resolution microscopes. In this context, transparent model organisms such as the zebrafish *Danio rerio* in which developing tissues and organs can be easily observed and imaged using fluorescent proteins have become very popular. One limiting factor however is the acquisition of a sufficient amount of data, in standardized and reproducible conditions, to allow robust quantitative analysis. One way to improve this is by developing mounting methods to increase the number of embryos that can be imaged simultaneously in near-to-identical orientation.

Results: Here we present an improved mounting method allowing semi-automated and high-content imaging of zebrafish embryos. It is based on a 3D-printed stamp which is used to create a 2D coordinate system of multiple μ-wells in an agarose cast. Each μ-well models a negative of the average zebrafish embryo morphology between 22 and 96 h-post-fertilization. Due to this standardized and reproducible arrangement, it is possible to define a custom well plate in the respective imaging software that allows for a semi-automated imaging process. Furthermore, the improvement in Z-orientation significantly reduces post-processing and improves comparability of volumetric data while reducing light exposure and thus photo-bleaching and photo-toxicity, and improving signal-to-noise ratio (SNR).

Conclusions: We present here a new method that allows to standardize and improve mounting and imaging of embryos. The 3D-printed stamp creates a 2D coordinate system of μ-wells in an agarose cast thus standardizing specimen mounting and allowing high-content imaging of up to 44 live or mounted zebrafish embryos simultaneously in a semi-automated, well-plate like manner on inverted confocal microscopes. In summary, image data quality and acquisition efficiency (amount of data per time) are significantly improved. The latter might also be crucial when using the services of a microscopy facility.

Keywords: Zebrafish, Reproducibility, Standardization, Automation, Quantitative imaging, High content imaging, 3D printed stamp, Multidimensional imaging

* Correspondence: Lecaudey@bio.uni-frankfurt.de
Department of Developmental Biology of Vertebrates, Institute for Cell biology and Neuroscience, Goethe University, Max-von-Laue-Str. 13, 60438 Frankfurt am Main, Germany

Background

Understanding how an organism develops from a unique, fertilized egg is largely based on observations. This includes observations at all scales to detect changes at the level of the whole organism, organs, tissue, cells and molecules. Recording data of biological processes in the mm to nm scale requires specialized and appropriate instruments that can magnify such small structures – the microscope. Therefore, since the emergence of microscopes, the fields of developmental biology and bioimaging have been developing in a coordinated manner [1–3]. Recent technological advances in microscopy, e.g. confocal spinning-disc and light-sheet microscopy, now allow for data acquisition with high spatiotemporal resolution, reduced phototoxicity and photobleaching [4].

Developmental biology evolved from a discipline largely based on qualitative methods. Due to always better microscopes, image analysis methods and computer technology, it is in a process of digital and quantitative transformation [5–8]. Given the noisy and variable character of biological systems and possibly small effect size, it is important to record a sufficient number of samples to obtain a quantitative and representative view of a biological process. Furthermore, to process biological samples of whole organisms in a high-content manner, it is important to have a standardized way of sample mounting, data acquisition, data processing and analysis. Therefore, high-resolution imaging of multiple samples in standardized conditions is of key importance.

As high-resolution imaging and quantitative biology became more and more important for developmental biologists, the zebrafish *Danio rerio* was established as an ideal model organism for vertebrate development [9–13]. Beside its genetic tractability and robustness, two of its main advantages in respect to modern light microscopy and bio-image data analysis are (i) the transparency of the embryo making it ideal for live imaging and (ii) the high number of off-spring enabling high-content screening.

To get the most of these two advantages, it is however necessary to mount, image and analyse as many embryos as possible simultaneously and in conditions that are as similar as possible. To date, this is still largely limited by the classical way most scientists mount embryos for imaging. The factors limiting this standardization are summarized in Table 1 (left column).

Therefore, there is a need for methods to standardize sample mounting and image acquisition of multiple embryos at a spatiotemporal resolution suited for 3D segmentation and 2D tracking experiments. The protocol we describe here was designed to be used with XY scanning universal sample holders that usually come with any motorized-stage inverted microscope. Similar to previous

Table 1 Limitations of traditional zebrafish mounting techniques

Standard method	Limitations	Solutions	Improvement
Mounting in 1% low-melting point agarose (LMPA)	Polymerization speed limits the number of embryos that can be mounted in parallel Embryo growth is restricted by high-percentage LMPA	Use lower concentration of LMPA Use heating device to keep the LMPA liquid longer	• Mounting time extended •Growth and change in shape of the embryo allowed • Sample size increased • Embryo retrieval facilitated • Possibility to grow embryos to adulthood afterwards
No pre-defined positions	Positioning and alignment in XY are neither standardized nor reproducible	Mount embryos at pre-defined, identically oriented, equidistant positions (Figs. 1c, and 2d, f)	• Relative positions of embryos identical in all experiments • Easier setup of multi-dimensional imaging experiments • Easier navigation between different XY locations • Possibility for semi-automated imaging • Identification of individual embryo facilitated for downstream experiments (genotyping)
No µ-wells that model the average embryo shape at a defined developmental stage	Time-consuming orientation of the embryos Body axes not aligned to optical sectioning in Z due to the huge yolk sac (Fig. 1a)	Use µ-wells that model the average embryo shape (Fig. 1a')	• Orientation of individual embryos during mounting in LMPA much faster • No need to re-orient the embryos in XY plane post-imaging • Aligned morphological shapes in Z projections • Reduced stack and file size •Reduced photo-bleaching and -toxicity • Reduced post-processing • Increased scanning speed • Improved signal-to-noise ratio

approaches [14–17], it uses a 3D-printed stamping device to produce an agarose imprint with a diameter of 20 mm on the cover glass of a 35 mm µ-dish. The imprint consists of 44 equally spaced micro-wells (µ-wells), which are designed to fit the average morphology of a zebrafish embryo between 24 and 96 h-post-fertilization (hpf) (Fig. 1a-b).

The aim was to develop a standardized mounting method allowing us to: (i) mount many samples in parallel in a 2D coordinate system of rows and columns, (ii) reduce the acquisition time and thus photo-bleaching and photo-toxicity during imaging, (iii) semi-automatize the acquisition, (iv) reduce the post-processing steps, and (v) facilitate subsequent processing such as genotyping due to a 1:1 correlation between image data and specimen arrangement sequence.

Results

Our 3D-printed stamp improves two main aspects of specimen preparation: orientation in Z (Figs. 1a-a', 2a-a') and standardization of the embryo arrangements in X and Y (Figs. 1c, and 2d). To demonstrate the efficacy of this method, we used the zebrafish posterior lateral line (pLL) system as a model.

The lateral line is a mechano-sensory organ that enables aquatic vertebrates to sense water movements and pressure, and therefore receive information about their environment. Its functional units are the neuromasts (NMs), which are groups of about 30 cells organized around a core of hair cells surrounded by support cells [18]. In the adult, the lateral line comprises hundreds of NMs distributed over the entire body surface. The first NMs to form in the zebrafish embryo are deposited by two groups of about 130 epithelial cells, the posterior lateral line primordia (pLLP), which migrate from head to tail on both sides of the embryo during the second day of development (Fig. 2c, Additional file 1: Movie S1). At about 50 hpf the pLL spreads linearly along the lateral body flanks of the embryo (Fig. 2a-b'; ~2 mm in length). Using this mounting technique, lateral line development can be recorded over more than 20 h, in up to 44 positions, in a confocal Z-stack of less than 120 µm and a time interval of 5–10 min (depending on the number of channels and exposure time).

Each morphogenetic process demands unique imaging conditions and may present several challenges that have to be overcome. Table 2 highlights specific challenges concerning imaging of the lateral line development.

Figure 2c shows three timepoints of a single time-lapse recording with 34 positions, 2 stitched FOVs, 2 channels (transgenes *cldnb:GFP* and *cxcr4b:H2B-RFP*), imaged with a 10 min. interval over a period of 13 h (Additional file 1: Movie S1). As mentioned above, the most crucial factor for automation of zebrafish embryo imaging is (i) that the embryos are arranged at equidistant spaces and (ii) that they are oriented in the same way. Using our mounting protocol allows the user to image zebrafish embryos systematically with up to 5 times more embryos as compared to traditional mounting methods. Figure 2d shows 36 embryos in 2 channels (brightfield + transgene *cldnb:GFP*) imaged with a 10x objective after an overnight time-lapse, highlighting (1) the regularity in mounting and (2) embryo integrity after time-lapse imaging (see also Additional file 1: Movie S1). To quantify mounting precision and reproducibility, the

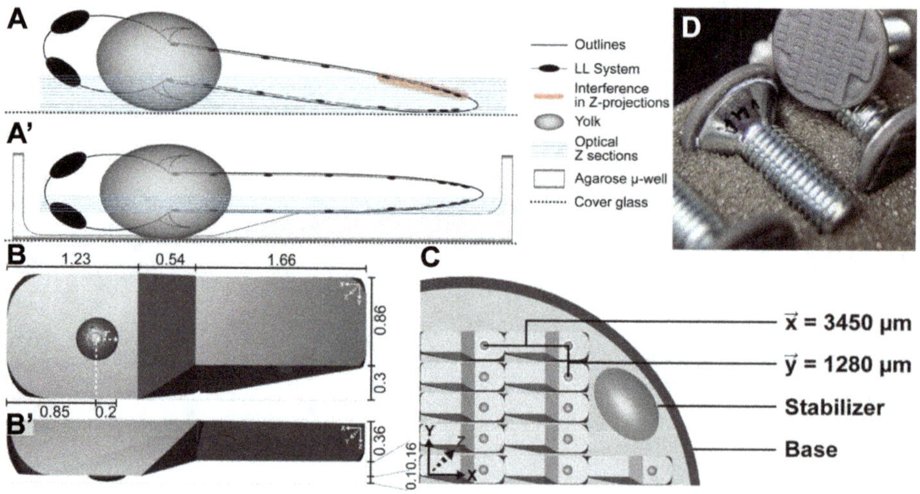

Fig. 1 Stamp and µ-well dimensions (**A, A'**) Scheme showing a zebrafish embryo mounted classically, directly on the coverslip (**A**) and with µ-well (**A'**). **B, B'** Top-view (**B**) and lateral view (**B'**) of a single µ-well with dimensions in mm. **C** Design and dimensions of the stamp. The base connects all parts and fits exactly a Ø 20 mm coverslip of a Ø 35 mm imaging dish. $\vec{x}$ and $\vec{y}$ are the distances between two adjacent µ-wells. Stabilizers were introduced as cornerstones and to make the structure more rigid. **D** For better handling, the stamp is mounted onto a countersunk screw

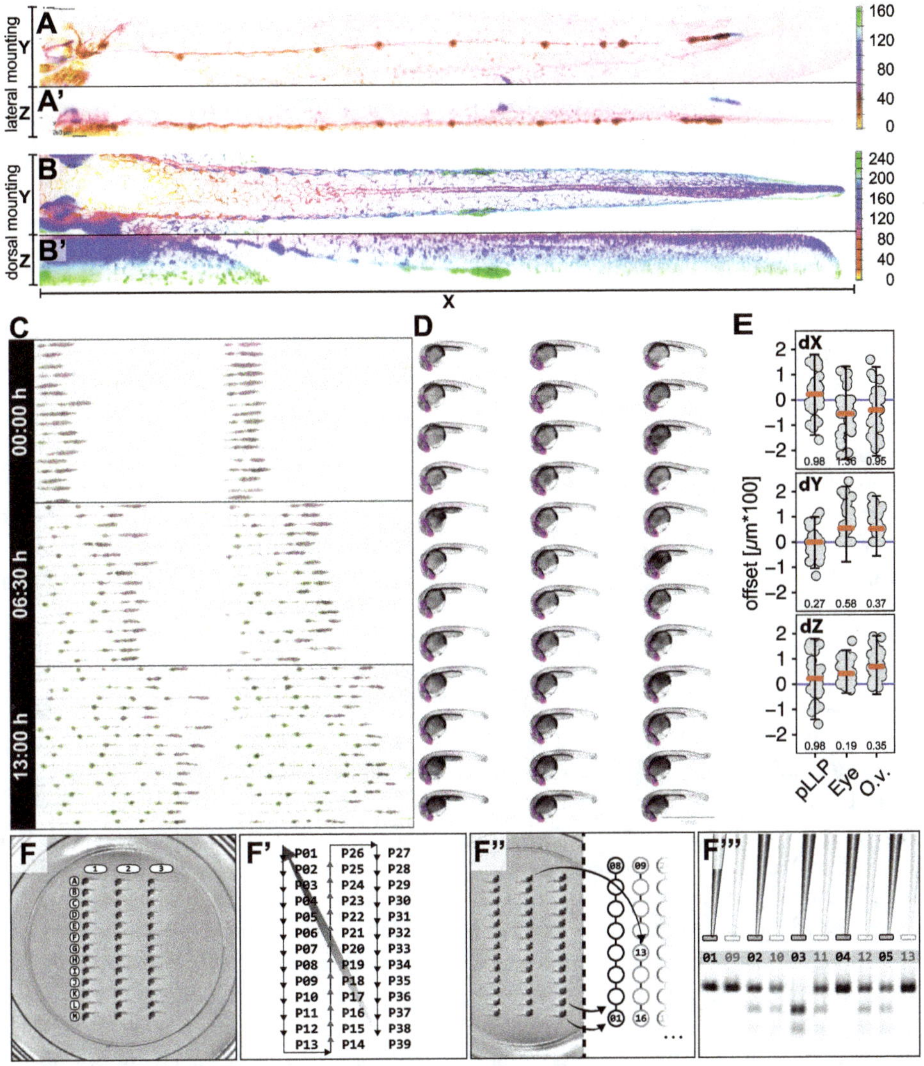

Fig. 2 Representative Results. **A** Maximum intensity projection (MIP) of a 50 hpf embryo mounted on its side (XY) and corresponding MIP of the orthogonal view (XZ) of the same embryo showing how flat the embryos are mounted in the plate (**A'**). **B** MIP of a 40 hpf embryo mounted dorsally (XY) and corresponding MIP of the orthogonal view (XZ) of the same embryo (**B'**). Depth colour encoding is indicated by colour scales on the right. **c** Multi-position (36), multi-channel (2) time-lapse recording (13 h duration; 15 min. Interval, see also Additional file 1: Movie S1). **D** Multichannel (2) Extended Depth of Focus (EDF) projections from widefield Z-stacks (recorded with 20x Objective). Scale Bar = 1 mm (**E**) Multi-point coordinates in X, Y and Z (recorded with 40x Objective). The *offset* describes the distance of each point from the mean of all points in X, Y and Z (See methods). Panel 1–3 (top to bottom) show dimensions X, Y and Z in comparison for the pLLP, the eye and the otic vesicle. The red line indicates the median, the blue line indicates zero offset, error bars indicate mean ± standard deviation. Numeric values indicate the variance. **F-F'''** Systematic retrieval for genotyping. **F** Mounted embryos in a 2-D coordinate system of rows (A-M) and columns (1–3). **F'** Imaging Sequence in a snake by column fashion. In a time-lapse setting, it starts at point 1 (P01) again to initiate the next timepoint. **F''** After imaging, the embryos are retrieved in the same sequence as they were imaged (snake by column, left panel). **F'''** Each genotyping result on the electrophoresis gel is easily correlated to one imaging dataset with defined X-Y coordinates

positions (in X, Y and Z) of three common zebrafish model organs (i) the pLLP, (ii) the eye and (iii) the otic vesicle were acquired and evaluated (Fig. 2e, see methods 'offset calculation').

Discussion

It has been more than 10 years since zebrafish has been recognized as a very powerful model organism for in vivo imaging and high content screening [9, 19]. Although efforts have been made to standardize zebrafish embryo mounting and imaging, most protocols were designed for high throughput widefield screening [16, 17, 20, 21] rather than high-resolution confocal microscopy [22–24]. In recent years a couple of studies tackled specifically this issue by developing standardized mounting methods using 3D-printed stamps [14, 15, 25, 26]. Here we provide a handy

Table 2 Specific challenges when imaging pLLP morphogenesis.

Challenges	Solutions
pLL morphogenesis lasts about 30 h: Embryos need to be immobilized but still survive and grow over a longtime period	Use of low concentration of LMPA containing Tricaine
pLLP migrates over almost 2 mm so the area to image is 2 to 3 times longer than one field of view (FOV): (i) If the embryos are not mounted parallel to the coverslip, the size of Z-stack increases significantly. (ii) The contra-lateral pLLP eventually enters the Z-volume and introduce interfering signal in post-processing Z-projections.	Multiple FOV need to be imaged and stitched The use of µ-wells with a defined shape orients the tails parallel to the coverslip (see Fig. 2a)' to both reduce the size of Z-stacks and prevent the interference of the contra-lateral pLLP.

and simple approach for a standardized and much more efficient way of mounting zebrafish for high resolution confocal microscopy.

Troubleshooting

The existing limitations in classical mounting methods, the solutions proposed by our new method and the improvements are summarized in Table 1. The most critical steps of the protocol are:

i. The detachment of the stamp from the solidified 1% Agarose (see 'Design and preparation of 3D stamp'), which can cause the formation of air inclusions between the cover glass and the agarose.
ii. The speed of polymerization of the 0,3% LMPA which should not be too fast to give enough time to place and orient each embryo in a µ-well.
iii. The application of the 0,5% LMPA to the embryos embedded in 0,3% LMPA (see 'Mounting' part of the Methods section) which should happen very carefully so that the embryos are not carried away.

Major improvements

Our method presents two major improvements over currently used mounting methods:

1) The low percentage LMPA used allows for:
   a. An extended timespan for mounting which is necessary to align the high number of embryos.
   b. Facilitated retrieval of the embryos after imaging and systematic identification of individual positions (e.g. for genotyping). For the latter, make sure your numbering for identification is in accordance with the sequence in which you imaged the embryos (Fig. 2f-f'').
   c. More freedom for the embryo to grow and elongate during long time-lapse imaging (Fig. 2c).

2) The mounting cast allows for:
   a. Standardized and reproducible positions of the embryos as shown for the lateral line primordium, the eye and the otic vesicle (Fig. 2e).
   b. Significant increase of number of embryos (more than 5 times more) imaged in a single experiment (Fig. 2c-d).
   c. Consistent and minimized distances between different positions (Fig. 1c)
   d. Identical and stage-aligned orientations of all embryos in XYZ (Figs. 1, and 2d-e and 5)
   e. Significant reduction of the Z-stack size and therefore of the illumination of the samples (Fig. 2a', e).

Flexibility

The design of the µ-well is not limited to a specific model organism or spatial orientation but different µ-wells may be designed exactly fitting the specific requirements. In fact, we have developed and successfully tested a stamp for dorsal mounting (Fig. 2b, b' – dorsal mounting) of zebrafish larvae. Also, it is possible to adopt the existing models for mounting other embryos including *O.latipes* (Medaka). All stamp models developed can be found at Github (https://github.com/LecaudeyLab/3DModels) and are posted as a technology offer at INNOVECTIS (http://innovectis.de/en/technologies/goethe-depository/3d-printed-stamp-for-standardized-mounting-and-high-content-confocal-imaging-of-zebrafish-embryos/) to ease transfer of technology.

Further improvements

While round dishes require a manual adjustment to align the embryo body axes to the microscope stage axes, using rectangular dishes together with a rectangular stamp would minimize the degrees of freedom when mounting the dish onto the sample holder.

Comparison to existing methods

In the past 5–10 years, several labs have developed methods for high-throughput imaging based on microfluidic devices. The limitations of such approaches is (i) that it is difficult and expensive to develop for non-specialized labs and (ii) it is perfectly suited for high-throughput screening, low-resolution purposes but less for long time-lapse or high resolution imaging since the embryos are generally not immobilized [20, 21, 27, 28]. Approaches based on 3D-printed stamps, like ours, are much more flexible and easier to establish in any lab, either using an in-house 3D printer or as we did it by outsourcing the printing at low costs. Recently, a multi-sample preparation protocol for long-term time-lapse imaging using a light-sheet fluorescence microscope (LSFM) has been published [29]. However the amount of embryos that can be imaged simultaneously was limited to five due to the specific sample mounting imposed by the orthogonal arrangement of the objectives in such systems. In principle, our 3D-stamp could however be combined with more recently-developed LSFM that are not based on orthogonal objectives and thus can use classical cover slip or petri dishes mounting method [30].

Conclusions

In conclusion, the main advantages of our 3D-printed stamp over existing ones are:

(i) the amount and the narrow arrangement of μ-wells allow for reduced stage movements and imaging of many embryos simultaneously (up to 44 in our case).
(ii) improved time resolution during time lapse imaging of multiple positions.
(iii) the specifically-designed form of the μ-well maintains the AP axis of the embryo parallel to the coverslip and thus allows to reduce Z-stack size, exposure time, photo-bleaching and photo-toxicity, and image reorientation post-processing and to improve signal-to-noise ratio.
(iv) the reduced concentration of LMPA to 0.3% gives the embryo more freedom to grow and change its shape [14, 15] while still being sufficiently immobilized.
(v) semi-automated imaging and further downstream sample processing, including genotyping significantly facilitated by the organization in a 2D coordinate system.

Methods

Zebrafish husbandry and transgenic lines

Animals used in this study were obtained from own breeding at the animal facility of the Goethe University Frankfurt according to the German animal welfare act and approved by the German authorities (veterinary department of the Regional Board of Darmstadt). Adult zebrafish were maintained under standard conditions and embryos were staged according to Kimmel et al. (1995). The Transgenic lines *Tg(−8.0cldnb:lynEGFP)*^{zf106} (*cldnb:gfp*) and *TgBAC (cxcr4b:H2B-RFP)*^{fu13} (*cxcr4b:H2B-RFP*) have been described previously [31, 32].

Design and preparation of 3D-stamp

To model a negative of an average zebrafish embryo between 24 and 96 hpf, the dimensions of different embryonic structures were measured on corresponding stage-match embryos:

(i) Embryo: Length / width (X, Y)
(ii) Yolk Sac: Diameter (in XY) and depth (in Z)
(iii) Trunk: Length / width / depth (X, Y, Z)

Using Microsofts '3D Builder', a single μ-well was assembled by transformation of basic shapes like cube, sphere and wedge. Afterwards, the maximum number of μ-wells fitting in a circle with a diameter of 20 mm were positioned. The 3D printing was then carried out by a commercial partner using a Formlabs 'Form1+' extrusion printer with the Formlabs material 'Tough' photopolymer. Using this material, it is possible to print structures at a maximum resolution of 25x50x100 μm, which is necessary to capture the intricate details of the embryonic morphology. To ease the handling, a zinc-plated countersunk screw with dimensions 8 × 20 mm (DIN7991) was attached to the back of the printout using superglue (Fig. 1d).

Preparation of the agarose mounting cast

Stamps were cleaned of dust or other particles with 70% Ethanol and pressured air. A solution of 1% agarose (w/v) was prepared in a clean 100 mL blue cap bottle by dissolving 200 mg of agarose in 20 mL of E3 in the microwave. 650 μL of this solution was then applied with a 1 mL pipette to the coverslip (Ø 20–21 mm) of a 35-mm imaging dish (e.g. ibidi μ-dish or MatTek microwell, see list of materials). The stamp was gently placed on it (Fig. 3a-b) and its position carefully adjusted to be in the centre of the coverslip. The dish was then gently rotated to distribute the excess agarose over the entire dish surface to stabilize the imprint once polymerized.

If bubbles emerge upon immersion of the pipette tip into the 1% agarose solution, the solution is too hot. Wait until no or almost no bubbles are appearing anymore.

The agarose was then allowed to polymerize for ~ 30 min. To remove the stamp, a clean preparation needle was slipped between the stamp and the polymerized agarose (Fig. 3b), and the stamp was then removed by holding the dish in place and carefully but swiftly lifting it. If necessary, air bubbles appearing between the cover glass and the agarose imprint were eliminated by puncturing with a preparation needle. The mounting cast (Fig. 3c) was either used immediately or stored at 4 °C for several days (with closed lid).

Preparation of mounting media

Two solutions of low-melting point agarose (LMPA) were prepared in clean 100 mL blue cap bottles by dissolving

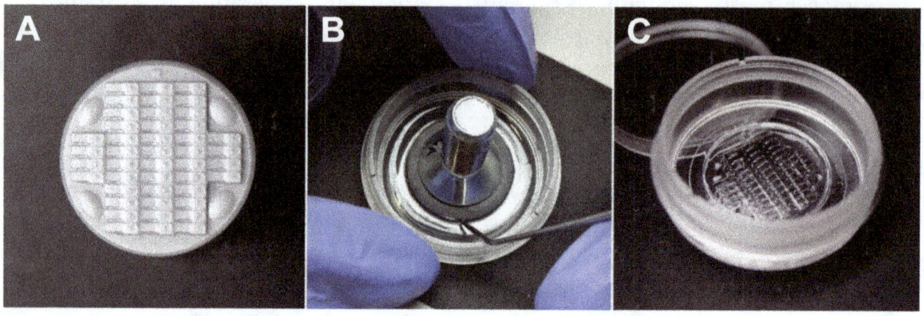

Fig. 3 Stamping procedure. **a** Clean stamp surface **b** Removal of the 3D-stamp from the polymerized agarose cast using a preparation needle **c** Ready-to-use mounting cast

60 mg and 100 mg LMPA in 16 mL of E3 in the microwave (0,375% and 0,625%, respectively). Per stamped cast, 2 aliquots of 1.6 mL were prepared in 2 mL tubes for each LMPA concentration, and placed in a heating block adjusted to 41 °C. For live imaging, 400 μL of 4,2 mg/mL Tricaine (25X) were added to keep the embryos anesthetized during the imaging session. Final concentrations of LMPA were therefore 0,3% and 0,5%, respectively. The LMPA containing Tricaine was prepared fresh for each mounting session.

Mounting

Before mounting, the quality of the imprint was examined under a stereo-microscope and deformation of single μ-well were corrected with a preparation needle, if necessary. Since the outlines of the stamp become transparent when adding the LMPA and to be able to still locate the embryos for alignment, the illumination contrast and mirror of the transmitted light base was adjusted to see the μ-wells again (Fig. 4a-c and Additional file 2: Movie S2).

The embryos were anesthetized in the petri dish they grew in with 4 or 5 drops of 4,2 mg/mL Tricaine (40 μg/mL in E3) added 4–5 min before usage. For mounting, the imprint was first gently filled from the border (Fig. 4a and Additional file 2: Movie S2) with 500 μL of 0.3% LMPA. Then, 44 embryos (for a 44 μ-well stamp) were collected from their petri dish with a glass Pasteur pipette (Additional file 2: Movie S2). To minimize the amount of liquid added to the LMPA, the embryos were allowed to sink to the air – liquid interface and immediately added in one drop to the liquid LMPA in the stamped cast.

Each embryo was moved to a separate μ-well with a preparation needle. The yolk was positioned within the half-spherical structure of each well, and the tail aligned horizontally with the shape of the well (Fig. 5 and Movie S2). The LMPA was allowed to polymerize for about 40 min. For time-lapse recording longer than 1 h, 1 mL of 0.5% LMPA was carefully added on top and allowed to polymerize for another 10 min to stabilize the structure.

The speed of polymerization of the LMPA is critical and should not be too fast to give enough time to place and orient each embryo in a μ-well. The temperature of the room should be around 23 °C. For indefinite time of embryo orientation, higher room temperature or a 5 V terrarium heating mat (at maximum temperature, ca. 38 °C) can be used. For this, a hole with the diameter of an imaging dish should first be cut in the middle of the heating mat. For mounting, the mat should be placed on the stereomicroscope stage with the dish in the hole.

Extra attention should be given when adding 0.5% LMPA on top of the 0.3% LMPA for time-lapse imaging.

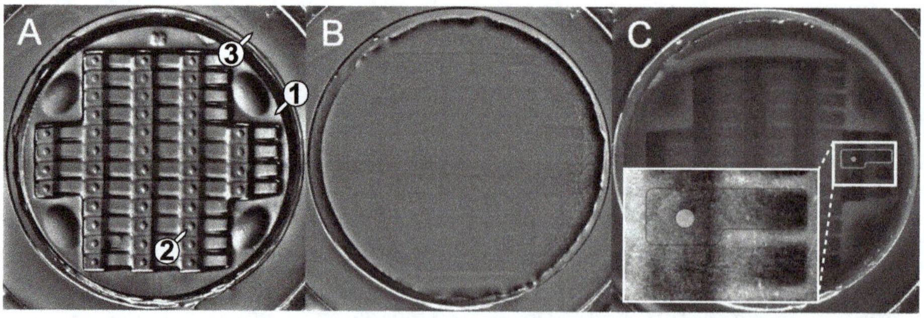

Fig. 4 Mounting cast under the stereo microscope (**a**) without LMPA and (**b, c**) with LMPA with different mirror and contrast adjustment in B (μ-wells not visible) and C (μ-wells visible) (1). inner well and mounting area (2) single μ-well (3) outer well. See also movie S2

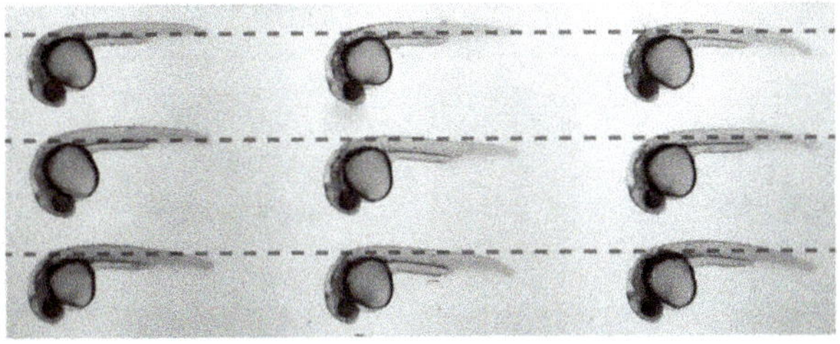

Fig. 5 Mounted and tail-aligned embryos (9 out of 44). Stripe lines indicate horizontal tail alignment

The LMPA will still be very fragile. Apply the 0.5% LMPA to the outer well first, then carefully raise the level.

Imaging setup

The dish was placed onto the sample holder of an inverted confocal spinning disc microscope so that the embryos aligned to the Y axis of the microscope stage. The stage was then moved to place the embryo at Position 01 (P01, top-left position) right above the objective (Fig. 6a).

The following steps describe the procedure on a fully motorized Nikon-Ti spinning disc system using NIS5.0, but can be performed with any commercial software:

1. Define positions (option **a** or **b**)
   a. Pre-defined points list in XY
      - Move the stage to P01 (Fig. 6a)
      - Define multipoint list ('Custom Multipoint Definition' in NIS 5.0) using the fixed distances (dX and dY) between embryos
      - dX/dY = 3450 / 1280 μm
      - Bring P01 in focus and offset all points in Z
      - Save the list to be able to re-use it in a future experiment.

To re-use a pre-existing point list: load the list, move to P01, bring it in focus and offset all points in X, Y and Z to align all points to reference P01.

   b. Custom well plate
      - Define a new custom well plate (Fig. 6C)
      - Calibrate the stage
      - Move to each position

2. Refine Positions
   - Move to the single positions one after another. If necessary, refine X, Y and Z and update the coordinates.
3. Start imaging

Retrieval

For further experiments such as genotyping, the embryos were retrieved from the agarose in the same sequence as they were imaged (snake by column, Fig. 2f-f'). For this purpose, a glass pipette was directed to the head region of an embryo and a gentle under pressure was applied till the embryo was gently sucked into the glass pipette (Fig. 7a). Each embryo was then placed in a single tube of an 8-tube PCR strip and lysed to extract genomic DNA. The genotyping PCR was then performed and analysed by gel electrophoresis using an 8x-multichannel pipette. When using a 34-well comb, the pipette tips will reach every second well of the agarose gel. Filling the wells staggered (offset by 1), one can load 4 × 8 wells in one row (Fig. 2f'''). Since each embryo has a defined position, it is straightforward to associate each genotype to the corresponding image data (Fig. 2f-f''). Since a single mismatch would mess up the entire experiment by resulting in a frameshift of the one-to-one correspondence, this is a very important feature. The imaging dish can be reused several times. For cleaning remove the agarose bed from the dish using a small scoop or preparation needle (Fig. 7b) and wipe it gently with a lint-free tissue soaked in ethanol.

Offset calculations

Multi-point coordinates of three independent experiments, each with information of 39 points and coordinates

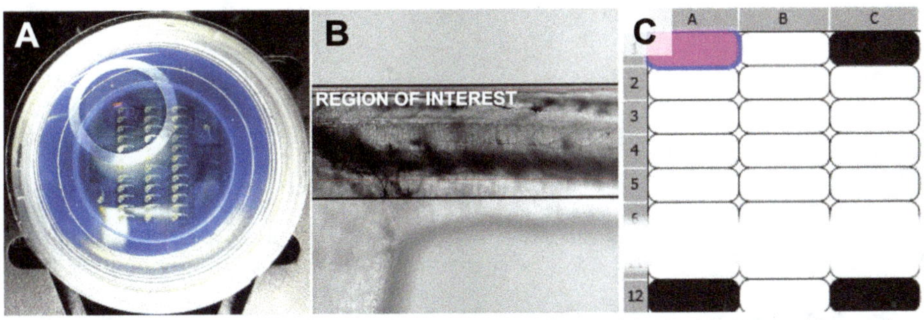

Fig. 6 Imaging setup. **a** Stage positioned on embryo at position 1 (P01 – upper left corner) **b** ROI definition and focus refinement in brightfield. **c** Navigation through the μ-wells by a custom defined well plate. Magenta well is current. Black wells are used to calibrate the well plate

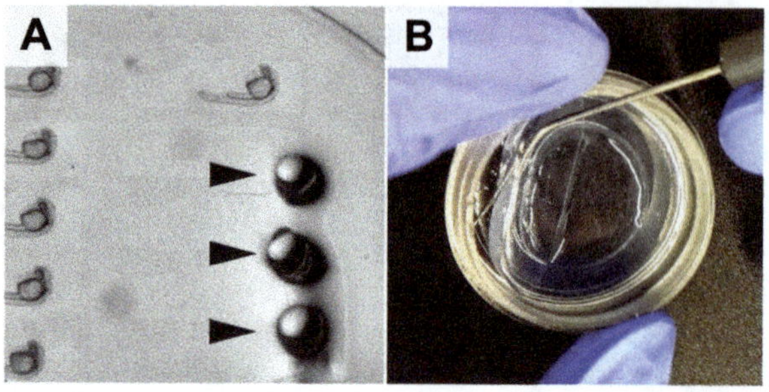

Fig. 7 Embryo retrieval. **a** Arrowheads point to incisions indicating locations where the pipette was introduced into the agarose to retrieve the embryos **b** Removal of the agarose bed with a preparation needle

in X, Y and Z, were saved and used to demonstrate mounting precision (Additional file 3: Dataset S3). For each set of 39 points the global means of X, Y and Z were calculated and used as references for distance calculation. For graphical representation ggplot2 for R was used [33].

1. Calculate global means

$$Reference(X) = \mu_X = \mathbf{X_0}$$
$$Reference(Y) = \mu_Y = \mathbf{Y_0}$$
$$Reference(Z) = \mu_Z = \mathbf{Z_0}$$

2. Calculate Offset

$$\boldsymbol{Offset}(\mathrm{X})_n = \mathrm{X}_n - \mathrm{X}_0$$
$$\boldsymbol{Offset}(\mathrm{Y})_n = \mathrm{Y}_n - \mathrm{Y}_0$$
$$\boldsymbol{Offset}(\mathrm{Z})_n = \mathrm{Z}_n - \mathrm{Z}_0$$

Materials

Description	Company	Catalog No.
µ-Dish	Ibidi	81,218–200
Stamp	Printed with a Formlabs 'Form1+' 3D printer. Resin = Formlabs 'Tough' photopolymer. Resolution = 25x50x100 µm	–
Preparation needles	VWR	USBE5470
Pasteur Pipettes	Roth	4518
Rubber or silicone bulb	VWR	612–2327
Microtubes 2 mL	Sarstedt	2691
Heating block	PeqLab	HX2
Microwave oven	Severin	MW7849
Stereo microscope	Leica	M165FC
Transmitted Light Base	Leica	MDG36
Agarose	Sigma	A9539
Low-melt Agarose	Roth	6351.2
16% Formaldehyde (w/v), Methanol-free	Thermo Fischer	28,908
Tricaine methane sulfonate	Sigma	A5040
NaCl	Roth	9265.2
KCl	Roth	P017.1
$CaCl_2$	Roth	886.1
$MgSO_4$*$7H_20$	Roth	T888.2
Countersunk screw DIN7991, 8 × 20 mm	Dresselhaus (Hornbach)	7,662,389
Superglue Blitzschnell Pipette	UHU (bueroshop24)	509,141
10X P-APO λ/0.45/4.0	Nikon	MRD00105
20X APO LWD λ-S/0.95/0.95	Nikon	MRD77200
40X APO LWD λ-S/1.15/0.6	Nikon	MRD77410
Eclipse Ti-E	Nikon	MEA53100
Motorized XY-table	Nikon	MHTIMOT-E
Piezo Z-Table (300 µm range)	Nikon	MHPIEZOZT
Universal Sample Holder	Nikon	MH00555001020
CSU-W1	Nikon	BIO1
ZYLA Plus	Nikon	BIO4

Supplementary information

: **Additional file 1: Movie S1.** Multi-position time-lapse. The movie shows a multi-position, multi-timepoint (15 h / 10 min. interval) dual-channel Z-projection of confocal-Z-stacks of about 50 slices (2.5 µm spacing). Two fields of view were stitched per embryo.

: **Additional file 2: Movie S2.** Mounting procedure. The video is divided in 5 consecutive episodes. (E1) Add mounting medium to the agarose cast (E2) Adjust Light path to make the µ-wells visible again (E3) Move mounting dish aside and collect embryos (E4) Apply embryos to the centre of the mounting dish (5) Distribute embryos.

: **Additional file 3: Dataset S3.** Multi-point coordinates. Single coordinates represent the centre of each respective organ in X, Y and Z as perceived by using *cldnb:lyn-gfp* transgenic embryos. The data is provided as '.xml' files, the native points-list data format of NIS-Elements. To parse the files and process the data R was used [34] [35]. The documentation with appropriate code comments is presented as '.html and '. Rmd' files.

Abbreviations

dX: Distance in X; dY: Distance in Y; E1–5: Episode 1–5; EDF: Extended Depth of Focus; FOV: Field of View; hpf: Hours post fertilization; LMPA: Low-melting point agarose; LSFM: Light-sheet fluorescence microscope; MIP: Maximum Intensity Projection; NM: Neuromast; P01: Position 1; pLL: Posterior lateral line; pLLP: Posterior lateral line primordium; SNR: Signal-to-Noise Ratio; µ-well: Micro-well

Acknowledgements

We are grateful to Denise Werner and Tore Dittrich for excellent fish care. Special thanks go to Christine Molenda, who was using and helping to mature this protocol. Furthermore, we thank Andreas Ebert, Felix Groden and Dmitri Baulin for rigorous testing of the method and Immersive Boardgames (Immersive Boardgames, 2018) for the excellent printout quality and short delivery time.

Authors' contributions

DSK designed the study, carried out the experiments and analysis of the results. VL provided the infrastructure and funding for performing the experiments. DSK and VL wrote the final manuscript. All authors read and approved the final manuscript.

Funding

This work was supported by the Deutsche Forschungsgemeinschaft (DFG-SPP1782 project number LE2681 to V. L and INST 161/896-1 FUGG to V.L.). Funding bodies were not involved in the design of the study nor in the collection, analysis, and interpretation of the data or in writing the manuscript.

Ethics approval and consent to participate

Animals used in this study were obtained from own breeding at the animal facility of the Goethe University Frankfurt according to the German animal welfare act and approved by the German authorities (veterinary department of the Regional Board of Darmstadt).

Competing interests

The authors declare that they have no competing interests.

References

1. Thorn K. A quick guide to light microscopy in cell biology. Mol Biol Cell. 2016;27(2):219–22.
2. Fischer RS, Wu Y, Kanchanawong P, Shroff H, Waterman CM. Microscopy in 3D: a biologist's toolbox. Trends Cell Biol. 2011;21(12):682–91.
3. Mavrakis M, Pourquié O, Lecuit T. Lighting up developmental mechanisms: how fluorescence imaging heralded a new era. Development. 2010;137(3):373–87.
4. Huisken J, Stainier DYR. Selective plane illumination microscopy techniques in developmental biology. Development. 2009;136(12):1963–75.
5. Davidson LA, Baum B. Making waves: the rise and fall and rise of quantitative developmental biology. Development. 2012;139(17):3065–9.
6. Rittscher J. Characterization of biological processes through automated image analysis. Annu Rev Biomed Eng. 2010;12:315–44.
7. Megason SG, Fraser SE. Digitizing life at the level of the cell: high-performance laser-scanning microscopy and image analysis for in toto imaging of development. Mech Dev. 2003;120(11):1407–20.
8. Keller PJ, Schmidt A, Wittbrodt J, Stelzer E. Reconstruction of zebrafish early embryonic development by scanned light sheet microscopy. Science. 2008; 322(5904):1065–69. https://doi.org/10.1126/science.1162493.
9. Beis D, Stainier DYR. In vivo cell biology: following the zebrafish trend. Trends Cell Biol. 2006;16(2):105–12.
10. Weber M, Mickoleit M, Huisken J. Multilayer mounting for long-term light sheet microscopy of zebrafish. J Vis Exp. 2014;84:e51119.
11. Mickoleit M, Schmid B, Weber M, Fahrbach FO, Hombach S, Reischauer S, et al. High-resolution reconstruction of the beating zebrafish heart. Nat Methods. 2014;11(9):919–22.
12. Montero J-A, Heisenberg C-P. Gastrulation dynamics: cells move into focus. Trends Cell Biol. 2004;14(11):620–7.
13. Höckendorf B, Thumberger T, Wittbrodt J. Quantitative analysis of embryogenesis: a perspective for light sheet microscopy. Dev Cell. 2012;23(6):1111–20.
14. Campinho P, Lamperti P, Boselli F, Vermot J. Three-dimensional microscopy and image analysis methodology for mapping and quantification of nuclear positions in tissues with approximate cylindrical geometry. 373. Phil Trans R Soc B. https://doi.org/10.1098/rstb.2017.0332.
15. Donoughe S, Kim C, Extavour CG. High-throughput live-imaging of embryos in microwell arrays using a modular specimen mounting system. Biology Open. 2018;7:bio031260. https://doi.org/10.1242/bio.031260. Published 9 July 2018.
16. Wittbrodt JN, Liebel U, Gehrig J. Generation of orientation tools for automated zebrafish screening assays using desktop 3D printing. BMC Biotechnol. 2014;14(1):36.
17. Yu T, Jiang Y, Lin S. A 3-dimensional (3D)-printed Template for High Throughput Zebrafish Embryo Arraying. J Vis Exp. 2018;(136):e57892. https://doi.org/10.3791/57892.
18. Ghysen A, Dambly-Chaudiere C. The lateral line microcosmos. Genes Dev. 2007;21(17):2118–30.
19. Kaufman CK, White RM, Zon LI. Chemical genetic screening in the zebrafish embryo. Nat Protoc. 2009;4(10):1422–32.
20. Pulak R. Tools for automating the imaging of zebrafish larvae. Methods. 2016;96:118–26.
21. Wielhouwer EM, Ali S, Al-Afandi A, Blom MT, Riekerink MBO, Poelma C, et al. Zebrafish embryo development in a microfluidic flow-through system. Lab Chip. 2011;11(10):1815–24.
22. Herrgen L, Schröter C, Bajard L, Oates AC. Multiple embryo time-lapse imaging of zebrafish development. Methods Mol Biol. 2009;546:243–54.
23. Swinburne IA, Mosaliganti KR, Green AA, Megason SG. Improved long-term imaging of embryos with genetically encoded α-Bungarotoxin. PLoS One. 2015;10(8):e0134005.
24. Hirsinger E, Steventon B. A Versatile Mounting Method for Long Term Imaging of Zebrafish Development. J Vis Exp. 2017;(119):e55210. https://doi.org/10.3791/55210.
25. Masselink W, Wong JC, Liu B, Fu J, Currie PD. Low-cost silicone imaging casts for zebrafish embryos and larvae. Zebrafish. 2014;11(1):26–31.
26. Alessandri K, Andrique L, Feyeux M, Bikfalvi A, Nassoy P, Recher G. All-in-one 3D printed microscopy chamber for multidimensional imaging, the UniverSlide. Sci Rep. 2017;7:42378.
27. Chronis N. Worm chips: microtools for C. elegans biology. Lab Chip. 2010;10(4):432–7.
28. Crane MM, Chung K, Stirman J. Lu H. microfluidics-enabled phenotyping, imaging, and screening of multicellular organisms. Lab Chip. 2010;10(12):1509–17.
29. Daetwyler S, Günther U, Modes CD, Harrington K, Huisken J. Multi-sample SPIM image acquisition, processing and analysis of vascular growth in zebrafish. Development. 2019;146:6.
30. Power RM, Huisken J. A guide to light-sheet fluorescence microscopy for multiscale imaging. Nat Methods. 2017;14(4):360–73.
31. Haas P, Gilmour DT. Chemokine signaling mediates self-organizing tissue migration in the zebrafish lateral line. Dev Cell. 2006;10(5):673–80.
32. Yakulov TA, Todkar AP, Slanchev K, Wiegel J, Bona A, Groß M, et al. CXCL12 and MYC control energy metabolism to support adaptive responses after kidney injury. Nat Commun. 2018;9(1):3660.
33. Wickham H. ggplot2. Springer; 2016. 1 p.
34. Wickham H, Hester J, Ooms J. xml2: Parse XML [Internet]. CRAN.R-project.org. 2018 [cited 2019 Apr 3]. Available from: https://CRAN.R-project.org/package=xml2
35. R Core Team. R: a language and environment for statistical computing [internet]. www.R-project.org. Vienna, Austria; 2018. Available from: https://www.R-project.org/. [cited 2019 Apr 3]

16

Enhanced Natamycin production by *Streptomyces natalensis* in shake-flasks and stirred tank bioreactor under batch and fed-batch conditions

Elsayed Ahmed Elsayed[1,2*], Mohamed A. Farid[2] and Hesham A. El-Enshasy[3,4]

Abstract

Background: Natamycin is an antifungal polyene macrolide antibiotic with wide applications in health and food industries. Currently, it is the only antifungal food additive with the GRAS status (Generally Regarded as Safe).

Results: Natamycin production was investigated under the effect of different initial glucose concentrations. Maximal antibiotic production (1.58 ± 0.032 g/L) was achieved at 20 g/L glucose. Under glucose limitation, natamycin production was retarded and the produced antibiotic was degraded. Higher glucose concentrations resulted in carbon catabolite repression. Secondly, intermittent feeding of glucose improved natamycin production due to overcoming glucose catabolite regulation, and moreover it was superior to glucose-beef mixture feeding, which overcomes catabolite regulation, but increased cell growth on the expense of natamycin production. Finally, the process was optimized in 7.5 L stirred tank bioreactor under batch and fed-batch conditions. Continuous glucose feeding for 30 h increased volumetric natamycin production by about 1.6- and 1.72-folds in than the batch cultivation in bioreactor and shake-flasks, respectively.

Conclusions: Glucose is a crucial substrate that significantly affects the production of natamycin, and its slow feeding is recommended to alleviate the effects of carbon catabolite regulation as well as to prevent product degradation under carbon source limitation. Cultivation in bioreactor under glucose feeding increased maximal volumetric enzyme production by about 72% from the initial starting conditions.

Keywords: Natamycin, Catabolite repression, Degradation, *Streptomyces natalensis*, Feeding, Bioreactor

Background

Natamycin is a polyene antibiotic produced by different *Streptomyces* species in aerobic submerged cultivations. It is also known as pimaricin, tennecetin or natacyn [1, 2]. It belongs to the tetraene macrolide antibiotics due to the presence of 4 conjugated double bonds in its lactone ring, which is in turn attached to a mycosamine unit [3, 4]. Major producing strains are *S. natalensis, S. gilvosporeus, S. chattanogenesis* and *S. lydicus* [5, 6]. Natamycin exhibits its antimicrobial effect by disrupting the ergosterol content in fungal cell walls, and therefore is characterized by its wide spectrum of antifungal activity against yeast, molds and protozoa [7, 8]. Accordingly, natamycin has found a wide range of applications in the treatment of skin infections, ophthalmic mycoses, oral candidiasis, vaginal candidiasis, bronchopulmonary infections and many systemic fungal infections [9, 10]. Moreover, natamycin is widely applied in many food industries as a preservative due to its low toxicity to mammalian cells compared to other antimicrobial compounds, and the unlikelihood development of bacterial resistance [1, 11]. It is considered one of the most important antibiotics added as food preservative approved by the US-FDA and the European Union. Moreover, it is currently the only antifungal food additive with a GRAS (Generally Regarded as Safe) status [12]. It is used in

* Correspondence: eaelsayed@ksu.edu.sa
[1]Bioproducts Research Chair, Zoology Department, Faculty of Science, King Saud University, 11451 Riyadh, Kingdom of Saudi Arabia
[2]Chemistry of Natural and Microbial Products Department, National Research Centre, Dokki, Cairo, Egypt

many cheese and sausage ripening processes, food packages or coatings due to its inability to diffuse into the product. It is also used to prevent fungal growth in beverages as well as in crop protection [13, 14].

The production of natamycin is usually carried out in submerged cultivations. Different strategies have been used to maximize natamycin production. These include optimization of medium composition and cultivation parameters [2, 5, 15, 16], engineering of the biosynthetic pathway [17], random mutation and selection [18] and genome shuffling and chromosomal integration [19]. Furthermore, addition of short-chain carboxylic or fatty acids has been found to improve natamycin production [5, 20]. The production process of antibiotics mainly depends on applying batch or fed-batch modes [21–24]. Fed-batch is characterized by feeding one or more of the components of the production medium, usually the limiting components, to avoid problems associated with their depletion from medium. Pathway engineering and genetic manipulation greatly improved natamycin production, where the improved strains produced about about 3.3 and 8.2 g/L natamycin in shake-flask and bioreactor levels, respectively [17, 19]. On the other hand, optimization of medium composition and cultivation parameters in shake-flask level resulted in the production of 1.5–1.6 g/L natamycin [15, 16]. Li et al. [20] were able to produce about 10.38 g/L natamycin in shake-flask cultivations with the addition of lower alcohols. On the other hand, combining butanol addition with random mutation resulted in the production of 1.62 and 2.8 g/L natamycin in shake-flask and bioreactor levels, respectively [18]. We have previously investigated the effect of addition of short-chain carboxylic acids on the production of natamycin in both shake-flask and bioreactor levels. The optimized cultivation and addition conditions resulted in the production of 3.0 and 3.98 g/L natamycin in both cultivation systems, respectively [5].

The production of polyene macrolides is limited by the application of glucose as a carbon source due to its integration in the antibiotic molecule [25]. However, glucose is a fast assimilated C-source and is known to repress the production of many antibiotics due to carbon catabolite regulation [26]. Despite prolonged industrial interest in natamycin, relatively little information has been published regarding production processes under the influence of glucose limitation.

In the present work, the effect of initial glucose concentration on the kinetics of cell growth, glucose consumption and natamycin production was investigated. Moreover, the effect of intermittent feeding of glucose or glucose-beef extract mixture on natamycin production was investigated in shake-flask cultivations. Finally, the process of natamycin production was studied in 7.5 L-stirred tank bioreactor under batch and continuous feeding conditions.

Results

Effect of initial glucose concentration on kinetics of cell growth and natamycin production in batch cultures

These experiments aimed at investigating the effect of different initial glucose concentration on cell growth and production kinetics in shake flask cultivations. The obtained results (Fig. 1, Table 1) clearly showed that initial glucose concentration significantly affected production process parameters ($p = 0.001$). Under all glucose concentrations, cells grew exponentially for the first 40 h, where they reached their maximal cell dry weight. After which, cells entered the stationary phase for the rest of the cultivation time. It can be seen that both maximal cell dry weight and cell growth rates increased proportionally with the initial glucose concentration. At 50 g/L glucose, a maximal cell dry weight of 5.66 ± 0.19 g/L was obtained, which was 73.6% higher than X_{max} obtained at 10 g/L. Furthermore, it can be seen that average cell growth rates at 20, 30, 40 and 50 g/L increased by about 33.0, 46.7, 60.0 and 88.0%, respectively, than the Q_x obtained at 10 g/L. Additionally, the maximal cell dry weights obtained at 30–50 g/L recorded comparable values by the end of cultivation (Table 1).

For glucose consumption, results showed that both 10 and 20 g/L glucose cultivations were characterized by complete glucose consumption at an average consumption rate of − 0.22 and − 0.41 g/L/h after 40 and 80 h, respectively. On the other hand, cultivations supplemented with higher glucose concentration (Q_s) did not suffer from glucose depletion, and their average glucose consumption rates increased with increasing glucose concentrations (− 0.44, − 0.51 and − 0.60 g/L/h, for 30, 40 and 50 g/L, respectively). By the end of the cultivation time, about 16.7, 29.8 and 29.9% of the initial glucose remained in the cultivation media supplemented with 30, 40 and 50 g/L glucose, respectively.

Concerning natamycin production, results showed that natamycin production started at all glucose concentrations after 8 h. At 10 g/L glucose, natamycin was produced with an average production rate of 0.011 g/L/h and reached its maximal volumetric production of 0.35 ± 0.04 g/L after 40 h. Increasing glucose concentration to 20 g/L has a pronounced effect on natamycin production, where its production rate increased by about 91% (0.02 g/L/h) and reached a maximal production of 1.58 ± 0.032 g/L after 80 h. On the other hand, further increase in glucose concentration (≥ 30 g/L) adversely affected natamycin production where the maximal volumetric productivity decreased by about 48.1, 72.8 and 72.8% at 30, 40 and 50 g/L, respectively, from the maximal production at 20 g/L. Moreover, the average production rate at these cultivations decreased and ranged from 0.006 to 0.013 g/L/h. The collective data in Table 1 shows that the growth and production coefficient yields recorded their maximal values at 20 g/L

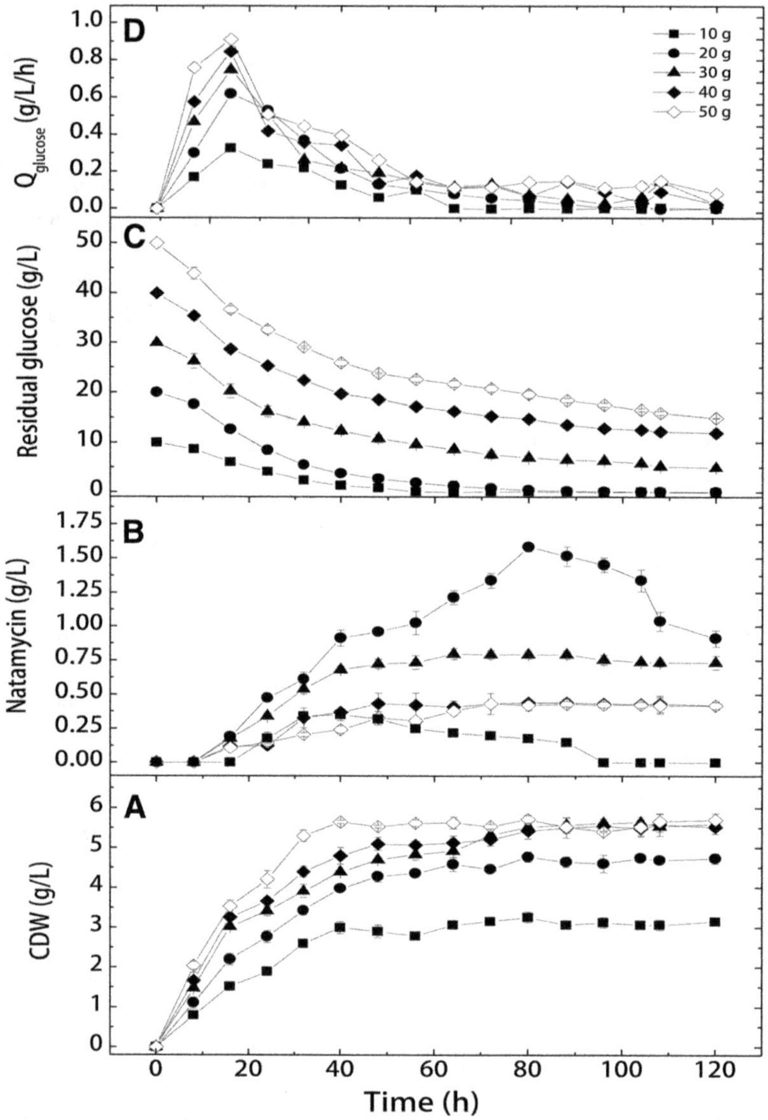

Fig. 1 Effect of different initial glucose concentrations on cell growth and natamycin production *by S. natalensis* (CDW, cell dry weight; $Q_{glucose}$, glucose consumption rate)

Table 1 Kinetic parameters of cell growth and natamycin production by *S. natalensis* NRRL 2651 as affected by initial glucose concentration

Concentration (g/L)	X_{max} (g/L)	P_{max} (g/L)	ΔS (%)	Q_x^a (g/L/h)	Q_p^a (g/L/h)	Q_s^a (g/L/h)	$Y_{x/s}$ (g/g)	$Y_{p/s}$ (g/g)	$Y_{p/x}$ (g/g)
10	3.26 ± 0.13	0.35 ± 0.04	100	0.075	0.011	- 0.22	0.35	0.038	0.121
20	4.78 ± 0.13	1.58 ± 0.03	100	0.099	0.021	- 0.41	0.28	0.083	0.332
30	5.66 ± 0.14	0.82 ± 0.03	85.8	0.11	0.013	- 0.39	0.39	0.044	0.185
40	5.59 ± 0.23	0.43 ± 0.02	72	0.12	0.006	- 0.38	0.31	0.027	0.122
50	5.66 ± 0.19	0.43 ± 0.07	70	0.14	0.006	- 0.50	0.31	0.015	0.102

X_{max}: maximal cell dry weight; P_{max}: maximal natamycin production; ΔS: % consumed glucose; Q_X: average cell growth rate; Q_P: average natamycin production rate; Q_S: average glucose consumption rate; $Y_{X/S}$: maximal yield of biomass per consumed glucose; $Y_{P/S}$: maximal yield of natamycin per consumed glucose; $Y_{P/X}$: maximal yield of natamycin per biomass. [a] Data are calculated at the end of the exponential growth phase

glucose. Furthermore, at 20 g/L, it can be seen that the substrate conversion yield into biomass was the least recorded (0.28 g cells/g consumed glucose), while natamycin yields per consumed glucose and produced cell mass were at their maximal values (0.083 g natamycin/g consumed glucose, and 0.33 g natamycin/g cells).

It can be further observed that at 10 and 20 g/L glucose the natamycin production decreased gradually with time reaching a minimum of 0.15 ± 0.02 and 0.913 ± 0.06 g/L at 88 and 120 h, respectively. Furthermore, in cultivation with 20 g/L glucose, natamycin completely disappeared in culture broth after 96 h, in other words, 16 h after complete glucose consumption. Therefore, the data representing the progress of natamycin production rates at different initial glucose concentrations were summarized in Fig. 2. It can be seen that the production rates increased initially at all glucose concentrations, and then started to slow down. At 10 and 20 g/L glucose, natamycin production rates started to decrease after 40 and 80 h and recorded negative rate values (– 0.0053 and – 0.0143 g/L/h, respectively) indicating that the produced natamycin was degraded in the medium (rates below the base line). On the other hand, at higher glucose concentrations, natamycin production rates decreased and were almost around the zero level (– 0.0012, – 0.0004 and – 0.0007 g/L/h at 30, 40 and 50 g/L, respectively), meaning that natamycin production was stopped. In such case, the produced natamycin still can be determined in the production medium at all three concentrations (30–50 g/L).

Effect of intermittent feeding of glucose-beef extract mixture on cell growth and natamycin production by *S. natalensis* in shake flasks

During these cultivations, a sterile feeding solution containing glucose and beef extract was fed to the growing cells at the same initial C:N ratio (before glucose depletion). Feeding started after 48 h of cultivation and was repeated every 24 h for 5 repeated batches. Obtained results (Fig. 3) showed that cells grew exponentially for the first 48 h with an average growth rate of 0.08 g/L/h, reaching 3.86 ± 0.021 g/L, and consumed glucose at an average consumption rate of – 0.27 g/L/h. Upon intermittent addition, cells continued to grow almost with the same growth rate (0.07 g/L/h) and reached a maximal cell growth of 12.25 ± 0.45 g/L at 168 h. However, during intermittent feeding of glucose-beef extract mixture, the average glucose consumption rate was increased by about 1.15-folds (– 0.311 g/L/h). Consequently, glucose started to accumulate in the production medium with each feeding batch, where a final residual concentration of 19.6 ± 1.27 g/L of glucose was found by the end of the cultivation (39% of the total added glucose).

For natamycin production, it can be seen that during the first normal batch (48 h), natamycin was produced by a normal batch production rate of 0.022 g/L/h, where volumetric production reached 1.057 ± 0.09 g/L. However, after the first feeding batch (at 72 h), natamycin concentration recorded 1.61 ± 0.02 g/L, a 20% increment from the 1.34 ± 0.05 g/L obtained in normal batch cultivation after 72 h (Fig. 1). After the 2nd feeding batch (at 96 h), natamycin concentration reached 1.709 ± 0.009 g/L, which corresponds to 17.5% increase from its corresponding batch value (1.454 ± 0.05 g/L

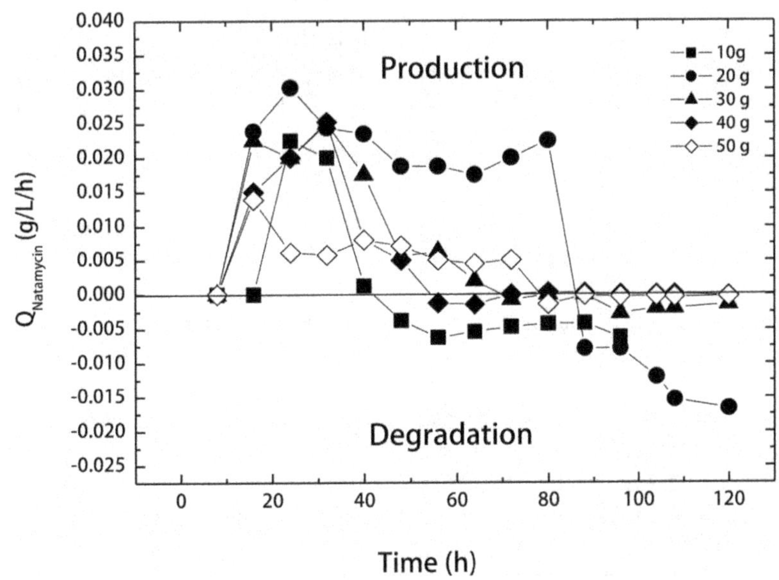

Fig. 2 Production and degradation rates of natamycin in batch culture as affected by different initial glucose concentrations ($Q_{natamyccin}$, natamycin production rate)

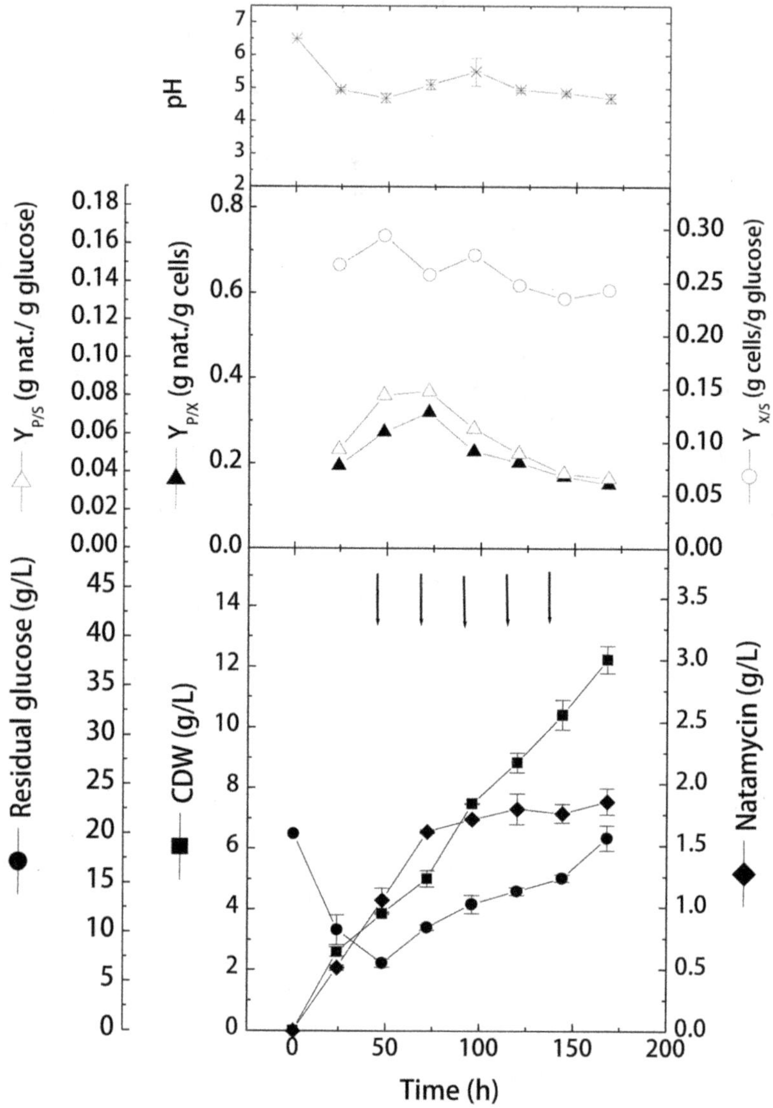

Fig. 3 Effect of intermittent feeding of glucose and beef extract on cell growth and natamycin production by *S. natalensis* in shake-flask cultivation (CDW, cell dry weight; $Y_{X/S}$: yield of biomass per consumed glucose; $Y_{P/S}$: yield of natamycin per consumed glucose; $Y_{P/X}$: yield of natamycin per biomass)

at 96). Although batch data (Fig. 1) showed that natamycin started to decompose in the production medium after 88 h, and reached a final of 0.913 ± 0.06 g/L by 120 h, however, the intermittent feeding significantly improved natamycin production ($p = 0.02$), where a final concentration of 1.855 ± 0.11 g/L was obtained at 168 h.

Yield coefficient results showed that *S. natalensis* cells continued to produce natamycin after the 1st addition batch with yield coefficients similar to those reported in the batch cultivation ($Y_{x/s}$, 0.26 g cells/g consumed glucose; $Y_{P/s}$, 0.083 g natamycin/g consumed glucose; $Y_{P/X}$, 0.32 g natamycin/g cells). However, yield coefficients decreased with further addition batches.

Effect of intermittent feeding of glucose on cell growth and natamycin production by *S. natalensis* in shake flasks

This part of the work was designed to investigate the effect of intermittent feeding of glucose only on the kinetics of cell growth and natamycin production. The cultivation was performed as previously described, except that the feeding solution contained only glucose at a concentration of 10 g/L. Results obtained (Fig. 4) showed that, similar to previous batch experiment, cells grew with an average growth rate of 0.085 g/L/h and a glucose consumption rate of – 0.31 g/L/h reaching 4.09 ± 0.16 g/L at 48 h. After starting the intermittent glucose feeding phase, cells continued to grow exponentially for the rest of the cultivation. However, it can be seen that the average growth rate decreased to 0.04 g/L/h, which

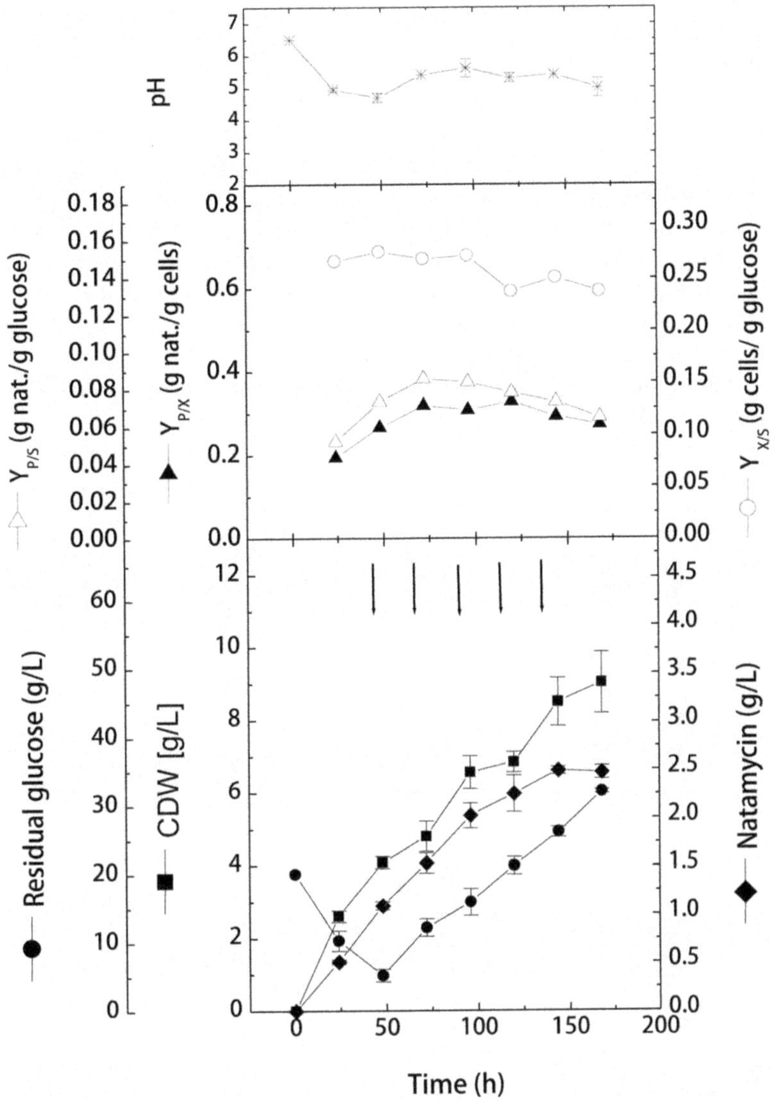

Fig. 4 Effect of intermittent feeding of glucose on cell growth and natamycin production by *S. natalensis* in shake-flask cultivation (CDW, cell dry weight; $Y_{X/S}$: yield of biomass per consumed glucose; $Y_{P/S}$: yield of natamycin per consumed glucose; $Y_{P/X}$: yield of natamycin per biomass)

is about 42.9% lower than the average growth rate recorded in case of feeding glucose-beef mixture (0.07 g/L/h). This lower growth rate was reflected on the maximal cell growth obtained by the end of cultivation (9.03 ± 0.82 g/L), which was lower by about 26.3% from the maximal growth obtained in glucose-beef feeding.

During batch -wise glucose feeding phase, the average glucose consumption decreased from − 0.31 to − 0.19 g/L/h, which is about 60% of the average glucose consumption rate obtained during batch-wise glucose-beef feeding (− 0.31 g/L/h). This decrease in substrate consumption rate highly increased glucose accumulated in the cultivation flasks, where it recorded 32 ± 0.29 g/L at 168 h, representing about 64% of the total fed glucose.

Looking at natamycin production kinetics, it can be observed that after the 1st addition batch, natamycin production increased by 1.4-folds recording 1.533 ± 0.11 g/L, and then increased with each glucose addition until reaching a maximal of 2.49 ± 0.04 g/L at 144 h. Additionally, it can be seen that after the 1st glucose addition, natamycin production increased by about 18, 25.7, 41.5 and 33.3% after the 2nd, 3rd, 4th and 5th addition, respectively, than those concentrations obtained after the corresponding glucose-beef additions. It is further worth to mention that the maximal produced natamycin in this experiment (2.49 ± 0.04 g/L) was 1.57-folds higher that the maximal concentration obtained in normal shake flask batch (1.58 ± 0.03 g/L). Yield coefficient results showed that $Y_{P/s}$ was slightly improved from 0.083 to 0.086 g natamycin/g consumed glucose for glucose-beef and glucose addition, respectively, suggesting that more glucose was converted into natamycin.

Cell growth and natamycin production 7.5 L-stirred tank bioreactor under batch cultivation mode

This part of the work aimed at investigating the effect of transferring natamycin production process from shake-flask level into stirred tank bioreactor level. *S. natalensis* cells were inoculated into 7.5 L-stirred tank bioreactor with 5 L working volume. The cultivation was firstly performed under batch conditions to monitor cell behavior and production kinetics. Results obtained (Fig. 5) showed that bioreactor cultivation significantly improved kinetics of cell growth and natamycin production in comparison to shake-flask batch cultivation ($p = 0.003$). During the exponential growth phase, cells grew with an average growth rate of 0.111 g/L/h, and reached 6.3 ± 0.04 g/L of cell growth. Although cells grew almost with the same growth rate recorded in shake-flask cultivation (0.099 g/L/h), however, in bioreactor cultivation, the exponential growth phase was extended to 56 h (14 h longer than that in shake-flask) before entering the stationary phase. This longer active growth was reflected on the higher cell concentration obtained before the onset of the stationary phase (1.97-folds higher than cell concentration in shake-flask culture). The dissolved oxygen (DO) concentration started to decrease sharply with time up to 20 h, at an average DO consumption rate of – 2.5%/h. After 20 h, DO concentration remained more or less constant until the end of cultivation, meaning that dissolved oxygen was not limiting. A final cell growth of 6.46 ± 0.04 g/L was obtained at 103 h.

Furthermore, in bioreactor cultivation, the average glucose consumption rate increase by about 23% (– 0.443 g/L/h) than that recorded in shake-flask cultivation (– 0.36 g/L/h). Such higher consumption rate can also be seen from the highest yield coefficient ($Y_{x/s}$), which recorded 0.73 g cells/g

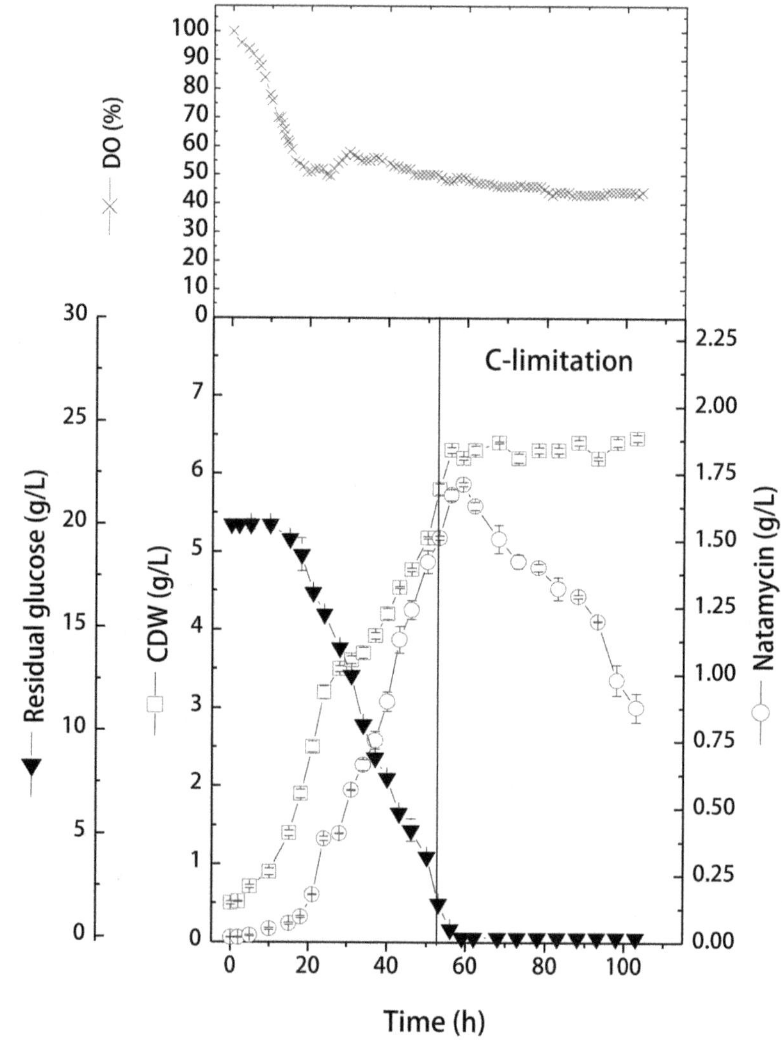

Fig. 5 Cell growth and natamycin production by *S. natalensis* in 7.5 L-stirred tank bioreactor under batch cultivation mode (CDW, cell dry weight; DO, dissolved oxygen)

consumed glucose (0.28 g cells/g consumed glucose in shake-flask cultivation).

Natamycin was produced during the stationary growth phase with an average production rate of 0.038 g/L/h, 1.8-folds higher than that recorded in shake-flask cultivation. Consequently, a maximal natamycin production of 1.71 ± 0.01 g/L was obtained after 59 h. However, similar to shake-flask cultivation, it can be seen that natamycin degradation started directly after glucose depletion (60 h), while in shake-flask cultivation, degradation started after 80 h. Also, in both cultivations, the natamycin degradation rates were similar (− 0.014 and − 0.019 g/L/h for shake-flask and bioreactor cultivations, respectively).

Kinetics of cell growth, glucose consumption and natamycin production by *S. natalensis* during fed-batch cultivation in 7.5 L-stirred tank bioreactor

The fed-batch bioreactor cultivation was designed to investigate the effect of continuous glucose feeding on the kinetics of cell growth and natamycin production in order to overcome the problems of glucose limitation. After inoculation, the cultivation was run as normal batch for the first 50 h. Based on the glucose consumption rate obtained from the bioreactor batch run (− 0.443 g/L/h), sterile glucose solution was continuously fed to the bioreactor at a feeding rate of 0.5 g/L/h. Feeding was started shortly before glucose becomes limited (at 3.8 g/L), and this phase lasted for the next 30 h. After 80 h, feeding was stopped and the cultivation mode was turned again into batch mode. Results (Fig. 6) showed that the first batch session proceeded as normal batch where cells reached 5.12 ± 0.04 g/L with a natamycin production of 1.4 ± 0.035 g/L. During this phase, growth and production kinetic parameters were similar to a great extent to those obtained in the previous batch cultivation.

Upon feeding glucose to the production medium, cells continued to grow exponentially for the first 14 h of feeding with a similar growth rate (0.103 g/L/h), and reached their maximal growth of 6.21 ± 0.05 g/L at 64 h. Afterwards, cell growth entered the stationary phase, and remained more or less constant. During the feeding phase, glucose consumption remained constant, where no accumulation was observed and the average available glucose concentration ranged from 4.0 ± 0.07 to 4.4 ± 0.04 g/L. Therefore, natamycin production continued during the feeding phase, where the average production rate increased up to 0.04 g/L/h (34.5 and 95.7% higher than the production rate in bioreactor and shake-flask batch cultivations, respectively). Consequently, natamycin production reached 2.52 ± 0.07 g/L by the end of the feeding phase, and continued to be produced during the second batch phase (at a smaller production rate), where a maximal production of 2.72 ± 0.01 g/L was obtained at 103 h. It can be seen that feeding increased maximal natamycin by 1.6- and 1.72-folds from the maximal production

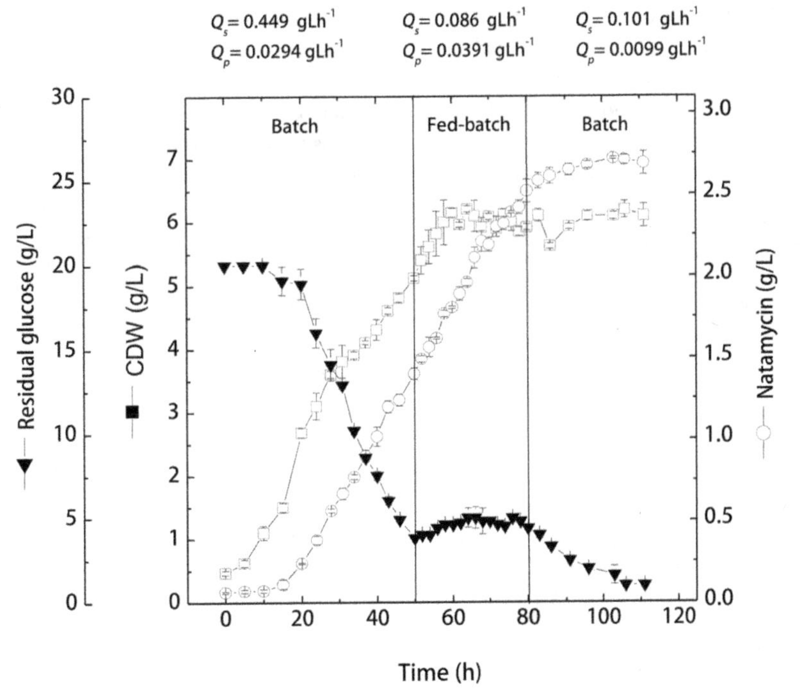

Fig. 6 Kinetics of cell growth, glucose consumption and natamycin production by *S. natalensis* during fed-batch cultivation in 7.5 L-stirred tank bioreactor (CDW, cell dry weight)

obtained in bioreactor and shake-flask batch cultivations, respectively.

The production yield coefficients in terms of consumed glucose and producing cells ($Y_{p/s}$, $Y_{p/x}$) were significantly improved in response to continuous glucose feeding. The $Y_{p/x}$ increased from 0.28 g natamycin/g cells (at 50 h) reaching a maximal of 0.42 g/g by 80 h, and increased further during the 2nd batch phase to reach 0.47 g/g at 86 h. Similarly, $Y_{p/s}$ increased from 0.09 g natamycin/g consumed glucose reaching its maximal at 80 h (0.16 g/g, 2-folds the yield before feeding).

Finally, Table 2 summarized different kinetic parameters obtained for cell growth and natamycin production under different cultivation conditions. It can be seen that the final cultivation in bioreactor under glucose feeding conditions significantly ($p = 0.003$) improved process parameters in terms of cell growth, substrate consumption, and natamycin production kinetics.

Discussion

Obtained results showed that glucose concentration has a pronounced effect on natamycin production by *S. natalensis*. Increasing glucose concentration up to 20 g/L increased natamycin production, and further increase in its concentration had remarkably decreased the concentration of natamycin. Furthermore, cell growth kinetics, glucose consumption and natamycin production rates, as well as yield coefficients, showed their maximal values at 20 g/L. These results are in accordance with those previously reported [5, 15], where highest natamycin concentration were obtained at lower glucose concentrations. Additionally, higher glucose concentrations showed higher cell growth levels. This can be explained on the basis of carbon catabolite repression and feedback regulation of higher glucose concentrations. Generally, actinomycetes respond negatively towards higher concentrations of glucose and similar low molecular readily metabolizable carbon sources [25]. Higher glucose concentrations usually stimulate higher growth and retarded the production of secondary metabolites. It has been reported that readily utilizable sugars repress or inhibit the enzymatic machinery responsible for the biosynthesis of many antibiotics, e.g. puromycin, kanamycin, penicillin and cephalosporin [27, 28].

Recently, it has been postulated that carbon catabolite repression regulates glucose metabolism by the phosphoenolpyruvate-dependent phospotransferase system affecting glucose kinase (Glk), which acts only on glucose and converts it into glucose-6-phosphate [25, 29]. It has been found that unknown glucose metabolites activate Glk, and that there is an unknown regulatory factor that binds to Glk and changes its function from catabolic one into regulatory one, leading finally to carbon catabolite repression phenomenon [30].

Furthermore, our results showed that natamycin concentrations decreased in 10–20 g/L glucose cultivations, while cultivations at higher glucose concentrations did not show such decrease in the volumetric natamycin production. This can be attributed to the cleavage of natamycin molecule by growing cells, present under glucose limitation or depletion, resulting in the release of glucose moiety from the macrolactone ring, and making it available for cells as maintenance energy [5, 15]. Similar results were reported on the decrease in natamycin concentration after reaching its maximal production [2, 18]. However, in both cases, the decrease in production was not attributed to natamycin hydrolysis, since the cultivation media contained other slowly metabolized carbon sources. Our results are also in good agreement with those reported by Marin and McDaniel [31], who suggested the same explanation for the decrease obtained in their production of candicidin after 70 h, when glucose was depleted from the medium. Additionally, it has been reported that amphotericin B has been used as

Table 2 Summarized kinetic parameters of cell growth and natamycin production by *S. natalensis* NRRL 2651 under different cultivation conditions applied

Parameter	Cultivation conditions				
	Shake-flask cultivation			Bioreactor cultivation	
	Batch (20 g/L)	Intermittent feeding (glucose+beef)	Intermittent feeding (glucose)	Batch (20 g/L)	Glucose feeding
X_{max} (g/L)	4.780 ± 0.13	12.25 ± 0.45	9.030 ± 0.84	6.460 ± 0.04	6.21 ± 0.05
P_{max} (g/L)	1.580 ± 0.03	1.855 ± 0.1	2.480 ± 0.04	1.710 ± 0.01	2.72 ± 0.01
$Q_x{}^a$ (g/L/h)	0.099	0.080	0.085	0.111	0.103[b]
$Q_p{}^a$ (g/L/h)	0.021	0.022	0.023	0.038	0.045[b]
$Q_s{}^a$ (g/L/h)	−0.41	−0.31	−0.220	−0.443	−0.448
$Y_{x/s}$ (g/g)	0.280	0.260	0.280	0.760	0.61
$Y_{p/s}$ (g/g)	0.083	0.083	0.086	0.090	0.162
$Y_{p/x}$ (g/g)	0.332	0.320	0.330	0.280	0.46

Xmax: maximal cell dry weight; Pmax: maximal natamycin production; QX: average cell growth rate; QP: average natamycin production rate; QS: average glucose consumption rate; YX/S: maximal yield of biomass per consumed glucose; YP/S: maximal yield of natamycin per consumed glucose; YP/X: maximal yield of natamycin per biomass. a Data are calculated at the end of the exponential growth phase. b Data are taken during the feeding phase

an endogenous substrate by the producing cells [32]. In such case, the produced antibiotic was transformed into the inactive form. The same authors found non-active forms of polifungin and candihexin lacking mycosamine moiety in the fermentation broth following glucose depletion, suggesting the degradation or modification of such antibiotics by the producing cells.

Another possible explanation for the decrease in natamycin concentration at higher glucose concentrations (30–50 g/L), is the fast glucose metabolism and the production of organic acids, which in turn decrease the pH of the medium. It is well known that pH shift towards acidic values results in the cleavage of aminosugar moiety of mycosamine and its release from the natamycin molecule [33]. Moreover, the propensity to autolysis is a well-known property of different actinomycetes in prolonged cultivations. This has been attributed to C-source depletion, which activates certain molecular mechanism, i.e. proteolytic cascades and nucleases, which in turn results in decreased cell proliferation, differentiation and cell death [34–36].

To overcome the carbon catabolite repression in actinomycetes, it is recommended to perform glucose cultivations with slow feeding of carbon source. Therefore, the second part of the work was conducted to investigate the effect of glucose feeding on the kinetics of cell growth and natamycin production. Firstly, intermittent repeated feeding of glucose alone or glucose-beef extract mixture was carried out on shake-flask levels. Results showed that both intermittent feeding strategies improved the production process in comparison to the preliminary batch cultivation. Volumetric natamycin production increased by about 1.3- and 1.57-folds in glucose-beef mixture or glucose feeding, respectively, than the corresponding batch cultivation. This suggests that slow feeding is useful to bypass the carbon catabolite repression regulation in *Streptomyces* in case of using glucose as a sole carbon source. Slow feeding of glucose allows higher respiration rates during antibiotic production phase (idiophase) as well as higher productivity [32]. Sugar addition has also been used for the increased production of other polyene antibiotics to overcome rapid glucose consumption [37, 38].

However, results showed that glucose feeding was superior to the feeding of glucose-beef mixture in terms of natamycin production. Also, cell growth was much higher in case of glucose-beef feeding. This can be accredited to the fact that the addition of beef extract tends to increase the concentration of organic nitrogen in the production medium. This will result in the conversion of the production phase (idiophase) into growth phase (trophophase), where the metabolism is directed towards cell growth and consequently lowered antibiotic production [28, 31]. This was accompanied by higher glucose consumption in the glucose-beef feeding, and therefore, lower amounts of glucose were accumulated by the end of the cultivation. Periodic addition of nitrogen source with the fed glucose increased growth over candidin production [28, 39]. Furthermore, the increase in cell growth in case of nitrogen source feeding can be due to the increase in phosphate content found in beef extract. Generally, complex nitrogen sources are used for their balanced contents of fatty acids, proteins, cations and low phosphate. However, the accumulation of beef extract in the medium can lead to increased levels of phosphate in the production medium, which in turns inhibits natamycin production. Phosphate effect on the production of antibiotics by streptomycetes is well known and reported [25, 40].

The production process was transferred to 7.5 L-stirred tank bioreactor to investigate the possibility of scaling-up the process from shake-flask into semi-industrial bioreactor level. Results showed that maximal natamycin production was reached at the onset of the stationary phase accompanied by complete glucose exhaustion. Additionally, natamycin production increased linearly and proportional to cell growth. Bioreactor cultivation is generally characterized by the presence of more controlled conditions in terms of pH, better oxygenation and mixing due to controlled aeration and agitation. Such conditions provide the growing cells with more adequate conditions promoting better growing environments and allowing maximal metabolic cellular activities [41, 42]. Our results also showed that oxygen consumption increased sharply for the first 20 h, representing cell growth in trophophase, and then oxygen consumption remained more or less stable for the rest of cultivation. These results are in agreement with those of Wang et al. [17], who reported on the enhancement of natamycin production by genetically manipulated *S. gilvosporeus*. They reported similar oxygen consumption rates for the first 12–20 h of cultivation, where the progress of cell growth was at its highest rates.

The batch bioreactor results showed also decreased concentrations of natamycin after glucose depletion from the cultivation medium. Accordingly, continuous feeding of glucose was carried out in bioreactor to investigate the kinetics of cell growth and natamycin production under these optimized conditions. Results showed that continuous glucose feeding significantly improved cell growth and natamycin production. Volumetric production increased by 1.6- and 1.72-folds from the concentration obtained in batch cultivations in bioreactor and shake-flasks, respectively. This was accompanied by improved specific production yields based on glucose consumption and growing cells. Also, glucose catabolite repression and product degradation were avoided and cells continue to grow and produce natamycin until the end of the cultivation. Furthermore, substrate consumption and natamycin production rates were kept nearly constant during batch and fed-batch phases. Although feeding glucose has not yet been explored for natamycin production, however, previous reports are found for the increased production of other polyene antibiotic using slow glucose feeding [28, 31]. Authors suggested that direct or indirect glucose metabolites are involved in the switch between primary and secondary metabolism. Generally, fed-batch cultivation is the choice for improving antibiotic production processes [21–23]. It is used in the production of secondary metabolites, which strictly require fast metabolizable carbon sources in order to avoid their fast metabolism and the conversion from idiophase into torphophase [37, 38]. Furthermore, Wang et al. [43] reported on the production of isovaleryispiramycin under glucose limitation by investigating the effect of different glucose concentrations on the activity of the rate limiting enzyme in the biosynthetic pathway, branched-chain α-keto acid dehydrogenase (BCKDH). They found that glucose depletion in the late production phase significantly decreased the activity of the BCKDH and consequently isovaleryispiramycin concentration. They suggested glucose feeding during this phase to keep glucose levels between 0.0 and 1.0 g/L, which significantly improved enzyme activity and hence, isovaleryispiramycin production.

Conclusions

From the aforementioned results, it can be inferred that initial glucose concentration significantly affects natamycin production through activation or inhibition of carbon catabolite repression mechanism. Slow glucose feeding was found to support better natamycin production and prevents the degradation of the product due to glucose limitation. Furthermore, the addition of glucose and beef extract shifted the glucose metabolism and the cultivation from the idiophase into the trophophase, favoring cell growth and retarding natamycin production. Finally, 7.5 L-stirred tank bioreactor was used to optimize the fed-batch cultivation for natamycin production. Such fed-batch resulted in the production of 2.72 ± 0.01 g/L of natamycin after 103 h, which corresponds to an increase of about 1.6- and 1.72-folds from the maximal production obtained in bioreactor and shake-flask batch cultivations, respectively.

Methods

Microorganism and inoculum preparation

Streptomyces natalensis (NRRL 2651) obtained from the National Center for Agricultural Utilization Research (NRRL, Peoria, Illinios, USA), was used throughout this work. Inoculum in the form of spores obtained from a densely sporulated culture grown on ISP agar medium for 7 days at 30 °C was used [5]. The arisen spores were harvested with a sterile physiological saline solution (0.9% NaCl, w/v), and were counted directly using a haemocytometer slide. The natamycin production medium was inoculated with spore suspension at a final concentration of 1×10^7 spores/mL.

Cultivation media

ISP agar medium [16], used for sporulation, was composed of (g/L) as follows: Malt extract (Difco), 10.0; Glucose (BDH), 4.0; and Yeast extract (Difco), 4.0. The final pH was adjusted to 7.2 before sterilization. Unless otherwise stated, natamycin production medium was composed as follows (g/L): Glucose (Difco), 20.0; Yeast extract (Difco), 2.0; Beef extract (Difco), 8.0 and DL-Asparagine (Fluka), 0.5. The final pH was adjusted to 7.2 before sterilization. Glucose was sterilized separately and added to the medium directly before inoculation. Cultivations were carried out using Erlenmeyer flasks with a total volume of 250 mL containing 50 mL medium. The flasks were cultivated on a rotary shaking incubator at 30 °C and 200 rpm (New Brunswick Scientific Co., Edison, New Jersey, USA).

Bioreactor cultivation

Submerged cultivations were carried out in a 7.5 L stirred tank bioreactor Microferm (MF-107, New Brunswick Scientific Co., New Brunswick, NJ, USA) with a working volume of 5 L. Sterilization was carried out at 121 °C for 30 min. and the separately sterilized glucose was added after cooling. The bioreactor vessel was inoculated as previously described [15], and agitation was performed using a three 4-bladed rushton turbine impellers ($d_{i(impeller\ diameter)} = 48$ mm; $d_{t(tank\ diameter)} = 143$ mm, $d_i d_t^{-1} = 0.34$) at 600 rpm. Aeration was performed by filtered sterile air (1 v/v/m). Dissolved oxygen concentrations were analyzed by polarographic electrode (Ingold, Germany). The pH was adjusted to 7.2 by the addition of 2.5 M NaOH and controlled during the cultivation. Foam was suppressed, when necessary, by the addition of silicon antifoam reagent (Fluka, Switzerland).

Feeding cultivations on shake-flask and bioreactor levels

This section was designed to evaluate the effect of feeding glucose (or glucose + beef extract) on the kinetics of cell growth and natamycin production by *S. natalensis* NRRL 2651. In shake-flask cultures, after inoculation and cell growth for 48 h, 5 mL of sterilized feeding solution (10 g/L glucose, or 10 g/L glucose + 4 g/L beef extract) were added to the growing cells under aseptic conditions. The feeding solution of glucose and beef extract was prepared at the same C:N ratio in the initial cultivation medium. Intermittent feeding was repeated every 24 h for 5 batches. At the end of each addition run, representative samples (2 flasks) were withdrawn and necessary analyses were performed. In case of bioreactor cultivations, cells were allowed to grow in a batch mode for 50 h, and then sterilized glucose solution was continuously fed at a feeding rate of 0.5 g/L/h (depending on the rate of glucose consumption calculated form the bioreactor batch experiments). Continuous glucose feeding was conducted for 30 h, and then the bioreactor cultivation mode was switched again to normal batch mode. Samples were taken and assayed at regular intervals.

Analyses

Sample preparation

During cultivation progress, samples were periodically withdrawn (a: shake flasks: 3 flasks, each 50 mL broth, or b: bioreactor: 10 mL broth). The volume of the samples was determined using pre-weighed sterile Falcon centrifuge tubes (BD Biosciences, Franklin Lakes, NJ, USA), and samples were then centrifuged at 5000 rpm for 20 min. A small fraction of the supernatant was frozen at – 20 °C for glucose and natamycin determination.

Determination of cell dry weight

The remaining pelleted cells in the Falcon tubes were washed thoroughly using saline solution, and were then centrifuged again. The final supernatant was discarded and the cell pellet was allowed to dry overnight at 80 °C until a constant weight was reached [44]. The final obtained cell dry weight was expressed as g/L.

Determination of natamycin concentration

The determination of natamycin concentration in the fermentation broth was carried out by two methods. Firstly, the fermentation broth was assayed biologically by standard agar plate diffusion method using *Saccharomyces cerevisiae* as the test organism [5]. Aliquots of the cultivation broth were loaded into wells bored in agar plates, which were previously seeded with the test organism at standard assay conditions. Plates were incubated at 30 °C. The diameters of the inhibition zones were recorded after 24 h. A biological standard curve was drawn between the logarithms of different concentrations of standard natamycin (Gist-Brocades, Delft, Netherlands) and the inhibition zone.

For confirmation, natamycin concentrations were determined using HPLC (Agilent 1200, Agilent Technologies, USA) as described in Wang et al. [17]. The HPLC system was equipped with a UV detector (303 nm) and a Zorbax Eclipse C18 column. The mobile phase contained NH_4Cl: acetonitrile: tetrahydrofuran (75:20:5) and it was run at a flow rate of 0.5 mL/min. A standard natamycin (Gist-Brocades, Delft, Netherlands) calibration curve was used to determine natamycin concentration.

Determination of glucose concentration

The concentration of residual glucose in the medium was detected spectrophotometrically. A glucose enzyme determination kit (Biocon Diagnostik GmbH, Burbach, Germany) was used.

Calculation of kinetic parametrs

$$Q_x \text{ Cell growth rate } (g/L/h) = \frac{X_2 - X_1}{t_2 - t_1} \quad (1)$$

where X_2 and X_1 are cell mass concentrations at time t_2 and t_1, respectively.

$$Q_s, \text{ Glucose consumption rate } (g/L/h) = \frac{C_2 - C_1}{t_2 - t_1} \quad (2)$$

where C_2 and C_1 are glucose concentrations at time t_2 and t_1, respectively.

$$Q_P, \text{ Natamycin production/degradation rate } (g/L/h) = \frac{P_2 - P_1}{t_2 - t_1} \quad (3)$$

where P_2 and P_1 are natamycin concentrations at time t_2 and t_1, respectively.

Statistical analysis and result reproducibility

Experiments were repeated thrice to ensure result reproducibility. In case of shake flask cultures, three flasks were simultaneously withdrawn at the same interval, and their contents were assayed separately. Results were analyzed using SPSS 9.0, and were expressed as mean ± SD for three replicates. ANOVA one-way analyses were used to analyze results, and statistical significance was defined as $p < 0.05$.

Abbreviations

ANOVA: Analysis of variance

Acknowledgments

Not Applicable.

Authors' contributions

EAE, MAF and HAE proposed the research concept and designed the experiments. They provided necessary tools for experiments, experimental instructions, analyzed and interpreted the data and wrote the manuscript. EAE and HAE carried out the experimental work. All authors read and approved the final manuscript.

Funding

This work was funded by the Deanship of Scientific Research, King Saud University through Vice Deanship of Scientific Research Chairs. Funding was obtained based on the proposed study and the funding bodies had no further role in data collection, analysis, interpretation or manuscript preparation.

Ethics approval and consent to participate

Not applicable.

Consent for publication

Not applicable.

Competing interests

The authors declare that they have no competing interests.

Author details
[1]Bioproducts Research Chair, Zoology Department, Faculty of Science, King Saud University, 11451 Riyadh, Kingdom of Saudi Arabia. [2]Chemistry of Natural and Microbial Products Department, National Research Centre, Dokki, Cairo, Egypt. [3]Institute of Bioproduct Development (IBD), Universiti Teknologi Malaysia (UTM), 81130 UTM Skudai, Malaysia. [4]City of Scientific Research and Technology Application, New Burg Al Arab, Alexandria, Egypt.

References

1. Aparicio JF, Barreales EG, Payero TD, Vicente CM, de Pedro A, Santos-Aberturas J. Biotechnological production and application of the antibiotic pimaricin: biosynthesis and its regulation. Appl Microbiol Biotechnol. 2016;100:67–78.
2. Zeng Z, Miao W, Zeng H, Zhao K, Zhou Y, Zhang J, Zhao Q, Tursun D, Xu D, Li F. Production of natamycin by *Streptomyces gilvosporeus* Z28 through solid-state fermentation using agro-industrial residues. Bioresour Technol. 2019;273:377–85.
3. Liu SP, Yuan PH, Wang YY, Liu XF, Zhou ZX, Bu QT, Yu P, Jiang H, Li YQ. Generation of the natamycin analogs by gene engineering of natamycin biosynthetic genes in *Streptomyces chattanoogensis* L10. Microbiol Res. 2015;173:25–33.
4. Martín JF, Aparicio JF. Enzymology of the polyenes pimaricin and candicidin biosynthesis. Methods Enzymol. 2009;459:215–42.
5. Elsayed EA, Farid MA, El-Enshasy HA. Improvement in natamycin production by *Streptomyces natalensis* with the addition of short-chain carboxylic acids. Process Biochem. 2013;48:1831–8.
6. Atta HM, El-Sayed AS, El-Desoukey MA, Hassan M, El-Gazar M. Biochemical studies on the Natamycin antibiotic produced by *Streptomyces lydicus*: fermentation, extraction and biological activities. J Saudi Chem Soc. 2015;19:360–71.
7. Raab W. Pimaricin, ein Antibioticum gegen Pilze und Trichomonaden. Arzneimittelforschung. 1967;17:538–43.
8. Sharma N, Aron N, Agarwal T, Sharma C. Antimicrobial agents for ocular use: bacterial, fungal, viral, and protozoal infections. In: Velpandian T (ed.) Pharmacology of ocular therapeutics, Springer International Publishing, Switzerland, 2016, pp. 285–332.
9. Shulka PK, Singh P, Yadav RK, Pandey S, Bhunia SS. Past, present, and future of antifungal drug development. Top Med Chem. 2018;29:125–68.
10. Hurst WJ, Finley JW, de Man JM. Additives and contaminants. In: deMan JM (ed.) Principles of food chemistry, Springer International Publishing, Switzerland, 2018, pp. 429–474.
11. Wu H, Liu W, Dong D, Li J, Zhang D, Lu C. *SlnM* gene overexpression with different promoters on natamycin production in *Streptomyces lydicus* A02. J Ind Microbiol Biotechnol. 2014;41:163–72.
12. Kallscheuer N. Engineered microorganisms for the production of food additives approved by the European Union - a systematic analysis. Front Microbiol. 2018;9:1–20 Article ID 1746.
13. Jiang T, Feng L, Zheng X, Li J. Physicochemical responses and microbial characteristics of shiitake mushroom (*Lentinus edodes*) to gum Arabic coating enriched with natamycin during storage. Food Chem. 2013;138:1992–7.
14. Fox PF, McSweeney PLH, Cogan TM, Guinee TP. Microbiology of cheese ripening. Fundamentals of Cheese Science. In: Fox PF, McSweeney PLH, Cogan TM, Guinee TP, editors. Fundamentals of cheese science. New York: Springer; 2017. p. 333–90.
15. Farid MA, El-Enshasy HA, El-Diwany AI, Elsayed EA. Optimization of the cultivation medium for natamycin production by *Streptomyces natalensis*. J Basic Microbiol. 2000;40:157–66.
16. El-Enshasy HA, Farid MA, Elsayed EA. Influence of inoculum type and cultivation conditions on natamycin production by *Streptomyces natalensis*. J Basic Microbiol. 2000;40:333–42.
17. Wang S, Liu F, Hou Z, Zong G, Zhu X, Ling P. Enhancement of natamycin production on *Streptomyces gilvosporeus* by chromosomal integration of the *Vitreoscilla* hemoglobin gene (*vgb*). World J Microbiol Biotechnol. 2014;30:1369–76.
18. Ge J, Wang C, Huang S, Du R, Liu K, Song G, Ping W. Biosynthesis regulation of natamycin production from *Streptomyces natalensis* HDMNTE-01 enhanced by response surface methodology. Prep Biochem Biotechnol. 2017;47:939–44.
19. Qi Z, Zhou Y, Kang Q, Jiang C, Zheng J, Bai L. Directed accumulation of less toxic pimaricin derivatives by improving the efficiency of a polyketide synthase dehydratase domain. Appl Microbiol Biotechnol. 2017;101:2427–36.
20. Li M, Chen S, Li J, Ji Z. Propanol addition improves natamycin biosynthesis of *Streptomyces natalensis*. Appl Biochem Biotechnol. 2014;172:3224–432.
21. Elsayed EA, Omar HG, El-Enshasy HA. Development of fed-batch cultivation strategy for efficient oxytetracycline production by *Streptomyces rimosus* at semi-industrial scale. Braz Arch Biol Biotechnol. 2015;58:676–85.
22. Omar H, Elsayed EA, Kenawy AA, Wadaan MA, El-Enshasy HA. Effect of cultivation scale and shear stress on cell growth and oxytetracycline production by *Streptomyces rimosus* in semi-defined medium. J Pure Appl Microbiol. 2014;8:121–8.
23. Boumehira AZ, El-Enshasy HA, Hacène H, Elsayed EA, Aziz R, Park EY. Recent progress on the development of antibiotics from the genus *Micromonospora*. Biotechnol Bioprocess Eng. 2016;21:199–223.
24. El-Enshasy HA, Beshay UI, El-Diwany AI, Omar HM, El-Kholy AE, El-Najar R. Improvement of rifamycins production by *Amycolatopsis mediterranei* in batch and fed-batch cultures. Acta Microbiol Polonica. 2003;51:301–13.
25. Wohlleben W, Bera A, Mast Y, Stegmann E. Regulation of secondary metabolites of *Actinobacteria*. In: Wink J, Mohammadipanah F, Hamedi J (eds.) Biology and biotechnology of *Actinobacteria*, Springer International Publishing, Switzerland, 2017, pp. 181–232.
26. Drew SW, Demain AL. Effect of primary metabolites on secondary metabolism. Ann Rev Microbiol. 1977;31:343–9.
27. Martin JF, Demain AL. Control of antibiotic biosynthesis. Microbiol Rev. 1980;4:230–51.
28. Martin JF, McDaniel LE. Submerged culture production of the polyene antifungal antibiotics candidin and candihexin. Dev Ind Microbiol. 1974;15:324–37.
29. van Wezel GP, König M, Mahr K, Nothaft H, Thomae AW, Bibb M, Titgemeyer F. A new piece of an old jigsaw: glucose kinase is activated posttranslationally in a glucose transport-dependent manner in *Streptomyces coelicolor* A3(2). J Mol Microbiol Biotechnol. 2007;12:67–74.
30. Görke B, Stülke J. Carbon catabolite repression in bacteria: many ways to make the most out of nutrients. Nat Rev Microbiol. 2008;6:613–24.
31. Martin JF, McDaniel LE. Kinetics of biosynthesis of polyene macrolide antibiotics in batch culture: cell maturation time. Biotechnol Bioeng. 1975;17:925–38.
32. Martin JF, McDaniel LE. Production of polyene macrolide antibiotics. Adv Appl Microbiol. 1977;21:1–52.
33. Belakhov VV, Garabadzhiu AV. Polyene macrolide antibiotics: mechanisms of inactivation, ways of stabilization, and methods of disposal of unusable drugs. Russ J Gen Chem. 2015;85:2985–3001.
34. Filippova SN, Vinogradova KA. Programmed cell death as one of the stages of Streptomycete differentiation. Microbiology. 2017;86:439–54.
35. Farid MA, El-Diwany AI, Daniel EN, El-Enshasy HA. Production of oxytetracycline and lysine using mixed-culture fermentation. Chem Mikrob Technol der Lebensm. 1995;17:139–44.
36. El-Enshasy HA, Farid MA, El-Diwany AI. Oxytetracycline production by free and immobilized cells of *Streptomyces rimosus* in batch and repeated batch cultures. Prog Biotechnol. 1996;11:437–43.
37. Siewert G, Kieslich K. Preparation and ultraviolet light-induced transformation of an antifungal mixture of heptaene antibiotics of *Streptomyces surinam*. Appl Microbiol. 1971;21:1007–10.
38. Tereshin IM. Polyene antibiotics: present and future. University of Tokyo Press, Tokyo, Japan. 1970.
39. Gil JA, Liras P, Martin JF. The polyenes: properties, biosynthesis, and fermentation. In: Vandamme EJ, editor. Biotechnology of industrial antibiotics. New York and Basel: Marcel Dekker Inc.; 1984. p. 513–29.
40. Barreiro C, Martinez-Castro M. Regulation of the phosphate metabolism in *Streptomyces* genus: impact on the secondary metabolites. Appl Microbiol Biotechnol. 2019;103:1643–58.
41. Dailin DJ, Elsayed EA, Othman NZ, Malek R, Phin HS, Aziz R, Wadaan M, El Enshasy HA. Bioprocess development for kefiran production by *Lactobacillus kefiranofaciens* in semi industrial scale bioreactor. Saudi J Biol Sci. 2016;23:495–502.
42. El-Enshasy HA, Elsayed EA, Suhaimi N, Malek RA, Esawy M. Bioprocess optimization for pectinase production using *Aspergillus niger* in a submerged cultivation system. BMC Biotechnol. 2018;18:71 https://doi.org/10.1186/s12896-018-0481-7.
43. Wang Y-H, Wu C-H, Chu J, Hao Y-Y, Zhuang Y-P, Zhang S-L. Regulation of branched-chain amino acid catabolism: glucose limitation enhances the component of isovalerylspiramycin for the bitespiramycin production. Bioprocess Biosyst Eng. 2010;33:257–65.
44. Elsayed EA, El-Enshasy HA. Effects of different aeration rates and feeding strategies on cell growth and invertase production kinetics by *Saccharomyces boulardii*. J Sci Ind Res. 2018;77:575–82.

Multiplex quantitative PCR for single-reaction genetically modified (GM) plant detection and identification of false-positive GM plants linked to *Cauliflower mosaic virus* (CaMV) infection

Aurélie Bak and Joanne B. Emerson*

Abstract

Background: Most genetically modified (GM) plants contain a promoter, P35S, from the plant virus, *Cauliflower mosaic virus* (CaMV), and many have a terminator, TNOS, derived from the bacterium, *Agrobacterium tumefaciens.* Assays designed to detect GM plants often target the P35S and/or TNOS DNA sequences. However, because the P35S promoter is derived from CaMV, these detection assays can yield false-positives from non-GM plants infected by this naturally-occurring virus.

Results: Here we report the development of an assay designed to distinguish CaMV-infected plants from GM plants in a single multiplexed quantitative PCR (qPCR) reaction. Following initial testing and optimization via PCR and singleplex-to-multiplex qPCR on both plasmid and plant DNA, TaqMan qPCR probes with different fluorescence wavelengths were designed to target actin (a positive-control plant gene), P35S, P3 (a CaMV-specific gene), and TNOS. We tested the specificity of our quadruplex qPCR assay using different DNA extracts from organic watercress and both organic and GM canola, all with and without CaMV infection, and by using commercial and industrial samples. The limit of detection (LOD) of each target was determined to be 1% for actin, 0.001% for P35S, and 0.01% for both P3 and TNOS.

Conclusions: This assay was able to distinguish CaMV-infected plants from GM plants in a single multiplexed qPCR reaction for all samples tested in this study, suggesting that this protocol is broadly applicable and readily transferrable to any interested parties with a qPCR platform.

Keywords: *Cauliflower mosaic virus*, CaMV, GMO, GM plant, Multiplex qPCR, Detection methods

Background

A genetically modified (GM) plant possesses genetic material that has been modified in order to introduce a new gene or trait to the plant, for example herbicide resistance, disease resistance, or insect tolerance [1]. Most of the engineered genetic constructs in GM plants are built with the 35S promoter (P35S) from *Cauliflower Mosaic Virus* (CaMV) and the NOS terminator (TNOS) derived from the soil-borne bacterium, *Agrobacterium tumefaciens.* Indeed, according to studies published in 2014–2015, P35S and TNOS are used in 65.7 and 53.49% of all commercialized GM crops, respectively, and either or both were used in 81.4% of all GM events, indicating that most GM plants can be detected by methods that target P35S and TNOS [2, 3]. To comply with GM crop legislation and respect consumer transparency, reliable GM plant detection methods have been developed, mostly focused on P35S and TNOS [4–8] (Bak A, Emerson JB. Towards distinguishing Cauliflower mosaic virus (CaMV) infection from genetic modification (GM) in crop plants: detection assays and biology, management, and food safety of CaMV. Submitted). However, it is known that CaMV infection of

* Correspondence: jbemerson@ucdavis.edu
Department of Plant Pathology, University of California, Davis, CA 95616, USA

non-GM plants can yield false-positive results from GM plant detection assays, due to the presence of the P35S region in both the CaMV genome and many GM crops [9, 10]. Therefore, a non-GM plant infected with CaMV could be incorrectly identified as a genetically modified organism (GMO).

Most GM plant detection methods follow DNA-based approaches, particularly polymerase chain reaction (PCR) and quantitative PCR (qPCR), because they allow for fast and specific GM plant screening [2, 11]. The PCR method uses DNA sequence-specific primers and a DNA polymerase enzyme to amplify new DNA strands from existing DNA strands (e.g., DNA from the sample of interest) as templates [12]. qPCR methods use, in addition to the same PCR reagents, a double-stranded deoxyribonucleic acid (dsDNA) fluorescent binding dye or a DNA sequence-specific fluorescent probe that allows for the quantification of the newly amplified DNA molecules by measuring the fluorescence produced during each PCR cycle [13, 14]. The number of amplification cycles required to reach a fixed fluorescence signal threshold (i.e., exceeding background levels to be considered detected) is called the Ct (cycle threshold) value. Two different methods of qPCR are extensively used, SYBR Green (non-specific dsDNA binding) and TaqMan [13, 15]. The TaqMan method uses a DNA sequence-specific hydrolysis probe labelled with a fluorophore and a quencher [16], and although the Taqman method is more specific than SYBR Green, both of these assays can be adapted to be relatively or absolutely quantitative [17–19].

New highly sensitive, reliable, and cost-effective GM plant detection methods continue to be developed [6, 20–22]. Nevertheless, few methods can discriminate between GM plants and CaMV infection, and all currently available methods would require multiple assays to identify false-positive GM plants. For example, in [23], the authors described PCR tests to detect P35S and the P3 gene of CaMV for differentiating between GM plants and CaMV infection in parallel reactions [23]. Here we sought to adapt these single-reaction PCR assays, together with PCR assays for TNOS, into a single multiplex qPCR assay to detect CaMV infection and GM plants that use constructs with P35S or TNOS, thereby minimizing reagent costs and processing time.

In this paper, we present an absolute multiplex TaqMan qPCR assay combining primers and TaqMan probes that can identify most GM events and can discriminate GM plants from CaMV infection. Four TaqMan qPCR probes with different fluorescence wavelengths were designed to target actin (a positive-control gene universal to plants), P35S, P3 (a CaMV specific gene), and TNOS. After the validation of the primers and probes using regular PCR and SYBR Green methods, the TaqMan method was used and multiplexed for the simultaneous detection of all four targets. The specificity of the primers and probe sets was tested using different combinations of plant DNA (non-GM and GM plants, infected or not with CaMV, and commercial and industrial samples) and the limit of detection (LOD) for each of the four targets was assessed. We present both the optimization process for developing this assay, along with tests to assess its robustness. For all samples tested, this assay was able to distinguish CaMV-infected plants from GM plants in a single multiplexed qPCR reaction.

Results

A workflow chart is presented in Fig. 1, showing the steps that we followed to develop the TaqMan multiplex qPCR assay for distinguishing true GM plants from false-positive GM plants due to CaMV infection. Briefly, literature-derived forward and reverse specific primer sets (Table 1) [23–25] were tested by PCR to confirm that they specifically amplified the corresponding target and template (Additional file 2: Figure S1). One primer from each set was then tested via qPCR with the probe as the second primer to ensure probe specificity, using a SYBR Green assay (Additional file 3: Figure S2). The forward and reverse primer pairs were then tested together with the fluorescent probe for each target, using a TaqMan assay, initially in singleplex (one target per reaction) and then progressively increasing the number of targets and probes by multiplexing two, three, and, eventually, all four sets in the same reaction. The efficacy of the TaqMan quadruplex (hereafter, multiplex) assay with all four targets (actin, P35S, P3, and TNOS) was confirmed by comparison of each target in TaqMan singleplex vs. multiplex qPCR assays (Fig. 2). After validation, the specificity of the multiplex assay was tested using different combinations of plant DNA (Fig. 3) and commercial and industrial plant DNA samples (Fig. 4 and Fig. 5). Finally, the limit of detection (LOD) was determined for each target (Table 2 and Additional file 4: Table S2). The following sections describe further details for each step of this process.

Confirmation of primer and plasmid efficacy by PCR

Forward and reverse specific primers (see Table 1) were tested by regular PCR to confirm amplification of the corresponding target DNA sequence, using plasmid and then plant DNA as templates (Additional file 2: Figure S1). After cloning the actin and P3 genes into separate pDonr207 vectors, two resulting plasmids (#1 and #2) for each target were tested by PCR using the actin and P3 primers, respectively. The PCR positively amplified the corresponding targets, which appeared as single gel electrophoretic bands of the appropriate size, with no observed non-specific amplification (Additional file 2: Figure S1a). The actin and P3 plasmids labeled #1 were

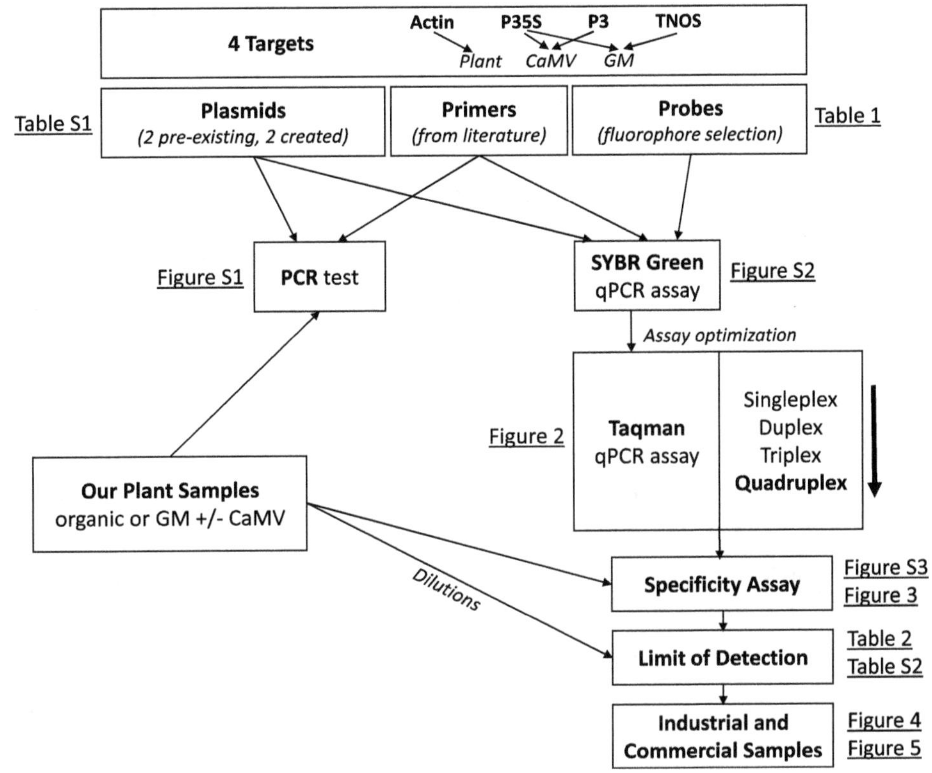

Fig. 1 Workflow chart. Diagram representing the different steps followed to develop the multiplex qPCR assay

then used for subsequent experiments. P35S and TNOS primers were similarly able to specifically amplify their targets in the pICH51277 and pICH41421 plasmids, respectively (Additional file 2: Figure S1a).

After primer efficacy was confirmed using the plasmid DNA templates, primers were similarly tested using plant DNA templates. The actin primers were confirmed by PCR using all three of our uninfected plant DNA extracts (GM canola, watercress, and non-GM canola) separately as templates (Additional file 2: Figure S1b). The efficacy of P35S and TNOS primers was confirmed by PCR using GM canola DNA as template (Additional file 2: Figure S1c). Finally, the CaMV infection success in all of our infected plants and our ability to specifically amplify CaMV targets in DNA from these CaMV-infected plants was confirmed by PCR using P3 and P35S primers on template DNA from CaMV-infected watercress, non-GM canola, and GM canola (Additional file 2: Figure S1d).

Table 1 Primer table. Table with the primers and probes used in the qPCR assays

Target	Primers	Sequences	Amplicon (bp)	References
Actin = Plant reference gene	Actin	Forward 5′-CAAGCAGCATGAAGATCAAGGT-3′	103	[24]
		Reverse 5′-CACATCTGTTGGAAAGTGCTGAG-3′		
		Probe 5′-HEX-CCTCCAATCCAGACACTGTACTTYCTCTC-BHQ-3′		
CaMV Promoter 35S	P35S	Forward 5′-CGTCTACAAAGCAAGTGGATTG-3′	79	[25]
		Reverse 5′-TCTTGCGAAGGATAGTGGGATT-3′		
		Probe 5′-FAM-TCTCCACTGACGTAAGGGATGACGCA-QSY-3′		
CaMV gene P3	P3	Forward 5′-TGAAATCCTCAGTGACCAAAAATC-3′	152	[23]
		Reverse 5′-TACAAGGACAATCATTGATGAGC-3′		
		Probe 5′-ABY-AAGCCGTTGCAGCGAAAATCGTTAATGA-QSY-3′		
A.tumefaciens nopaline synthase terminator	TNOS	Forward 5′-GTCTTGCGATGATTATCATATAATTTCTG-3′	151	[24]
		Reverse 5′-CGCTATATTTTGTTTTCTATCGCGT-3′		
		Probe 5′-JUN-AGATGGGTTTTTATGATTAGAGTCCCGCAA-QSY-3′		

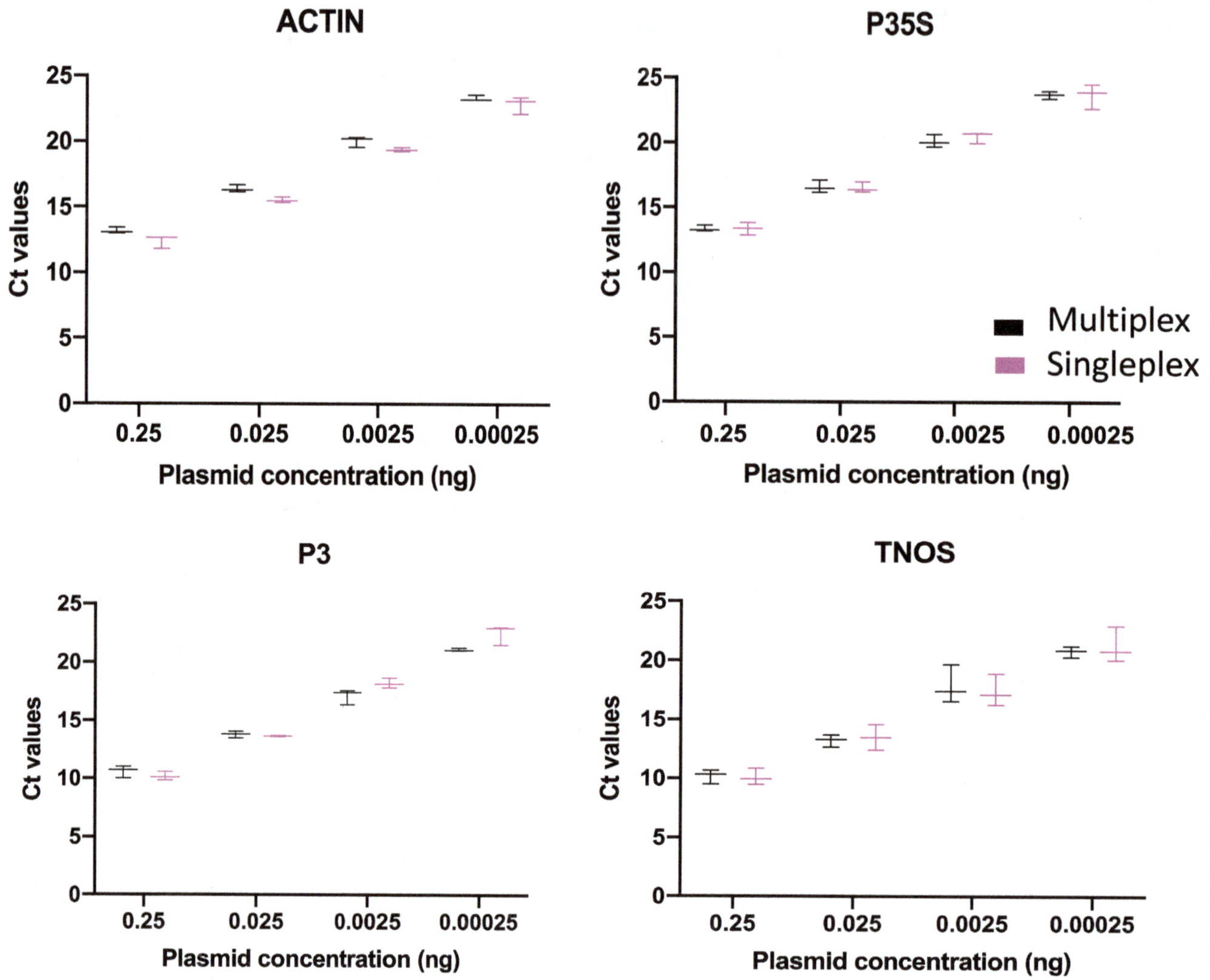

Fig. 2 Multiplex to singleplex qPCR Ct comparison. Graphics showing the comparison of the Ct values for each target (actin, P35S, P3, and TNOS) obtained from the multiplex assay (in black) and from the singleplex assay (in pink). The error bars show results from three independent experiments. ($N = 3$, no significant differences between any pair of multiplex vs. singleplex Ct values for the same plasmid concentration and target, Student's t test, $P < 0.05$)

Confirmation of primer and probe efficacy using SYBR green qPCR assays

Forward and reverse specific primers for each target were used to confirm amplification of the target, using a SYBR Green assay and the corresponding plasmid as template DNA. Subsequently, one primer from each pair was replaced with the probe (without the fluorescent dye) for that target and used with the second primer (see Table 1) in a SYBR Green assay to ensure that the probe sequences were effective prior to purchasing probes with fluorophores (Additional file 3: Figure S2). For both of these tests, we generated melting curves to track double-stranded DNA dissociation with rising temperature in the SYBR Green qPCR reaction, leading to an increase in the fluorescence absorbance intensity. The temperature at which 50% of the DNA is denatured is known as the melting temperature, and this is manifested as a peak in the melting curve. The SYBR Green method allowed us to generate a melting curve for each primer set, including both forward-reverse and primer-probe pairs, which allowed us to ensure that each primer set yielded a specific PCR product (one peak at one temperature) without non-specific amplification or primer-dimers, either of which would have resulted in peaks at different temperatures [26]. Indeed, we confirmed the specificity of all of our primer and probe sets with the formation of a single peak in the melting curve (Additional file 3: Figure S2).

Development and validation of the multiplex qPCR assay

The fluorophores added to the probes were chosen to minimize spectral profile overlap so that each amplified

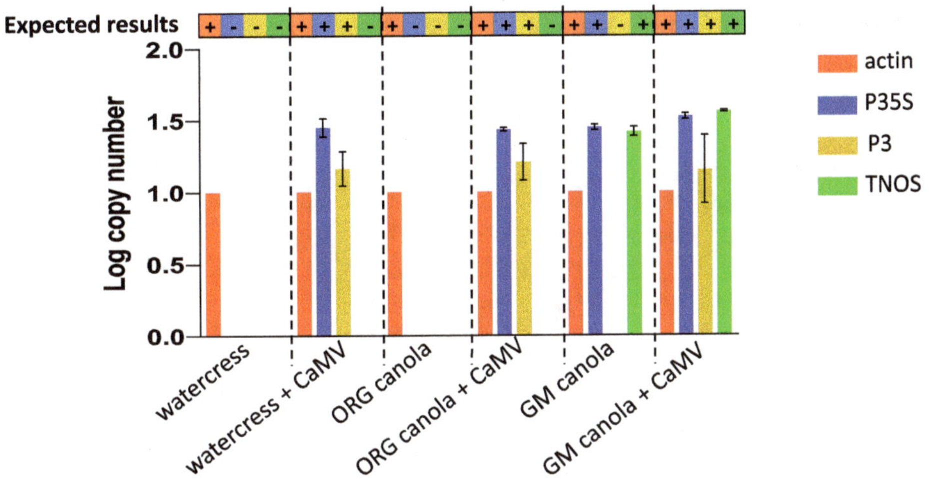

Fig. 3 Specificity of the multiplex qPCR assay. Log copy number of each target (actin, P35S, P3, and TNOS) for organic (watercress and ORG Canola) or GM (GM canola) plants infected or not with CaMV. The expected results are presented in the bar above the graphic (+ indicates expected amplification, – indicates no expected amplification). The error bars show results from three independent experiments, and no error bars are shown for actin because all copy numbers were normalized to actin

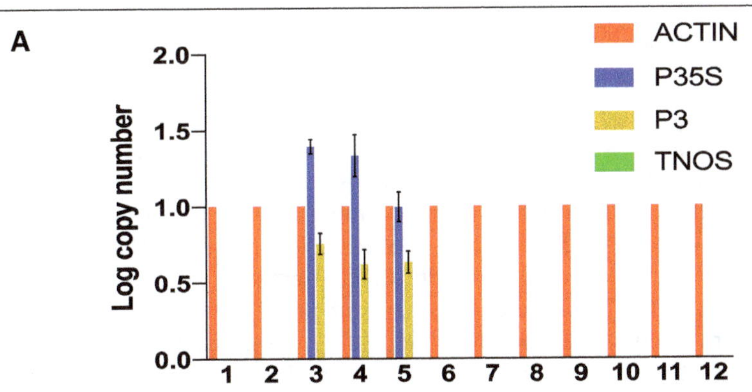

B

Samples #	Plant	Location	CaMV infected	GMO	3rd party
1	Watercress	Jalisco, Mexico	-	-	-
2	Watercress	Jalisco, Mexico	-	-	-
3	Watercress	Ubajara, Brazil	+	-	+
4	Watercress	Ubajara, Brazil	+	-	+
5	Watercress	Ubajara, Brazil	+	-	+
6	Watercress	Jalisco, Mexico	-	-	Nm
7	Watercress	Jalisco, Mexico	-	-	Nm
8	Watercress	Jalisco, Mexico	-	-	Nm
9	Watercress	Jalisco, Mexico	-	-	Nm
10	Watercress	Jalisco, Mexico	-	-	Nm
11	Broccoli	Jalisco, Mexico	-	-	Nm
12	Broccoli	Jalisco, Mexico	-	-	Nm

Fig. 4 Multiplex qPCR assay on industrial samples. **a** Log copy number of each target (actin, P35S, P3, and TNOS) for different plant samples provided by Nutrilite by Amway. The error bars show results from three independent experiments, and no error bars are shown for actin because all copy numbers were normalized to actin. **b** Sample information for results in (a), + (detected) or – (not detected) indicates the result of our multiplex assay. The final column, "3rd party," has GM plant detection results (+/–) from an independent 3rd party assay that could not distinguish CaMV infection from GM plant detection. GMO = genetically modified organism, Nm = not measured

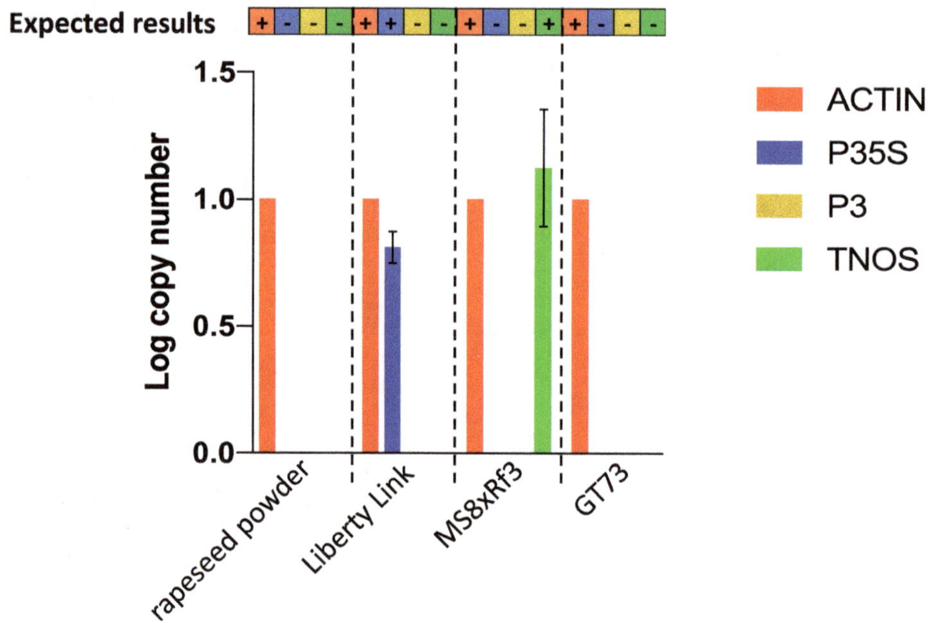

Fig. 5 Multiplex qPCR assay on commercially available samples. Log copy number of each target (actin, P35S, P3, and TNOS) for different purchased plant and/or DNA samples. The expected results are presented in the bar above the graphic (+ indicates expected amplification, – indicates no expected amplification). Liberty Link has P35S, MS8xRf3 has TNOS, and GT73 has neither P35S nor TNOS. The error bars show results from three independent experiments, and no error bars are shown for actin because all copy numbers were normalized to actin

target would emit fluorescence in a different wavelength, allowing the four targets to be detected in the same qPCR reaction. The four chosen fluorophores were FAM™, HEX®, ABY® and JUN®, with emission spectra peaks at 517 nm, 551 nm, 580 nm, and 617 nm, respectively. To develop the TaqMan multiplex qPCR assay, a plasmid dilution series was tested by TaqMan qPCR. All four targets were first tested in singleplex reactions, then a duplex reaction with actin plus P35S was tested, followed by a triplex reaction with actin, P35S and P3 before the quadruplex reaction with all four targets (actin, P35S, P3, and TNOS). The primer and probe concentrations of the reaction mix were optimized at each step for the best amplification of the four targets (see Methods for the final, optimized protocol).

Once the multiplex protocol was established, the efficacy of the multiplexed assay was validated using a plasmid dilution series tested in singleplex (reactions containing a single set of primers and the probe for one specific target) with Ct values compared to Ct values for the same target in multiplex (reactions containing primer and probe sets for all four of the targets) (Fig. 2) [27, 28]. No significant difference in sensitivity or efficiency was found between the singleplex and multiplex qPCRs for any of the four targets ($P < 0.05$). Indeed, the Ct values for each target in the multiplex reaction were very similar to those obtained from the singleplex reaction (< 1 difference in the Ct values).

Table 2 Limit of Detection (LOD) of the multiplex qPCR assay. The LOD for each target (actin, P35S, P3, and TNOS) is the percentage of plant DNA containing the target of interest diluted in relevant background DNA (i.e., background not containing the target of interest, see Methods) at the lowest concentration in which the target was detectable; 10 samples were tested for each dilution, and in each case, all 10 were positive (detected) at the LOD

	LOD (%)
actin	1
P35S	0.001
P3	0.01
TNOS	0.01

Primer efficiency and generating standard curves for the multiplex qPCR assay

A serial dilution of the four plasmids mixed together in UltraPure water was used to determine the primer efficiency and generate standard curves. The amplification efficiencies for each of the four targets ranged between 92 and 96%, which is well within the acceptable range of 80 to 120% (the ideal amplification efficiency is 100%, assuming that the PCR product concentration doubles every cycle during the exponential phase of amplification [29, 30]). For each multiplex qPCR assay, a standard curve for each primer set was generated, which could be used for placement of unknown targets to determine their log copy numbers and Ct values to assess whether or not a given target was detected. The Y-intercept, slope, and R^2 data for the standard curves associated with the primer efficiency tests are summarized in Additional file 4: Figure S3.

Most qPCR assays use standard curves based on plasmid dilution in pure water, as we did above, and therefore, we used plasmid dilutions for all of the standard curves and multiplex qPCR measurements reported here. However, we recognize that most actual targets from samples of interest will be amplified from a "noisy" plant DNA background (as in, the target sequence will be within the plant genome, as opposed to in a relatively small plasmid), so we were curious to test the difference in amplification efficiency when plasmids were diluted in relevant DNA background instead of just water. To test this, a serial dilution of the four plasmids mixed together and diluted in plant DNA (for P35S, TNOS, and P3) or bacterial DNA (for actin) was used in multiplex qPCR assays to see how much the primer efficiency differed from results from plasmids in water. The amplification efficiencies for each of the four targets was between 80 and 89% when the plasmids were diluted in background DNA, which was 7–13% lower than in pure water but still within the acceptable amplification efficiency range of 80–120% (Additional file 4: Figure S3). As dilution in organismal DNA (here, plant or bacterial DNA, as opposed to plasmid DNA) is a non-standard approach, we did not use these methods for any of our downstream calculations.

Specificity of the multiplex qPCR assay

Our different plant DNA samples (GM and non-GM plants with or without CaMV infection) were used to test the specificity of the multiplex qPCR assay. We wanted to ensure that all expected targets amplified and no non-specific amplification was observed. For these measurements, primer efficiencies were adjusted to 100% for the calculation of the log copy number, which was normalized to the reference target, actin (see Methods). All expected targets amplified, and no cross-reactions were observed between the different targets in the multiplex qPCR assay (Fig. 3). Specifically, TNOS did not amplify in any non-GM samples, CaMV P3 did not amplify in any uninfected plants, and P35S did not amplify in any uninfected, non-GM plants. As expected, actin amplified well from all plant DNA samples, TNOS amplified well from GM plant DNA, P3 amplified well from CaMV-infected plant DNA, and P35S amplified well from all tested GM and/or CaMV-infected plant DNA. These results indicate that the multiplex qPCR assay has 100% specificity.

Sensitivity (limit of detection) of the multiplex qPCR assay

In order to determine the limit of detection (LOD) for the P35S, P3, and TNOS targets via the multiplex qPCR method, different concentrations of CaMV-infected GM canola DNA (which would be positive for P35S, P3, and TNOS) were diluted in uninfected, non-GM canola DNA (i.e., a complex plant DNA background that did not contain P35S, P3, or TNOS). The LOD of actin was determined using different concentrations of plant DNA diluted in bacterial DNA. Based on the specificity of the multiplex assay results, the Ct value 28 was the threshold Ct considered to be positive. Ten replicates were used for each assay, and the LOD of P3 and TNOS was found to be 0.01%, the LOD of P35S was 0.001%, and the LOD of actin was 1% (Table 2). These percentages represent the concentration of the target that was detectable within the background DNA and are approximately the concentration of target expected to be detectable by this method. The LOD of actin, which appears to be very high compared to the LOD of P3, P35S, and TNOS, is probably due to the poor efficacy of the actin dye-quencher combination (the least sensitive of the four fluorescent dyes was assigned to the actin probe specifically because sensitivity to the actin control was the least important, relative to sensitivity to P3, P35S, and TNOS targets). Similarly, differences in the LOD of the other targets are likely due to differences in the efficacy of the dye-quencher combination for each target.

To evaluate the sensitivity of detection of CaMV infection in a GM plant DNA background, the Ct values obtained for each target (actin, P35S, P3, and TNOS) for different percent dilutions (10, 5, 2, 1, 0.1 and 0.01%) of CaMV-infected GM plant DNA in uninfected GM plant DNA were evaluated and compared to the Ct values obtained from uninfected GM plant DNA alone (Additional file 5: Table S2). As shown in Additional file 5: Table S2, the Ct values of the different targets changed only slightly between the different concentrations of CaMV-infected GM plant DNA tested, suggesting that the assay is not effective at determining specific CaMV concentrations in a GM plant background. However, CaMV was detected (via the amplification of P3) in a GM plant background even at extremely low concentrations of CaMV infection, suggesting that the assay is highly sensitive to CaMV infection (Additional file 5: Table S2). Together, these results mean that the multiplex qPCR assay described here can be used to detect CaMV infection with high sensitivity but should not be used to quantify the amount of infection.

Applicability to test samples

The multiplex qPCR assay was tested on different plant samples provided by Nutrilite by Amway (Fig. 4). Initially, we (the study authors) were blind to the expected results from each of these samples, and the results from prior testing of the same samples by a third party were provided to us after we generated our results in Fig. 4. We were informed that the tested samples were broccoli and watercress from organic (and non-GM plant-containing) farms in Brazil and Mexico (Fig. 4b) and that three samples (numbers 3, 4, and 5) had previously tested positive for GM plants by a proprietary third-party assay. We were informed that false-positive GM plant detection due to CaMV infection was strongly suspected for these three samples (and none of the others), but that CaMV infection had never been previously confirmed or disambiguated from GM plant detection. In our multiplex qPCR assay, those three plant samples (3, 4, and 5) tested positive for CaMV infection (positive for P3 and P35S) but negative for GM plants (no amplification of TNOS) (Fig. 4a). Though proprietary, we infer that the third-party GM plant detection assay was based on detection of P35S, which would be present in a CaMV-infected non-GM plant, and therefore that the third-party assay could not disambiguate GM plant detection from a false-positive GM plant due to CaMV infection. Our addition of P3 (which was positive for these three samples, indicating CaMV infection) and TNOS (which was negative for these three samples, indicating that a GM plant was not likely, though our assay does not consider all possible terminators) allowed us to disambiguate a very likely false-positive GM plant due to CaMV infection. All of the remaining samples were GM plant- and CaMV-negative (no amplification of P35S, TNOS, or P3), consistent with the organic source of the samples and indicative of a lack of detectable CaMV infection (Fig. 4a).

The multiplex qPCR assay was also tested on commercial samples purchased from Sigma-Aldrich (GMO Genomic DNA Standard Set, Cat. No 55231) to further ensure the specificity of the assay. The canola DNA LIBERTY LINK™ Falcon GS40/90, MS8xRf3, and GT73 ROUNDUP READY™ samples gave the expected results using the multiplex qPCR assay, specifically, amplification of P35S for LIBERTY LINK™ Falcon GS40/90, amplification of TNOS for MS8xRf3, and no amplification of TNOS or P35S for GT73 ROUNDUP READY™ (Fig. 5). A non-GM plant, non-CaMV-infected rapeseed (also called canola) powder was also used as a negative control after DNA extraction (Rapeseed, Cat. # ERMBF434A - ERM® certified Reference Material, nominal 0% GMO), and as expected, only actin amplified from that sample (Fig. 5).

Discussion

Here we have described a method that allows, in a single reaction, for the detection of a very low level of CaMV infection (0.01%) in both GM and non-GM plants. The ability of this assay to detect CaMV infection is due to the detection or non-detection of the CaMV-specific gene, P3. Although the assay can clearly show when even a trace amount of CaMV infection is present, it cannot be used to quantify the amount of CaMV infection. The method also allows for the detection of GM plants that contain P35S and/or TNOS, accounting for most known GM plants. Nevertheless, the method cannot detect GMOs that do not use at least one of these two markers. This multiplex qPCR assay can distinguish between CaMV infection and most common GM plants, as follows: detection of P35S and/or TNOS but not P3 would indicate an uninfected GM plant, detection of TNOS (regardless of the other results) would indicate a GM plant, and detection of P35S and P3 but not TNOS would indicate CaMV infection of a likely non-GM plant. There is still the potential for false-positive GM plant detection in this assay, due to potential *Agrobacterium tumefaciens* infection, which could be overcome by considering non-GMO targets in the *A. tumefaciens* genome, as discussed elsewhere [9] (Bak A, Emerson JB. Towards distinguishing Cauliflower mosaic virus (CaMV) infection from genetic modification (GM) in crop plants: detection assays and biology, management, and food safety of CaMV. Submitted). Also, the efficacy of the P3 and P35S primers on divergent CaMV strains is to be determined. For example, a recent in silico test designed to evaluate the specificity of common CaMV primers to a wide diversity of CaMV strains detected 100% sequence homology for the P3 primers used here to 54% (forward) and 78% (reverse) of 96 tested CaMV strains, with the number and locations of SNPs variable among strains [10]. Although these P3 primers were expected to perform the best among the pre-existing primers in that study, primers that target a more diverse range of CaMV strains could be considered for future improvements to this assay, for example, newly designed CaMV ORFV primers with three degenerate base-pairs per primer to target all 96 of the tested strains [10]. Finally, the positive control actin primers should theoretically result in positive PCR products for all plants [24]; this was the case for all plants in our study, but we did not attempt validation of the actin primers beyond the species studied here.

Conclusions

PCR and qPCR methods have remained the primary GM plant detection techniques. However, most GM plant detection methods do not allow for discrimination between CaMV infection and GM plants, and to our knowledge, there is no currently available technique that allows for rapid, efficient, and relatively affordable disambiguation between CaMV infection and GM plant detection. Here we have described an optimized multiplex qPCR assay for GM plant detection that allows for the detection of most GM plants, along with the identification of false-positive GM plants linked to *Cauliflower mosaic virus* (CaMV) infection in a single reaction. This method should be compatible with and relatively easily transferable to any diagnostic (or other laboratory) facility with a qPCR machine.

Methods

Plant material

Non-GM canola (*Brassica napus*) seeds and non-GM watercress (*Nasturtium officinale*) seeds were purchased from Amazon (www.amazon.com). GM canola (*Brassica napus* cv. HyCLASS 969 Roundup-Ready® which has P35S and TNOS) seeds were provided by Stephen Kaffka (UC Davis). The plants were grown in a greenhouse at 20–25 °C with 14 h light period.

CaMV infection

Dried turnip tissue infected with wild-type CaMV strain W260 was obtained from James E. Schoelz at the University of Missouri. To prepare inoculum (virus sap) for our plants, dried CaMV-infected turnip tissue was ground in two volumes of 20 mM phosphate buffer (pH 7.2). One-week-old canola and watercress plants were mechanically inoculated with the virus sap and used for experiments at 3 weeks post infection. Two leaves per plant were dusted with carborundum (Sigma-Aldrich, St Louis, MO) to facilitate penetration and rub-inoculated with the virus sap using a cotton-stick, as previously described [31, 32]. Three weeks post infection, two plants from each variety were combined together and their DNA was extracted.

Plant DNA extraction

For all plant DNA extractions (including those from our own fresh plant tissues, from dried tissue received either as coarse cut or fine ground from Nutrilite by Amway, or from powder (Rapeseed, Cat. # ERMBF434A - ERM® certified Reference Material, nominal 0% GMO)), plant tissues were ground in liquid nitrogen prior to DNA extraction. Plant DNA was then extracted using the DNeasy plant mini kit (Qiagen) following the manufacturer's recommendations. Following the measurement of the DNA concentration using the Thermo Scientific™ NanoDrop™ OneC Microvolume UV-Vis spectrophotometer, DNA extracts were diluted to a final concentration of 10 ng/μL in nuclease-free water. DNA was stored at − 20 °C until further use.

Plasmid DNA

Two control plasmids, one containing a single copy of the P35S promoter and the other containing a single copy of the TNOS transcriptional terminator (plasmids pICH51277 and pICH41421, respectively) were provided by Gitta Coaker (UC Davis). Two more control plasmids, one for actin and one for P3 (a CaMV-specific gene), were constructed using the Gateway® cloning technology, according to the manufacturer's instructions (Invitrogen, Carlsbad, CA, USA), including the design and synthesis of Gateway primers containing the attB and attP sites (Additional file 1: Table S1, att sites added to P3 primers from ref. [23] and actin primers from ref. [24]) to facilitate insertion into the vector. The actin genetic region was amplified by PCR, using DNA extracted from turnip leaves (see method above) as the template and the gene-specific gateway primers (Gtw-Actin) listed in Additional file 1: Table S1. The P3 CaMV gene was amplified by PCR using CaMV-infected turnip DNA (extracted from infected leaves as described above) as the template and the gene-specific gateway primers (Gtw-P3) listed in Additional file 1: Table S1. The PCR products were then inserted into the entry vector pDONR207 via Gateway® BP recombinant reaction between attB and attP sites. After transformation in *Escherichia coli* DH5alpha, the clones were selected on LB agar plates supplemented with gentamycin (50 μg/ml). The plasmids were purified using the Zyppy™ plasmid miniprep kit according to the manufacturer's recommendations. Finally, the plasmid DNA concentrations were measured using the Thermo Scientific™ NanoDrop™ OneC Microvolume UV-Vis spectrophotometer and diluted to 10 ng/μL in nuclease-free water. Two plasmids (derived from different bacterial colonies) were tested by PCR (Additional file 2: Figure S1) and the plasmids labeled #1 were used for further experiments.

DNA from each of the four plasmids (containing actin, P35S, P3, and TNOS, respectively) was serially diluted at different concentrations (0.25 ng/μL; 0.025 ng/μL; 0.0025 ng/μL and 0.00025 ng/μL), and, depending on the experiment, mixed together (multiplex assay) or used separately (singleplex assay). DNA was stored at − 20 °C until further use.

Oligonucleotides and probes

Primers and TaqMan® probes used in this experiment are presented in Table 1. The primers were produced by Sigma-Aldrich, and the TaqMan® probes were synthesized by either Sigma-Aldrich (MilliporeSigma, Life Science business of Merck KGaA, Darmstadt, Germany) for HEX®-Actin and FAM™-P35S or Applied Biosystems (Thermo Fisher Scientific, Waltham, Massachusetts, U.S) for ABY®-P3 and JUN®-TNOS. The primers and probes used in this paper were selected because they are widely cited in the literature and known to work on various types of plants and samples in singleplex PCR and/or qPCR reactions [23–25]. The four fluorophores (FAM™, HEX®, JUN®, and ABY®) for the TaqMan probes were selected to minimize overlap of their spectral profiles.

PCR settings

PCR was performed using forward and reverse specific primers (Table 1) and GoTaq (Promega, Madison, WI, USA) according to the manufacturer's instructions and using the following program for 35 cycles: denaturation at 95 °C for 1 min, annealing at 55 °C for 1 min, and extension at 72 °C for 12 s. The PCR reaction was then loaded on a 1.5% agarose gel, separated by gel electrophoresis, and visualized with a UV transilluminator to identify PCR products.

SYBR green qPCR method

SYBR Green qPCR was tested using each plasmid DNA as template with the specific primer set for each target. The qPCR was performed on a QuantStudio 6 Flex instrument (Thermo Fisher Scientific) equipped with a 384-well block. The reaction consisted of a 10 µL amplification mix containing 5 µL of PowerUP™ SYBR™ Green Master Mix (Thermo Fisher Scientific), 400 nM of each primer, and 4 µL of plasmids (at different concentrations: 0.25 ng/µL, 0.025 ng/µL, 0.0025 ng/µL, and 0.00025 ng/µL). No template controls (NTCs) were used as negative controls. All reactions were tested in triplicate. The reactions were performed in standard mode. They consisted of a first step at 50 °C for 2 min, followed by a step at 95 °C for 10 min, followed by 40 cycles of a step at 95 °C for 15 s and a step at 55 °C for 1 min. The fluorescent signals and the melting curves were analyzed using QuantStudio Real-Time PCR software version 1.1 (Thermo Fisher Scientific) with a manual threshold.

TaqMan qPCR

TaqMan singleplex and multiplex qPCR runs were performed on a QuantStudio 6 Flex instrument (Thermo Fisher Scientific) equipped with a 384-well block. Each multiplex reaction consisted of a 15 µL amplification mix containing 7.5 µL of Multiplex Master Mix (Thermo Fisher Scientific), 400 nM of each primer, 250 nM of FAM™-P35S probe and 300 nM of HEX®-Actin, ABY®-P3 and JUN®-TNOS probes, and 4 µL of sample DNA (40 ng) or 4 µL of plasmids (at different concentrations: 0.25 ng/µL, 0.025 ng/µL, 0.0025 ng/µL, and 0.00025 ng/µL). Singleplex reactions were the same but only included one set of primers and one probe, adjusting the final volume to 15 µL using UltraPure water. Mustang Purple™ dye was used as a passive reference for normalization. No template controls (NTCs) were used as negative controls. All reactions were tested in triplicate. The distribution of samples and controls was not randomized within each PCR plate, but samples and controls were in multiple blocks throughout each plate, meaning that groups of control wells were interspersed with groups of sample wells, and vice versa.

The reactions were performed in standard mode. They consisted of a first step at 50 °C for 2 min, followed by 95 °C for 10 min, followed by 40 cycles of: 95 °C for 15 s then 55 °C for 1 min. The fluorescent signals were analyzed using QuantStudio Real-Time PCR software version 1.1 (Thermo Fisher Scientific) with a manual threshold. Ct values below 28 were considered positive [33].

Specificity trial

To evaluate the specificity of this multiplex assay, high percentages (≥1% (w/w)) of non-GM and GM plant materials, infected or not infected with CaMV, were tested in triplicate [18]. Specifically, the six different samples tested were: non-GM watercress (there are currently no widely produced GM watercress plants), CaMV-infected non-GM watercress, non-GM canola, CaMV-infected non-GM canola, GM canola, and CaMV-infected GM canola. The assay was considered specific when it amplified all targets expected to be in the sample and no targets not expected to be in the sample.

Sensitivity trial

To determine the limit of detection (LOD) of the different targets, 10-fold serial dilutions from 1% down to 0.0001% were tested for each target. To obtain these low concentrations of DNA and to test the LOD of P3, P35S, and TNOS, CaMV-infected GM canola DNA was serially diluted in non-GM Canola DNA. Similarly, to test the LOD of actin, plant DNA was serially diluted in bacterial DNA [34, 35].

Copy number calculation

For absolute quantification, standard curves were obtained for each target based on a dilution series of all four of the DNA plasmids containing the target genes mixed together. The slopes of the corresponding standard curves were used to calculate the amplification efficiency percentage of each primer set using the mathematical formula: $10^{(-1/\text{slope of the standard curve})}-1$. The DNA concentrations from the standard curve were converted to the number of copies, using the following formula:

Number of copies = (DNA amount (g) $*6022*10^{23}$) / (plasmid length (bp)*660) where $6022*10^{23}$ is Avogadro's number and 660 is the average weight of a base pair (Kamau, 2013).

The Ct values of each target obtained from the multiplex qPCR assays of the plant samples were interpolated as unknowns from the linear regression standard curves to determine the log copy numbers, using the following formula: (Ct − intercept)/slope. The log copy numbers were then adjusted to a 100% primer efficiency, depending on the primer set percentage efficiency from the standard curve. Finally, the log copy numbers were normalized with the actin reference gene, using the following formula: log copy number of the target/log copy number of actin [24].

Statistical analyses

The statistical analyses were performed using GraphPad Prism version 8.00 for Mac (GraphPad Software, La Jolla California USA, www.graphpad.com) and data were analyzed by Student's t-tests. All experiments were repeated at least three times, with the number of repeats indicated in the text and/or figure(s).

Supplementary information

Additional file 1: Table S1. Gateway primers used for cloning actin and P3 (Gtw-Actin and Gtw-P3).

Additional file 2: Figure S1. Primer sets tested by PCR. a Specific primers tested by PCR using the corresponding plasmids as template. After the cloning of the actin and P3 amplification products, two different plasmids (1 and 2) were tested. b Actin primers tested by PCR using plant DNA extract as template (GM Canola, Watercress and Non-GM Canola). c P35S and TNOS specific primers tested by PCR using GM Canola DNA as template. d P3 and TNOS specific primers tested by PCR on CaMV-infected plants (GM Canola, Non-GM Canola and Watercress).

Additional file 3: Figure S2. Melting curves for each primer set and probe obtained by SYBR Green assay. Specific primers were tested by SYBR Green qPCR using the corresponding plasmid as template. Melting curves for each set of primers (Forward + Reverse) or using one Primer + the Probe are shown with the temperature on the x-axis and the derivative reporter (ΔRn) on the y-axis. The derivative reporter is calculated as the negative first derivative of the normalized fluorescence (Rn) generated by the reporter during PCR amplification. It allows for visualization of the maximum rate of change in fluorescence during the temperature ramp.

Additional file 4: Figure S3. Standard curves for each target. A serial dilution of the four plasmids mixed together and diluted in water (blue) or in plant DNA (green) for P35S, P3, and TNOS or in bacterial DNA (green) for actin were tested in multiplex qPCR to determine primer efficiency and standard curves for each primer set.

Additional file 5: Table S2. CaMV-infected GM plant DNA versus uninfected GM plant DNA. Ct values for each target (actin, P35S, P3, and TNOS) for different dilution percentages (10, 5, 2, 1, 0.1 and 0.01%) of CaMV-infected GM plant DNA in GM plant DNA, compared to the Ct values obtained from GM plant DNA alone. Here, GMO is short for "GM plant."

Abbreviations

CaMV: Cauliflower mosaic virus; Ct: Cycle threshold; GM: Genetically Modified; GMO: Genetically Modified Organism; LOD: Limit of detection; NTC: No Template Control; P35S: Cauliflower mosaic virus 35S promoter; PCR: Polymerase Chain Reaction; qPCR: quantitative Polymerase Chain Reaction; TNOS: Nopaline synthase terminator

Acknowledgements

We acknowledge Tera Pitman and Amisha Poret-Peterson for helpful discussions about qPCR. We acknowledge Mysore Sudarshana for the use of his QuantStudio 6 Flex instrument (Thermo Fisher Scientific). We acknowledge Gitta Coaker, Stephen Kaffka, and James E. Schoelz for sharing research materials with us, as detailed in the Methods section.

Authors' contributions

JBE conceived the project. AB and JBE designed the research. AB performed research and analyzed the data. AB and JBE wrote the article. Both authors read and approved the final manuscript.

Funding

This work was funded by Nutrilite by Amway from a grant awarded to JBE. The project was designed solely by JBE and AB, who provided periodic progress reports to Nutrilite by Amway via conference calls. The funding body played no part in directing the research, but they did offer insights from an industry perspective that helped us to test the robustness of the assay, for example by including commercially available products in our tests (Fig. 5). They also provided samples for us to test "blind" (specifically, samples in Fig. 4), which we initially received with no information about expected results, and for which results from a 3rd party GMO diagnostics facility were revealed to us after we ran our own tests, as reported in Fig. 4.

Ethics approval and consent to participate

Nothing to declare.

Consent for publication

Not applicable.

Competing interests

The authors declare that they have no competing interests.

References

1. Oliver MJ. Why we need GMO crops in agriculture. Mo Med. 2014;111(6):492–507.
2. Wu Y, Wang Y, Li J, et al. Development of a general method for detection and quantification of the P35S promoter based on assessment of existing methods. Sci Rep 4. 2014;8(4):7358.
3. Fu W, Zhu P, Wang C, et al. A highly sensitive and specific method for the screening detection of genetically modified organisms based on digital PCR without pretreatment. Sci Rep. 2015;5:12715.
4. Ramessar K, Capell T, Twyman RM, et al. Trace and traceability - a call for regulatory harmony. Nat Biotechnol. 2008;26:975–8.
5. Davison J. GM plants: science, politics and EC regulations. Plant Sci. 2010;178(2):94–8.
6. Maghari BM, Ardekani AM. Genetically modified foods and social concerns. Avicenna J Med Biotechnol. 2011;3(3):109–17.
7. Wong AYT, Chan AWK. Genetically modified foods in China and the United States: a primer of regulation and intellectual property protection. Food Sci Human Wellness. 2016;5(3):124–40.
8. Fraiture MA, Herman P, Taverniers I, et al. Current and new approaches in GMO detection: challenges and solutions. Biomed Res Int. 2015;392872.
9. Wolf C, Scherzinger M, Wurz A, et al. Detection of *Cauliflower mosaic virus* by the polymerase chain reaction: testing of food components for false-positive 35S-promoter screening results. Eur Food Res Technol. 2000;210:367–72.
10. Becker R, Ulrich A. Improved detection and quantification of *Cauliflower mosaic virus* in food crops: assessing false positives in GMO screening based on the 35S promoter. Eur Food Res Technol. 2018;244:1861.
11. Holden MJ, Levine M, Scholdberg T, et al. The use of 35S and Tnos expression elements in the measurement of genetically engineered plant materials. Anal Bioanal Chem. 2010;396(6):2175–87.
12. Garibyan L, Avashia N. Polymerase chain reaction. J Invest Dermatol. 2013;133(3):1–4.

13. VanGuilder HD, Vrana KE, Freeman WM. Twenty-five years of quantitative PCR for gene expression analysis. BioTechniques. 2008;44(5):619–26.
14. Emerson JB, Adams RI, Román CMB, et al. Schrödinger's microbes: tools for distinguishing the living from the dead in microbial ecosystems. Microbiome. 2017;5(1):86.
15. Hou Y, Zhang H, Miranda L, Lin S. Serious overestimation in quantitative PCR by circular (supercoiled) plasmid standard: microalgal pcna as the model gene. PLoS One. 2010;5:e9545.
16. Tajadini M, Panjehpour M, Javanmard SH. Comparison of SYBR green and TaqMan methods in quantitative real-time polymerase chain reaction analysis of four adenosine receptor subtypes. Adv Biomed Res. 2014;3:85.
17. Singh M, Bhoge RK, Randhawa G. Crop-specific GMO matrix-multiplex PCR: a cost-efficient screening strategy for genetically modified maize and cotton events approved globally. Food Control. 2016;70:271–80.
18. Cottenet G, Blancpain C, Sonnard V, et al. Two FAST multiplex real-time PCR reactions to assess the presence of genetically modified organisms in food. Food Chem. 2019;15(274):760–5.
19. Cottenet G, Blancpain C, Sonnard V, et al. Development and validation of a multiplex real time PCR method to simultaneously detect 47 targets for the identification of genetically modified organisms. Anal Bioanal Chem. 2013; 405(21):6831–44.
20. Fraiture MA, Herman P, Papazova N, et al. An integrated strategy combining DNA walking and NGS to detect GMOs. Food Chem. 2017;232:351–8.
21. Salisu IB, Shahid AA, Yaqoob A, et al. Molecular approaches for high throughput detection and quantification of genetically modified crops: a review. Front Plant Sci. 2017;8:1670.
22. Demeke T, Dobnik D. Critical assessment of digital PCR for the detection and quantification of genetically modified organisms. Anal Bioanal Chem. 2018;410(17):4039–50.
23. Chaouachi M, Fortabat MN, Geldreich A, et al. An accurate real-time PCR test for the detection and quantification of *Cauliflower mosaic virus* (CaMV): applicable in GMO screening. Eur Food Res Technol. 2008;227(3):789–98.
24. Scholtens IMJ, Molenaar B, van Hoof RA, et al. Semiautomated TaqMan PCR screening of GMO labelled samples for (unauthorised) GMOs. Anal Bioanal Chem. 2017;409(15):3877–89.
25. Fernandez S, Charles-Delobel C, Geldreich A, et al. Quantification of the 35S promoter in DNA extracts from genetically modified organisms using real-time polymerase chain reaction and specificity assessment on various genetically modified organisms, part I: operating procedure. J AOAC Int. 2005;88(2):547–57.
26. Pryor RJ, Wittwer CT. Real-time polymerase chain reaction and melting curve analysis. Methods Mol Biol. 2006;336:19–32.
27. Kamau E, Alemayehu S, Feghali KC, et al. Multiplex qPCR for detection and absolute quantification of malaria. PLoS One. 2013;8(8):e71539.
28. Llewellyn S, Inpankaew T, Nery SV, et al. Application of a multiplex quantitative PCR to assess prevalence and intensity of intestinal parasite infections in a controlled clinical trial. PLoS Negl Trop Dis. 2016;10(1): e0004380.
29. Bustin SA, Benes V, Garson JA, et al. The MIQE guidelines: minimum information for publication of quantitative real-time PCR experiments. Clin Chem. 2009;55:611–22.
30. Zhang Q, Wang J, Deng F, et al. TqPCR: a touchdown qPCR assay with significantly improved detection sensitivity and amplification efficiency of SYBR green qPCR. PLoS One. 2015;10(7):e0132666.
31. Bak A, Gargani D, Macia JL, et al. Virus factories of *Cauliflower mosaic virus* are virion reservoirs that engage actively in vector transmission. J Virol. 2013;87(22):12207–15.
32. Bak A, Cheung AL, Yang C, et al. A viral protease relocalizes in the presence of the vector to promote vector performance. Nat Commun. 2017;8:14493. https://doi.org/10.1038/ncomms14493.
33. Nybo K. qPCR: technical replicate variation. BioTechniques. 2011;50(1):23–5.
34. Arumuganathan K, Earle ED. Nuclear content of some important plant species. Plant Mol Biol Report. 1991;9:208–18.
35. Broeders S, Huber I, Grohmann L, et al. Guidelines for validation of qualitative real-time PCR methods. Trends Food Sci Technol. 2014;37:115–26.

18

The next generation of biopanning: next gen sequencing improves analysis of bacterial display libraries

Sarah D. Stellwagen[1,2], Deborah A. Sarkes[1*], Bryn L. Adams[1], Mia A. Hunt[1,3], Rebecca L. Renberg[1,3], Margaret M. Hurley[1] and Dimitra N. Stratis-Cullum[1]

Abstract

Background: Bacterial surface display libraries are a popular tool for novel ligand discovery due to their ease of manipulation and rapid growth rates. These libraries typically express a scaffold protein embedded within the outer membrane with a short, surface-exposed peptide that is either terminal or is incorporated into an outer loop, and can therefore interact with and bind to substrates of interest.

Results: In this study, we employed a novel bacterial peptide display library which incorporates short 15-mer peptides on the surface of *E. coli*, co-expressed with the inducible red fluorescent protein DsRed in the cytosol, to investigate population diversity over two rounds of biopanning. The naive library was used in panning trials to select for binding affinity against 3D printing plastic coupons made from polylactic acid (PLA). Resulting libraries were then deep-sequenced using next generation sequencing (NGS) to investigate selection and diversity.

Conclusions: We demonstrated enrichment for PLA binding versus a sapphire control surface, analyzed population composition, and compared sorting rounds using a binding assay and fluorescence microscopy. The capability to produce and describe display libraries through NGS across rounds of selection allows a deeper understanding of population dynamics that can be better directed towards peptide discovery.

Keywords: Bacterial display, Directed evolution, Peptide display, Biopanning, NGS

Background

Natural selection has resulted in elegant solutions to the pressures that organisms face. On a macro scale, examples include the cryptic shape and coloration of stick insects, or the specialized feeding organs of mosquitoes, and microscopically, ligand binding and enzymes. Artificially harnessing the power of natural selection has not only provided useful organisms at a larger scale that have shaped culture and diet (e.g. agricultural varieties of plants and animals), but has also contributed to discovering molecular tools that have influenced disciplines from medicine to materials research. Selection of particular molecular interactions is key to natural and engineered antibody-antigen recognition for diagnostics, new materials, and an understanding of chemical interactions [1–4]. Peptide discovery that exploits artificial selection pressures to shape diverse peptide libraries have led to a wide variety of unique sequences for targeted use [5–9]. Understanding library diversity, how peptide populations evolve during the discovery process, as well as methods to analyze diversification on a deeper scale are important for improving discovery techniques that rapidly produce unique molecules for purpose-driven benefit.

Surface display libraries employ polypeptides (including peptides, scaffold proteins, and antibody fragments) typically presented at the surface of phage coats or bacterial, yeast, or mammalian cell membranes. Each organism expresses a unique, random sequence with the potential to interact with specific antigenic or molecular targets [10]. After incubating a library with a substrate of interest, the unbound members are washed away while the remaining bound members are amplified and used for further

*Correspondence: deborah.a.sarkes.civ@mail.mil
[1]Biotechnology Branch, CCDC US Army Research Laboratory, 2800 Powder Mill Rd, Adelphi, MD, 20783 USA
Full list of author information is available at the end of the article

screening under more stringent conditions or analyzed to understand binding interactions. While phage display has been the traditional system used for ligand screening, bacterial display libraries provide advantages due to ease of manipulation, as phage reinfection steps are eliminated, and fast growth rates, which are important factors for rapid discovery of binding peptides to assess new and emergent threats [6, 8, 9].

Historically, peptide libraries were assumed to be highly diverse, unbiased sequence compilations ready for selection against a target material, and examination of the enriched peptides was limited by Sanger sequencing of a few hundred sequences. High throughput next generation sequencing (NGS) technologies now provide an avenue to investigate populations at a much broader scale [11–13]. We use a newly generated library (Fig. 1) to assess diversity and follow population evolution across rounds of biopanning against polylactic acid (PLA) 3D printing plastic. This library relies on a previously engineered outer membrane scaffold protein (eCPX) that co-presents random 15-mer peptides for screening on the N-terminus, and a fixed P2X peptide at the C-terminus to monitor expression via binding to the fluorescent protein YPet-Mona [14–16]. If a stop codon were present in the random peptide, P2X binding by YPet-Mona (Fig. 1b and c) would not occur because the remainder of the eCPX scaffold would not be produced or shuttled to the outer membrane. This interaction can be monitored using Fluorescence-Activated Cell Sorting (FACS) [17]. Additionally, we incorporated inducible DsRed fluorescence into the vector cassette (Fig. 1) allowing visualization to compare binding efficiency as an alternative method to indirect binding assays [6]. Indirect binding assays measure the cell proliferation of bound cells after a recovery stage to extrapolate and relatively quantify the number of bound library members. Serial dilution and plating of the resulting culture of amplified cells allows counting of colony forming units, while visualizing and quantifying cells directly bound to nonconductive materials by scanning electron microscopy (SEM) would require sputter coating with a metal such as gold. Here, we analyzed peptide sequences from the naive library and after each round of biopanning using Illumina NGS technology (Fig. 2) to understand the overall dynamics of population selection, and then visualized the bound cells directly on the PLA surface using the incorporated DsRed protein with fluorescence microscopy. The biopanning process is described in detail in Fig. 3, and is similar to previously described methods for isolating aluminum binders using a related eCPX-based library [6], except that here we additionally co-expressed the DsRed protein for direct visualization.

Results

Analysis of library diversity

We used Illumina next generation sequencing technology to assess peptide diversity for a newly created bacterial display library and after 2 rounds of biopanning against PLA 3D printing plastic (Figs. 2 and 3). The PLA

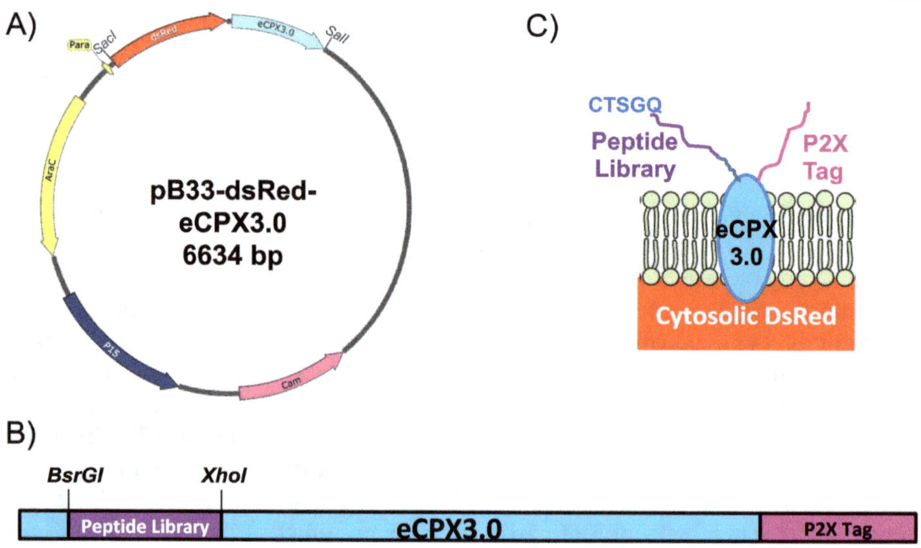

Fig. 1 pB33-dsRed-eCPX3.0 Peptide Library Design. **a** Plasmid map of DsRed-peptide display expression plasmid showing relative location and orientation of genes. **b** Diagram of eCPX3.0 gene from 5′ to 3′ indicating location of 15-mer peptide display library sequence, restriction sites flanking the library, and the location of the C-terminal P2X tag. **c** Schematic of mature eCPX3.0 display scaffold embedded in the cell membrane with N-terminal peptide library (preceded by CTSGQ) and C-terminal P2X tag displayed at the cell surface, and DsRed accumulating in the cytosol for cell visualization. The N-terminal peptide library contains random sequences, while the fixed C-terminal P2X tag allows expression monitoring through binding to the fluorescent protein YPet-Mona

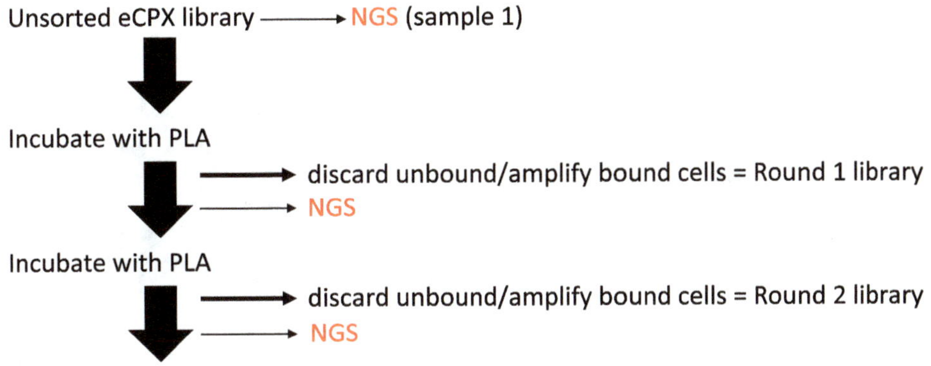

Fig. 2 Flow Chart for Biopanning and NGS Sequencing. The naive, unsorted library was both sequenced and used for biopanning against PLA 3D printing plastic. After washing the PLA sample and amplifying bound library members, the Round 1 library was sequenced and again panned against fresh PLA. The final, Round 2 library was sequenced along with the naive and Round 1 libraries

coupons were prepared by melting filaments together as described in Fig. 4 and the experimental methods. Illumina NGS sequencing resulted in 1,212,302 reads from the naive library, 1,418,018 reads from the Round 1 library, and 1,405,367 reads from the Round 2 library (Additional file 1: Table S1). The strategy for filtering sequences from this data is outlined in Fig. 5, showing examples of valid, useable sequence, and invalid sequences containing stop codons, frame shifts, or no peptide insert. We define 'unique' sequences as the total complement of different amino acid sequences for any category. For example, a valid insert with more than one representative is counted as a single representative of the 'unique' category for valid sequences. Valid inserts (inserts in the

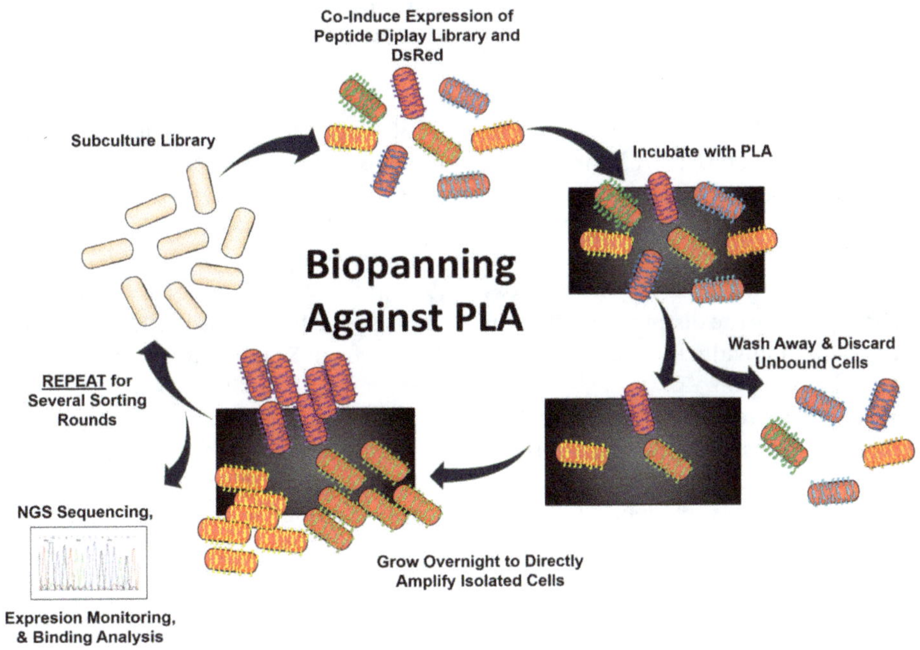

Fig. 3 Biopanning experimental design schematic. Biopanning is a cyclical process that begins with subculturing the naive, unsorted library. Co-expression of cytosolic DsRed and membrane-localized peptides from the pB33-dsRed-eCPX3.0 library is achieved by induction with L-arabinose during log phase of cell growth. The induced library is then incubated with PLA 3D printing plastic for 15 min with shaking. Next, the PLA coupons were washed vigorously with PBS 1% Tween for 30 min to remove unbound cells, and then placed in LB +Cm with 2% D-Glucose media to proliferate and recover the bound library. The presence of glucose halts the production of new copies of membrane-displayed peptides, which are slowly diluted during cell proliferation, along with the cytosolic DsRed. The process is repeated for sorting rounds 1 and 2, changing only the starting library material (using the previous sorting round). After each round, NGS sequencing was performed and peptide expression level monitored. Bound cells can be visualized directly on the PLA using fluorescence microscopy due to the presence of DsRed, and binding level can also be indirectly compared using regrowth assays and cell counting

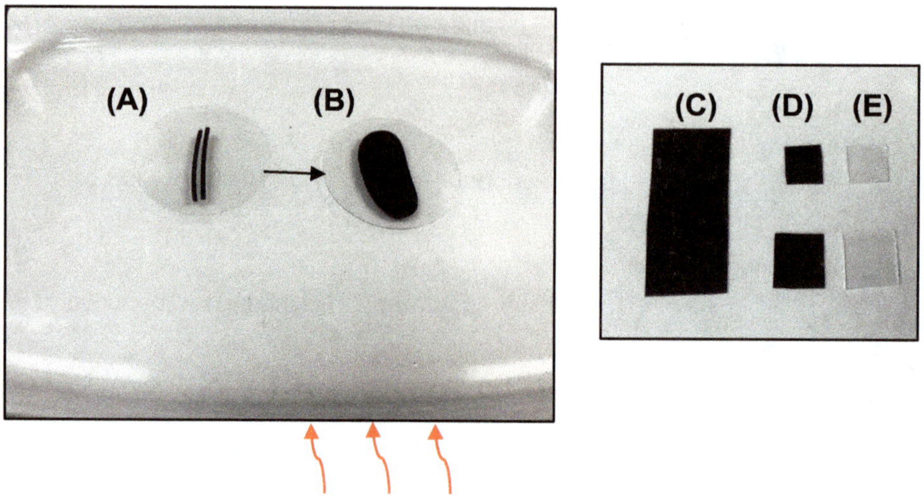

Fig. 4 Polylactic Acid (PLA) and sapphire material sample preparation. Two pieces cut from a spool of PLA printing filament ~4 cm in length were placed on a Pyrex baking dish under a sapphire wafer (**a**) and then heated with a hot plate until melted (**b**). After melting, PLA pieces were removed from the Pyrex dish and sapphire wafers, then cut into coupons for biopanning trials (**c**), or cut into smaller pieces used for binding quantification (**d**) with equally sized sapphire pieces (**e**)

correct reading frame without stop codons) comprised 38% of all sequences in the naive library, 85% of which were unique (or 33% of the total library; Fig. 6). After one round of selection, valid inserts increased to represent 42% of the library population; however, 11% of these (5% of the total library) were unique. The second round of sorting resulted in more than half the library's representation as valid inserts (55%), with an even further decrease in unique valid inserts (3% of valid inserts were unique or 2% of the total library). Library members with inserts that resulted in a frame shift, contained stop codons, or that had an empty scaffold made up 27%, 21%, and 13% of the naive library population, respectively. A general reduction was observed for members from these categories across panning rounds, accompanied by the distinct reduction in the representation of unique individuals, as compared to valid sequences (Fig. 6).

The most frequent sequences represented in the naive, unsorted library generally consisted of single-residue inserts (Table 1), although the bulk of the unsorted library consisted of 15-mers (375,287 out of 396,641 unique sequences, or 95%). Leucine alone had the highest representation and was identified 32 times (0.003%), while the most frequent 15-mer (PRKTLKGTLTVPSYI) was identified 8 times (0.0007%). The most common insert sequence discovered in the Round 1 library, a 15-mer (VLPQTSFFAATCRRS), was represented 115 times (0.01%), which increased to 283 in Round 2. 15-mers were again the main component of the library in Round 1, consisting of 64,507 out of 67,885 sequences (95%). The most common insert sequence discovered in Round 2, also a 15-mer (CHISPEKRRLIVCAD), was represented 4,619 times (0.38%), which increased from 28 in Round 1. In Round 2, 15-mers comprised 20,903 of 21,931 unique sequences (95%). The frequency of the population's top sequence increased 4-fold from the naive library to Round 1, and 40-fold from Round 1 to Round 2.

Representation of all 20 amino acids as a decimal of the overall (i.e. all unique sequences and not just 15-mer) library character was generated round by round, both including and excluding frequency of sequence occurrence (Additional file 1: Figure S1). This represents a

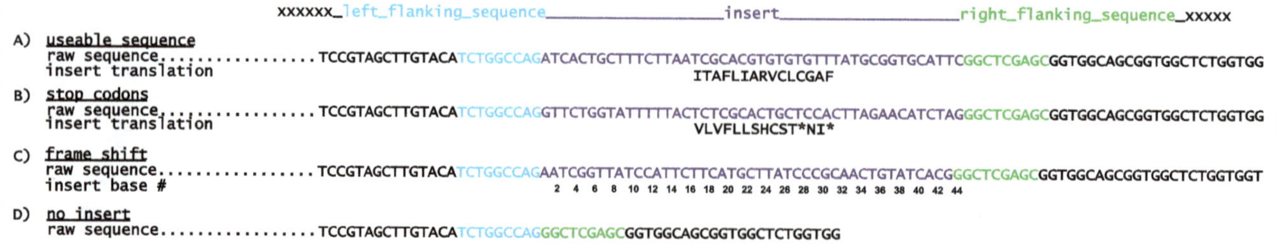

Fig. 5 Strategy for filtering NGS sequences. A valid insert sequence consists of correct flanking regions, an insert within the correct reading frame (typically 45 bp), and without stop codons (**a**). Sequences that translate into an insert with stop codons (**b**), result in a frame shift (**c**), or that have no insert (**d**) are filtered into separate categories

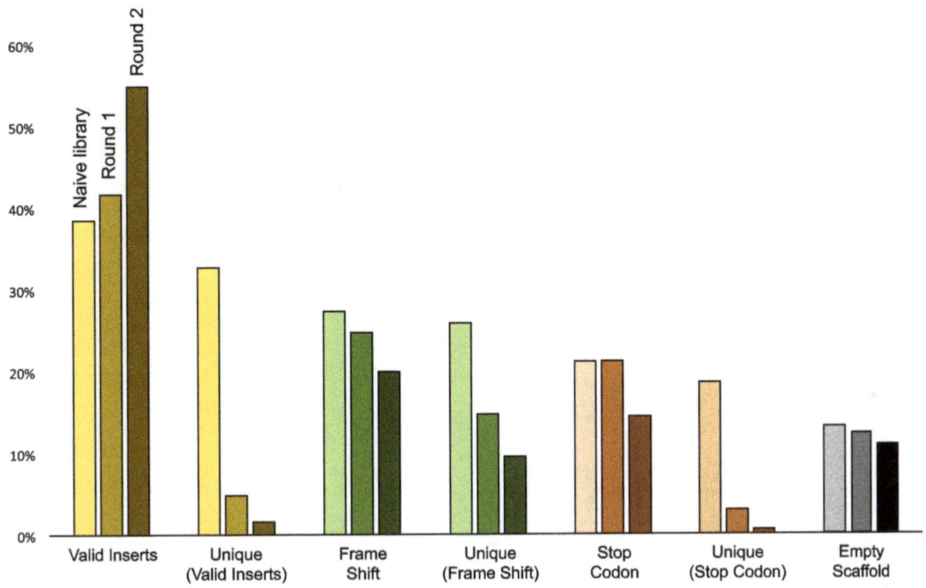

Fig. 6 Percent representation of each insert type from the naive and sorted libraries after NGS sequencing. Colors are grouped by total representation of the library followed by frequency of each unique insert for each category, and are in the order Naive library, Round 1, and Round 2. The Empty Scaffold category does not have a frequency representation, as members cannot be tracked across panning rounds. Raw sequence counts can be found in Additional file 1: Table S1

flat numeric analysis of current sequencing data with no correction for background probabilities from previous genomic analyses. Despite that, this information is useful as a depiction of both static and dynamic characteristics of the library. There is a consistently high representation of hydroxyl containing residues S, T, basic residue R, and hydrophobic residue L. Additionally there is underrepresentation from acidic residues D and E and hydrophilic residues N and Q. Round by round modifications to the distribution are subtle, but there is an increase in hydrophilic residues C and Y, an increase in hydrophobic residues M and L, an increase in aromatic residues F and W, and a decrease in D, E, N and Q. These trends are visible regardless of the incorporation of sequence frequency.

Analysis of sequence motifs in round by round 15-mer sequence character by the pLogo generator [18] further examines the inclusion of background probability statistics for *E. coli* K12 and provides an assessment of spatial organization within the sequence (Additional file 1: Figure S2). The trends seen from the rough numbers in Additional file 1: Figure S1 are accentuated in this representation. There is an up-regulation of hydrophilic residues S, T, C, Y, up-regulation of aromatic residue W and basic residue R, down-regulation of acidic residues D and E, and down-regulation of nonpolar residues A, G and L. The increase in M, C, and W by round is more visible in this representation, as is the down-regulation by round of E and A. By round 2, dominance of key residues C, W, S, M and R are clearly visible. There is very little spatial dependence on this up- (or down-) regulation, as befits interactions with a surface and with sequences which are largely expected to be loose coil in structure (as can be seen from the list in Table 1). There are further differences in sequence character in round 2 when comparing the pLogo generated motif of the entire 15-mer population of the library (Additional file 1: Figure S3).

The increase in cysteine after sorting is particularly interesting since it could have both positive and negative effects on the peptide library. Cysteine could help constrain any structural component of the peptide that aids in binding affinity or specificity by creating disulfide bonds that increase rigidity [19]. However, when several cysteine residues are present in the library, unwanted intermolecular disulfide bonding could also occur, which could cause clumping of the displayed cells and interfere with isolation of individual peptides within the population for characterization. The peptides fused by disulfide bonds could have cooperative binding effects, which is not a problem in and of itself, but could be difficult to understand and replicate off-cell. In this study it was shown that the percentage of peptide sequences with zero cysteines decreases round by round, and the percentage of sequences containing more than a single cysteine changes only slightly (Additional file 1: Table S2). Since the increase in cysteine is primarily due to an increase in peptides containing a single cysteine, intra-sequence disulfide binds are not significantly changing and probably don't have much effect on peptide rigidity.

Table 1 Top 40 most frequent sequences from the naive library and after 2 rounds of biopanning

	Naive Library	No. Seqs	Round 1	No. Seqs	Round 2	No. Seqs
1	L	32	VLPQTSFFAATCRRS	115	CHISPEKRRLIVCAD	4619
2	P	31	SNRLRGENLTFRRHK	110	YPLCSTYGNVLVGGS	3989
3	R	30	DRRRKIERYDAQISV	89	PLKQGIYETTSTKGT	3321
4	S	22	T	87	YGSRTFPVCSTVHRR	2185
5	A	19	S	84	YRLMEMEEYHKDVFA	1985
6	T	13	PAKGSIKNPILLGIR	74	KHLLCVMQGGTAFRG	1783
7	E	12	QPLSESLEDKTSIDV	74	RIFGFLSCKPNIIKL	1094
8	H	11	SS	73	IVNRMWCLISVISQA	921
9	SP	11	THNARRLMPCIDASD	73	FWGLYPLFCCTGCSP	909
10	Y	11	YPGRMEALFAHVSNA	73	SKLFSSWVLCACWHC	778
11	Q	10	RDPSWRAETVNCGNL	69	AVLAIYPNSLYWLQA	656
12	TT	10	RHQNPQEEI	69	YIAMILHGFSQAFPG	650
13	V	10	YTYHGRNSERLLRMK	69	LLFSFFNCFTGNDGQ	636
14	F	9	TQDETRRGGMSTTSS	68	NKWSPRTDFNTESVR	623
15	GPCLETCLTA	9	HTYCVHPSMATCWHY	66	HAPWECLISVLASAH	616
16	ERSVG	8	ANLPDSHHIAHNKDT	65	ARGLLTAISFYAFQC	602
17	LT	8	AMPDFVILDHGVTRR	64	RLWMRSPTSTAGGIR	601
18	PRKTLKGTLTVPSYI	8	KNSLTCCYG	63	AERLHQITVMLLRSF	595
19	SG	8	RRNYIHIGAKAAGRS	63	LSNLLHYLYALGQPG	595
20	SKAPLRKTNMKLR	8	DVLAKYSSLERQDLP	62	PILLCYVLYCEYASS	587
21	TS	8	SVAATCWRYWYDKPL	62	SWTMLICQLHSYLYT	583
22	ATYYLQPYVPISEET	7	RPPCEEMSINCPGVT	61	PRALWEASMVVTLAC	581
23	DQLGVL	7	TYPCIYA	61	AVCKPVGPKAPWRDR	575
24	GPWATWSSYDKWHNR	7	VSLIDQCDCTMQSNP	61	SSRSLQQNSISRLSM	571
25	GT	7	HRRRKPIHLPEYPLP	60	IRKQWDFLCLGFLLS	559
26	KS	7	RHSICPRFRGLLPHS	60	SAFPACGLASYFPWF	554
27	PCNFISKHFVPHRRH	7	RML	60	GVVRHCPAWQKSPWP	532
28	PIVGVYK	7	ARTSG	59	HSALDVLSWVISSLM	532
29	PR	7	DWMKFHVEA	59	GLRIQAYALCYWEPV	526
30	SFLL	7	GPANWTCAWQAVTAF	59	NRAIIGEIAGLIKFL	522
31	TD	7	LLVKLGP	59	PAAAFFTQVMSLLRQ	503
32	WD	7	RGVGTSHPQSVCVKP	59	RLHCLERYINLAFLP	503
33	YT	7	CIT	58	RMHHLMFRDSHSHTD	502
34	AMTSLPSTCRQLHSN	6	WLHGVLLNFKKHGVC	58	TQKHKTLCFSIRKPG	494
35	AQDRVRLKSFGRASL	6	IHSAGRMSINLCVGI	57	LRYGIWQVSHIALTS	492
36	CQFNVWANLARIAIG	6	IRRTLLGRPVWFFEE	57	QSFVMESVLSFGFSF	492
37	EDGPTRLWPLSRAAW	6	RSSRSVDPLTYRRNA	57	DRRRKIERYDAQISV	486
38	FSLTVFVMDQENVNR	6	YQPWREDRRSSAGSA	57	PHKLWRLGCFVAMDL	485
39	GDSEGLRMSTLGSSS	6	GLVRVFWPTPERGHR	56	KTVFCLTVLPRFEAA	482
40	GTIGFLDLPRKTVMR	6	AVKAYSTFRQSPTRL	55	RHLVSWLCTVLHPDP	481

Binding assays

We compared a binding assay using direct visualization that measures fluorescent pixel coverage of microscopy images (Figs. 7a & 8) with an indirect binding assay that quantifies plated CFUs to estimate library density (Fig. 7b). We incubated PLA and sapphire samples with the naive library, libraries produced after each round of sorting, and negative control cultures. Both assays showed

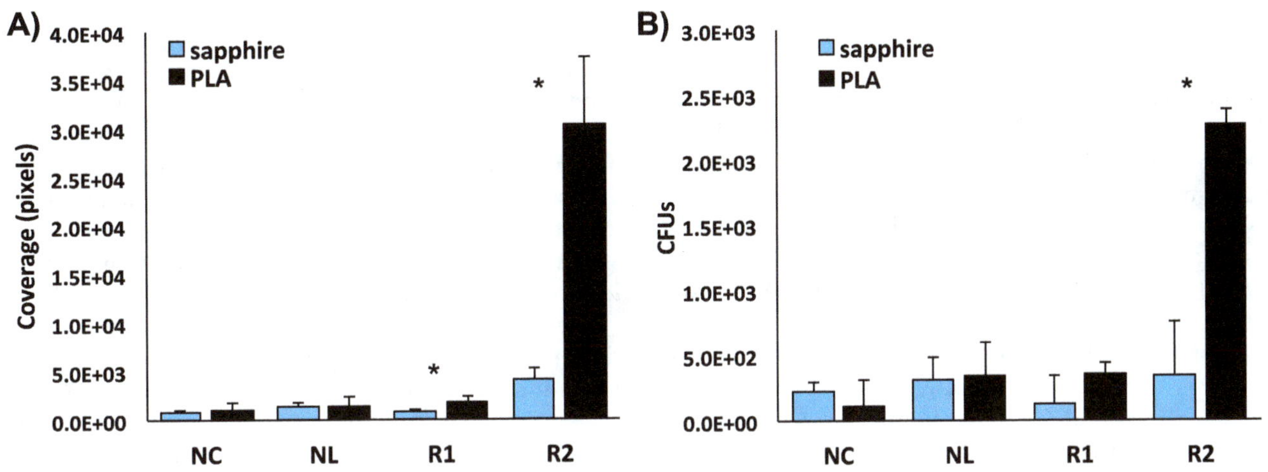

Fig. 7 Binding quantification for each assay type. After incubation of libraries and negative control with PLA and sapphire, the number of red pixels were quantified using fluorescent microscopy (**a**) and colony forming units quantified using indirect binding methods (**b**). NC = negative control, NL = naive library, R1 = Round 1, R2 = Round 2. Mean ± standard error

similar trends, and binding was significantly higher on the PLA coupons than sapphire pieces after Round 2 for both visualization ($P < 0.0001$) and indirect ($P = 0.0015$) assays, and after Round 1 using the visualization assay ($P < 0.0001$). There were no differences between the naive library, sorting rounds, and negative control when comparing growth curves (Additional file 1: Figure S4).

Along with the quantified results in Fig. 7, the fluorescence microscopy images shown in Fig. 8 demonstrate that after two rounds of sorting against PLA, cells displaying peptides from the 15mer library are able to bind the desired material and remain bound after vigorous washing (Fig. 8a). Furthermore, the binding level is higher for PLA (Fig. 8a) as compared to binding to a sapphire wafer (Fig. 8b). The enrichment of binding for PLA over sapphire indicates that the peptides are not just "sticky" in general, but have demonstrated affinity for the material of interest. There is some evidence of cell clumping in Fig. 8a, which may be caused by intermolecular disulfide bonds, discussed above, as a consequence of enrichment for cysteine residues in the library. Additionally, a peptide-free negative control culture, which still displays the empty eCPX scaffold and produces cytosolic DsRed, shows very little binding (Figs. 7 and 8c), indicating that the enriched peptide library is responsible for binding, rather than being caused by either of these over-expressed proteins, or by the cells used to generate the library.

Discussion

Next generation sequencing technologies allow a deeper understanding of peptide selection from display libraries. For the first time, we used NGS to analyze libraries that employ bacterial display across rounds of biopanning to enrich libraries for surface binding. We sequenced populations from a newly developed, unsorted library (naive = N), and after two rounds of selection (Round 1 = R1, Round 2 = R2) against PLA 3D printing plastic (Figs. 5 and 6). Sequencing results confirm population diversity of the

Fig. 8 Fluorescent microscopy images of bound cells. **a** Round 2 library on PLA, **b** Round 2 library on sapphire wafer, and **c** negative control on PLA after 10 min incubation with sample materials and 10 min washing

naive library and a reduction in this diversity across panning rounds (Fig. 6). Visualization of the Round 2 library demonstrates enrichment of PLA binding peptides after selection (Figs. 7 and 8). Valid, unique inserts represented 33% of the total naive bacterial display library (Fig. 6), more than double what was observed after deep sequencing of a naive 7-mer phage display library produced using NNK saturation mutagenesis, which recovered only 15% of sequences with valid, potential binders [20]. The representation of valid inserts increased through successive rounds of biopanning (N: 38%, R1: 42%, R2: 55% of the total library); however, the frequency of individual representatives decreases dramatically after just the first round (N: 33%, R1: 5%, R2: 2% of the total library), as adherent members are carried forward while weakly bound members are washed away in increasingly stringent washing steps.

The total fraction of nonbinding library members that display an empty scaffold, or contain sequence that results in a frame shifts or stop codon decreased 3–7% from the naive library across rounds of sorting (Fig. 6). Interestingly however, the frequency of individual sequences that contain frame shifts or stop codons (we cannot track empty scaffold individuals across rounds as they lack unique insert sequences) decreased more dramatically, reducing the diversity of these sequences 16% (frame shift) or 18% (stop codon). It is expected that after applying selective binding pressure, some individuals exhibiting binding not related to peptides or the display scaffold, or that are carried over for other reasons that are less understood (i.e. genomic mutations, growth rate advantages, etc.), would be enriched to some extent due to the natural evolution process. However, unlike the overall population increase observed for individuals that display a valid peptide (Fig. 6), the representation of nonbinding library members does not increase (note binding increase with sorting rounds in Fig. 7), suggesting a stronger association of valid inserts with the PLA surface. Nonbinding library population diversity does exhibit a similar dramatic decrease as compared to valid insert population diversity.

The library used in this study was created to reduce stop codons by employing 15 x NNS saturation mutagenesis, where N represents any base and S represents G or C. This technique produces a library of inserts with a representation of all 20 amino acids, but with an incidence rate of 3% of the codons in the naive library as stops. With 15-mer insert lengths, this translates to an expected 45% of sequences containing stop codons if stop codons occur singularly within an insert. Only 24% percent of our naive library scaffolds with a 15-mer contain at least one stop codon; a further 20% of these contain at least 2 or more stops. Moreover, a percentage of inserts with frame shifts (27% of the total library) also must contain stop codons, although it is not possible to distinguish a correct reading frame in order to quantify their frequency. There is a tradeoff between insert length and stop codon occurrence, as longer inserts with more residues have a higher chance of incorporating a stop. For example, libraries with a 7-mer insert length would be expected to have only a 21% stop codon occurrence. Library generation must balance the benefits of longer inserts that present secondary structure for enhanced binding with maintaining a diverse library of valid inserts free of stop codons.

The amino acid representation within the library does show subtle changes round by round as expected (Additional file 1: Figures S1 and S2). However, we note that the overall shape of the profile stays roughly the same (Additional file 1: Figure S1) and that diversity may be diminished but not extinguished (Fig. 6, Unique Valid Inserts). There is a general increase in residues capable of hydrogen bond donation consistent with the fact that PLA is a chemically generic aliphatic polyester capable of hydrogen bond acceptance through the polymer backbone. We also note an increase in the sulfur containing residues cysteine and methionine. $n \rightarrow \pi^*$ interactions have been noted between cysteine sulfurs and protein backbone carbonyl groups [21] and similar interactions with the PLA backbone may contribute to the up-regulation of C and M clearly shown in the motif representation. Trends shown from the entire library to the top 40 most frequent sequences to the top 10 most frequent sequences in round 2 are also subtle. It is clear that the predominance of W in the library as a whole is less important in the context of the top 10 PLA binding sequences, while the cysteine interactions mentioned above are clearly in play for all. No true spatial dependence is flagged in any sequence analysis. We note from visual inspection of the top 10 binding sequences in Round 2 that hydrogen bond donors such as R, C, S, Y and T are highly represented.

Assessing populations for binding enrichment on nonantigenic material surfaces using bacterial display libraries in the absence of fluorescence has depended on methods that require additional steps, such as SEM sputter coating for nonconductive surfaces or indirect binding assays [6, 22]. Indirect binding assays require an additional growth step after material incubation to amplify individual binders, followed by serial dilution and replicated plating of library populations. Our new library incorporates an inducible DsRed-Express2 fluorescent protein to allow direct visualization and quantification of enrichment on sample surfaces to eliminate SEM sample preparation or indirect assays that consume time and laboratory supplies, and avoids potential errors caused by cell growth rate differences. After incubating PLA and sapphire pieces with induced libraries and subsequent washing, binding was quantified by measuring either fluorescent pixel coverage, or the number of CFUs from an indirect assay (Fig. 7).

Similar trends were observed for each method of quantification, and in addition to significant differences between PLA and sapphire binding from the Round 2 library, the visualization assay also uncovered differences in Round 1 binding not found in the indirect assay (Fig. 7a).

While we demonstrate the enrichment of peptides with affinity for PLA in this work, and that the enriched peptide library is not purely "sticky" since it does not bind to sapphire with similar affinity (Fig. 8b), the primary goal of this work was to demonstrate how NGS and co-expression of fluorescent proteins can be used to enhance data analysis of display libraries in general, rather than to provide an optimized peptide sequence for binding PLA. To further optimize affinity and specificity of peptides for PLA, it could be useful to employ a constrained peptide library, as they have been shown to improve both affinity and specificity [19]. Specificity for PLA, for instance over other plastics, can also be improved by incorporation of negative sorting steps in future studies. Strategies including negative sorting have previously been shown to provide specificity over structurally similar biological targets [5, 17, 23], and are expected to analogously aid in biopanning peptides towards inorganic materials.

Conclusions

Bacterial display is a valuable tool for rapidly screening peptide libraries. Our results demonstrate that the library generated for this study is diverse, contains a high proportion of valid inserts compared to phage libraries, and can be used successfully for material surface binding enrichment. Binding sequence motif analysis shows little spatial dependence and a general increase in the representation of hydrogen bond donors to interact with the polyester backbone of PLA. The DsRed-Express2 fluorescent protein provides an effective means for visualizing and quantifying adherent library members as an alternative method to indirect binding assays. Similar co-expression strategies with fluorescent proteins and substrate-specific peptides could also be employed for Förster resonance energy transfer (FRET) or other uses [24]. High-throughput next generation sequencing technology allows researchers to verify initial library diversity, gain a better understanding of the effects of selective pressure applied to library populations, and provide a baseline analysis for improving library design and application.

Methods

Bacterial strains, plasmids, materials and growth conditions

All experiments were performed with *E. coli* MC1061 F' Electrocompetent cells strain SS320 (Lucigen; cat.no. 60512-2) [25]. Bacterial cells were routinely cultured in Luria-Bertani (LB) Miller broth (BD Difco) supplemented with 34 μg/ml chloramphenicol (LB+Cm). Agar (15 g/L) was added for solid media and cultures maintained at 37 °C. Liquid cultures were grown in 14 ml culture tubes (Falcon) and shaken at 250 rpm. The eCPX peptide display scaffold was encoded on plasmid pBAD33-eCPX [16] and is available from Addgene. The plasmid was modified by replacing eCPX with eCPX3.0, which includes a P2X peptide displayed on the C-terminus [16]. Additional modification to the Addgene plasmid sequence included mutation of an *Sfi*I site upstream of the peptide insertion site to create a *Bsr*GI site for ease of library insertion. DsRed-Express2 was PCR amplified from commercial vector pCMV-DsRed-Express2 (Takara) and inserted directly upstream of eCPX. The backbone vector was constructed by co-inserting DsRed-Express2 and the modified eCPX 3.0 sequence into pBAD33-eCPX using *Sac*I and *Sal*I restriction sites to create pB33-dsRed-eCPX3.0 (Fig. 1).

To construct the library, the pB33-dsRed-eCPX3.0 backbone vector was digested with *Bsr*GI and *Xho*I and purified using DNA Clean and Concentrator kit (Zymo Research). A DNA library of peptide inserts in the form of NNS within the eCPX3.0 gene was commercially synthesized (GeneArt), similarly digested and ligated into the vector backbone using electroligase (New England Biolabs). A total of 1.5 μg of newly assembled constructs were then electroporated into 250 μl electrically competent cells in batches of 1 μl DNA with 25 μl cells using a 0.1 cm gap, pre-chilled electroporation cuvette, and pulsed at 1.8 kV, 10 μF, and 100 Ω. The cuvette was rinsed three times with 1 ml warm SOC and transferred to a 125 ml vented flask. This process was repeated until the entire volume of pB33-dsRed-eCPX3.0 library was transformed ($\sim$10 times). Approximately 70 ml warm SOC was added to the flask and incubated at 37 °C for 1 h. After recovery in SOC, the library transformants were added to 500 ml of warm LB+Cm for 4 h. The culture was then split into five flasks of 1 L LB+Cm and grown overnight. The following day, each liter was pelleted and resuspended in 10 ml LB+15% glycerol. From this, 1 ml frozen library stocks were prepared and stored at -80 °C, with each stock containing $\sim 10^{11}$ cells.

PLA coupons

A 1 kg spool of 1.75 mm 'True Black' PLA 3D Printer Filament from HATCHBOX® was used to make 1 x 3 cm coupons for use in binding trials. PLA plastic filament for 3D printing is delivered in spools, and must be melted into thin pieces between smooth-surfaced materials. Two $\sim$4 cm pieces of PLA filament were placed side-by-side on the bottom of a Pyrex baking dish (which maintains a smooth surface and can withstand rapid changes in heat) and set onto a hot plate heated to level 6/10, or $\sim$90C. A 5 cm diameter sapphire (aluminum oxide) wafer was placed on top of the PLA filaments (Fig. 4). After several minutes, as the PLA melted, a small 200 gram weight was used to gently press the sapphire wafer, spreading the melting plastic underneath into a flat sample approximately 0.5 mm thick. Once fully melted, the Pyrex dish was immediately moved to a benchtop and allowed to cool for exactly 7 min. After cooling, a razor blade was used to carefully lift the PLA sample from the Pyrex dish and remove the sapphire wafer. Scissors sterilized with ethanol were used to cut the sample into 1 x 3 cm coupons for biopanning trials, 0.5 x 0.5 cm coupons for indirect assays, or 0.75 x 0.75 cm coupons for visualization. Sapphire wafer was also used to as an alternate material for comparison after binding trials (described below and depicted in Fig. 3). Sapphire has a smooth, characterized surface and can be cut to size using an etching tool.

Biopanning

Biopanning rounds consist of several steps: 1) incubating the target material with the naive library culture, 2) washing unbound bacteria from the material surface, 3) placing the washed material with bound library members into fresh media, 4) allowing the bound members to amplify (resulting in the Round 1 library), and 5) repeating steps 1–4 for two total panning rounds. The pB33-dsRed-eCPX3.0 library was prepared for panning as previously described [6, 17, 26] with modifications depicted in Fig. 3 for the current study. In brief, peptide expression was induced overnight using 0.4% L-arabinose after 250 ml overnight stocks diluted 1:50 in LB+Cm reached an OD600 0.50–0.60. Induced cells were chilled on ice for 15 min and transferred to a sterilized crystalizing dish containing four 1 x 3 cm PLA samples sterilized with 70% ethanol. Samples were kept at 4 °C for 15 min with shaking to allow bacteria to bind. PLA coupons were removed from the library culture, briefly dipped into sterile phosphate buffered saline (PBS) to remove the bulk of unbound bacteria, and transferred to 20 ml of PBS supplemented with 1% Tween. Coupons were washed for 30 min at 330 rpm after incubation with the naive library, and for 1 h after incubation with the Round 1 library. After washing, bound cells were recovered by placing the PLA coupons into LB+Cm media supplemented with 2% D-Glucose (LB+Cm/Glu) and grown at 37 °C with shaking overnight. Fifty randomly selected colonies from Round 2 were sequenced using the pBAD Reverse universal primer (Genewiz).

Visualization binding assay

In order to visually compare adhesion across samples after biopanning against PLA, 0.5 x 0.5 cm PLA coupons and sapphire pieces were incubated with induced 5 mL cultures of the naive library, sorting rounds, or negative control (which consists of a library member with an empty eCPX scaffold, i.e. no peptide). Cultures were prepared from freezer stocks grown overnight in LB+Cm. Fresh LB was inoculated 1:50 with the overnight stocks and brought to an OD of 0.5-0.6, then induced overnight with 0.4% arabinose as described above. PLA and sapphire pieces were incubated for 10 min with each culture at 230 rpm, then washed for an additional 10 min in fresh culture tubes containing 4 ml of PBS 1% Tween at 330 rpm to remove weakly bound cells. A final dip in 4 ml of fresh PBS ensured any unbound bacteria in the wash solution would be completely removed. Material samples were placed on microscope slides and visualized using an epifluorescent microscope (Nikon Eclipse TE2000-E confocal) at 40X and a red fluorescent filter (AT-TRITC/CY3). Samples were prepared independently in triplicate and ten images were captured from each sample and analyzed for the number of red fluorescent pixels using ImageJ [27]. Student's T-tests were used to statistically verify differences in fluorescent pixel density.

Indirect binding assay

Indirect binding assays were performed to quantify the number of cells bound to PLA and sapphire pieces as previously described [6]. In brief, incubation of sample materials with libraries and the negative control was the same as described above, however, after washing, the material samples were placed in 6 ml of fresh LB+Cm/Glu and grown for 1 h at 37 °C with shaking. Serial dilutions of amplified cells from each sample were plated on LB+Cm agar plates. Samples were prepared independently in triplicate and the averages and standard deviations determined. Student's T-tests were used to statistically verify differences in relative binding results.

Next generation sequencing

Three library culture samples were prepared for Illumina next generation sequencing: the naive library and libraries recovered after each of two rounds of biopanning. Plasmids were extracted from overnight cultures using a ZymoPure plasmid maxiprep kit (Zymo Research) and prepared for sequencing according to the Illumina 16S amplicon protocol using custom forward (AGTTCTGGCTTTCACCGCAG) and reverse (CCGTAGTACTGGTTTTGTTGTAGTC) primers corresponding to eCPX scaffold regions adjacent to the peptide insert and run on a NextSeq500. Primers also included standard Illumina platform adapter overhangs and multiplex barcodes (not shown).

Illumina fastq files were analyzed using modified Matlab scripts based on the format presented in Matochko et al. (2012)[12]. Sequences were parsed into categories based on: 1) valid sequences: inserts that had both correct flanking sequences, were in the correct reading frame, contained no stop codons, and had all bases above the quality threshold (sequences with an Illumina quality score above Q20, i.e. bases that have a 99% or better chance of having been correctly assigned), 2) inserts containing nucleotides with incorrect reading frames, which resulted in a nonfunctional eCPX, 3) inserts in the correct reading frame that contained stop codons, and 4) empty eCPX scaffolds (scaffolds that did not contain any insert; Fig. 5). Valid sequences were then sorted to discover unique sequences. Additional sequence analysis was performed with in-house scripts, and the pLogo probability logo generator [18] was used to study sequence motifs.

Supplementary information

Additional file 1: Figure S1 Amino acids represented within unique sequences. Amino acid representation of all unique sequences within each library **a** excluding and **b** including the effects of frequency of sequence occurrence for Naive (blue), Round 1 (orange) and Round 2 (grey). **Figure S2** Unique 15-mer sequence motifs from all libraries. Visualization of unique 15-mer sequence motifs calculated against background probabilities for *E. coli* K12 in **a** Naive **b** Round 1 and **c** Round 2 libraries, as generated by the pLogo software (http://plogo.uconn.edu) [18]. **Figure S3** Unique 15-mer sequence motifs from the Round 2 library and subsets of the Round 2 library. Visualization of unique 15-mer sequence motifs calculated against background probabilities for *E. coli* K12 in the Round 2 library: **a** all 15-mers **b** 40 most frequently occurring 15-mers and **c.** 10 most frequently occurring 15-mers, as generated by the pLogo software (http://plogo.uconn.edu) [18]. **Figure S4** Library growth curves. Growth curves over 6 h for the naive library, both sorting rounds, and the negative control after a 1:100 dilution from overnight cultures in fresh media. **Table S1** Raw count of total and unique sequences for all sequence categories of each library. **Table S2** Raw count and decimal of total sequences in library containing 0–4 cysteines, round by round.

Abbreviations

NC: Negative control; NGS: Next generation sequencing; NL: Naive library; PLA: Polylactic acid; R1: Round 1; R2: Round 2

Acknowledgements

We thank Jeffrey Rice for technical discussions and advice.

Authors' contributions

SDS led experimental design, collected data, performed analyses, and led manuscript preparation. DAS aided in experimental design, data collection, analysis, manuscript preparation, and submission. BLA conceived of the DsRed panning library and aided in experimental design and manuscript preparation. MAH generated the panning library. RLR collected all next generation sequencing data. MMH contributed to data analysis and manuscript preparation. DNS contributed to experimental design, manuscript preparation, and supervised the work. The final manuscript was read and accepted by all co-authors.

Funding

This work was funded entirely by the U.S. Army Combat Capabilities Development Command Army Research Laboratory (CCDC ARL), in part through contracts with other entities for personnel: the U.S. Army Research Laboratory Postdoctoral Fellowship Program administered by the Oak Ridge Associated Universities and General Technical Services, LLC. As such, all authors were Federal Government employees or contractors when this research was completed. The study design, the collection, analysis, and interpretation of data, as well as the writing of the manuscript, were performed with CCDC ARL funds, facilities, and research personnel. The views expressed here are those of the authors and do not represent the opinions of the U.S. Army, the U.S. Department of Defense, or the U.S. Government.

Ethics approval and consent to participate

Not applicable.

Consent for publication

Not applicable.

Competing interests

The authors declare that they have no competing interests.

Author details

[1]Biotechnology Branch, CCDC US Army Research Laboratory, 2800 Powder Mill Rd, Adelphi, MD, 20783 USA. [2]Department of Biological Sciences, University of Maryland Baltimore County, 1000 Hilltop Circle, Baltimore, MD, 21250 USA. [3]General Technical Services, Suite 301, 1451 Route 34 South, Wall Township, NJ, 07727 USA.

References

1. Ibsen KN, Daugherty PS. Prediction of antibody structural epitopes via random peptide library screening and next generation sequencing. J Immunol Methods. 2017;451:28–36.
2. Molek P, Strukelj B, Bratkovic T. Peptide phage display as a tool for drug discovery: Targeting membrane receptors. Molecules. 2011;16(1):. https://doi.org/10.3390/molecules16010857.
3. Martins IM, Reis RL, Azevedo HS. Phage display technology in biomaterials engineering: Progress and opportunities for applications in regenerative medicine. ACS Chem Biol. 2016;11(11):2962–80.
4. Ploss M, Facey SJ, Bruhn C, Zemel L, Hofmann K, Stark RW, Albert B, Hauer B. Selection of peptides binding to metallic borides by screening m13 phage display libraries. BMC Biotechnol. 2014;14(1):12.
5. Sarkes DA, Hurley MM, Stratis-Cullum DN. Unraveling the roots of selectivity of peptide affinity reagents for structurally similar ribosomal inactivating protein derivatives. Molecules. 2016;21(11):. https://doi.org/10.3390/molecules21111504.
6. Adams BL, Finch AS, Hurley MM, Sarkes DA, Stratis-Cullum DN. Biomaterials: Genetically engineered peptides for inorganics: Study of an unconstrained bacterial display technology and bulk aluminum alloy (adv. mater. 33/2013). Adv Mater. 2013;25(33):4530.
7. Bessette PH, Rice JJ, Daugherty PS. Rapid isolation of high-affinity protein binding peptides using bacterial display. Protein Eng Des Sel. 2004;17(10):731–9.
8. Kenrick SA, Daugherty PS. Bacterial display enables efficient and quantitative peptide affinity maturation. Protein Eng Des Sel. 2010;23(1):9–17.
9. Kogot JM, Pennington JM, Sarkes DA, Kingery DA, Pellegrino PM, Stratis-Cullum DN. Screening and characterization of anti-seb peptides using a bacterial display library and microfluidic magnetic sorting. J Mol Recogn. 2014;27(12):739–45.
10. Smith GP, Petrenko VA. Phage display. Chem Rev. 1997;97(2):391–410.
11. Matochko W, Derda R. In: Derda R, editor. Next-Generation Sequencing of Phage-Displayed Peptide Libraries. Methods in Molecular Biology, vol. 1248. Chap. 17: Springer; 2015, pp. 249–66. https://doi.org/10.1007/978-1-4939-2020-4_17.
12. Matochko WL, Chu K, Jin B, Lee SW, Whitesides GM, Derda R. Deep sequencing analysis of phage libraries using illumina platform. Methods. 2012;58(1):47–55.
13. 't Hoen PAC, Jirka SMG, ten Broeke BR, Schultes EA, Aguilera B, Pang KH, Heemskerk H, Aartsma-Rus A, van Ommen GJ, den Dunnen JT. Phage display screening without repetitious selection rounds. Anal Biochem. 2012;421(2):622–31.
14. Daugherty PS. Protein engineering with bacterial display. Curr Opin Struct Biol. 2007;17(4):474–80.
15. Rice JJ, Schohn A, Bessette PH, Boulware KT, Daugherty PS. Bacterial display using circularly permuted outer membrane protein ompx yields high affinity peptide ligands. Protein Sci Publ Protein Soc. 2006;15(4):825–36.

16. Rice JJ, Daugherty PS. Directed evolution of a biterminal bacterial display scaffold enhances the display of diverse peptides. Protein Eng Des Sel. 2008;21(7):435–42.
17. Sarkes DA, Jahnke JP, Stratis-Cullum DN. Semi-automated biopanning of bacterial display libraries for peptide affinity reagent discovery and analysis of resulting isolates. JoVE (J Visualized Exp). 2017;130:56061. https://doi.org/10.3791/56061.
18. O'Shea JP, Chou MF, Quader SA, Ryan JK, Church GM, Schwartz D. plogo: a probabilistic approach to visualizing sequence motifs. Nat Methods. 2013;10(12):1211.
19. Bosma T, Rink R, Moosmeier MA, Moll GN. Genetically encoded libraries of constrained peptides. ChemBioChem. 2019;20(14):1754–8.
20. Ryvkin A, Ashkenazy H, Smelyanski L, Kaplan G, Penn O, Weiss-Ottolenghi Y, Privman E, Ngam PB, Woodward JE, May GD, Bell C, Pupko T, Gershoni JM. Deep panning: Steps towards probing the igome. PLoS ONE. 2012;7(8):41469.
21. Kilgore HR, Raines RT. n $\rightarrow \pi^*$ interactions modulate the properties of cysteine residues and disulfide bonds in proteins. J Am Chem Soc. 2018;140(50):17606–11.
22. Adams BL, Hurley MM, Jahnke JP, Stratis-Cullum DN. Functional and selective bacterial interfaces using cross-scaffold gold binding peptides. JOM. 2015;67(11):2483–93.
23. Sarkes DA, Hurley MM, Coppock MB, Farrell ME, Pellegrino PM, Stratis-Cullum DN. Rapid discovery of peptide capture candidates with demonstrated specificity for structurally similar toxins. In: Proc. SPIE 9863, Smart Biomedical and Physiological Sensor Technology XIII, 986305. International Society for Optics and Photonics; 2016. p. 1–10. https://doi.org/10.1117/12.2224799.
24. Erickson MG, Moon DL, Yue DT. Dsred as a potential fret partner with cfp and gfp. Biophys J. 2003;85(1):599–611.
25. Casadaban MJ, Cohen SN. Analysis of gene control signals by dna fusion and cloning in escherichia coli. J Mol Biol. 1980;138(2):179–207.
26. Kogot JM, Zhang Y, Moore SJ, Pagano P, Stratis-Cullum DN, Chang-Yen D, Turewicz M, Pellegrino PM, de Fusco A, Soh HT, Stagliano NE. Screening of peptide libraries against protective antigen of bacillus anthracis in a disposable microfluidic cartridge. PLoS ONE. 2011;6(11):26925.
27. Schneider CA, Rasband WS, Eliceiri KW. Nih image to imagej: 25 years of image analysis. Nat Methods. 2012;9(7):671–5.

19

Engineering integrative vectors based on phage site-specific recombination mechanism for *Lactococcus lactis*

Innanurdiani Koko[1], Adelene Ai-Lian Song[2,3*], Mas Jaffri Masarudin[1,3] and Raha Abdul Rahim[1,3,4*]

Abstract

Background: Site-specific integration system allows foreign DNA to be integrated into the specific site of the host genome, enabling stable expression of heterologous protein. In this study, integrative vectors for secretion and surface display of proteins were constructed based on a lactococcal phage TP901–1 integrating system.

Results: The constructed integration system comprises of a lactococcal promoter (P_{nisA} or P_{170}), phage attachment site (attP) from bacteriophage TP901–1, a signal peptide (USP45 or SPK1) for translocation of the target protein, and a $PrtP_{344}$ anchor domain in the case of the integrative vectors for surface display. There were eight successfully constructed integrative vectors with each having a different combination of promoter and signal peptide; pS1, pS2, pS3 and pS4 for secretion, and pSD1, pSD2, pSD3 and pSD4 for surface display of desired protein. The integration of the vectors into the host genome was assisted by a helper vector harbouring the integrase gene. A nuclease gene was used as a reporter and was successfully integrated into the *L. lactis* genome and Nuc was secreted or displayed as expected. The signal peptide SPK1 was observed to be superior to USP45-LEISSTCDA fusion in the secretion of Nuc. As for the surface display integrative vector, all systems developed were comparable with the exception of the combination of P_{170} promoter with USP45 signal peptide which gave very low signals in whole cell ELISA.

Conclusion: The engineered synthetic integrative vectors have the potential to be used for secretion or surface display of heterologous protein production in lactococcal expression system for research or industrial purposes, especially in live vaccine delivery.

Keywords: *Lactococcus lactis*, Integrative vector, Secretion, Surface display, Signal peptide

Background

Lactococcus lactis, a lactic acid bacteria (LAB) that has been conventionally known as a starter culture in food fermentations such as cheese and yoghurt, was granted with the Generally Recognized as Safe (GRAS) status by the US FDA. It is the most well studied LAB strain due to its well characterised genome which allows researchers to manipulate this bacterium for desired applications. *L. lactis* is also known as the model LAB and a promising bacteria in recombinant protein technology that has been extensively used as cell factory for the production of industrially important protein such as enzymes, compounds, antigens, allergens, and cytokines [1–5].

Due to its extensive applications, numerous cloning and expression vectors have been constructed for this bacterium. However, the instability of the replicating plasmid DNA is a drawback where it can be lost under non-selective conditions [6]. Plasmid copy numbers often varies with growth, resulting in different expression levels of the genes they carry [7]. Besides, over expression of heterologous protein causes metabolic load which burdens the host and affect the protein integrity [8]. Due to these limitations, constructions of more stable expression systems were attempted through

* Correspondence: adelene@upm.edu.my; raha@upm.edu.my
[2]Department of Microbiology, Faculty of Biotechnology and Biomolecular Sciences, University Putra Malaysia, 43400 UPM, Serdang, Selangor, Malaysia
[1]Department of Cell and Molecular Biology, Faculty of Biotechnology and Biomolecular Sciences, University Putra Malaysia, 43400 UPM, Serdang, Selangor, Malaysia

genomic integration of integrative vectors harbouring the desired gene in *L. lactis*. DNA integrating system has been demonstrated in other LAB such as *Lactobacillus* for vaccine construction, for instance, the expression of surface anchored llama heavy chain antibody against rotavirus [9–12] and other Gram-positive bacteria such as *Bacillus* sp. using spore surface display system to express target protein on the outer surface of bacteria [13]. The integrating system has also been long developed for surface display of target protein for vaccine production in mycobacteria [14, 15] and other applications [16–18]. The most well characterised, easy manipulated bacteria, *Escherichia coli* has a long lists of integrating systems developed to express various kinds of recombinant protein for numerous applications [19–24].

As reported by Leenhouts et al., (1990), chromosomal integration in *L. lactis* mediated by replacement recombination and Campbell-like mechanism [25] resulted in unstable integration and the integrated DNA fragment were gradually lost under non-selective condition. The instability of the integrated plasmid was due to its replicative activity and the direct repeats flanking the inserted DNA sequences resulted from the Campbell-like integration permitted intra-molecular homologous recombination by the host system. These conditions caused the integrated plasmid to be cleaved out from the genome, losing the integrated plasmid [26–28].

A site-specific recombination system for *L. lactis* has been developed to generate stable, single-copy chromosomal integration based on the lactococcal temperate bacteriophage TP901–1. The system requires phage-encoded elements, which are the phage attachment site, *attP* and an integrase protein which facilitates recombination between the *attP* site and the bacterial attachment site, *attB* on the host genomic DNA at a high frequency. For the recombination to occur, integrase recognises a short sequence of the site (43 bps) which makes the system applicable in numerous *L. lactis* strains [29–34]. Through utilisation of this established integration system, we developed a variety of integrative vectors for secretion and surface display of targeted proteins and compared their efficiency using nuclease as a reporter.

L. lactis possess only one exported housekeeping protease (HtrA), and secrete only one major extracellular protein, USP45, thus minimizing protein degradation by the extracellular proteolytic system, simplifying downstream purification [35]. Secretion is preferable in protein delivery when the target location is the digestive tract of humans or animals, because it facilitates interaction between the expressed protein and its target [36]. On the other hand, surface display enables expression of recombinant protein on the surface of bacteria for various applications such as live vaccine development, antibody production, peptide library screening, and production of whole-cell biocatalysis and bioadsorbents for environmental applications [37].

In this report, we describe the construction of a single-copy integration vector for *L. lactis* based on the TP901–1 site-specific recombination. The integrative vectors were suicide vectors for *L. lactis* host, designed to replicate in *Escherichia coli* for easy DNA manipulations. When it is transformed into the *L. lactis*, the vector integrates into the *attB* site of the genome and the expressed target protein is either secreted or surface displayed on the outer part of the cells. Different combinations of promoters and signal peptides were used and protein expression and secretion efficiency were compared between systems.

Results

Helper plasmid (pNZint)

L. lactis harbouring the helper plasmid, pNZint was analysed for the expression of integrase protein as integration of the foreign DNA is only possible with co-expression of the integrase gene. The integrase protein was tagged with 6 histidine residues to allow detection via Western blotting. Protein expression was analysed for 12 h post-induction (20 ng/ml nisin) with 2 h intervals. The expected size of the desired protein was 55.6 kDa where expression of integrase was observed from 2 h onwards (Fig. 1).

Vector constructions and integration

A total of eight vectors, pS1–4 and pSD1–4 were successfully constructed as listed in Table 1. The constructed secretion and surface display integrative vectors were subsequently cloned with *nuc* gene, thereby named pS1–4nuc and pSD1–4nuc. The integrative vectors harbouring *nuc* were then transformed into *L. lactis* NZ9000 harbouring pNZint, and the positive integrants which would obtain erythromycin resistance were then tested for their stability to stay integrated in the attB site of the genome. After 100 generations grown in non-selective conditions, all random colonies chosen which were sub-cultured on to erythromycin plates had positive growth, indicating that the integrated plasmid was still inside and has not been excised. In addition, PCR using the attB flanking primers (F-attB/R-attB) on genomic DNA extracted from the integrants further confirmed presence of the integrated plasmid.

Figure 2 (a) and (b) show an example of PCR verification of selected integrants using primers flanking the attB site on the genome after 100 generations with the expected DNA size of 3944 bps, 3834 bps, 3927 bps and 3902 bps for pS1–4nuc, respectively, whereby the empty genomic DNA (negative control) was 300 bps. Figure 3 shows the gel electrophoresis image of the PCR amplified integrated surface displayed integrative plasmids

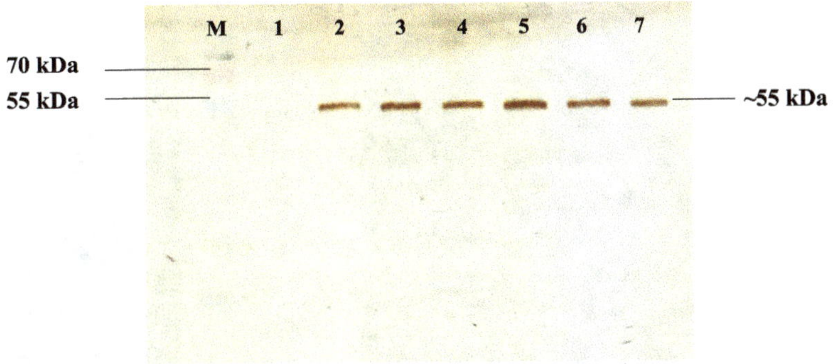

Fig. 1 Western blot showing time optimization for the expression of integrase protein. Lane M) Pageruler prestained protein ladder, 1–7) protein sample at 0 h, 2 h, 4 h, 6 h, 8 h, 10 h and 12 h respectively

(pSD1nuc, pSD2nuc, pSD3nuc, and pSD4nuc) with the expected DNA size of 4916 bps, 4875 bps, 4912 bps and 4943 bps respectively. However, some unspecific bands were also observed which is probably due to unspecific binding of the flanking primers in the genome

Nuclease expression

The secreted nuclease from the integrated secretion plasmids were determined by Toluidine Blue DNA (TBD) assay through colony overlay method as described by Lachica et al., (1971) [38]. The *L. lactis* strains harbouring integrated *nuc* were observed to form bright pink halo zone as compared to the negative control, a non-integrated *L. lactis*. Figure 4 shows the representation of halo zone formed around the bacteria culture on the TBD plate. The TBD assay concluded that *L. lactis* integrated with pS4nuc (driven by P_{nisA} promoter and SPK1 signal peptide) has the highest Nuc secretion with mean 0.575 cm followed by pS2nuc, pS3nuc and pS1nuc with mean 0.50 cm, 0.25 cm, 0.20 cm respectively (Fig. 5, raw data set in Additional file 1: Table S1). The ANOVA test indicates there is no significant difference in the expression of Nuc driven by P_{170} and P_{nisA} promoters. However, SPK1 showed better secretion of Nuc compared to USP45-LEISSTCDA fusion with significance of P-value < 0.05.

Immunofluorescence analysis

The surface displayed nuclease on *L. lactis* NZ9000 was analysed to determine the bio-functionality of the cell wall anchoring motif, PrtP, an LPXTG anchoring system, via immunofluorescence microscope. It was observed that *L. lactis* integrated with pSDnuc1–4 were able to translocate and anchor the expressed Nuc onto the surface of the bacteria proven by the emission of green fluorescence signals as compared to the negative control, a non-integrated *L. lactis* (Fig. 6).

Whole cell ELISA assay

Quantitative expression of surface anchored nuclease was analysed by whole-cell ELISA assay with nuclease specific antibody. Among the surface displayed vector constructs, strain harbouring pSD1nuc showed extremely low signals, almost as low as the controls while pSD2-4nuc have almost the same signals intensities with no significance difference. Although it was observed that pSD1nuc and pSD3nuc strains driven by P_{170} and P_{nisA} promoter, respectively, show a highly significant difference with the latter showing greater signal intensities.

Table 1 Summary of secretion and surface display integrative vector components

Plasmid constructs	Size (bps)	Promoter	Signal peptide	Propeptide	LPXTG anchor domain	Function
pS1	3111	P_{170}	USP45	LEISSTCDA	–	Secretion
pS2	3072	P_{170}	SPK1	–	–	Secretion
pS3	3144	P_{nisA}	USP45	LEISSTCDA	–	Secretion
pS4	3086	P_{nisA}	SPK1	–	–	Secretion
pSD1	4152	P_{170}	USP45	LEISSTCDA	$PrtP_{344}$	Surface display
pSD2	4107	P_{170}	SPK1	–	$PrtP_{344}$	Surface display
pSD3	4185	P_{nisA}	USP45	LEISSTCDA	$PrtP_{344}$	Surface display
pSD4	4146	P_{nisA}	SPK1	–	$PrtP_{344}$	Surface display

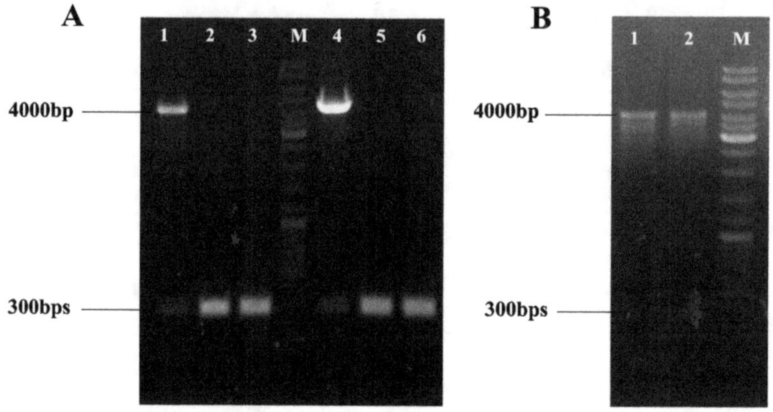

Fig. 2 PCR verification of integrated plasmid in the *L. lactis* genome using attB flanking primers. **a** M) Generuler DNA ladder mix 1) integrated pS1nuc plasmid, 2) false positive clones, 3) negative control; 4) integrated pS2nuc plasmid, 5) false positive clone, 6) negative control. **b** 1) integrated pS3nuc, 2) integrated pS4nuc, M) Generuler DNA ladder mix

We cannot conclude that P_{nisA} is a stronger promoter for surface displaying of proteins than P_{170} as this result was not consistent when comparing pSD2 and pSD4. Similarly, inconclusive results were also observed on the surface display of nuclease controlled by USP45 and SPK1 signal peptides as pSD1nuc and pSD2 showed significant intensity differences while pSD3 and pSD4 did not. Therefore, we can only conclude that the combination of P_{170} with USP45 was inferior compared to the other systems in the surface display integrative vectors (Fig. 7, raw data set in Additional file 2: Table S2).

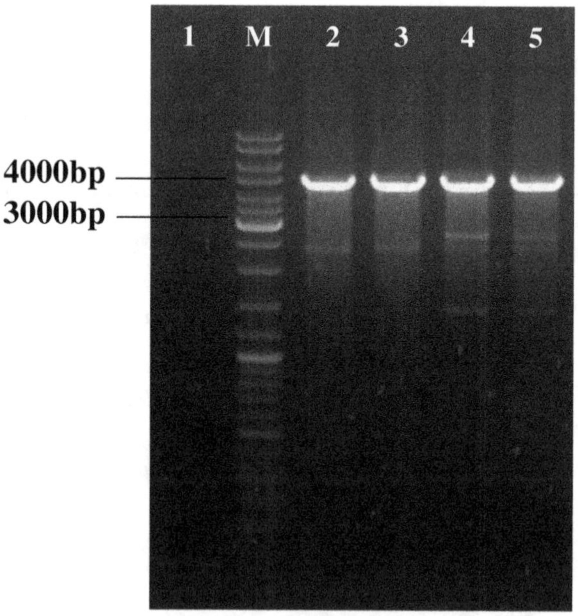

Fig. 3 Gel electrophoresis of PCR verification using attB flanking primers of integrated surface displayed integrative plasmids in the *L. lactis* genome. M) Generuler DNA ladder mix, 1) negative control, 2) pSD1nuc, 3) pSD2nuc, 4) pSD3nuc, and 5) pSD4nuc

Discussion

In this work, two types of *Lactococcus lactis* integrative vectors utilizing the bacteriophage TP901–1 integrating system were successfully constructed. The TP901–1 is a temperate bacteriophage that resembles resolvase recombination mechanism. The TP901–1 attB was found to be located in a putative open reading frame encoded for competence protein, ComGC (Accession number: DJ61374.1) [39]. The integrative plasmid carrying the attP site recombines with the bacteria's DNA sequence

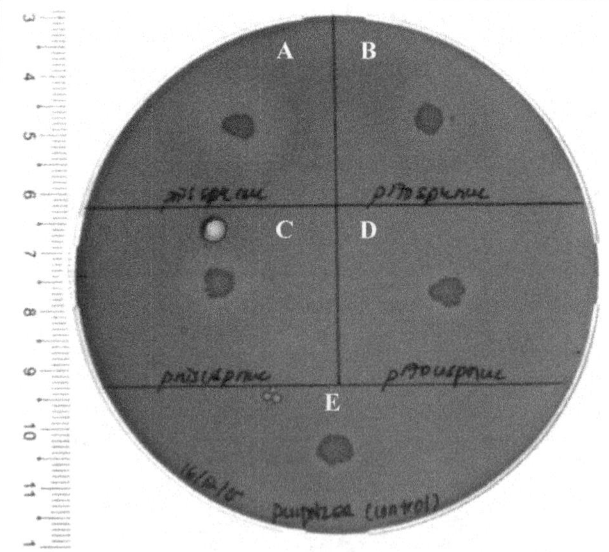

Fig. 4 TBD test on integrated secretion plasmids clones, image taken using UV light gel documentation. **a**) pS4nuc, **b**) pS2nuc, **c**) pS3nuc, **d**) pS1nuc, and **e**) negative control

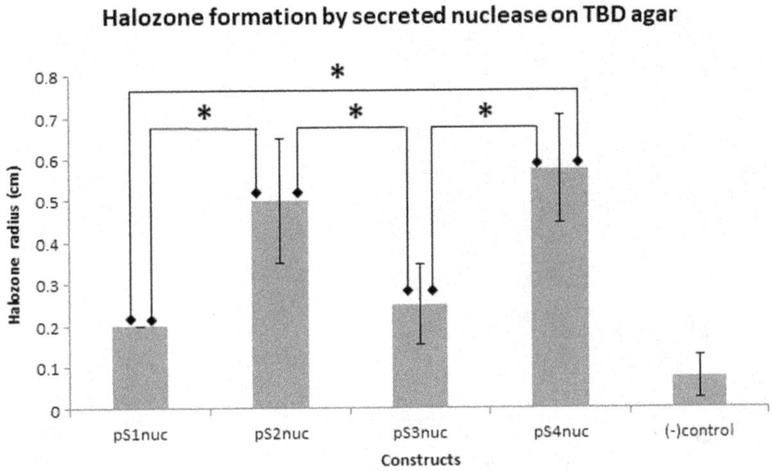

Fig. 5 TBD assay of secreted nuclease by *L. lactis* harbouring integrated *nuc*. Data shown is an average of the radius of halozone of three replicates and *indicates significance $P < 0.05$. All samples are significant compared to the negative control

on the attB site, disrupting the open reading frame. However, up to date, there is no reports regarding the effect of this ORF disruption on the physiology of the cells [39, 40]. *Bacillus subtilis*, a naturally competent bacterium, utilises the proteins to take up of extracellular DNA into the cell. However, *L. lactis* and other dairy isolated bacteria were reported to be unable to take foreign DNA into the intracellular environment naturally. Thus, disruption of the protein might not have any significant effect.

The site-specific integrating system was designed to integrate the desired plasmid into *L. lactis* genome and express the gene of interest into the extracellular environment, where it is either fully secreted into the medium or surface displayed on *L. lactis* cell wall. The secretion integrative vectors, pS and the surface display integrative vectors, pSD were engineered with different combinations of promoters and signal peptides; P_{170} versus P_{nisA}, and USP45 fused with LEISSTCDA versus SPK1 signal peptides alone, resulting in pS1, pS2, pS3 and pS4 secretion integrative vectors, and, pSD1, pSD2, pSD3 and pSD4 surface display integrative vectors, respectively (Table 2). Nuclease was used as a reporter to determine the ability of the bacteriophage integrating system to integrate into the *L. lactis* genome and the best combination of vector constructs to secrete or surface display the desired protein.

The data obtained showed that all the integrative vectors constructed were able to integrate into the genomic DNA of *L. lactis* and the integrants were stably integrated in the host genome. However, in this study, only PCR was used to verify integration of the plasmid. Further verification of the integration could possibly employ an alternative method which may include whole genome-sequencing to confirm that integration has taken place at the expected site and that ectopic integration has not taken place. However, it is unlikely that ectopic integration has taken place in the current study as the mechanism for integration using the attP/attB site is very specific, unlike homologous recombination where recombination may occur at different positions depending on the degree of similarity of the sequences.

The integrants constructed in this study were able to either secrete or surface display the reporter protein. The secreted Nuc results showed that pS2nuc and pS4nuc gave the highest secretion based on the radius of halo zone on TBD agar plates. Both pS2 and pS4 were driven by SPK1 signal peptide which demonstrated that the SPK1 signal peptide was superior over USP45 even though the USP45 was fused with synthetic propetide LEISSTCDA which was reported to enhance protein secretion [43]. This is despite USP45 being an endogenous signal peptide and the most commonly used signal peptide for secretion of heterologous protein in *L. lactis*. On the other hand, SPK1 is a signal peptide isolated from *Pediococcus pentasoceus* which was previously shown to possess comparative secretion efficiency to USP45 although total protein was at times much lower than in USP45 [44, 45].

From the analysis, there was no significant difference on the Nuc expression controlled by P_{nisA} and P_{170} promoters where the *P*-value for both pS1 against pS3 and pS2 against pS4 were above 0.05. Therefore, we can conclude that P_{nisA} and P_{170} promoters have comparable strength although the data showed a slight increase in the expression of Nuc by P_{nisA} promoter. The well-established NICE system with P_{nisA} is known for its strong and tightly regulated promoter [46–48] induced by nisin, whereas P_{170} is a lactococcal auto-inducible promoter by the accumulation of lactic acid in the

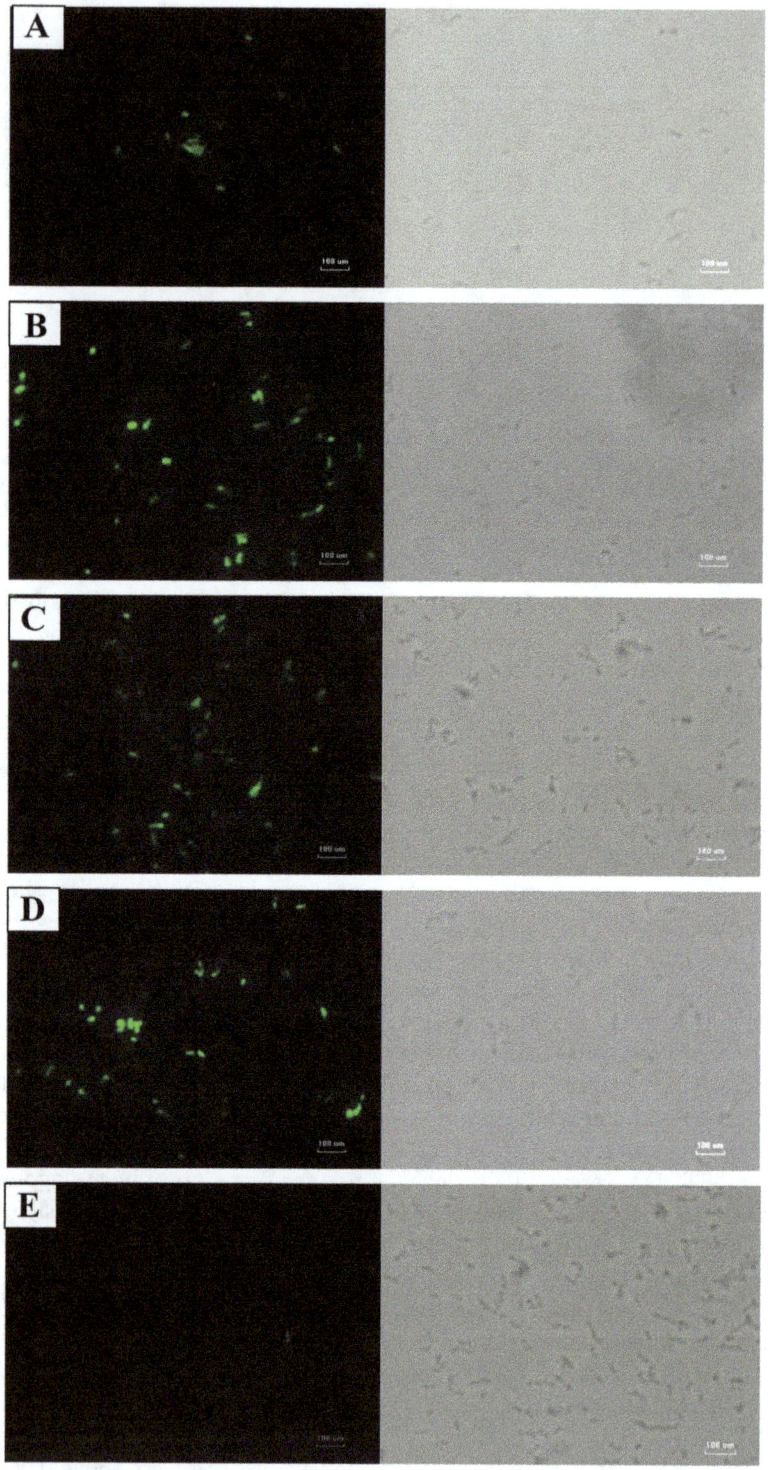

Fig. 6 Fluorescent microscopy of *L. lactis* harbouring integrated pSDnuc1–4. The fluorescence images indicate surface anchored nuclease protein expressed from integrated plasmids; Left Panel: Green filter, Right Panel: Phase contrast; **a** pSD1nuc, **b** pSD2nuc **c** pSD3nuc, **d** pSD4nuc and non-integrated *L. lactis* NZ9000 as negative control (**e**)

culture medium which has the potential to be used in food-grade expression system [49–51]. In the industrial application where nisin induction will be costly for mass production and the safety concern of nisin in pharmaceutical applications made P_{170} promoter as a good alternative.

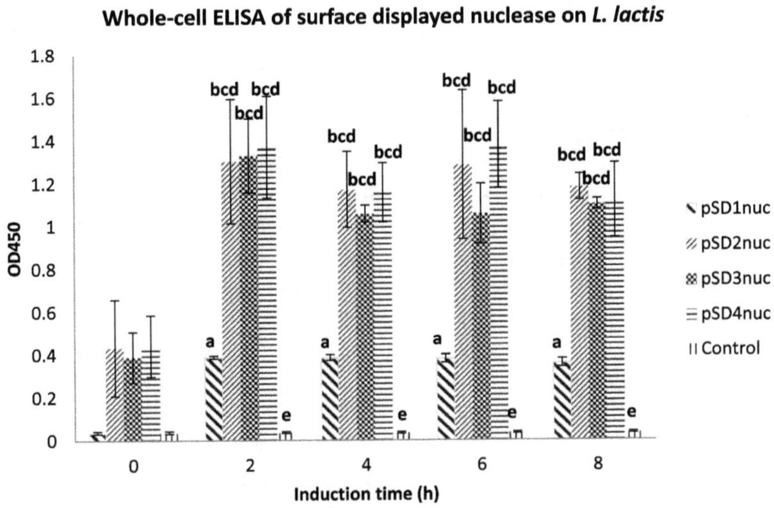

Fig. 7 Whole-cell ELISA of induction time optimisation (from 0 to 8 h induction) for surfaced displayed nuclease on different integrated *L. lactis* constructs. [abcde] common letter on the bars mean no significance difference on the expression of nuclease by the respective integrative vectors ($P < 0.05$)

Table 2 Bacterial strains and plasmids used in this study

Strain or plasmid	Relevant phenotypes	References
Strains		
E. coli Top 10		Invitrogen
L. lactis NZ9000	MG1363 pepN::nisRK, host strain for all lactococcal plasmids	[41]
L. lactis pNZint	*L. lactis* harbouring pNZint plasmid	This study
Plasmid		
pCRTM-Blunt II-TOPO®	Zero Blunt® TOPO® PCR cloning Kit	Invitrogen
pUC19	2686 bps, Amp^R, Cloning vector	New England Biolabs Inc. (NEB)
pNZ8048	3348 bps, Cm^R, Nisin inducible expression vector	[41]
pMG36e	3611 bps, Erm^R	[42]
pNZSPKNuc	Modified pNZ8048 carrying SPK1 signal peptide, *nuc* gene from *S. aureus*	Microbial Biotech Lab, UPM
pNZ170Gus	Modified pNZ8008, P_{nisA} was replaced with P_{170}	Microbial Biotech Lab, UPM
pNZint	Modified pNZ8048 carrying integrase gene of TP901−1	This study
[a]pS1	Secretion integrative vector harboring *attP* of T901−1, P_{170}, *E. coli* rep origin, USP45-LEISSTCDA, Em^R	This study
[a]pS2	Secretion integrative vector harboring *attP* of T901−1, P_{170}, *E. coli* rep origin, SPK1, Em^R	This study
[a]pS3	Secretion integrative vector harboring *attP* of T901−1, P_{nisA}, *E. coli* rep origin, USP45-LEISSTCDA, Em^R	This study
[a]pS4	Secretion integrative vector harboring *attP* of T901−1, P_{nisA}, *E. coli* rep origin, SPK1, Em^R	This study
[a]pSD1	Surface display integrative vector harboring *attP* of T901−1, P_{170}, *E. coli* rep origin, USP45-LEISSTCDA, PrtP anchor domain, Em^R	This study
[a]pSD2	Surface display integrative vector harboring *attP* of T901−1, P_{170}, *E. coli* rep origin, SPK1, $PrtP_{344}$ anchor domain, Em^R	This study
[a]pSD3	Surface display integrative vector harboring *attP* of T901−1, P_{nisA}, *E. coli* rep origin, USP45-LEISSTCDA, $PrtP_{344}$ anchor domain, Em^R	This study
[a]pSD4	Surface display integrative vector harboring *attP* of T901−1, P_{nisA}, *E. coli* rep origin, SPK1, $PrtP_{344}$ anchor domain, Em^R	This study

[a] Plasmids had variations where nuclease reporters were cloned into the MCS and denoted with suufix-nuc in text

In the case of surface display integrative vectors, pSD, the integrated *nuc* was successfully expressed and anchored on the cell wall of *L. lactis* as shown in the immunofluorescence images (Fig. 6). The anchor domain $PrtP_{344}$, which has the well-known LPXTG cell wall anchoring motif, successfully anchored the Nuc onto the surface of *L. lactis* [52–54]. Whole cell ELISA assay showed that pSD1nuc which had the combination of P_{170} promoter with USP45 signal peptide gave very low signals in the ELISA assay, although fluorescent signals were still detectable in immunofluorescence assay (Fig. 7). The expressed Nuc on the other clones which harboured pSD2nuc, pSD34nuc and pSD4nuc were comparable, although pSD2nuc and pSD4nuc, both which had the signal peptide SPK1 gave slightly better signals, although values were non-significant.

Conclusions

In conclusion, the synthetic integrative vectors constructed in this study based on site-directed recombination via the attP/attB site were able to integrate the vectors into the lactococcal chromosome facilitated by integrase enzyme. The secretion integrative vectors, pS1, pS2, pS3 and pS4 were able to secrete Nuc protein into the extracellular region, while surface displayed integrative vectors; pSD1, pSD2, pSD3 and pSD4 were able to display Nuc reporter protein on the cell wall via the $PrtP_{344}$ anchor domain. SPK1 signal peptide was shown to be superior over USP45 in secretion while the combination of P_{170} and USP45 was consistently inferior in both secretion and surface display integrative vectors. These constructed integrative expression vectors can be used to secrete and surface display recombinant protein and polypeptides stably for various applications, especially in live vaccine delivery.

Methods

Bacterial strains, plasmid and culture conditions

Escherichia coli Top 10 (Invitrogen, USA) cells were grown at 37 °C with agitation (250 rpm) in Luria Bertani (LB) broth or LB agar (Merck) (1.5% w/v bacteriological agar) with 50 μg/ml of kanamycin, 100 μg/ml of ampicillin or 75 μg/ml of erythromycin when necessary. *Lactococcus lactis* subsp. *cremoris* NZ9000 (MoBiTech GmbH, Germany) was grown in M17 (Merck) broth or agar supplemented with 0.5% (w/v) glucose as standing culture at 30 °C. Whenever required, erythromycin (10 μg/ml) and chloramphenicol (7.5 μg/ml) were used for the selection of recombinant *L. lactis* cultures. *E. coli* transformation was carried out using $CaCl_2$ method [55] while *L. lactis* competent cell preparation and transformation were performed according to the protocol described by [56]. The electroporated cells were grown in SGM17 broth containing 0.5% (w/v) glucose and 0.5 M sucrose supplemented with chloramphenicol and/or erythromycin at a final concentration of 7.5 μg/ml and 10 μg/ml respectively and incubated at 30 °C for 2 days.

All plasmids used and developed in this study are listed in Table 2. *L. lactis* NZ9000 was used as the host for site-specific integration, providing the chromosomal attB site for the lactococcal phage TP901–1. NZ9000 was transformed in advanced with pNZint, which is a modified pNZ8048 harbouring the integrase gene (*int*), encoding the TP901–1 integrase (Genbank accession number: AF304433.1), to facilitate site-specific recombination. The constructed secretion integrative vectors pS1, pS2, pS3, pS4 and surface display integrative vectors pSD1, pSD2, pSD3, pSD4 are suicide vectors harbouring the P_{170} promoter or P_{nisA}, USP45-LEISSTCDA or SPK1 signal peptide, MCS, TP901–1 attP site, *E. coli* origin of replication, erythromycin resistance gene (Em^R). The integrative vectors were then transformed into the earlier *L. lactis* NZ9000 harbouring pNZint which is subsequently integrated into the attB site of the host genome.

PCR amplifications and DNA manipulations

PCR amplification procedures were performed using either *Taq* DNA polymerase, for analytical purposes, or *Pfu* DNA polymerase, for cloning and sequencing (Fermentas, USA). The integrated DNA was verified via PCR with KOD FX Neo (TOYOBO, Japan). PCR primers were designed based on the known DNA sequences and relevant restriction enzymes (RE) were introduced via primers when needed (Table 3) which were synthesized by Integrated DNA Technologies (Singapore). DNA fragments were purified using GeneJET PCR purification Kit and GeneJET Gel Extraction Kit (Thermofisher, USA). Plasmids were isolated using Favorpep™ Plasmid DNA Extraction Mini Kit (Favorgen, Taiwan). Conventional and FastDigest RE and T4 DNA ligase were obtained from Thermofisher, USA.

Plasmid constructions

The respective genes and DNA fragments were PCR amplified using primers listed in Table 3. The integrative vectors were constructed based on the pNZ8048 plasmid as backbone. A cassette consisting of USP45 fused with LEISSTCDA propeptide to increase secretion efficiency, pNZ8048 multiple cloning sites (MCS) modified to include more RE sites, pNZ8048 terminator (T), and TP901–1 bacteriophage attachment site (attP) was synthesized by Integrated DNA Technologies (Singapore) and denoted as attP cassette.

The vectors constructions strategy was divided into four phases; Phase I) Construction of PZatt cassette composed of the P_{170} promoter, USP45 signal peptide, LEISSTCDA propeptide and attP sequence; Phase II) Construction of the secretion integrative vectors (pS1–

Table 3 Primers used in this study

Primers	Sequences	RE
F-attB	CTACTGCTGCTTCACCAGTTT	–
R-attB	GTATGCAGCGATGTTGTTACCC	–
F-attP	GGAC **CTCGAG**TCCAACTCGCTTAATTGC	XhoI
R-attP	GGAC**GGATCC**GCTAAAACGTCTCAGAAA	BamHI
F-Int	GACG**CCATGG**CACATCATCATCATCATCATATGACTAAGAAAGTAGCAATCTATAC	NcoI
R-Int	GCAC**GAGCTC**TTAAGCGAGTTGGAATTTAAATATG	HindIII
F-PnisA	GCG**AGATCT**AGTCTTATAACTATAC	BglII
R-PnisA	GCA**GATATC**GTGAGTGCCTCCTTATAAT	EcoRV
F-P170	GCGAC**AGATCT**GAACTATGAATATC	BglII
R-P170	GCAC**GATATC**AACTGTTCTTTTTTAATTTTT	EcoRV
F-USP45	GACG**GATATC**ATGAAAAAAAAGA	EcoRV
R-USP45	CTGA**GTCGAC**GATATTTCGAGAGC	SalI
F-SpkNuc	GCGT**GATATC**ATGAAAAAAAATATTAACGTTGGTAT	EcoRV
R-NucH	GGC**GAGCTC**TTAGTGGTGATGATGGTGA	SacI
F-NucEV	GCGT**GATATC**ATGAAAAAAAATATTAACGTTGGTAT	EcoRV
F-NucNI	CGAG**CCATGG**CTATGAAAAAAAATATTAACG	NcoI
R-NucEI	CGCG**GAATTC**TTGACCTGAATCAGCGTTGTCTTCG	EcoRI
F-Em	GACG**GGATCC**AATCAGGCTTGATCCCCAGTAAGTC	BamHI
R-Em	GACG**AGATCT**AGGATGAGGAGGCAGATTGCCTTG	BglII
F-PrtP344	GAGC**GAATTC**TGGCTCTAGAGGATCCAAGTC	EcoRI
R-PrtP344	TAAACTAT**GCGGCCGC**GACAGGCTATTCTTC	NotI
F-E. coli Ori	GGAG**GGATCC**GGCGTAATCATGGTCATAG	BamHI
R-E. coli Ori	AGGC**GGATCC**CTGTCAGACCAAGTTTACTC	BamHI

RE sites are in bold

4); (a) Plasmid circularisation by incorporation of erythromycin resistant gene and *E. coli* replication origin into the PZatt cassette, denoted as pS1; (b) Replacement of P_{170} promoter with P_{nisA} promoter to produce pS3 plasmid; (c) Replacement of USP45-LEISSTCDA fusion with SPK1 signal peptide into each pS1 and pS3 plasmids to produce pS2 and pS4 plasmids (Additional file 3: Figure S1); Phase III) Construction of the surface display integrative vectors (pSD1–4) by incorporation of $PrtP_{344}$ anchor domain into each secretion integrative vector; Phase IV) Cloning of reporter gene, nuclease (Additional file 4: Figure S2).

In Phase I, the attP cassette was ligated with P_{170} promoter to produce the PZatt cassette through EcoRV RE. In Phase II (a), the PZatt cassette was ligated with the erythromycin resistant gene (Em^R) that was isolated from pMG36e to produce the PZER cassette using BglII RE. The PZER cassette was then fused with colE1 origin of replication from the pUC19 plasmid via BamHI RE to produce the functional circular plasmid, denoted as pS1. At each step of cassette construction, the cassettes were subcloned into pCRTM-Blunt II-TOPO® for maintenance. The P_{170} promoter of pS1 was cleaved out using BglII and EcoRV and was substituted with a similarly digested P_{nisA} promoter to produce pS3 plasmid (Phase II-b). On the other hand, the pS2 plasmid was constructed by replacing USP45 signal peptide with SPK1 using EcoRV and KpnI RE. To construct pS4 plasmid, USP45 in pS3 plasmid was cleaved out and was replaced with SPK1 signal peptide using NcoI and KpnI RE (Phase II-c). These resulted in secretion integrative plasmids (pS1–4) constructs.

The surface display integrative plasmids, pSD1–4 were constructed by insertion of PCR amplified $PrtP_{344}$ into the respective secretion integrative vectors using EcoRI and NotI RE. All the successfully constructed integrative vectors were inserted with nuclease reporter gene, *nuc* in the MCS via NcoI and EcoRI sites to determine the functionality of the plasmids and to analyse the level of expression of each integrative vectors. Separately, the helper vector, pNZint (plasmid carrying integrase gene to assist integration) was constructed by cloning the integrase gene into pNZ8048 plasmid using NcoI and SacI RE.

Cloning and integration of nuclease gene

The integration of foreign DNA into the host genome only happens with the presence of the integrase protein.

Thus, *L. lactis* NZ9000 competent cells were first transformed with pNZint helper plasmid and the positive transformants were verified by restriction enzyme digestion and sequencing. The integration of the newly constructed integrative vectors harbouring the nuclease reporter was conducted in a few steps. First, *L. lactis* NZ9000 harbouring pNZint was made competent using protocol described by Holo and Nes (1989) with addition of 7.5 µg/ml of chloramphenicol into the culture media. Next, the competent cells were transformed with each integrative vector constructed according to the mentioned protocol with minor modifications. During the transformation process, after an hour of incubation, 20 ng/ml of nisin was added into the transformants to allow the expression of integrase protein and integration of the plasmid harbouring the nuclease gene. Transformants were then plated on SGM17 agar plate supplemented with 10 µg/ml of erythromycin and incubated at 30 °C for 2–3 days. The colonies were then verified with PCR using attB flanking primers (F-attB and R-attB) and were sent for sequencing. The positive colonies were used for further analysis.

Stability tests

Doubling time of *L. lactis* was first determined using the *L. lactis* growth profile which was plotted prior. From the doubling time, cells were grown up to 100 generations by subculturing the cells to fresh media before it reaches stationary phase. After roughly 100 generations, cells were plated onto non-selective GM17 media. Then, colonies were transferred onto selective GM17 media containing 10 µg/ml of erythromycin due determine if the cells still possess antibiotic resistance. For further verification, genomic DNA was extracted from the integrants and verified using PCR with attB flanking primers to determine the presence of the integrated plasmid.

Expression of heterologous protein

The expressed heterologous protein was determined by Western blotting. The grown culture of *L. lactis* harbouring integrated *nuc* was harvested. The secreted Nuc in the medium was harvested using 100% trichloroacetic acid, TCA (Merck, Germany) and the cell pellet was sonicated to harvest intracellular Nuc. Briefly, the harvested protein was added with loading buffer and boiled for 10 min. Treated samples were ran on SDS-polyacrylamide gel electrophoresis and then transferred to polyvinylidene difluoride (PVDF) membrane (ThermoFisher, USA). After blocking with 2% BSA at room temperature for 2 h, the membrane was incubated with the nuclease specific primary antibody, *Staphylococcus aureus* anti-nuclease antibody (LSBio, USA) at 1:10000 dilution followed by a secondary antibody goat anti-rabbit IgG H&L (HRP) (Abcam, USA) at 1:3000 dilution for another hour. The labelled protein on the membrane was detected using DAB substrate (Amresco, USA). On the other hand, the integrase protein was detected using 1:1000 dilution of primary antibody, Pierce™ 6X-His epitope tag antibody (ThermoFisher, USA) and 1:1000 dilution of secondary antibody, Peroxidase Conjugated Affinity Purified Goat anti-Mouse IgG (OriGene, USA).

Nuclease assay (TBD assay)

Toluidine Blue DNA Agar was used for the detection of thermostable deoxyribonuclease activity. The agar was prepared according to the manufacturers protocol (HiMedia, India). Nuclease activity was detected using colony overlay as described by Lachica et al. (1970) [38] with minor modifications. *L. lactis* strain harboring *nuc* gene was spotted into the Brain Heart Infusion (BHI) agar (Oxoid, UK) and incubated overnight at 30 °C. Then, 15 ml of boiled TBD agar that has been cooled to 50 °C was poured onto the overnight cultured BHI agar. The overlaid agar was incubated at 37 °C for 4 h and bright pink halos observed.

Immunofluorescence microscopy analysis

Cell wall anchoring protein was determined by immunofluorescence microscopy based on fluorescence conjugated secondary antibody tagging. The bacteria suspected to surface display nuclease were harvested, washed with PBS, applied to microscope slides, and fixed with 4% (v/v) formaldehyde. The cells were blocked with 3% (w/v) of BSA for 30 min at RT and washed thrice. The slides were incubated for 1 h with anti-nuclease antibody (diluted in 1:500 in PBS and 1% BSA) (LSBio, USA) on a slide chamber at RT, washed thrice with PBS. Then, the cells were labelled with polyclonal anti-nuclease primary antibody raised in rabbit (LSBio, USA) at 1:500 dilution and incubated for 1 h at RT. The slide was washed thrice with 1X PBS for 10 min and then incubated with anti-rabbit secondary antibody conjugated with Alexa Fluor 488 (Merck, USA) at 1:1000 dilution with 1 h incubation at RT followed by washing. The samples were analysed using Zeis Confocal Microscope (Carl Zeis, Germany) fixed with green filter.

Whole cell ELISA assay

The anchored recombinant protein was quantified using ELISA assay with the specific antibody. The overnight cultures were transferred into fresh media and incubated at 30 °C until reached OD_{600} 0.5 and was induced with 20 ng/ml nisin. The cultures were harvested and normalized at constant OD_{600}. They were then fixed with 4% (w/v) paraformaldehyde at RT, washed with 1X PBS and then blocked with 3% (w/v) BSA before washed again. The cells were then incubated with primary polyclonal nuclease anti-rabbit antibody (LSBio, USA), washed and then incubated with HRP conjugated secondary anti-rabbit antibody (Abcam, USA). The treated cells were transferred into 96 wells plate and incubated with substrate at RT in the dark. The reactions were stopped with 1 M H_2SO_4 (Sigma, USA). The samples were then analysed using ELISA reader at OD_{450} (Magellan for F50, USA).

Supplementary information

Additional file 1: Table S1. Data set of TBD assay of secreted nuclease.

Additional file 2: Table S2. Data set of whole-cell ELISA of surface displayed nuclease.

Additional file 3: Figure S1. Schematic representations of the construction strategy of the integrative vectors. Phase I) constructions of PZatt cassette backbone; Phase II) construction of secretion integrative vectors, (a) pS1 plasmid, (b) pS3 plasmid, (c) pS2 and pS4.

Additional file 4: Figure S2. Schematic representations of the construction strategy the integrative vectors. Phase III) construction of surface display integrative vectors, (pSD1–4); Phase IV) cloning of *nuc* gene into each integrative vectors.

Abbreviations

ANOVA: Analysis of variance; attB: Bacteria attachment site; attP: Phage attachment site; ELISA: Enzyme-linked immunosorbent assay; GRAS: Generally Recognised as Safe; LAB: Lactic Acid Bacteria; pS: Secretion integrative vector; pSD: Surface display integrative vector; RE: Restriction Enzyme; TBD: Toluidine Blue DNA

Acknowledgements

The authors wish to thank Nur Aqlili Riana Alias for providing the *nuc* gene.

Authors' contributions

IK designed the research, conducted the reported experiments, data analysis and wrote the manuscript. AAS assisted in the writing of the manuscript and data analysis. RAR proposed the research concept and assisted in design of experiments. MJM contributed in the experimental design. All authors read and approved the final manuscript.

Funding

This project was funded by Malaysian Ministry of Science, Technology and Innovation (02-01-04-SF1608). The funding body had no role in the design of the study, the collection, analysis, and interpretations of data and in writing the manuscript.

Ethics approval and consent to participate

Not applicable.

Consent for publication

Not applicable.

Competing interests

The authors declare that they have no competing interests.

Author details

[1]Department of Cell and Molecular Biology, Faculty of Biotechnology and Biomolecular Sciences, University Putra Malaysia, 43400 UPM, Serdang, Selangor, Malaysia. [2]Department of Microbiology, Faculty of Biotechnology and Biomolecular Sciences, University Putra Malaysia, 43400 UPM, Serdang, Selangor, Malaysia. [3]Institute of Bioscience, University Putra Malaysia, 43400 UPM Serdang, Selangor, Malaysia. [4]Chancellory, Universiti Teknikal Malaysia, 76100 Durian Tunggal, Melaka, Malaysia.

References

1. Song AAL, In LLA, Lim SHE, Rahim RA. A review on *Lactococcus lactis*: from food to factory. Microb Cell Factories. 2017;16:1–15.
2. Pontes DS, de Azevedo MSP, Chatel J-M, Langella P, Azevedo V, Miyoshi A. *Lactococcus lactis* as a live vector: heterologous protein production and DNA delivery systems. Protein Expr Purif. 2011;79:165–75.
3. Souza RD, Pandeya DR, Hong S. Review article *Lactococcus Lactis* : an efficient gram positive cell factory for the production and secretion of recombinant protein. Biomed Res. 2012;23:1–7.
4. Bahey-El-Din M, Gahan CGM, Griffin BT. *Lactococcus lactis* as a cell factory for delivery of therapeutic proteins. Curr Gene Ther. 2010;10:34–45.
5. Morello E, Bermúdez-Humarán LG, Llull D, Solé V, Miraglio N, Langella P, et al. *Lactococcus lactis*, an efficient cell factory for recombinant protein production and secretion. J Mol Microbiol Biotechnol. 2008;14:48–58.
6. Jannitire L, Niaudet B, Pierre E, Ehrlich SD. Stable gene amplification in the chromosome of *Bacillus subtilis*. Gene. 1985;40:47–55.
7. Mckay LL. Functional properties of plasmids in lactic streptococci. Antonie Van Leeuwenhoek. 1983;49:259–74.
8. Glick BR. Metabolic load and heterologous gene expression. Biotechnol Adv. 1995;13:247–61.
9. Martín MC, Pant N, Ladero V, Günaydin G, Andersen KK, Alvarez B, et al. Integrative expression system for delivery of antibody fragments by lactobacilli. Appl Environ Microbiol. 2011;77:2174–9.
10. Coddeville M, Ritzenthaler P. Control of directionality in bacteriophage mv4 site-specific recombination: functional analysis of the Xis factor. J Bacteriol. 2010;192:624–35.
11. Martin MC, Alonso JC, Sua JE, Alvarez MA, La P. Generation of food-grade recombinant lactic acid bacterium strains by site-specific recombination. Appl Environ Microbiol. 2000;66:2599–604.
12. Gosalbes MJ, Esteban CD, Luis J, Pérez-martínez G, Gala L. Integrative food-grade expression system based on the lactose regulon of *Lactobacillus casei*. Appl Environ Microbiol. 2000;66:4822–8.
13. Iwanicki A, Pi I, Grela A, Łę T, Obuchowski M. A system of vectors for *Bacillus subtilis* spore surface display. Microb Cell Factories. 2014;13:1–9.
14. Stover BCK, Bansal GP, Hanson MS, Burlein JE, Palaszynski SR, Young JF, et al. Protective immunity elicited by recombinant *Bacille Calmette-Guerin* (BCG) expressing outer surface protein a (OspA) lipoprotein : a candidate Lyme disease vaccine. J Exp Med. 1993;178:197–209.
15. Joseph J, Fern R, Cardona P, Mothe B, Gatell JM. Molecular characterization of heterologous HIV-1gp120 gene expression disruption in *Mycobacterium bovis* BCG host strain : a critical issue for engineering mycobacterial based-vaccine vectors. J Biomed Biotechnol. 2010;2010:357370.
16. Hatfull GF. Mycobacteriophages. Microbiol Spectr. 2018;6:1-32.
17. Van Kessel JC, Marinelli LJ, Hatfull GF. Recombineering mycobacteria and their phages. Perspectives (Montclair). 2008;6:851–7.
18. Murry J, Sassetti CM, Moreira J, Lane J, Rubin EJ. A new site-specific integration system for *Mycobacteria*. Tuberculosis. 2005;85:317–23.
19. Hook CD, Samsonov VV, Ublinskaya AA, Kuvaeva TM, Andreeva EV, Gorbacheva LY, et al. A novel approach for *Escherichia coli* genome editing combining in vivo cloning and targeted long-length chromosomal insertion. J Microbiol Methods. 2016;130:83–91.
20. Zucca S, Pasotti L, Politi N, Gabriella M, De Angelis MG, Magni P. A standard vector for the chromosomal integration and characterization of BioBrick ™ parts in *Escherichia coli*. J Biol Eng. 2013;7:1–13.
21. Yeom S-J, Kim YJ, Lee J, Kwon KK, Han GH, Kim H, et al. Correction : long-term stable and tightly controlled expression of recombinant proteins in antibiotics-free conditions. PLoS One. 2017;12:e0175955.
22. Datta S, Costantino N, Court DL. A set of recombineering plasmids for gram-negative bacteria. 2006;379:109–15.
23. Sabri S, Steen JA, Bongers M, Nielsen LK, Vickers CE. HL HA2 Knock-in / Knock-out (KIKO) vectors for rapid integration of large DNA sequences , including whole metabolic pathways, onto the *Escherichia coli* chromosome at well-characterised loci. 2013;12:1–1.
24. Chung M, Yeh I, Sung L, Wu M, Chao Y. Enhanced integration of large DNA into *E . coli* chromosome by CRISPR / Cas9. Biotechnol Bioeng. 2017;114:172–83.

25. Leenhouts KJ, Kok J, Venema G. Stability of integrated plasmids in the chromosome of *Lactococcus lactis*. Appl Environ Microbiol. 1990;56:2726–35.
26. Leenhouts KJ, JAN K, Venema G. Campbell-Like Integration of Heterologous plasmid DNA into the Chromosome of *Lactococcus lactis* subsp. lactis. Appl Environ Microbiol. 1989;55:394.
27. Chopin MC, Chopin A, Rouault A, Galleron N. Insertion and amplification of foreign genes in the *Lactococcus lactis* subsp. lactis chromosome. Appl Environ Microbiol. 1989;55:1769–74.
28. Zeigler DR. Integration vectors for gram-positive bacteria. *Bacillus* Genetic Stock Center, Catalog Strains, Seventh Edition; 2002. p. 4.
29. Stoll SM, Ginsburg DS, Calos MP. Phage TP901-1 site-specific integrase functions in human cells. J Bacteriol. 2002;184:3657.
30. Brøndsted L, Østergaard S, Pedersen M, Hammer K, Vogensen FK. Analysis of the complete DNA sequence of the temperate bacteriophage TP901–1 : evolution, structure , and genome organization of lactococcal bacteriophages. Virology. 2001;283:93–109.
31. Petersen KV, Martinussen J, Jensen PR, Solem C. Repetitive, marker-free, site-specific integration as a novel tool for multiple chromosomal integration of DNA. Appl Environ Microbiol. 2013;79:3563–9.
32. Brøndsted L, Hammer K. Use of the integration elements encoded by the temperate lactococcal bacteriophage TP901-1 to obtain chromosomal single-copy transcriptional fusions in *Lactococcus lactis*. Appl Environ Microbiol. 1999;65:752.
33. Christiansen B, Brøndsted L, Vogensen FK, Hammer K. A resolvase-like protein is required for the site-specific integration of the temperate lactococcal bacteriophage TP901-1. J Bacteriol. 1996;178:5164–73.
34. Christiansen B, Johnsen MG, Stenby E, Vogensen FK, Hammer K. Characterization of the lactococcal temperate phage TP901-1 and its site-specific integration. J Bacteriol. 1994;176:1069–76.
35. Ng DTW, Sarkar CA. Engineering signal peptides for enhanced protein secretion from *Lactococcus lactis*. Appl Environ Microbiol. 2013;79:347–56.
36. Le Loir Y, Azevedo V, Oliveira SC, Freitas DA, Miyoshi A, Bermúdez-Humarán LG, et al. Protein secretion in *Lactococcus lactis* : an efficient way to increase the overall heterologous protein production. Microb Cell Fact. 2005;4:2.
37. Yang H, Kim Y, Chang H. Construction of an integration-proficient vector based on the site-specific recombination mechanism of enterococcal temperate phage φ FC1 construction of an integration-proficient vector based on the site-specific recombination mechanism of Enterococcal T. J Bacteriol. 2002;184:1859.
38. Lachica RV, Genigeorgis C, Hoeprich PD. Metachromatic agar-diffusion methods for detecting staphylococcal nuclease activity. Appl Microbiol. 1971;21:585–7.
39. Breüner A, Brøndsted L, Hammer K. Resolvase-like recombination performed by the TP901-1 integrase. Microbiology. 2001;147(Pt 8):2051–63.
40. Mulder J, Wels M, Kuipers OP, Kleerebezem M. Unleashing natural competence in *Lactococcus lactis* by induction of the competence regulator ComX. Appl Environ Microbiol. 2017;83:1–13.
41. Kuipers OP, de Ruyter PGG, Kleerebezem M, de Vos WM. Quorum sensing-controlled gene expression in lactic acid bacteria. J Biotechnol. 1998;64:15–21.
42. van de Guchte M, van der Vossen JM, Kok J, Venema G. Construction of a lactococcal expression vector: expression of hen egg white lysozyme in *Lactococcus lactis* subsp. lactis. Appl Environ Microbiol. 1989;55:224–8.
43. Loir YLE, Nouaille S, Commissaire J, Bretigny L, Gruss A, Langella P. Signal peptide and propeptide optimization for heterologous protein secretion in *Lactococcus lactis*. Appl Environ Microbiol. 2001;67:4119–27.
44. Baradaran A, Sieo CC, Illias RM, Rahim RA, Yusoff K, Foo HL. Cloning and in silico characterization of two signal peptides from *Pediococcus pentosaceus* and their function for the secretion of heterologous protein in *Lactococcus lactis*. Biotechnol Lett. 2013;35:233–8.
45. Subramaniam M, Baradaran A, Rosli MI, Rosfarizan M, Khatijah Y, Raha AR. Effect of signal peptides on the secretion of β-cyclodextrin glucanotransferase in *Lactococcus lactis* NZ9000. J Mol Microbiol Biotechnol. 2012;22:361–72.
46. Kuipers OP, Beerthuyzen MM, de Ruyter PG, Luesink EJ, de Vos WM. Autoregulation of Nisin Biosynthesis in *Lactococcus lactis* by Signal Transduction. J Biol Chem. 1995;270:27299–304.
47. de Ruyter PG, Kuipers OP, Beerthuyzen MM, van Alen-Boerrigter I, de Vos WM. Functional analysis of promoters in the nisin gene cluster of *Lactococcus lactis*. J Bacteriol. 1996;178:3434–9.
48. NICE. Expression System for *Lactococcus lactis*. 2010.
49. Madsen SM, Arnau J, Vrang A, Givskov M, Israelsen H. Molecular characterization of the pH-inducible and growth phase-dependent promoter P170 of *Lactococcus lactis*. Mol Microbiol. 1999;32:75–87.
50. Madsen SM, Hindré T, Le Pennec J-P, Israelsen H, Dufour A. Two acid-inducible promoters from *Lactococcus lactis* require the cis-acting ACiD-box and the transcription regulator RcfB. Mol Microbiol. 2005;56:735–46.
51. Jørgensen CM, Vrang A, Madsen SM. Recombinant protein expression in *Lactococcus lactis* using the P170 expression system. FEMS Microbiol Lett. 2014;351:170–8.
52. Leenhouts K, Buist G, Kok J. Anchoring of proteins to lactic acid bacteria. Antonie Van Leeuwenhoek. 1999;76:367–76.
53. Hansson M, Samuelson P, Gunneriusson E, Ståhl S. Surface display on gram positive bacteria. Comb Chem High Throughput Screen. 2001;4:171–84.
54. Habimana O, Bellon-Fontaine M-N, Buist G, Juillard V, Le Goff C, Briandet R, et al. Positive role of cell wall anchored proteinase PrtP in adhesion of lactococci. BMC Microbiol. 2007;7:36.
55. Dagert M, Ehrlich SD. Prolonged incubation in calcium chloride improves the competence of *Escherichia coli* cells. Gene. 1979;6:23–8.
56. Holo H, Nes I. Transformation of *Lactococcus* by electroporation. Methods Mol Biol. 1989;47:195–9.

PERMISSIONS

 Every chapter published in this book has been scrutinized by our experts. Their significance has been extensively debated. The topics covered herein carry significant findings which will fuel the growth of the discipline. They may even be implemented as practical applications or may be referred to as a beginning point for another development.

The contributors of this book come from diverse backgrounds, making this book a truly international effort. This book will bring forth new frontiers with its revolutionizing research information and detailed analysis of the nascent developments around the world.

We would like to thank all the contributing authors for lending their expertise to make the book truly unique. They have played a crucial role in the development of this book. Without their invaluable contributions this book wouldn't have been possible. They have made vital efforts to compile up to date information on the varied aspects of this subject to make this book a valuable addition to the collection of many professionals and students.

This book was conceptualized with the vision of imparting up-to-date information and advanced data in this field. To ensure the same, a matchless editorial board was set up. Every individual on the board went through rigorous rounds of assessment to prove their worth. After which they invested a large part of their time researching and compiling the most relevant data for our readers.

The editorial board has been involved in producing this book since its inception. They have spent rigorous hours researching and exploring the diverse topics which have resulted in the successful publishing of this book. They have passed on their knowledge of decades through this book. To expedite this challenging task, the publisher supported the team at every step. A small team of assistant editors was also appointed to further simplify the editing procedure and attain best results for the readers.

Apart from the editorial board, the designing team has also invested a significant amount of their time in understanding the subject and creating the most relevant covers. They scrutinized every image to scout for the most suitable representation of the subject and create an appropriate cover for the book.

The publishing team has been an ardent support to the editorial, designing and production team. Their endless efforts to recruit the best for this project, has resulted in the accomplishment of this book. They are a veteran in the field of academics and their pool of knowledge is as vast as their experience in printing. Their expertise and guidance has proved useful at every step. Their uncompromising quality standards have made this book an exceptional effort. Their encouragement from time to time has been an inspiration for everyone.

The publisher and the editorial board hope that this book will prove to be a valuable piece of knowledge for researchers, students, practitioners and scholars across the globe.

LIST OF CONTRIBUTORS

Olivier Albagli, Alicia Maugein, Lukas Huijbregts, Géraldine Carlier and Raphaël Scharfmann
INSERM U1016, CNRS UMR8104, Institut Cochin, Université Paris Descartes, 123 Boulevard de Port-Royal, 75014 Paris, France

Delphine Bredel
INSERM U1016, CNRS UMR8104, Institut Cochin, Université Paris Descartes, 123 Boulevard de Port-Royal, 75014 Paris, France
Laboratoire de Recherche Translationnelle en Immunothérapie, Institut Gustave Roussy, 114 Rue Edouard Vaillant, 94800 Villejuif, France

Patrick Martin
Université Côte d'Azur, CNRS UMR7277 INSERM U1099, iBV (Institut de Biologie Valrose), Université Nice Sophia Antipolis, Bâtiment Sciences Naturelles; UFR Sciences, Parc Valrose, 28, avenue Valrose, 06108 Nice Cedex 2, France

Alessandro Ascione, Claudia Arenaccio, Alessandra Mallano, Michela Flego, Mara Gellini, Mauro Andreotti, Stefano Vella and Maurizio Federico
National Center for Global Health, Istituto Superiore di Sanità (ISS), Viale Regina Elena 299, 00161 Rome, Italy

Craig Fenwick and Giuseppe Pantaleo
Service of Immunology and Allergy, Lausanne University Hospital, University of Lausanne, Lausanne, Switzerland

Syukri Shukor, Alfred Tamayo and Lorenzo Tosi
Center for Surgery, Innovation, and Bioengineering, Department of Surgery, Massachusetts General Hospital, Harvard Medical School and the Shriners Hospitals for Children, 51 Blossom Street, Boston, MA 02114, USA

Biju Parekkadan
Center for Surgery, Innovation, and Bioengineering, Department of Surgery, Massachusetts General Hospital, Harvard Medical School and the Shriners Hospitals for Children, 51 Blossom Street, Boston, MA 02114, USA
Harvard Stem Cell Institute, Cambridge, MA 02138, USA
Department of Biomedical Engineering, Rutgers University, Piscataway, NJ 08854, USA

H. Benjamin Larman
Division of Immunology, Department of Pathology, Johns Hopkins University, Baltimore, MD, USA

Majid Haddad Momeni and Maher Abou Hachem
Department of Biotechnology and Biomedicine, Technical University of Denmark, Søltofts Plads, 2800 Kgs, Lyngby, Denmark

Lo Gorton
Department of Biochemistry and Structural Biology, Lund University, P.O. Box 124, 221 00 Lund, Sweden

Paolo Bollella
Department of Biochemistry and Structural Biology, Lund University, P.O. Box 124, 221 00 Lund, Sweden
Department of Chemistry and Drug Technologies, Sapienza University of Rome, Piazzale Aldo Moro 5, 00185 Rome, Italy

Roberto Ortiz and Esben Thormann
Department of Chemistry, Technical University of Denmark, Kemitorvet 207, 2800 Kgs, Lyngby, Denmark

Qiancheng Zhang
School of Chemical Engineering, Inner Mongolia University of Technology, Hohhot 010051, Inner Mongolia, China

Zhanying Liu and Jian Pang
School of Chemical Engineering, Inner Mongolia University of Technology, Hohhot 010051, Inner Mongolia, China
Inner Mongolia Energy Conservation and Emission Reduction Engineering Research Center in Fermentation Industry, Hohhot 010051, China

Qingsheng Qi and Junshu Wang
State Key Laboratory of Microbial Technology, Shandong University, Qingdao 266237, China

Yassir El Alaoui and Yahya Bensouda
Laboratory of Pharmaceutics, Faculty of Medicine and Pharmacy, Mohammed V University, Rabat, Morocco

Soumaya El Bouchikhi
Laboratory of Pharmaceutics, Faculty of Medicine and Pharmacy, Mohammed V University, Rabat, Morocco
Laboratory of Medical Biotechnology, Faculty of Medicine and Pharmacy, Mohammed the Vth University in Rabat, Rabat, Morocco

Azeddine Ibrahimi
Laboratory of Medical Biotechnology, Faculty of Medicine and Pharmacy, Mohammed the Vth University in Rabat, Rabat, Morocco

Philippe Pagès
PhP Stats, Création et analyse d'information, Conseil, études et formations en statistique, 19, rue Pasteur, 94170 Le Perreux, France

Qi Zhang and Pi-wu Wang
Biotechnology Center of Jilin Agricultural University, Jilin Agricultural University, Changchun 130118, People's Republic of China

Di Qin
Biotechnology Center of Jilin Agricultural University, Jilin Agricultural University, Changchun 130118, People's Republic of China
School of Biosciences and Veterinary Medicine, University of Camerino, MC, Camerino, Italy

Cristina Miceli
School of Biosciences and Veterinary Medicine, University of Camerino, MC, Camerino, Italy

Xiao-Yi Liu
College of Life Sciences, Kyung Hee University, 1732, Deogyeong daero, Giheung-gu, Yongin-si, Gyeonggi-do 17104, Republic of Korea

Chuan Zhang, Bao-Cheng Yang, Wen-Ting Liu, Zhong-Yuan Li, Ya-Jian Song, Tong-Cun Zhang and Xue-Gang Luo
Key Laboratory of Industrial Fermentation Microbiology of the Ministry of Education & Tianjin Key Lab of Industrial Microbiology, College of Biotechnology, Tianjin University of Science and Technology, Tianjin 300457, China
State Key Laboratory of Food Nutrition and Safety, Tianjin 300457, China

Taj Arndell
CSIRO, Agriculture and Food, Canberra, ACT, Australia

Niharika Sharma
New South Wales Department of Primary Industries, Research Excellence, Orange, NSW, Australia

Peter Langridge, Ute Baumann and Nathan S. Ryan Whitford
School of Agriculture, Food & Wine, The University of Adelaide, Waite Campus, Urrbrae, SA 5064, Australia

Watson-Haigh
Bioinformatics Hub, School of Biological Sciences, The University of Adelaide, Adelaide, SA 5005, Australia

Henrique C. T. Veras
Grupo Engenharia de Biocatalisadores, Universidade de Brasília - UnB, Campus Darcy Ribeiro, Instituto de Ciências Biológicas, Bloco K, 1° andar, Asa Norte, Brasilia 70.790-900, Brazil
Empresa Brasileira de Pesquisa Agropecuária, EMBRAPA Agroenergia, Brasília-DF, Brazil

Nádia S. Parachin
Grupo Engenharia de Biocatalisadores, Universidade de Brasília - UnB, Campus Darcy Ribeiro, Instituto de Ciências Biológicas, Bloco K, 1° andar, Asa Norte, Brasilia 70.790-900, Brazil
Programa de Pós-Graduação em Biologia Microbiana, Instituto de Biologia, Universidade de Brasília - UnB, Brasilia, Brazil

Patrícia V. Abdelnur and Christiane G. Campos
Empresa Brasileira de Pesquisa Agropecuária, EMBRAPA Agroenergia, Brasília-DF, Brazil
Instituto de Química, Universidade Federal de Goiás - UFG, Goiânia, Brazil

João R. M. Almeida
Empresa Brasileira de Pesquisa Agropecuária, EMBRAPA Agroenergia, Brasília-DF, Brazil
Programa de Pós-Graduação em Biologia Microbiana, Instituto de Biologia, Universidade de Brasília - UnB, Brasilia, Brazil

Igor F. Nascimento
Programa de Pós-Graduação em Administração, Universidade de Brasília - UnB, Brasília, Brazil

Eugeny Y. Bragin, Dmitry V. Dovbnya and Marina V. Donova
Institute of Biochemistry and Physiology of Microorganisms, Federal Research Center "Pushchino Center for Biological Research of the Russian Academy of Sciences", Nauki, 5, Pushchino, Russian Federation 142290
Pharmins Ltd., Institutskaya, 4, Pushchino, Russian Federation 142290

Mikhail I. Schelkunov
Skolkovo Institute of Science and Technology, Nobelya, 3, Moscow, Russian Federation 121205
Institute for Information Transmission Problems, Russian Academy of Sciences, Bolshoy Karetny, 19, build. 1, Moscow, Russian Federation 127051

Victoria Y. Shtratnikova
A.N. Belozersky Research Institute of Physico-Chemical Biology, M.V. Lomonosov Moscow State University, Leninskye gory, 1, building 40, Moscow, Russian Federation 119992

Masayuki Machida
Bioproduction Research Institute, National Institute of Advanced Industrial Science and Technology (AIST), 2-17-2-1 Tsukisamu-Higashi, Toyohira-ku, Sapporo, Hokkaido 062-8517, Japan

Koichi Tamano
Bioproduction Research Institute, National Institute of Advanced Industrial Science and Technology (AIST), 2-17-2-1 Tsukisamu-Higashi, Toyohira-ku, Sapporo, Hokkaido 062-8517, Japan
AIST-Waseda University Computational Bio Big-Data Open Innovation Laboratory (CBBD-OIL), National Institute of Advanced Industrial Science and Technology (AIST), 5-20, Building 63, Nishi-waseda campus, Waseda University, 3-4-1 Okubo, Shinjuku-ku, Tokyo 169-8555, Japan

Myco Umemura
AIST-Waseda University Computational Bio Big-Data Open Innovation Laboratory (CBBD-OIL), National Institute of Advanced Industrial Science and Technology (AIST), 5-20, Building 63, Nishi-waseda campus, Waseda University, 3-4-1 Okubo, Shinjuku-ku, Tokyo 169-8555, Japan
Bioproduction Research Institute, National Institute of Advanced Industrial Science and Technology (AIST), Central 6, 1-1-1 Higashi, Tsukuba, Ibaraki 305-8566, Japan

Mahoko Kuninaga and Hideaki Koike
Bioproduction Research Institute, National Institute of Advanced Industrial Science and Technology (AIST), Central 6, 1-1-1 Higashi, Tsukuba, Ibaraki 305-8566, Japan

Naoshi Kojima
Biomedical Research Institute, National Institute of Advanced Industrial Science and Technology (AIST), Central 6, 1-1-1 Higashi, Tsukuba, Ibaraki 305-8566, Japan

Patrick Hardinge and James A. H. Murray
Cardiff School of Biosciences, Cardiff, Museum Avenue, CF10 3AX Cardiff, UK

Yajie Wang and Jingjing Cai
Department of Pharmacy, Anhui Medical College, Hefei 230061, China

Xiaoping Song
Department of Pharmacy, Anhui Medical College, Hefei 230061, China
Anhui Engineering Research Center of Recombinant Protein Pharmaceutical Biotechnology, Hefei 230022, China

Yugang Guo
Institute of advanced technology, University of Science and Technology of China, Hefei 230031, China
Anhui Engineering Research Center of Recombinant Protein Pharmaceutical Biotechnology, Hefei 230022, China

Changsheng Shao
Anhui Engineering Research Center of Recombinant Protein Pharmaceutical Biotechnology, Hefei 230022, China

David Simon Kleinhans and Virginie Lecaudey
Department of Developmental Biology of Vertebrates, Institute for Cell biology and Neuroscience, Goethe University, Max-von-Laue-Str. 13, 60438 Frankfurt am Main, Germany

Elsayed Ahmed Elsayed
Bioproducts Research Chair, Zoology Department, Faculty of Science, King Saud University, 11451 Riyadh, Kingdom of Saudi Arabia
Chemistry of Natural and Microbial Products Department, National Research Centre, Dokki, Cairo, Egypt

Mohamed A. Farid
Chemistry of Natural and Microbial Products Department, National Research Centre, Dokki, Cairo, Egypt

Hesham A. El-Enshasy
Institute of Bioproduct Development (IBD), Universiti Teknologi Malaysia (UTM), 81130 UTM Skudai, Malaysia
City of Scientific Research and Technology Application, New Burg Al Arab, Alexandria, Egypt

Aurélie Bak and Joanne B. Emerson
Department of Plant Pathology, University of California, Davis, CA 95616, USA

Deborah A. Sarkes, Bryn L. Adams, Margaret M. Hurley and Dimitra N. Stratis-Cullum
Biotechnology Branch, CCDC US Army Research Laboratory, 2800 Powder Mill Rd, Adelphi, MD, 20783 USA

Sarah D. Stellwagen
Biotechnology Branch, CCDC US Army Research Laboratory, 2800 Powder Mill Rd, Adelphi, MD, 20783 USA
Department of Biological Sciences, University of Maryland Baltimore County, 1000 Hilltop Circle, Baltimore, MD, 21250 USA

Mia A. Hunt and Rebecca L. Renberg
Biotechnology Branch, CCDC US Army Research Laboratory, 2800 Powder Mill Rd, Adelphi, MD, 20783 USA
General Technical Services, Suite 301, 1451 Route 34 South, Wall Township, NJ, 07727 USA

Innanurdiani Koko
Department of Cell and Molecular Biology, Faculty of Biotechnology and Biomolecular Sciences, University Putra Malaysia, 43400 UPM, Serdang, Selangor, Malaysia

Mas Jaffri Masarudin
Department of Cell and Molecular Biology, Faculty of Biotechnology and Biomolecular Sciences, University Putra Malaysia, 43400 UPM, Serdang, Selangor, Malaysia
Institute of Bioscience, University Putra Malaysia, 43400 UPM Serdang, Selangor, Malaysia

Raha Abdul Rahim
Department of Cell and Molecular Biology, Faculty of Biotechnology and Biomolecular Sciences, University Putra Malaysia, 43400 UPM, Serdang, Selangor, Malaysia
Institute of Bioscience, University Putra Malaysia, 43400 UPM Serdang, Selangor, Malaysia
Chancellory, Universiti Teknikal Malaysia, 76100 Durian Tunggal, Melaka, Malaysia

Adelene Ai-Lian Song
Department of Microbiology, Faculty of Biotechnology and Biomolecular Sciences, University Putra Malaysia, 43400 UPM, Serdang, Selangor, Malaysia
Institute of Bioscience, University Putra Malaysia, 43400 UPM Serdang, Selangor, Malaysia

Index

S

T

V

X

Z

www.ingramcontent.com/pod-product-compliance
Lightning Source LLC
LaVergne TN
LVHW081024210926
848485LV00004B/942